# ADVANCES IN DATA, METHODS, MODELS AND THEIR APPLICATIONS IN GEOSCIENCE

Edited by **Dongmei Chen**

**Advances in Data, Methods, Models and Their Applications in Geoscience**
Edited by Dongmei Chen

**Published by InTech**
Janeza Trdine 9, 51000 Rijeka, Croatia



**Publishing Process Manager** Dejan Grgur

**Technical Editor** Teodora Smiljanic

**Cover Designer** InTech Design Team

Additional hard copies can be obtained from orders@intechopen.com

Advances in Data, Methods, Models and Their Applications in Geoscience,
Edited by Dongmei Chen ,
p. cm.
ISBN 978-953-307-737-6

For EU product safety concerns:
IN TECH d.o.o., Prolaz Marije Krucifikse Kozulić 3, 51000 Rijeka, Croatia,
info@intechopen.com or visit our website at intechopen.com.

# Meet the editor

Prof. Dongmei Chen is currently an associate professor at the Department of Geography, Queen's University in Kingston, Canada. She is also crossed-appointed at the Department of Environmental Science. She received her BA from Peking University, China, her MS from the Institute of Remote Sensing Application, Chinese Academic of Science, and her PhD in Geography from the joint doctoral program of San Diego State University and University of California at Santa Barbara, USA. Prof. Chen worked at the Environmental Systems Research Institute (ESRI), California as a GIS product specialist for two years. Her research interests focus mainly on understanding and modeling of interactions between human activities and physical environment by using geographic information science, remote sensing, spatial analysis and modeling technologies. She has worked on more than twenty funded environmental- and health-related research projects, and published over fifty journal articles and book chapters. Prof. Chen has received research awards from American Society for Photogrammetry and Remote Sensing, Canada Foundation of Innovation, Chinese Academic of Science and International Geographic Information Foundation.

# Contents

# Preface

Geoscience is one of the most broad and rapidly developed fields in last three decades. With the rapid diffusion of geospace technologies and growing attention on global environmental and climate change, many new data, methods and models have been applied to various aspects of geoscience. The focus of this book is on the recent advances in data, analysis methods and their applications in geoscience and environmental change studies. One of the problems associated with such a rich selection of themes is the organization challenge of bringing together a variety of techniques, applications, study areas and data. After the review process, a collection of sixteen articles written by scholars from thirteen countries has been selected for this book. These articles cover a wide range of data analysis methods and modeling techniques, and provide an excellent profile of their applications in geology, geophysics, hydrology and environmental change in different regions of the world.

One of the most obvious technological revolutions in last half century is the development and application of computer information technology. Modern geoscience research widely relies on information technologies, especially geographic information systems (GIS). In the first chapter, Rückemann shows how geoscience research greatly benefits from modern development of information and computing systems, especially high performance computing. He reviews the general information technology advances and discusses issues from hardware, system architecture, software, legal and collaborational aspects that prevent interest groups in geoscience and natural science from building complex integrated computing systems on a long term base. At the end of the chapter, he also presents successful case studies of integrated computing systems used in geoscience and natural science.

Many quantitative methods are widely used in geoscience. Chapters 2 and 3 are review papers on two common analysis methods for geoscience and natural science. In Chapter 2, Likkason covers different spectral analysis methods and tools that can be used to analyze and interpret geophysical data obtained in the field. He discusses the detailed treatment and analysis of periodic and aperiodic functions in Fourier methods, as well as the sampling theorem and Fourier analysis. These Fourier-based methods have found usefulness in the analysis of geophysical data. In Chapter 3, Liu et al. review spatial interpolation algorithms that are widely used to estimate values of properties at unsampled locations for spatial continuous data. There are many factors affecting the performance of spatial interpolation methods and growing concerns on the accuracy obtained from different interpolation methods. They propose a data-independent GIS-based evaluation process flow to quantitatively analyze and evaluate different spatial interpolation models.

The following chapters 4 to 11 consist of eight articles on different data, methods, models and their applications used in geophysics and geology studies. Ariyibi and Joshua (Chapter 4) summarize the basic principles of using data acquired by the dual frequency global positioning system (GPS) receivers to understand variation and dynamics of the ionosphere. They use the total electron content (TEC) and S4 index data recorded from GPS receiver at a low-latitude station of Ile – Ife to study the ionospheric condition during low geomagnetic activity period and during geomagnetic storm events.

The airborne Gamma Ray (GR) measurements have been used in geophysical research for decades. The information provided by a gamma-ray spectrometer can be used for tracing the amounts of mineral deposits in rocks. Gaci et al. (Chapter 5) conducts a mono (two)-dimensional fractal analysis of natural radioactivity measurements recorded over the Hoggar's area of Algeria. They found that the fractal behavior of the GR measurement is not affected by pre-processing operations.

Core samples are commonly used to study the evolution of climate, species and sedimentary composition from geologic history. Gaci and Zaourar (Chapter 6) overview different numerical analysis methods used for the study and investigation of core images. They used a new fractal-based model with the two-dimensional multifractional Brownian motion (2D-mBm) in order to explore the spatial evolution of the local regularity in digitized core images. They demonstrated that the regularity maps obtained by 2D multiple filter techniques from the digitized core images could characterize heterogeneities of the analyzed core.

The use of seismic data to study sedimentary rocks and their forming processes is the focus of seismic sedimentology. Yang et al. (Chapter 7) present a case study on 3D seismic sedimentology of nearshore subaqueous fans from Dongying Depression, Eastern China. In Chapter 8, Bilal presents a study on mapping and analyzing the volcano activities and seismicity characteristics to understand the nature and composition of the underlying lithospheric mantle along the Syrian rift.

The microstructures of geomaterials and their evolution under different environmental conditions are critical for understanding their mechanical behavior and performance in engineering environments. In Chapter 9, Sorgi and De Gennaro present results from a research program to evaluate the mechanical behavior of chalk in the shallow underground mine of Estreux, Northern France. Different techniques are used to analyze the process at the site, laboratory and microscopic scales.

Diamond mining and oil field planning are always hot topics in geoscience because of their high economic implication. Due to the close association of Kimberlite rocks with diamonds, Kimberlite magmas are invaluable to scientific and exploration communities. Kamenetsky et al. (Chapter 10) use data from the newly developed study of the diamondiferous Udachnaya-East kimberlite pipe in Siberia to understand chemical and physical characteristics of kimberlite magma. To optimize oil reservoir development and production planning, geological models that can capture uncertainty are needed for hydrocarbon field development.

Maucec et al. (Chapter 11) review recent advances in technology for building high-resolution geological models and its role in state-of-the-art and future reservoir management decisions for hydrocarbon field development. They introduce some basic tools and concepts of geostatistical spatial analysis and modeling, and present a fundamentally novel method to resolve most of the common geocellular modeling issues.

Chapters 12 to 14 present studies in hydrological models and their applications. Sediment yield exported by a basic (or catchment) over a period of time is important for the size of sediment storage capability of a dam and estimate of a reservoir life. Therefore, sediment yield prediction from catchments is very important for many dam plans and designs in places where water resources sedimentation is a serious problem. Preksedis Ndomba (Chapter 12) reports a study of sediment yields based on the relationship between sediment yield and various climatic regions in Tanzania. They develop simple sediment yield-fill equations on each catchment area and recommend them as simple and efficient planning tools for water-supply schemes of small reservoirs with limited local data. In Chapter 13, Ndomba et al. evaluate the suitability of a commonly used hydrological model, the Soil and Water Assessment Tool (SWAT) for sediment yield estimation in three catchments of Eastern Africa.

Many geo-objects or phenomena do not have clear boundaries and are often mixed together. Fuzzy set theories have been developed to model the uncertainty in transition or mixed regions. In Chapter 14, Barreto-Neto develops a fuzzy rule-based hydrologic model within a GIS environment to predict runoff in the Quilomo River watershed located in South of the State of São Paulo, Brazil. They found that the incorporation of the fuzzy theory to the hydrologic model made a better run off estimation than the traditional Boolean model, and confirm the advantage of fuzzy theory in the representation of natural phenomena.

Environmental change has been causing global concerns, and remote sensing technology has played an important role in monitoring environmental change. Chapter 15 written by Zhang et al. provides a good example of how remotely sensed images can be used to detect changes on the earth surface. They use time series MODIS satellite images to get the classified land cover classification maps to assess the rock desertification problems in Guangxi Province of China.

The evolution of species on earth is the topic of the last chapter (Chapter 16). Using fossil records in the Lower Eocene crustancean burrow system in the southwestern

Israel, Lewy et al. demonstrate the change of breeding mode from K-type to r-type during the end-Cretaceous biological crisis.

As a collection of edited research and review papers, this book attempts to create a sense of the current state of art in methods, technology and their applications in geoscience, as well as the direction currently pursued by the research community around the world. I hope that readers will find this book of benefit in understanding the current developments and trends in the merging areas of analysis methods, models, techniques, and applications in geoscience. However, considering the wide range of themes and rapid technology progress in geoscience, the topics covered in this book are only a small part of methods, models, techniques used in geoscience.

Above all, I would like to thank Mirna Cvijic, Dejan Grgur, and other publishing managers at InTech for their support and help, as well as acknowledge the efforts of the authors. All credits and responsibilities of each included article belong to the corresponding contributors

**Dongmei Chen**
Associate Professor at the Department of Geography,
Queen's University in Kingston,
Canada

# 1

# Queueing Aspects of Integrated Information and Computing Systems in Geosciences and Natural Sciences

Claus-Peter Rückemann
*Westfälische Wilhelms-Universität Münster (WWU),*
*Leibniz Universität Hannover,*
*North-German Supercomputing Alliance (HLRN)*
*Germany*

## 1. Introduction

Modern geosciences widely rely on information science technologies. In most cases of scientific research applications, tools, and state of the art hardware architectures are infamously neglected and therefore their development is not pressed ahead and not documented, a fault for continuous and future developments. Information Systems and Computing Systems up to now live an isolated life, rarely integrated and mostly lacking essential features for future application. Although in general technology advances and new tools arise, there is a number of aspects that prevent interest groups from building complex integrated systems and components on a long term base. These issues, from hardware and system architecture aspects to software, legal, and collaborational aspects, are top in the queue for realisation show-stoppers. This chapter presents the current status of integrated information and computing systems. It discusses the most prominent technical and legal aspects for applications in geosciences and natural sciences. Todays state of the art information systems provide a plethora of features for nearly any field of application. Present computing systems can provide various distributed and high end compute power. Compute resources in most cases have to be supported by highly performing storage resources. The most prominent disciplines on up to date resources are natural sciences like geosciences, geophysics, physics, and many other fields with theoretical and applied usage scenarios. For geosciences both information systems as well as computing resources are essential means of day to day work. The most immanent limitation is that there are only a very few facilities with these systems combining the information systems features with powerful compute resources. The goal we have to work on for the next years is to facilitate this integration of information and computing systems. Modern information systems can provide various information and visualise context for different purposes beyond standard Geoscientific Information Systems (GIS). Fields demanding for handling, processing, and analysing geoscientific data are manyfold. Geophysics as well as applied sciences provide various methods as to name magnetic methods, gravity methods, seismic methods, tomography, electromagnetic methods, resistivity methods, induced polarisation, radioactivity methods, well logging and various assisting and integrated methods and techniques. Integrating these methods with information systems and the support of remote sensing, cartography, depth imaging, and infrastructural and social sources a more and more holistic view on the earth system will be possible. This will help to gain insight in the fields of seismology, meteorology, climatology, prospection and exploration, medical geology, epidemiology, environmental planning and many more disciplines. The resulting information systems and applications are not only used for scientific research but for public information, education, disaster management, expert systems and many more. In various application areas the surplus value arises with intelligent combination of information. A cartographic system only displaying spatial data is of less significance for a seismological disaster management application if there are no additional information and features provided. Provisioning these information will in many case result in interactive computation. For some use cases requests for points of interest, dynamical cartography, event programming, flexible event triggering, long-term monitoring like in seismology, catastrophe management and meteorology are necessary, for others simulation or modeling of scenarios are essential. All these fields of application contain tasks that cannot be handled in extend for large and complex systems on one local compute and storage resource only. Processes like processing jobs, visualisation, traveling salesman problems, and multimedia production have to be transferred to systems with the capacity necessary for multiple requests at the same time. We should not isolate scholarly research from long term information science concepts and architectures used in geosciences disciplines. Therefore the overall goal is to integrate systems, concepts, software, hardware and other components on a higher level of collaboration and strategical decision. As many application scenarios arise from geosciences and natural sciences, a number of examples are given based on implementations from these disciplines and case studies done over the last decade. Present activities and future work to be done on development and strategies level are presented to help overcome the stagnancy in the evolution of integrated systems. This chapter will show the components that in most other cases are discussed independently and presents a basic concept for integrating systems as successfully used with geosciences and natural sciences case studies.

## 2. Capability versus capacity

The high end computing world vastly used by geosciences researchers can be described with capability computing on the one hand and capacity computing on the other hand which provide complex tools to expand the means and methods of research by continuously expanding the limits of feasibility.

### 2.1 Capability computing

Capability computing means to target the grand challenge problems in certain fields. This will for example be the case with earthquake simulation, multi-dimensional modeling of the earth's underground, tornado simulation, atmosphere simulation, galaxy cluster simulation, non-linear computation of complex structures, life-sciences and epidemiological simulation, archaeological and architectural simulation. For the foreseeable future complex information and computing systems will be a topic on this list as soon as resources evolve. Systems for capability computing have to provide capacity for a few large jobs. Job processing for single-job scenarios can handle larger problems or faster solution. For these single jobs have main potential for insight on their own, merely not from context. Capability computing will usually not be the first association with information systems.

### 2.2 Capacity computing

The more interesting along with the development of the next generation of system architectures is the opening of capacity computing for complex information systems. These systems provide capacity for many small or medium sized jobs. Job processing of several parallel jobs is suitable for these resources. The single job itself may be used for parameter studies, design alternatives, exploring pre-development stages, and in general these jobs on their own have less potential for insight. Making use of capacity computing resources many instances of compute jobs will run on a resource. This is what we need to enliven conventional information system implementations with new features, leading to new insights.

## 3. Computing resources

### 3.1 Paradigms

There is a number of different paradigms and resources that can be considered for this purpose and used from various geosciences disciplines. The topmost category for High End Computing (HEC) are High Performance Computing (HPC) and Supercomputing resources. These will in nearly all use cases be used in a non distributed manner. The lower end, Distributed Computing (DC) and services computing resources, are for example built on base of paradigms like Sky Computing, Cloud Computing, Grid Computing, Cluster Computing, Mobile Computing.

### 3.2 Performance pyramid

What do we have to expect to be the machines behind these paradigms? The answer is a performance pyramid. The performance pyramid for computing resources shows the following structure.

Top sector:

- International supercomputers,
- National supercomputers,
- Regional supercomputers.

Medium and bottom sector:

- Local and dedicated compute servers and clusters,
- Workstations,
- Mobile devices.

## 4. Computing obstacles

The obstacles for operating complex resources with interactive systems with efficient and effective operation can be overcome while reacting on several levels, architectures, hardware, frameworks and middleware, applications, energy consumption, competence resources, consultancy, and support.

As far as High End Computing (HEC) being a genus for High Performance Computing and various other ambitioned computing paradigms is an issue of national interest for most countries, reliability and security are the most important factors for operating these services. Science and Research is depending on the results of their computations. Just with this, everyone is depending on systems and operating systems used. So problems most imminent arise especially with the

- Large number of cores,
- Large number of nodes,
- Distributed memory usage,
- Large number of large hard disks,
- Read and write speed of storage.

With the increasing number of requests and interactivity the communication size, size of data, transfer band width, scalability, and mean times for failure get more important. An intelligent arrangement and configuration of system components and an overall management of system components gets into the focus.

The most prominent problem with the next generation of resources is quantity of components. The handling of quantity leads –besides many other challenges– to increased demands for encryption, IO, PCI, on-chip features, error correction (ECC), research and development, scientific and academic staff and supporting maintenance, operative and administrative staff, as well as for secondary dependencies like energy resources and unbreakable power supplies.

### 4.1 Consumption

The most prominent problem with quantity, besides the computing obstacles, is consumption. State of the art power and energy measures are for example Low Voltage memory (LV DIMM), Light Load Efficiency Mode (LLEM), multiple Power Supplies, watercooler chassis & air conditioning, higher temperature cooling, hot water cooling, hybrid cooling systems, Energy and Power Manager (Active Energy Manager, AEM and others), application/energy frequency optimisation, energy reduced low frequency Processors, Power Management, and Energy Management.

### 4.2 Shortcommings

Besides that modular, dynamical applications are rare, even in geosciences, shortcommings regarding application context and how to handle these aspects are obvious:

- Architectures (CPU, GPU, GPGPU, FPGA, . . .),
- Languages (high level languages, CUDA, . . .),
- Memory,
- Fast and broad band Networks,
- Efficiency,
- Manageability, . . .

## 5. Hardware resources

Most scientific projects consider software and hardware issues to be treated separately. This would most likely be a problem for developing integrated systems on a solid holistic base. As for overall costs, for example with power consumption and staff, only very few institutions will be able to operate and develop those systems. The more complex these systems get, the less can the distinction between infrastructure resources and systems resources be recognised. As for understanding the complexity of these issues to be inseparable in the dimension of future integrated systems, the following paragraphs will illustrate some most important hardware components.

### 5.1 Resources infrastructure

An unabdicable premise for safe and reliable operation complex and large systems are concepts and implementation of power resources, unbreakable power supplies, air conditioning, electronics, physical security and many more infrastructure components. Figure 1 shows infrastructure components necessary to operate a larger computer installation for the purpose of scientific computing: generator, air conditioning, power supplies, and electronical and physical security measures.

Fig. 1. Infrastructure components necessary to operate larger computer installations.

### 5.2 Resources cooling

Besides the infrastructure, various measures associated directly with the computer systems are necessary and this will depend on the type of installation. Figure 2 shows one type of rack water-cooling (SGI, 2011).

Fig. 2. Water-cooling in rack.

### 5.3 System core resources

For the main purpose of computing, large numbers of compute nodes are needed. The main system resources are cores and memory. The first two images in Figure 3 show a rack with compute nodes (SGI, 2011) and some thousand memory sticks needed for one supercomputer installation.

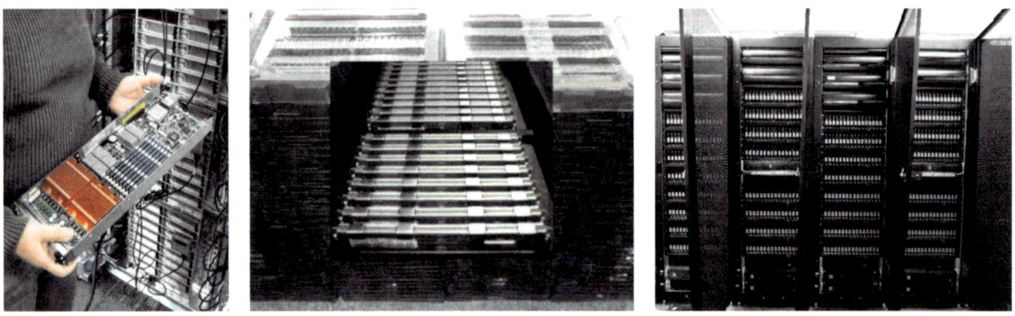

Fig. 3. Core resources and storage: compute nodes, memory, and disk storage unit.

### 5.4 System networks

With the increase of core resources, the more networks infrastructure is needed to operate these resources and make them accessible as a system. Figure 4 shows cabling and switches. Currently the significance of networks is rapidly increasing. Fibre optics are used to efficiently and effectively implementing networks. No wonder that in the year 2009 the physics Nobel Prize was dedicated to fibre optics, for the ground breaking achievements concerning the transmission of light in fibers for optical communication.

### 5.5 System storage

Besides cores, memory, and networks large storage capabilities are necessary for permanent storage. Figure 3 shows a disk storage unit consisting of several racks of hard disk drives, controllers, and servers.

Fig. 4. Networks, cabling, switches.

### 5.6 Systems connection

Linking high end resources is an important factor for ensuring economical use and enabling for access. In many cases these connections are built for redundancy, in order to switch connection for maintenance or emergencies.

### 5.7 System redundancy

Not only systems connections can be created using fallbacks. Figure 5 shows a network switch connecting some compute resources on redundant pathes. With large numbers of components the rate of failure increases. Redundancy and appropriate concepts will minimise the risk due to component failures. Large resources, as we have seen, not only need redundant cores and memory but various additional redundancies. Figure 5 further shows redundant rack power supplies, redundant disk drive enclosures with redundant disks, for example with appropriate RAID level, and redundant meta data storage servers.

Fig. 5. Redundancy with linking resources, rack power supplies, disk drive enclosures and disks, meta data storage servers.

### 5.8 System operating

System operating will have to ensure local system access, component and services monitoring as well as hardware, physical and logical maintenance. Figure 6 illustrates operating access using local console, remote access, monitoring and physical maintenance.

Fig. 6. Operating using local console access, remote access, monitoring and physical maintenance.

## 6. Resources prerequisites

The complexity of these aspects shows why the integration of computing resources has takes so much time and why this is on the turn now. When we want do the planning for future resources and consumptive prerequisites there are essential requirements for technical and competence resources. For the technical resources we have to implement efficient and effective general purpose system installations, for loosely as well as for massively parallel processing:

- Architecture (e.g. MPP Massively Parallel Processing / SMP Symmetric Multi-Processing),
- Accessing Computing Power (MPI / OpenMP / loosely coupled interactive),
- Efficiency (Computing Power / Power Consumption),
- Storage and Archive.

For the competence resources a sustainable infrastructure has to be built, regarding research, scientific consulting, staff, operation, systems management, technical consulting, and administrative measures. Goals with using these resources are dynamically provisioning of secondary information, calculation and computation results, modeling and simulation. For exploiting the existing and future compute and storage resources, the basic "trust in computing" and "trust in information" requirements have therefore to be implemented in complex environments. For many scenarios this leads to international collaborations for data collection and use as well as to modular development and operation of components. The geosciences as other natural sciences cannot fulfill these requirements without interdisciplinary research and collaboration.

## 7. Software needs hardware

The next sections will present some examples on how these resources are used in geosciences and geoinformatics with integrated systems. It will show the complexity of the next generation of system architectures and usage that arises from integrating the necessary components and resources.

### 7.1 Geoexploration and integrated systems

These sections present information system use cases and geoscientific application components that will profit from integrated information and computing systems using high end resources. It discusses features, concepts and frameworks, legal issues, technical requirements, and techniques needed for implementing these integrated systems. High Performance Computing has been recognised as one of the key technologies for the 21st century. It shows up with the problems existing today with implementing and using computing resources not only in batch mode but in combination with quasi interactive applications and it gives an outlook for future systems, components, and frameworks, and the work packages that have to be done by geoscientific disciplines, services and support, and resources providers from academia and economy.

There are two main objectives for interfacing modular complex integrated information and computing systems: "trust in computing" and "trust in information". This goal for a long-term strategy means to concentrate on implementing methods for flexible use of envelope interfaces for use with integrated information and computing systems for managing objects and strengthen trust with systems from natural and geosciences, spatial sciences, and remote sensing as to be used for application, e.g., in environment management, healthcare or archaeology. Spatial means are tools, for the sciences involved. Therefore processing and computing is referred to the content which is embedded and used from the visual domain. Over the last years a long-term project, Geo Exploration and Information (GEXI) (GEXI, 1996, 1999, 2011) for analysing national and international case studies and creating as well as testing various implementation scenarios, has shown the two trust groups of systems, reflected by the collaboration matrices (Rückemann, 2010a). It has examined chances to overcome the deficits, built a collaboration framework and illuminated legal aspects and benefits (EULISP, 2011; Rückemann, 2010b). The information and instructions handled within these systems is one of the crucial points while systems are evolving by information-driven transformation (Mackert et al., 2009).

For computing and information intensive systems the limiting constraints are manyfold. Recycling of architecture native and application centric algorithms is very welcome. In order to reuse information about these tasks and jobs, it is necessary to enable users to separate the respective information for system and application components. This can be done by structured envelope-like descriptions containing essential workflow information, algorithms, instructions, data, and meta data. The container concept developed has been called Compute Envelope (CEN). The idea of envelope-like descriptive containers has been inspired by the good experiences with the concept of Self Contained applications (SFC) (Rückemann, 2001). Envelopes can be used to integrate descriptive and generic processing information. Main questions regarding the topics of computing envelope interfaces are: Which content can be embedded or referenced in envelopes? How will these envelope objects be integrated into an information and computing system and how can the content be used? How can the context and environment be handled?

### 7.2 Geoscientific Information Systems and Active Source

One of the most essential components for integrating geoexploration components and the access to computing resources are dynamical Geoscientific Information Systems (GIS) (Rückemann, 2001). Some screenshots of a dynamical GIS (Rückemann, 2009) illustrate the facettes: Using dynamical data, raster, vector, secondary data, and events (Figure 7a), embedding dynamical components into components (Figure 7b), interacting with external components (Figure 7c), extending the user interface (Figure 7d).

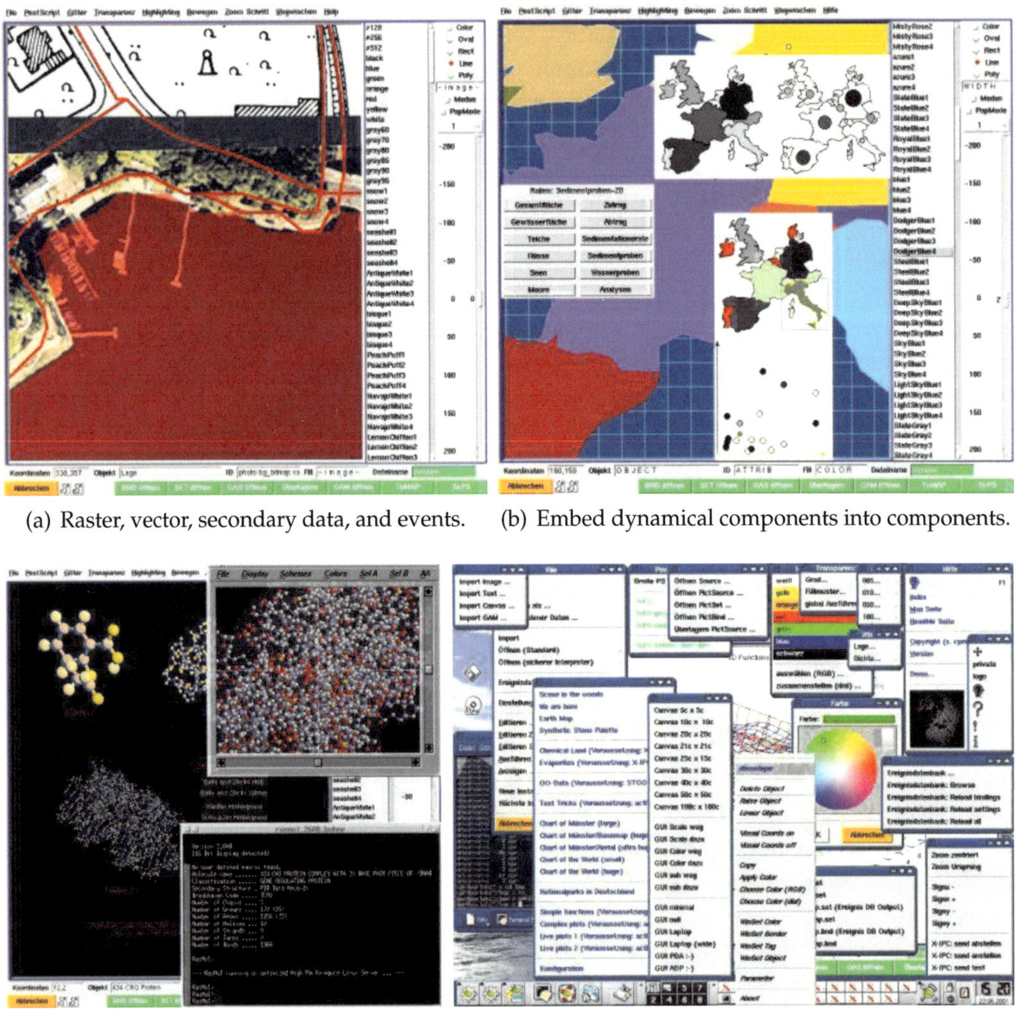

(a) Raster, vector, secondary data, and events.

(b) Embed dynamical components into components.

(c) Interact with external components.

(d) Extend the user interface.

Fig. 7. Facettes of an Integrated Information System.

### 7.3 Information databases

Information databases are the base for many components in Information Systems. There are various concepts built on classical database architectures, special information structures, and data collections. These architectures can be centralised or distributed. For the different purposes the implementations will use database software, file systems structures, meta data collection or combinations of more than one mechanism. In any case there should be a flexible and standardised way to interface and access the information. The information created within the LX Project has been subject of a long-term research initiative. It covers and combines for example educationary treatises, individual and definitions and descriptions, scientific results, as well as Point of Interest (POI) information. It is capable of handling categorisation, ergonomic multi-lingual representations as well as information alternatives for various different purposes. It integrates typesetting issues and publishing, including formulas, customised sorting, indexing, and timeline functions for the information objects. Listing 1 shows a simple example for an LX Encyclopedia entry.

```
Caldera        %-GP%-XX%---: Caldera           [Vulkanologie, Geologie]:
               %-GP%-DE%---:                   \lxidx{Schüsselkrater},
               %-GP%-DE%---:                   \lxidx{Chaldera},
               %-GP%-DE%---:                   \lxidx{Caldera},
               %-GP%-DE%---:                   \lxidx{cauldron},
               %-GP%-DE%---:                   \lxidx{Kessel}.
               %-GP%-DE%---:                   ...
               %-GP%-EN%---:                   \lxidx{Chaldera},
               %-GP%-EN%---:                   \lxidx{Schüsselkrater},
               %-GP%-EN%---:                   \lxidx{Caldera},
               %-GP%-EN%---:                   \lxidx{cauldron},
               %-GP%-EN%---:                   \lxidx{Kessel}.
               %-GP%-EN%---:                   ...
               %-GP%-DE%---:                   s. auch Capping stage
               %-GP%-EN%---:                   s. also Capping stage
               %-GP%-XX%---:                   %%SRC: ...
               %-GP%-XX%---:                   %%NET: http://...
```

Listing 1. Example LX Encyclopedia entry.

### 7.4 High End Computing

The integration of interactive and batch resources usage with dynamical applications does need flexibility in terms of interfaces and configuration. Various use cases studied distributed resources like capacity resources with Condor for example on ZIVcluster resources and on the other hand with High Performance Computing resources for example on ZIVSMP and HLRN (HLRN, 2011; ZIVGrid, 2008; ZIVHPC, 2011; ZIVSMP, 2011).

## 8. Integration framework – resources, services, disciplines

As well as analysing and separating the essential layers for building complex integrated systems, it is essential that these allow a holistic view on the overall system, for operation, development, and strategies level. The framework developed (Rückemann, 2010b) and studied for integrated information and computing is the Grid-GIS house (Figure 8) (Rückemann, 2009). An implementation of this kind is very complex and can only be handled with a modular architecture being able to separate tasks and responsibilities for resources (HEC, HPC), services, and disciplines. Figure 9 illustrates the logical components for integrated monitoring, accounting, and billing architecture for distributed resources and data usage. The legal aspects of combined usage of geo applications with services and distributed

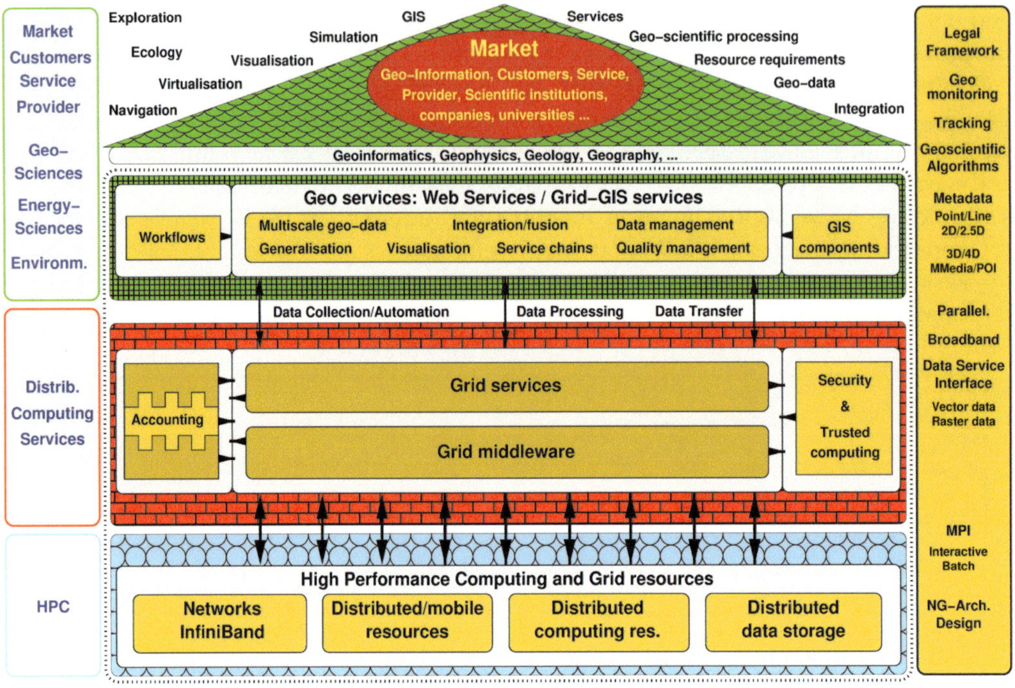


Fig. 8. Framework for integrated information and computing.

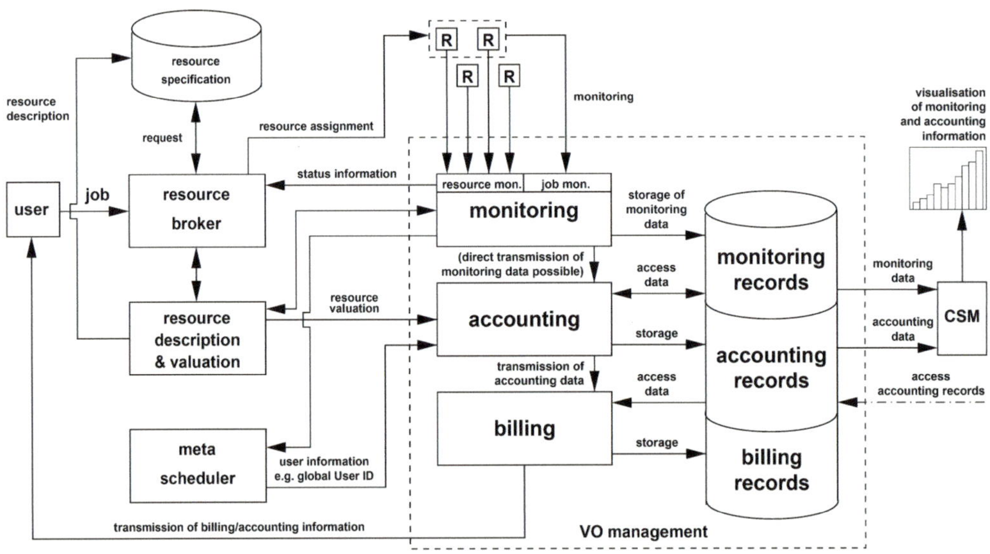


Fig. 9. Integrated monitoring, accounting, and billing architecture.

resources (Rückemann, 2010a;b) is shown in Figure 10. As the various components need dynamical and user-space interfaces, scripting mechanisms are unabdicable. The following sections discuss an example of a dynamical application using different distributed resources via flexible methods. Dynamical components and interfaces are implemented with Tcl and

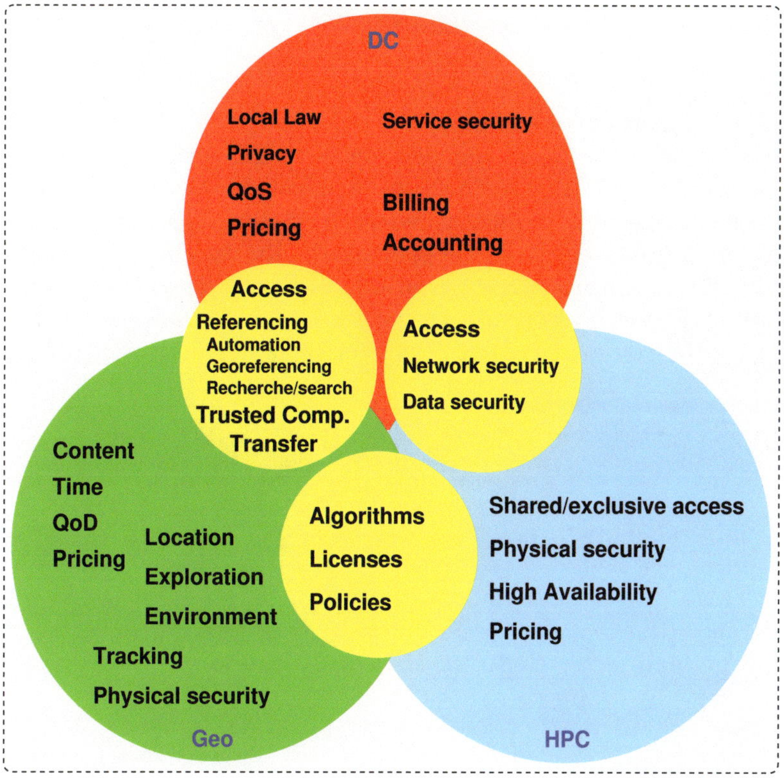


Fig. 10. Legal aspects of combined geo applications, services, and distributed resources usage.

derivatives and C (Tcl Developer Site, 2010), like the actmap application and Active Map procedures and data.

## 9. Integrated – InfoPoints using distributed resources

Using auto-events, dynamical cartography, and geocognostic aspects, views and applications using distributed compute and storage resources can be created very flexibly. As with the concept presented resources available from Distributed Systems, High Performance Computing, Grid and Cloud services, and available networks can be used. The main components are:

- interactive dynamical applications (frontend),
- distributed resources, compute and storage, configured for interactive and batch use,
- parallel applications and components (backend), as available on the resources,
- a framework with interfaces for using parallel applications interactively.

Besides the traditional visualisation a lot of disciplines like geo-resources and energy exploration, archaeology, medicine, epidemology and for example various applications within the tourism industry can profit from the e-Science components. These e-Science components can be used for Geoscientific Information Systems for dynamical InfoPoints and multimedia,

Points of Interest based on Active Source (Active POI), dynamical mapping, and dynamical applications.

### 9.1 InfoPoints and dynamical cartography

With the integration of interactive dynamical components and dynamical cartography various surplus values can be used. Figure **??**a shows an interactive Map of México (Rückemann, 2009). The yellow circle is an event sensitive Active Source object containing a collection of references for particular objects in the application. This type of object has been named InfoPoint. InfoPoints can use any type of start and stop routines triggered by events. Figure **??**b shows a defined assortment of information, a view set, fetched and presented by triggering an event on the InfoPoint. The information has been referenced from within

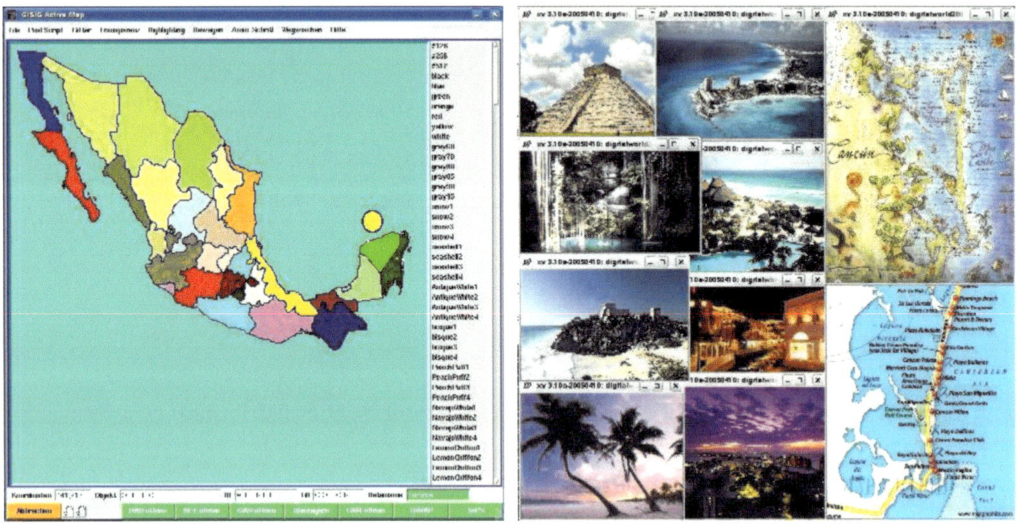

(a) Interactive México with InfoPoint Yucatán. (b) Sample view set of InfoPoint Yucatán.

Fig. 11. Integrated interactive dynamical components and InfoPoints.

the World Wide Web in this case. InfoPoints can depend on the cognitive context within the application as this is a basic feature of Active Source: Creating an application data set it is for example possible to define the Level of Detail (LoD) for zoom levels and how the application handles different kinds of objects like Points of Interest (PoI) or resolution of photos in the focus area of the pointing device.

### 9.2 Inside InfoPoints

The following passages show all the minimal components necessary for a fully functional InfoPoint. The example for this case study is mainly based on the Active Source framework. Triggered program execution ("Geoevents") of applications is shown with event bindings, start and stop routines for the data.

### 9.3 InfoPoints bindings and creation

Listing 2 shows the creation of the canvas for the InfoPoint and loading of the Active Source via bindings.

```
#
# actmap example -- (c) Claus-Peter R\"uckemann, 2008, 2009
#

#
# Active map of Mexico
#

erasePict
$w configure -background turquoise

pack forget .scale .drawmode .tagborderwidth \
  .poly .line .rect .oval .setcolor
pack forget .popupmode .optmen_zoom

openSource    mexico.gas
removeGrid

##EOF:
```

Listing 2. Example InfoPoint Binding Data.

This dynamical application can be created by loading the Active Source data with the `actmap` framework (Listing 3).

```
/home/cpr/gisig/actmap_sb.sfc mexico.bnd
```

Listing 3. Example creating the dynamical application.

### 9.4 InfoPoints Active Source

The following Active Source code (Listing 4) shows a tiny excerpt of the Active Source for the interactive Map of México containing some main functional parts for the InfoPoint Yucatán (as shown in Figure ??).

```
#BCMT-------------------------------------------------
###EN \gisigsnip{Object Data: Country Mexico}
###EN Minimal Active Source example with InfoPoint:
###EN Yucatan (Cancun, Chichen Itza, Tulum).
#ECMT-------------------------------------------------
proc create_country_mexico {} {
global w
#  Yucatan
$w create polygon 9.691339i 4.547244i 9.667717i \
  4.541732i 9.644094i 4.535433i 9.620472i 4.523622i \
  9.596850i 4.511811i 9.573228i 4.506299i 9.531496i \
...
 -outline #000000 -width 2 -fill green -tags {itemshape province_yucatan}
}

proc create_country_mexico_bind {} {
global w
$w bind province_yucatan <Button-1> {showName "Province␣Yucatan"}
$w bind province_quintana_roo <Button-1> \
                                   {showName "Province␣Quintana␣Roo"}
```

```
}

proc create_country_mexico_sites {} {
global w
global text_site_name_cancun
global text_site_name_chichen_itza
global text_site_name_tulum
set text_site_name_cancun         "Cancún"
set text_site_name_chichen_itza   "Chichén␣Itzá"
set text_site_name_tulum          "Tulum"

$w create oval 8.80i 4.00i 9.30i 4.50i \
  -fill yellow -width 3 \
  -tags {itemshape site legend_infopoint}
$w bind legend_infopoint <Button-1> \
  {showName "Legend␣InfoPoint"}
$w bind legend_infopoint <Shift-Button-3> \
  {exec browedit$t_suff}

$w create oval 9.93i 4.60i 9.98i 4.65i \
  -fill white -width 1 \
  -tags {itemshape site cancun}
$w bind cancun <Button-1> \
  {showName "$text_site_name_cancun"}
$w bind cancun <Shift-Button-3> \
  {exec browedit$t_suff}

$w create oval 9.30i 4.85i 9.36i 4.90i \
  -fill white -width 1 \
  -tags {itemshape site chichen_itza}
$w bind chichen_itza <Button-1> \
  {showName "$text_site_name_chichen_itza"}
$w bind chichen_itza <Shift-Button-3> \
  {exec browedit$t_suff}
...
}

proc create_country_mexico_autoevents {} {
global w
$w bind legend_infopoint <Any-Enter> {set killatleave \
   [exec ./mexico_legend_infopoint_viewall.sh $op_parallel ] }
$w bind legend_infopoint <Any-Leave> \
   {exec ./mexico_legend_infopoint_kaxv.sh }

$w bind cancun <Any-Enter> {set killatleave \
                     [exec $appl_image_viewer -geometry +800+400 \
                     ./mexico_site_name_cancun.jpg $op_parallel ] }
$w bind cancun <Any-Leave> {exec kill -9 $killatleave }

$w bind chichen_itza <Any-Enter> {set killatleave \
                     [exec $appl_image_viewer -geometry +800+100 \
                     ./mexico_site_name_chichen_itza.jpg $op_parallel ] }
$w bind chichen_itza <Any-Leave> {exec kill -9 $killatleave }
...
}
```

```
proc create_country_mexico_application_ballons {} {
global w
global is1
gisig:set_balloon $is1.country "Notation␣of␣State␣and␣Site"
gisig:set_balloon $is1.color "Symbolic␣Color␣od␣State␣and␣Site"
}

create_country_mexico
create_country_mexico_bind
create_country_mexico_sites
create_country_mexico_autoevents
create_country_mexico_application_ballons
scaleAllCanvas 0.8
##EOF
```

Listing 4. Example InfoPoint Active Source data.

The source contains a minimal example with the active objects for the province Yucatán in México. The full data set contains all provinces as shown in Figure **??**. The functional parts depicted in the source are the procedures for:

- `create_country_mexico`:
  The cartographic mapping data (polygon data in this example only) including attribute and tag data.
- `create_country_mexico_bind`:
  The event bindings for the provinces. Active Source functions are called, displaying province names.
- `create_country_mexico_sites`:
  Selected site names on the map and the active objects for site objects including the InfoPoint object. The classification of the InfoPoint is done using the tag `legend_infopoint`. Any internal or external actions like context dependent scripting can be triggered by single objects or groups of objects.
- `create_country_mexico_autoevents`:
  Some autoevents with the event definitions for the objects (Enter and Leave events in this example).
- `create_country_mexico_application_ballons`:
  Information for this data used within the Active Source application.
- Call section: The call section contains function calls for creating the components for the Active Source application at the start of the application, in this case the above procedures and scaling at startup.

Any number of groups of objects can be build. This excerpt only contains Cancun, Chichen Itza and Tulum. A more complex for this example data set will group data within topics, any category can be distinguished into subcategories in order to calculate specific views and multimedia information, for example for the category `site` used here:

- `city` (México City, Valladolid, Mérida, Playa del Carmen),
- `island` (Isla Mujeres, Isla Cozumel),

- `archaeological` (Cobá, Mayapan, Ek Balam, Aktumal, Templo Maya de Ixchel, Tumba de Caracol),
- `geological` (Chicxulub, Actun Chen, Sac Actun, Ik Kil),
- `marine` (Xel Há, Holbox, Palancar).

Objects can belong to more than one category or subcategory as for example some categories or all of these as well as single objects can be classified `touristic`. The data, as contained in the procedures here (mapping data, events, autoevents, objects, bindings and so on) can be put into a database for handling huge data collections.

### 9.5 Start an InfoPoint

Listing 5 shows the start routine data (as shown in Figure **??**). For simplicity various images are loaded in several application instances (`xv`) on the X Window System. Various other API calls like Web-Get `fetchWget` for fetching distributed objects via HTTP requests can be used and defined.

```
xv -geometry +1280+0   -expand 0.8 mexico_site_name_cancun_map.jpg &
xv -geometry +1280+263 -expand 0.97 mexico_site_name_cancun_map_hot.jpg &

xv -geometry +980+0    -expand 0.5 mexico_site_name_cancun.jpg &
xv -geometry +980+228  -expand 0.61 mexico_site_name_cancun_hotel.jpg &
xv -geometry +980+450  -expand 0.60 mexico_site_name_cancun_mall.jpg &
xv -geometry +980+620  -expand 0.55 mexico_site_name_cancun_night.jpg &

xv -geometry +740+0    -expand 0.4 mexico_site_name_chichen_itza.jpg &
xv -geometry +740+220  -expand 0.8 mexico_site_name_cenote.jpg &
xv -geometry +740+420  -expand 0.6 mexico_site_name_tulum_temple.jpg &
#xv -geometry +740+500  -expand 0.3 mexico_site_name_tulum.jpg &
xv -geometry +740+629  -expand 0.6 mexico_site_name_palm.jpg &
```

Listing 5. Example InfoPoint event start routine data.

### 9.6 Stop an InfoPoint

Listing 6 shows the stop routine data. For simplicity all instances of the applications started with the start routine are removed via system calls.

```
killall -9 --user cpr --exact xv
```

Listing 6. Example InfoPoint event stop routine data.

Using Active Source applications any forget or delete modes as well as using Inter Process Communication (IPC) are possible.

### 9.7 Integration and trust

Integrating components for mission critical systems does expect methods for handling "Trust in computation" and "Trust in information". This is what Object Envelopes (OEN) and Compute Envelopes (CEN) have been developed for (Rückemann, 2011). Listing 7 shows a small example for a generic OEN file.

```
<ObjectEnvelope><!-- ObjectEnvelope (OEN)-->
<Object>
<Filename>GIS_Case_Study_20090804.jpg</Filename>
<Md5sum>...</Md5sum>
<Sha1sum>...</Sha1sum>
<DateCreated>2010-08-01:221114</DateCreated>
<DateModified>2010-08-01:222029</DateModified>
<ID>...</ID><CertificateID>...</CertificateID>
<Signature>...</Signature>
<Content><ContentData>...</ContentData></Content>
</Object>
</ObjectEnvelope>
```

Listing 7. Example for an Object Envelope (OEN).

An end-user public client application may be implemented via a browser plugin, based on appropriate services. With OEN instructions embedded in envelopes, for example as XML-based element structure representation, content can be handled as content-stream or as content-reference. The way this will have to be implemented for different use cases depends on the situation, and in many cases on the size and number of data objects. Listing 8 shows a small example for an OEN file using a content DataReference.

```
<ObjectEnvelope><!-- ObjectEnvelope (OEN)-->
<Object>
<Filename>GIS_Case_Study_20090804.jpg</Filename>
<Md5sum>...</Md5sum>
<Sha1sum>...</Sha1sum>
<DateCreated>2010-08-01:221114</DateCreated>
<DateModified>2010-08-01:222029</DateModified>
<ID>...</ID><CertificateID>...</CertificateID>
<Signature>...</Signature>
<Content><DataReference>https://doi...</DataReference></Content>
</Object>
</ObjectEnvelope>
```

Listing 8. OEN referencing signed data.

### 9.8 Implemented solution for integrated systems with massive resources requirements

For most interactive information system components a configuration of the distributed resources environment was needed. In opposite to OEN use, making it necessary to have referenced instead of embedded data for huge data sets, for CEN it should be possible to embed the essential instruction data. So there is less need for minimising data overhead and communication. Envelope technology is meant to be a generic extensible concept for information and computing system components (Rückemann, 2011). Figure 12 shows the workflow with application scenarios from GEXI case studies (Rückemann, 2010b). Future objectives for client components are:

- Channels for limiting communication traffic,
- Qualified signature services and accounting,
- Using signed objects without verification,
- Verify signed objects on demand.

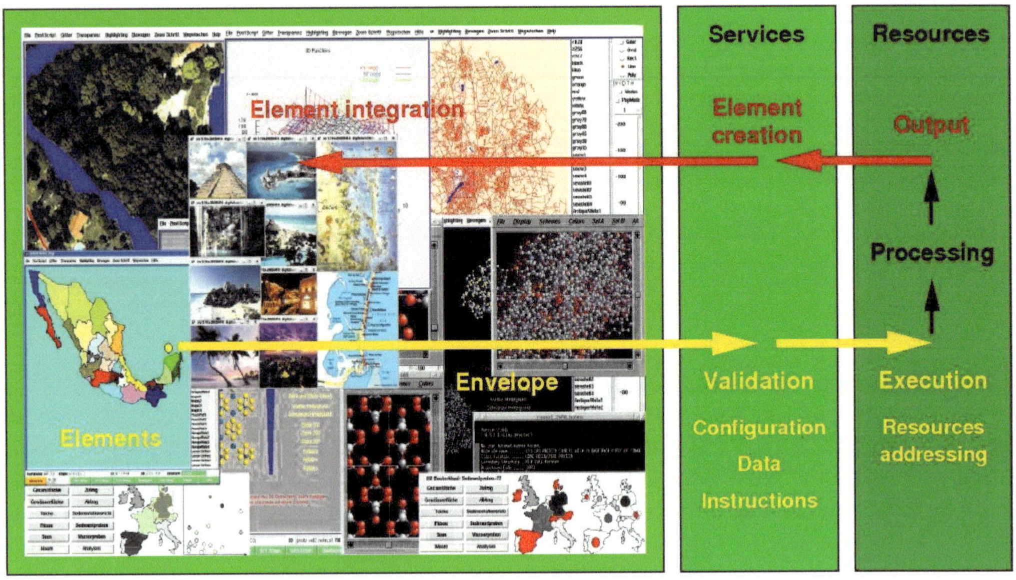


Fig. 12. Workflow with application scenarios from the GEXI case studies.

The tests done for proof of concept have been in development stage. A more suitable solution has now been created on a generic envelope base. An end-user public client application may be implemented via a browser plugin, based on appropriate services. The current solution is based on CEN files containing XML structures for handling and embedding data and information. This is so important because even in standard cases we easily have to handle hundreds of thousands of compute request from these components. In the easiest case it will be static information from information databases, for advanced cases it is for example conditional processing and simulation for thousands of objects like multimedia data or borehole depth profiling.

### 9.9 Integrated components in practice

When taking a look onto different batch and scheduling environments one can see large differences in capabilities, handling different environments and architectures. In the last years experiences have been gained in handling simple features for different environments for High Throughput Computing like Condor (Condor, 2010), workload schedulers like LoadLeveler (IBM, 2005) and Grid Engine (SGE, 2010), and batch system environments like Moab / Torque (Moab, 2010; Torque, 2010). Batch and interactive features are integrated with Active Source event management (Rückemann, 2001). Listing 9 shows a small example of a CEN embedded into an Active Source component.

```
#BCMT-----------------------------------------------
###EN \gisigsnip{Object Data: Country Mexico}
#ECMT-----------------------------------------------
proc create_country_mexico {} {
global w
#  Sonora
$w create polygon 0.938583i 0.354331i 2.055118i ...
#BCMT-----------------------------------------------
```

```
###EN \gisigsnip{Compute Data: Compute Envelope (CEN)}
#ECMT---------------------------------------------------
#BCEN <ComputeEnvelope>
##CEN <Instruction>
##CEN <Filename>Processing_Bat_GIS515.torque</Filename>
##CEN <Md5sum>...</Md5sum>
##CEN <Sha1sum>...</Sha1sum>
##CEN <Sha512sum>...</Sha512sum>
##CEN <DateCreated>2010-09-12:230012</DateCreated>
##CEN <DateModified>2010-09-12:235052</DateModified>
##CEN <ID>...</ID><CertificateID>...</CertificateID>
##CEN <Signature>...</Signature>
##CEN <Content>...</Content>
##CEN </Instruction>
#ECEN </ComputeEnvelope>
...
proc create_country_mexico_autoevents {} {
global w
$w bind legend_infopoint <Any-Enter> {set killatleave \
   [exec ./mexico_legend_infopoint_viewall.sh $op_parallel ] }
$w bind legend_infopoint <Any-Leave> \
   {exec ./mexico_legend_infopoint_kaxv.sh }
$w bind tulum <Any-Enter> {set killatleave \
   [exec $appl_image_viewer -geometry +800+400 \
   ./mexico_site_name_tulum_temple.jpg $op_parallel ] }
$w bind tulum <Any-Leave> \
   {exec kill -9 $killatleave }
} ...
```

Listing 9. CEN embedded with Active Source.

Interactive applications based on Active Source have been used on Grid, Cluster, and HPC (MPP, SMP) systems.

### 9.10 Resources interface

Using CEN features, it is possible to implement resources access on base of validation, verification, and execution. The sources (Listing 10, 11) can be generated semi-automatically and called from a set of files or can be embedded into an actmap component, depending on the field of application.

```
<ComputeEnvelope><!-- ComputeEnvelope (CEN)-->
<Instruction>
<Filename>Processing_Batch_GIS612.pbs</Filename>
<Md5sum>...</Md5sum>
<Sha1sum>...</Sha1sum>
<Sha512sum>...</Sha512sum>
<DateCreated>2010-08-01:201057</DateCreated>
<DateModified>2010-08-01:211804</DateModified>
<ID>...</ID>
<CertificateID>...</CertificateID>
<Signature>...</Signature>
<Content><DataReference>https://doi...</DataReference></Content>
<Script><Pbs>
<Shell>#!/bin/bash</Shell>
```

```
<JobName>#PBS -N myjob</JobName>
<Oe>#PBS -j oe</Oe>
<Walltime>#PBS -l walltime=00:10:00</Walltime>
<NodesPpn>#PBS -l nodes=8:ppn=4</NodesPpn>
<Feature>#PBS -l feature=ice</Feature>
<Partition>#PBS -l partition=hannover</Partition>
<Accesspolicy>#PBS -l naccesspolicy=singlejob</Accesspolicy>
<Module>module load mpt</Module>
<Cd>cd $PBS_O_WORKDIR</Cd>
<Np>np=$(cat $PBS_NODEFILE | wc -l)</Np>
<Exec>mpiexec_mpt -np $np ./dyna.out 2>&1</Exec>
</Pbs></Script>
</Instruction>
</ComputeEnvelope>
```

Listing 10. Embedded Active Source MPI script.

Examples for using High Performance Computing and Grid Computing resources include batch system interfaces and job handling. Job scripts from this type will on demand (event binding) be sent to the batch system for processing. The Actmap Computing Resources Interface (CRI) is an example for an actmap library (`actlcri`) containing functions and procedures and even platform specific parts in a portable way. CRI can be used for handling computing resources, loading Tcl or TBC dynamically into the stack (Tcl Developer Site, 2010) when given `set behaviour_loadlib_actlib "yes"`.

```
<ComputeEnvelope><!-- ComputeEnvelope (CEN)-->
<Instruction>
<Filename>Processing_Batch_GIS612.pbs</Filename>
<Md5sum>...</Md5sum>
<Sha1sum>...</Sha1sum>
<Sha512sum>...</Sha512sum>
<DateCreated>2010-08-01:201057</DateCreated>
<DateModified>2010-08-01:211804</DateModified>
<ID>...</ID>
<CertificateID>...</CertificateID>
<Signature>...</Signature>
<Content><DataReference>https://doi...</DataReference></Content>
<Script><Condor>
<Environment>universe = standard</Environment>
<Exec>executable = /home/cpr/grid/job.exe</Exec>
<TransferFiles>should_transfer_files = YES</TransferFiles>
<TransferInputFiles>transfer_input_files = job.exe,job.input
</TransferInputFiles>
<Input>input = job.input</Input>
<Output>output = job.output</Output>
<Error>error = job.error</Error>
<Log>log = job.log</Log>
<NotifyMail>notify_user = ruckema@uni-muenster.de</NotifyMail>
<Requirements>
requirements = (Memory >= 50)
requirements = ( ( (OpSys=="Linux")||(OpSys=="AIX"))&&(Memory >= 500) )
</Requirements>
<Action>queue</Action>
</Condor></Script>
```

```
</Instruction>
</ComputeEnvelope>
```

Listing 11. Embedded Active Source Condor script.

With Actmap CRI being part of Active Source, calls to parallel processing interfaces, e.g., using InfiniBand, can be used, for example MPI (Message Passing Interface) and OpenMP, already described for standalone job scripts for this purpose, working analogical (Rückemann, 2009).

### 9.11 Service and operation

With the complexity of the high level integration of disciplines, services, and resources there are various aspects that cannot be handled in general as they will depend on scenario, collaboration partners, state of current technology, and many other. Based on the collaboration framework operation can integrate Service Oriented Architectures (SOA) and Resources Oriented Architectures (ROA). Based on the pre-implementation case studies and application scenarios it will be necessary to define agreements on the low level of services (S-Level) and operation (O-Level). For the S-Level integrated systems will need to collect Service Level Requests (SLR), define Service Level Specifications (SLS), and specify appropriate Service Level Agreements (SLA). According to these, on the O-Level Operational Level Agreements (OLA) have to be arranged. In all practical cases these agreements together with the underlying Service Level Management (SLM) should clearly take up less than two to five percent of the overall capacity for the collaboration for an efficient and effective system. The most important mostly non-technical factor for planning complex integrated systems therefore is to limit the dominance and growth of management, administrative, and operational tasks. Economic target centred contracts can be a solution to set limits to possible usuriousness, for example to restrict against "hydrocephalic" reporting and auditing.

## 10. Evaluation

The case studies demonstrated that integrated systems can be successfully implemented with enormous potential for flexible solutions. Disciplines, services, and resources level can be handled under one integrated concept. Interactive dynamical information systems components have been enabled to use an efficient abstraction and to handle thousands of subjobs for parallel processing, in demanding cases without the disadvantages of distributed systems. With the results of the case studies we can answer one additional question: What are the essential key factors for long-term use of integrated components? The academic and industry partners involved in the case studies emphasised that the key factors are:

- Information, instructions, and meta data have to be self explanatory.
- Multi-lingual information need appropriate interfaces in order to be editable and processable along with each other.
- Information has to be stored in a common non proprietary way.
- Tools for processing and interfacing the information have to be available without restrictions.
- Component atoms need to be recyclable.
- Information and components have to be widely portable.

## 11. Outlook

There are a number of aspects that have to be addressed in future work. These are mostly not only on the technology but on collaborational, organisational, and funding level with integrated information and computing systems. For geosciences and natural sciences algorithms and concepts for processing, visualisation, and extended use of data and information are available. The grand challenge with the wisdom will be to succeed in overcoming the fate of decision makers on scientific funding, that gathering a critical mass of acceptance in the society is vital.

## 12. Conclusion

This chapter has shown some prominent aspects of the complexity of the next generation of system architectures that arises from integrating the necessary components and computing resources, used in geosciences and geoinformatics. With technology advances new tools arise for geosciences research and Information Systems and Computing Systems will become more widely available. Due to the complexity and vast efforts necessary to implement and operate these systems and resources there is a strong need to enable economic and efficient use and operation. Hardware and software system components cannot be neglected anymore and viewed isolated. System architecture issues to software, legal, and collaborational aspects are in the focus and must be handled for operation, development, and strategies level. Various application scenarios from geosciences and natural sciences profit from the new means and concepts and will help to push the development not only of Geoscientific Information Systems and Computing Systems but of Information Systems and Computing Systems for the geosciences.

## 13. Acknowledgements

I am grateful to all national and international academic and industry partners in the GEXI cooperations for the innovative constructive work and case study support as well as to the colleagues at the Leibniz Universität Hannover, at the Institut für Rechtsinformatik (IRI), the North-German Supercomputing Alliance (HLRN), the Westfälische Wilhelms-Universität (WWU) Münster, at the Zentrum für Informationsverarbeitung (ZIV) Münster, in the German Grid Initiative D-Grid and the participants of the postgraduate European Legal Informatics Study Programme (EULISP) and of the WGGEOSP work group as well as the colleagues at the last years GEOWS, GEOProcessing, CYBERLAWS, ICDS, and INFOCOMP international conferences for prolific discussion of scientific, legal, and technical aspects as well as to the staff at ZIV and the partner institutes and associated HPC companies for supporting this work by managing and providing HEC resources over the years. I am especially grateful to Hans-Günther Müller, SGI, for fruitful discussion and providing support with hardware photos.

## 14. References

Condor (2010). Condor, High Throughput Computing. URL: `http://www.cs.wisc.edu/condor/` [accessed: 2010-10-10].

EULISP (2011). *Fundamental Aspects of Information Science, Security, and Computing (Lecture)*, EULISP Lecture Notes, European Legal Informatics Study Programme, Institute for

Legal Informatics, Leibniz Universität Hannover (IRI / LUH). URL: http://www.eulisp.de [accessed: 2011-01-01].

GEXI (1996, 1999, 2011). Geo Exploration and Information (GEXI). URL: http://www.user.uni-hannover.de/cpr/x/rprojs/en/index.html#GEXI (Information) [accessed: 2010-05-02].

HLRN (2011). HLRN, North-German Supercomputing Alliance (Norddeutscher Verbund für Hoch- und Höchstleistungsrechnen). URL: http://www.hlrn.de [accessed: 2011-07-10].

IBM (2005). IBM Tivoli Workload Scheduler LoadLeveler. URL: http://www-03.ibm.com/systems/software/loadleveler/ [accessed: 2010-10-10].

Mackert, M., Whitten, P. & Holtz, B. (2009). *Health Infonomics: Intelligent Applications of Information Technology*, IGI Global, pp. 217–232. Chapter XII, in: Pankowska, M. (ed.), Infonomics for Distributed Business and Decision-Making Environments: Creating Information System Ecology, ISBN: 1-60566-890-7, DOI: 10.4018/978-1-60566-890-1.ch010.

Moab (2010). Moab: Admin Manual, Users Guide. URL: http://www.clusterresources.com [accessed: 2010-10-10].

Rückemann, C.-P. (2001). *Beitrag zur Realisierung portabler Komponenten für Geoinformationssysteme. Ein Konzept zur ereignisgesteuerten und dynamischen Visualisierung und Aufbereitung geowissenschaftlicher Daten*, Diss., Westfälische Wilhelms-Universität, Münster, Deutschland. 161 (xxii+139) S., URL: http://wwwmath.uni-muenster.de/cs/u/ruckema/x/dis/download/dis3acro.pdf [accessed: 2009-11-16].

Rückemann, C.-P. (2009). Dynamical Parallel Applications on Distributed and HPC Systems, *International Journal on Advances in Software* **2**(2). ISSN: 1942-2628, URL: http://www.iariajournals.org/software/ [accessed: 2009-11-16].

Rückemann, C.-P. (2010a). Integrating Future High End Computing and Information Systems Using a Collaboration Framework Respecting Implementation, Legal Issues, and Security, *International Journal on Advances in Security* 3(3&4): 91–103. Savola, R., (ed.), VTT Technical Research Centre of Finland, Finland, URL: http://www.iariajournals.org/security/sec_v3_n34_2010_paged.pdf [accessed: 2011-07-10], URL: http://www.iariajournals.org/security/ [accessed: 2011-07-10].

Rückemann, C.-P. (2010b). Legal Issues Regarding Distributed and High Performance Computing in Geosciences and Exploration, *Proceedings of the Int. Conf. on Digital Society (ICDS 2010), The Int. Conf. on Technical and Legal Aspects of the e-Society (CYBERLAWS 2010), February 10–16, 2010, St. Maarten, Netherlands Antilles*, IEEE Computer Society Press, IEEE Xplore Digital Library, pp. 339–344. ISBN: 978-0-7695-3953-9, URL: http://ieeexplore.ieee.org/stamp/stamp.jsp?tp=&arnumber=5432414 [accessed: 2010-03-28].

Rückemann, C.-P. (2011). Envelope Interfaces for Geoscientific Processing with High Performance Computing and Information Systems, *Proceedings International Conference on Advanced Geographic Information Systems, Applications, and Services (GEOProcessing 2011), February 23–28, 2011, Gosier, Guadeloupe, France / GEOProcessing 2011, ICDS 2011, ACHI 2011, ICQNM 2011, CYBERLAWS 2011, eTELEMED 2011, eL&mL 2011, eKNOW 2011 / DigitalWorld 2011*, XPS, Xpert Publishing

Solutions, pp. 23–28. Rückemann, C.-P., Wolfson, O. (eds.), 6 pages, ISBN: 978-1-61208-003-1, URL: http://www.thinkmind.org/download.php?articleid=geoprocessing_2011_2_10_30030 [accessed: 2011-07-10].

SGE (2010). Sun Grid Engine. URL: http://gridengine.sunsource.net/ [accessed: 2010-10-10].

SGI (2011). SGI, Silicon Graphics GmbH. URL: http://www.sgi.com [accessed: 2011-07-10].

Tcl Developer Site (2010). Tcl Developer Site. URL: http://dev.scriptics.com/ [accessed: 2010-10-10].

Torque (2010). Torque Admin Manual. URL: http://www.clusterresources.com/torquedocs21/ [accessed: 2010-10-10].

ZIVGrid (2008). ZIV der WWU Münster – ZIVGrid. URL: http://www.uni-muenster.de/ZIV/Server/ZIVGrid/ [accessed: 2008-12-23].

ZIVHPC (2011). ZIVHPC, HPC Computing Resources. URL: https://www.uni-muenster.de/ZIV/Technik/ZIVHPC/index.html [accessed: 2011-02-20].

ZIVSMP (2011). ZIVSMP, SMP Computing Resources. URL: https://www.uni-muenster.de/ZIV/Technik/ZIVHPC/ZIVSMP.html [accessed: 2011-02-20].

# 2

# Spectral Analysis of Geophysical Data

Othniel K. Likkason
*Physics Programme,
Abubakar Tafawa Balewa University, Bauchi,
Nigeria*

## 1. Introduction

The usefulness of a geophysical method for a particular purpose depends, to a large extent, on the characteristics of the target proposition (exploration target) which differentiate it from the surrounding media. For example, in the detection of structures associated with oil and gas (such as faults, anticlines, synclines, salt domes and other large scale structures), we can exploit the elastic properties of the rocks. Depending on the type of minerals sought, we can take advantage of their variations with respect to the host environment, of the electric conductivity, local changes in gravity, magnetic, radioactive or geothermal values to provide information to be analysed and interpreted that will lead to parameter estimation of the deposits.

This chapter deals with some tools that can be used to analyse and interpret geophysical data so obtained in the field. We shall be having in mind potential field data (from gravity, magnetic or electrical surveys). For example, the gravity data may be the records of Bouguer gravity anomalies (in milligals), the magnetic data may be the total magnetic field intensity anomaly (in gammas) and data from electrical survey may be records of resistivity measurements (in ohm-metre). There are other thematic ways in which data from these surveys can be expressed; depicting some other attributes of the exploration target and the host environments.

Potential fields for now are mostly displayed in 1-D (profile form), 2-D (map form) or 3-D (map form and depth display). Whichever form, the 1-D and 2-D data are usually displayed in magnitudes against space (spatial data). When data express thematic values against space (profile distance) or thematic values against time, they are called time-series data or time-domain data. The changes or variations in the magnitudes (thematic values) with space and/or time may reflect significant changes in the nature and attributes of the causative agents. We therefore use these variations to carefully interpret the nature and the structural features of the causative agents.

Most at times the picture of the events painted in the time-domain is poor and undiscernable possibly because of noise effects and other measurement errors. A noise in any set of data is any effects that are not related to the chosen target proposition. Even where such noise and measurements errors are minimized, some features of the data need to be gained/enhanced for proper accentuation. The only recourse to this problem is to make a transformation of the time-domain data to other forms or use some time-domain tools to analyse the data for improve signal to noise ratio; both of which must be

accomplished without compromising the quality of the original data. Even where the picture of the time-domain data 'looks good", we can perform further analyses on the signal for correlation, improvement and enhancement purposes. There are many time-domain and frequency-domain tools for these purposes. This is the reason for this chapter.
We shall be exploring the uses of some tools for the analyses and interpretations of geophysical potential fields under the banner of Spectral Analysis of Geophysical Data.
The first section of the chapter covers the treatment and analysis of periodic and aperiodic functions by means of Fourier methods, the second section develops the concept of spectra and possible applications and the third section covers spectrum of random fields; ending with an application to synthetic and real (field) data.

## 2. Periodic and aperiodic functions

A periodic function of time, t can be defined as f(t) = f(t+T), where T is the smallest constant, called the period which satisfies the relation. In general f(t) = f(t+NT), where N is an integer other than zero. As an example, we can find the period of a function such as f(t) = cos t/3 + cos t/4. If this function is periodic, with a period, T, then f(t) = f(t+T). Using the relation $\cos\theta = \cos(\theta + 2\pi m)$, m = 0, ±1, ±2, ..., we can compute the period of this function to be T = 24π.
Aperiodic function, on the other hand, is a function that is not periodic in the finite sense of time. We can say that an aperiodic function can be taken to be periodic at some infinite time, where the time period, T = ∞.
A function f(t) is even if and only if f(t) = f(-t) and odd if f(t) = -f(-t).

## 3. Sampling theorem

A function, f which is defined in some neighbourhood of c is said to be continuous at c provided (1) the function has a definite finite value f(c) at c and (2) as t approaches c, f(t) approaches f(c) as limit, i.e. $\lim_{t\to c} f(t) = f(c)$. If a function is continuous at all points of an interval a≤t≤b (or a<t<b, etc), then it is said to be continuous on or in that interval. A function $f(t) = t^2$ is a continuous function that satisfies the two conditions above.
A graph of a function that is continuous on an interval a≤t≤b is an unbroken curve over that interval. In practical sense, it is possible to sketch a continuous curve by constructing a table of values (t, f(t)), plotting relatively few points from this table and then sketching a continuous (unbroken) curve through these points.
A function that is not continuous at a point t = c is said to be discontinuous or to have a discontinuity at t = c. The function, f(t) = 1/t is discontinuous at t = 0 as the two requirements for continuity above are violated at t = 0. This discontinuity cannot be removed. On the other hand, a function, $f(t) = \frac{\sin t}{t}$ has a removable discontinuity at t = 0, even though the formula is not valid at t = 0. We can therefore extend the domain of f(t) to include the origin, as f(0) = 1.
A continuous function, f(t) can be sampled at regular intervals such that the value of the function at each digitized point is $f_n$, with n = 0, ±1, ±2, ±3, ... (Fig. 1).
It will appear as if the time domain function f(t) versus t is transformed into a digitized form domain ($f_n$ versus n).

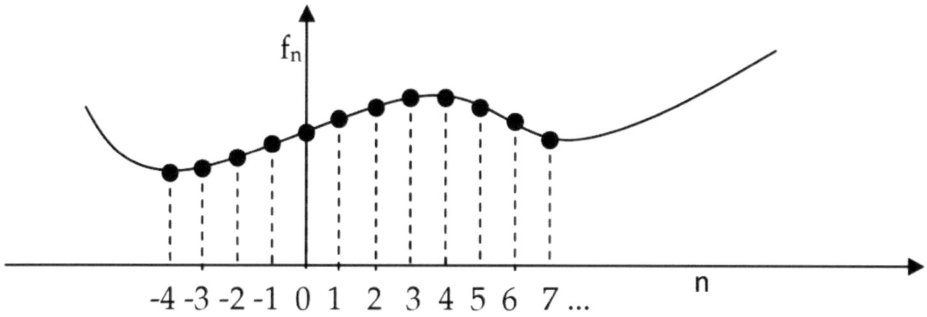


Fig. 1. Sampling a continuous signal $f_n$

To recover a function f(t) from its digital form, the sampling points must be sufficiently close to each other. There is actually a maximum sampling period particular to the function concerned, with which the complete recovery may be achieved.

According to the sampling theorem (or Shannon theorem), a function can be fully recovered by the sampling process provided (a) it is a reasonably well-behaved function and (b) band-limited (Bath, 1974).

Condition (a) implies that the function be continuous with no abnormal behaviour such as discontinuities (sharp breaks or singularities). This condition is practically always fulfilled in the case of functions representing natural processes (such as Laplacian fields). The band-limitedness of a function as a second condition, refers to a function which possesses a Fourier transform of non-zero value outside it. Most functions involving natural processes also fulfill this condition.

If a function is digitized with equal sampling interval, τ, the period present in that function which can be recovered by the process is 2τ, since we need a minimum of two sampling intervals to define one period. The equivalent frequency ($f_N = \frac{1}{2\tau}$) has a special significance in the subject of digitization. The frequency, $f_N$ is referred to as the folding or Nyquist frequency. The parameter, τ represents the maximum limit for the sampling period with which we can fully specify a function whose lowest period is 2τ. However, the sampling cut-off frequency, $f_C$ ($=\frac{1}{\tau}$) and so $f_C = 2f_N$.

It has been shown that the continuous function, f(t) which is band-limited can be reconstructed from the digital values $f_n$ by using the formula (Papoulis, 1962; Bath, 1974)

$$f(t) = \sum_{n=-\infty}^{\infty} f_n \frac{\sin \pi[(^t/_\tau)-n]}{\pi[(^t/_\tau)-n]} = \sum_{n=-\infty}^{\infty} f_n \frac{\sin \omega_c(t-n\tau)}{\omega_c(t-n\tau)}$$

Where $\omega_c = 2\pi f_c$. We see that to recover the continuous function, f(t) from the digital version, the $n^{th}$ sample is replaced by the sinc function (i.e. $[\sin\omega_c t]/\omega_c t$) which is scaled by the sample value $f_n$ and placed at time, nτ. The scaled and shifted sinc functions are then added together to give the original time function, f(t). Complete recovery also means running values from n = -∞ to +∞, which is practically impossible.

Aliasing is a kind of spectrum distortion which is brought about as a result of too coarse sampling. Fine sampling (implying $f_N > f_c$) and critical sampling (implying $f_N = f_c$), produce no aliasing effects. Coarse sampling (implying $f_N < f_c$) is undersampling and there will be considerable overlap between adjacent spectra in the recovered analogue function. Thus to avoid aliasing effect, we must make the sampling frequency sufficiently high to ensure making the Nyquist frequency at least equal to the cutoff frequency of the original signal.

## 4. Fourier analysis

Fourier analysis is the theory of the representation of a function of a real variable by means of a series of sines and cosines. The discussion of Fourier analysis starts with a statement of Fourier theorem. Fourier (1768 – 1830) stated without proof and used in developing a solution of the heat equation, the so-called Fourier theorem.

### 4.1 Fourier theorem

Any single-valued function, f(x) defined in the interval [-π, π] may be represented over this interval by the trigonometric series, i.e.

$$f(x) = \frac{a_0}{2} + \sum_{n=1}^{\infty}(a_n \cos nx + b_n \sin nx) \quad (1)$$

The expansion coefficients, $a_n$ and $b_n$ are determined by use of Euler's formulas:

$$a_n = \frac{1}{\pi}\int_{-\pi}^{\pi} f(x) \cos nx \, dx \; (n = 0, 1, 2, 3 \ldots), \; b_n = \frac{1}{\pi}\int_{-\pi}^{\pi} f(x) \sin nx \, dx \; (n = 1, 2, 3 \ldots)$$

Fourier investigated many special cases of the above theorem, but he was unable to develop a logical proof of it.

Dirichlet (1805 – 1859) in 1829 formulated the restrictions under which the theorem is mathematically valid. These restrictions are normally called Dirichlet conditions, and are summarized as follows: that for the interval [-π, π], the function, f(x) must (1) be single-valued, (2) be bounded, (3) have at least a finite number of maxima and minima, (4) have only a finite number of discontinuities: piece-wise continuous and (5) periodic, i.e. f(x + 2π) = f(x). For values of x outside of [-π, π], the series in equation (1) above converges to f(x) at values of x for which f(x) is continuous and to $\frac{1}{2}[f(x+0) + f(x-0)]$ at points of discontinuity. The quantities f(x+0) and f(x-0) refer to the limits from the right and left respectively of the point of discontinuity. The coefficients are still given by the Euler's formulas.

### 4.2 Different forms of Fourier series

According to the Fourier theorem, a function such as f(t), which satisfies Dirichlet's conditions can be represented by the following infinite series, the Fourier series, as

$$f(t) = \frac{a_0}{2} + \sum_{n=1}^{\infty}(a_n \cos nt + b_n \sin nt) \quad (2)$$

Where the constants $a_0$, $a_n$ and $b_n$ are given by

$$\left.\begin{aligned} a_0 &= \frac{1}{\pi}\int_{-\pi}^{\pi} f(t)\, dt \\ a_n &= \frac{1}{\pi}\int_{-\pi}^{\pi} f(t) \cos nt \, dt \; (n = 1, 2, 3 \ldots) \\ b_n &= \frac{1}{\pi}\int_{-\pi}^{\pi} f(t) \sin nt \, dt \; (n = 1, 2, 3 \ldots) \end{aligned}\right\} \quad (3)$$

These coefficients ($a_0$, $a_n$ and $b_n$) are called Fourier coefficients and the determination of their values is called Fourier or harmonic analysis.

The three expressions in equation (3) are respectively obtained by multiplying both sides of equation (2) by 1 (i.e. cos 0t), cos nt and sin nt and integrating with respect to t over the period length 2π. We then use the orthogonality properties of sine and cosine functions to eliminate some of the expressions.
The Fourier series in equation (2) will exactly represent the function f(t) only when an infinite number of terms are included in the summation, i.e. when n runs from 1 to infinity (∞). If a finite number of terms is taken, the sum will shoot beyond the value of f(t) in the neighbourhood of a discontinuity (boundaries of the function). The overshoot oscillates about the value with a decreasing amplitude as we move away from the discontinuity. Increasing the number of terms does not influence the error magnitude at the discontinuity, but only leads to a better approximation for the continuous part of f(t). The overshoot and oscillatory behaviour of f(t) at the vicinity of a discontinuity is known as Gibb's phenomenon.
If f(t) is an even function, then equation (2) becomes

$$f(t) = \frac{a_0}{2} + \sum_{n=1}^{\infty}(a_n \cos nt) \tag{4}$$

Since $b_n = 0$ for even f(t), Similarly for odd f(t), $a_0$, $a_n = 0$ and so

$$f(t) = \sum_{n=1}^{\infty} \mathrm{b_n} \sin nt) \tag{5}$$

To solve many physical problems, it is necessary to develop a Fourier series that will be valid over a wider interval. To obtain an expansion valid in the interval [-T, T], we let $f(t) = \frac{a_0}{2} + \sum_{n=1}^{\infty}(a_n \cos \varphi t + b_n \sin \varphi t)$ and determine $\varphi$ such that f(t) = f(t + 2T). In this case, $\varphi = \frac{n\pi}{T}$, hence we obtain:

$$f(t) = \frac{a_0}{2} + \sum_{n=1}^{\infty}(a_n \cos \frac{n\pi t}{T} + b_n \sin \frac{\mathrm{n\pi t}}{\mathrm{T}}), \ \text{-T} \leq \text{t} \leq \text{T} \tag{6}$$

If f(t) satisfies the Dirichlet conditions in this interval, then the Fourier coefficients $a_0$, $a_n$ and $b_n$ can be computed in a similar fashion as in equation (3).
The complex form of the Fourier series in equation (6) can be obtained by expressing $\cos \frac{n\pi t}{T}$ and $\sin \frac{n\pi t}{T}$ in the exponential using the Euler identity cos θ + sin θ = $e^{i\theta}$, where i = √-1 is a complex number. Thus the complex form of the Fourier series can be written as

$$f(t) = \sum_{n=-\infty}^{\infty} C_n e^{in\pi t/T} \ [\text{-T, T}] \tag{7}$$

Where $C_0 = a_0/2 = F(0)$, $C_n = (a_n - ib_n)/2 = F(n)$ and $C_{-n} = (a_n + ib_n)/2 = F(-n)$
Equation (7) is the complex form of the Fourier series. On multiplying both sides of this equation by $e^{-in\pi t/T}$ and integrating with respect to t, we obtain

$$C_n = \frac{1}{2T}\int_{-T}^{T} f(t) e^{-in\pi t/T} dt \tag{8}$$

Note that the amplitude spectrum, $|F(n)| = |F(-n)| = \frac{\sqrt{a_n^2+b_n^2}}{2}$ is symmetrical. From the identity in complex number representation: $a + ib = re^{i\theta}$, with $r = \sqrt{a^2 + b^2}$ and $\tan \theta = b/a$, we have $F(n) = |F(n)|e^{i\vartheta(n)}$ , $F(-n) = |F(-n)|e^{i\vartheta(-n)}$ with $tan[\vartheta(n)] = \frac{-b_n}{a_n}$ , $tan[\vartheta(-n)] =$

$\frac{b_n}{a_n}$ and therefore $tan[\vartheta(-n)] = -tan[\vartheta(n)]$, indicating that the phase spectrum, $\vartheta(n)$ is antisymmetrical.

Equation (7) is the time domain Fourier series and equation (8) is the frequency domain Fourier series. The two equations are an example of a Fourier transform pair, or we can say f(t) is the inverse Fourier transform of F(n) or $C_n$.

### 4.3 Application of Fourier series

Fourier series, as we have already seen, is used as an effective tool in the analysis of periodic functions. We have also noted that any periodic function having a period, T (and satisfying Dirichlet conditions) can be represented by the infinite series of trigonometric functions. Thus f(t) is represented by the addition of sinusoidal and cosinusoidal waves whose frequencies are integral multiples of some fundamental unit of frequency in the signal, f(t). The frequency, $\pi/T$ is called the fundamental and its integral multiples are called the harmonics.

If we plot the Fourier coefficients $a_n$ and $b_n$ as functions of frequency, we obtain a number of discrete spectral lines located at fixed spacing of $\pi/T$. As $T\to\infty$, the spacing of the spectral lines approaches zero. Analysis of the spectrum of a field gives the energy content of the field and may represent phenomenal changes in the attributes of the causative agents of the field. This allows systems to be analyzed in the frequency domain so as to find out the frequency response of the system from the impulse response and vice-versa. It can be used as an intermediate step in more elaborate signal processing techniques (e.g. the fast Fourier transform). Fourier integrals (next section) is useful in the analysis of limited-duration (transient) signals.

### 4.4 Fourier integrals

If we look back at equation (6) and decide to put the expressions for $a_0$, $a_n$ and $b_n$ that followed inside it and use the dummy variable, $\lambda$, we can write equation (6) in a more compact form as

$$f(t) = \frac{1}{2T}\int_{-T}^{T} f(\lambda)d\lambda + \sum_{n=1}^{\infty}\frac{1}{T}\int_{-T}^{T} f(\lambda)d\lambda \cos\frac{n\pi}{T}(t-\lambda)d\lambda \qquad (9)$$

We let $\omega_n = n\pi/T$ (the n[th] angular frequency), $\omega_{n-1} = (n-1)\pi/T$ and therefore $\Delta\omega = \omega_n - \omega_{n-1} = \pi/T$ and substituting all these in equation (9), we obtain

$$f(t) = \frac{1}{2T}\int_{-T}^{T} f(\lambda)d\lambda + \sum_{n=1}^{\infty}\frac{\Delta\omega}{\pi}\int_{-T}^{T} f(\lambda)d\lambda \cos\omega_n(t-\lambda)d\lambda \qquad (10)$$

If we let $T\to\infty$, the following changes will occur in equation (10)

a. The first part of the expression in the RHS will vanish to zero.
b. The increment $\Delta\omega$ becomes very small and in the limit $\Delta\omega\to d\omega$. Thus the digitally increasing $\omega_n$ becomes a continuous variable, $\omega$.
c. The summation can be replaced by an equivalent integral with appropriate limits.

Thus equation (10) becomes

$$f(t) = \frac{1}{\pi}\int_{0}^{\infty} d\omega \int_{-\infty}^{\infty} f(\lambda)\cos\omega(t-\lambda)d\lambda \qquad (11a)$$

This is the Fourier integral. We can further analyze equation (11a) by using the fact that

$\cos \omega(t-\lambda) = \cos \omega t \cos \omega\lambda + \sin \omega t \sin \omega\lambda$ and so

$$f(t) = \frac{1}{\pi}\int_0^{\infty} d\omega \int_{-\infty}^{\infty} f(\lambda) \cos \omega\lambda \cos \omega t \, d\lambda + \frac{1}{\pi}\int_0^{\infty} d\omega \int_{-\infty}^{\infty} f(\lambda) \sin \omega\lambda \sin \omega t \, d\lambda$$

$$= \int_0^{\infty}\Big[\frac{1}{\pi}\int_{-\infty}^{\infty} f(\lambda) \cos \omega\lambda \, d\lambda\Big] \cos \omega t \, d\omega + \int_0^{\infty}\Big[\frac{1}{\pi}\int_{-\infty}^{\infty} f(\lambda) \sin \omega\lambda \, d\lambda\Big] \sin \omega t \, d\omega$$

Or

$$f(t) = \int_0^{\infty} A(\omega) \cos \omega t \, d\omega + \int_0^{\infty} B(\omega) \sin \omega t \, d\omega$$

$$= \int_0^{\infty} [A(\omega) \cos \omega t + B(\omega) \sin \omega t] d\omega \tag{11b}$$

Where $A(\omega) = \frac{1}{\pi}\int_{-\infty}^{\infty} f(\lambda) \cos \omega\lambda \, d\lambda$ and $B(\omega) = \frac{1}{\pi}\int_{-\infty}^{\infty} f(\lambda) \sin \omega\lambda \, d\lambda$

Equation (11b) is the Fourier integral: a Riemann sum of integral consisting of the first of the RHS of the equation, called the Fourier cosine integral and the second part called the Fourier sine integral. This integral equation will converge to f(t) when f(t) is continuous and converges even at points of discontinuity just like a Fourier series.
Note that the function is suppose to be defined from -∞ to +∞, but because of the parity of the function, we only need the function from 0 to ∞. This also means that if we only are interested in the range of 0 to ∞, we can define the function from -∞ to any where we want, then we can have either cosine integral or sine integral by extending the function into negative range either in an even or odd form. Thus Fourier cosine and sine integrals are equivalent to the half-range expansion of Fourier series.

### 4.5 Fourier transforms and theorems

From the complex form of Fourier series given in equation (7) and expression for the coefficients given in equation (8), we make a transition $T\to\infty$ and introduce again the variable, $\omega = n\pi/T$, with $\frac{\Delta\omega}{\pi}T = 1$, since $\Delta n = 1$. Hence equations (7) and (8) would be written as

$f(t) = \sum_{-\infty}^{\infty} C_T(\omega)e^{i\omega t}\Delta\omega$ and $C_T(\omega) = \frac{1}{2\pi}\int_{-T}^{T} f(t)e^{-i\omega t}dt$, where $C_T(\omega) = TC_n/\pi$. If we let $T\to\infty$, $C_T(\omega)\to C(\omega)$, and so

$$C(\omega) = \frac{1}{2\pi}\int_{-\infty}^{\infty} f(t)e^{-i\omega t}dt \tag{12}$$

$$f(t) = \int_{-\infty}^{\infty} C(\omega)e^{i\omega t}d\omega \tag{13}$$

There are other ways of expressing equations (12) and (13) which are the Fourier transforms of each other. If we let $F(\omega) = 2\pi C(-\omega)$, then

$$F(\omega) = \frac{1}{2\pi}\int_{-\infty}^{\infty} f(t)e^{-i\omega t}dt \tag{14}$$

$$f(t) = \frac{1}{2\pi}\int_{-\infty}^{\infty} F(\omega)e^{i\omega t}d\omega \tag{15}$$

Equations (14) and (15) are called the Fourier transform pair. If f(t) satisfies the Dirichlet conditions and the integral $\int_{-\infty}^{\infty}|f(t)|dt$ is finite, then $F(\omega)$ exists for all $\omega$ and is called the

Fourier transform of f(t). The function f(t) in equation (15) is called the Fourier transform (inverse transform) of F(ω) and the pair may be represented by a notation f(t)↔F(ω). F(ω) is called the amplitude of the time domain field, f(t).
So far we have used as variables $t$ and $\omega$, representing time and angular frequency, respectively. Mathematics will, of course, be the same if we change the names of these variables. In describing the spatial variations of a wave, it is more natural to use either $r$ or $x$, $y$, and $z$ to represent distances. In a function of time, the period $T$ is the time interval after which the function repeats itself. In a function of distance, the corresponding quantity is called the wavelength $\lambda$, which is the increase in distance that the function will repeat itself. Thus, if $t$ is replaced by $r$, then the angular frequency $\omega$, which is equal to $2\pi/T$, should be replaced by a quantity equal to $2\pi/\lambda$, which is known as the wave number, $k$. In three dimensions, we can define a Fourier transform pair as

$$F(\vec{k}) = \frac{1}{(2\pi)^{3/2}} \iiint_{-\infty}^{\infty} f(\vec{r}) e^{-i\vec{k}.\vec{r}}\, d^3r \text{ and } f(\vec{r}) = \frac{1}{(2\pi)^{3/2}} \iiint_{-\infty}^{\infty} F(\vec{k}) e^{i\vec{k}.\vec{r}}\, d^3k$$

Where $\vec{r}$ and $\vec{k}$ are radius and wave number vectors respectively. If the wave number, $\vec{k}$ is along the z-axis of the coordinate space, then $\vec{k}.\vec{r}$ = kr cosθ and $d^3r = r^2 \sin\theta d\theta dr d\phi$ and so the Fourier transform of f(r) becomes, $F(\vec{k}) = \frac{1}{k}\sqrt{\frac{2}{\pi}} \int_0^{\infty} f(r) r \sin kr\, dr$ as ϕ runs from 0 to 2π, r runs from 0 to ∞ and θ runs from 0 to π.

Again, how to split $1/(2\pi)^3$ between the Fourier transform and its inverse is somewhat arbitrary. Here we split them equally to conform to most standard expressions.
There are number of useful theorems that connect the Fourier transform pair in equations (14) and (15) and these can be found in standard texts on the subject. We shall only mention a few ones that are easily provable as follows:

a. The Convolution Theorem - If $f_1(t) \leftrightarrow F_1(\omega)$ and $f_2(t) \leftrightarrow F_2(\omega)$ are Fourier pairs, then

$$f_1(t).f_2(t) \leftrightarrow \frac{1}{2\pi} F_1(\omega) * F_2(\omega) \text{ and } f_1(t) * f_2(t) \leftrightarrow F_1(\omega).F_2(\omega)$$

This a very important theorem which has a wide field application. It simply says that the spectrum of a product of two time functions is the convolution of their individual spectra. The theorem can be extended to any number of functions as

$$[f_1(t).f_2(t).f_3(t) \ldots f_n(t)] \leftrightarrow \left(\frac{1}{2\pi}\right)^{n-1} [F_1(\omega) * F_2(\omega) * F_3(\omega) \ldots F_n(\omega)].$$

Note that in general a convolution is expressed as

$$f_1(t) * f_2(t) = \int_{-\infty}^{\infty} f_1(\tau) f_2(t-\tau) d\tau$$

and

$$F_1(\omega) * F_2(\omega) = \int_{-\infty}^{\infty} F_1(p) f_2(\omega - p) dp.$$

The mathematical operation of convolution consists of the following steps:

1. Take the mirror image of $f_2(\tau)$ about the coordinate axis to create $f_2(-\tau)$ from $f_2(\tau)$.
2. Shift $f_2(-\tau)$ by an amount $t$ to get $f_2(t-\tau)$. If $t$ is positive, the shift is to the right, if it is negative, to the left.

3. Multiply the shifted function $f_2(t-\tau)$ by $f_1(\tau)$.
4. The area under the product of $f_1(\tau)$ and $f_2(t-\tau)$ is the value of convolution at $t$.
   The convolution in the time domain is multiplication in the frequency domain and vice versa. Convolution operation has commutative, associative and distributive properties, like most linear systems. We encounter convolution operations in filtering processes, truncating lengthy data using window functions and sampling a signal with a comb function (digitization).

b. The Multiplication Theorem - If $f_1(t)\leftrightarrow F_1(\omega)$ and $f_2(t)\leftrightarrow F_2(\omega)$ are Fourier pairs, then

$$\int_{-\infty}^{\infty} f_1(t) f_2(t) dt = \frac{1}{2\pi}\int_{-\infty}^{\infty} F_1^*(\omega) F_2(\omega)$$

Where $F_1^*(\omega)$ is the complex conjugate of $F_1(\omega)$. The product $F_1^*(\omega)F_2(\omega)$ is called the cross power spectrum. If we put

$$f_1(t) = f_2(t) = f(t)$$

and

$$F_1(\omega) = F_2(\omega) = F(\omega),$$

then the previous equation reduces to

$$\int_{-\infty}^{\infty} [f(t)]^2 dt = \frac{1}{2\pi}\int_{-\infty}^{\infty} [F(\omega)]^2 d\omega$$

This is called Parseval's theorem and the real quantity $[F(\omega)]^2$ represents the power spectrum (or energy spectrum) of the function f(t). Thus, it is interesting to note that if the amplitude spectrum $F(\omega)$ of a given is known, it is possible to compute its power spectrum $[F(\omega)]^2$ and its total energy,

$$E_T = \frac{1}{2\pi}\int_{-\infty}^{\infty} [F(\omega)]^2 d\omega.$$

c. The Correlation Theorem - The Fourier transform of the cross correlation function $\phi_{12}(\tau)$ is the cross power spectrum, $E_{12}(\omega)$ and that of the auto-correlation function, $\phi_{11}(\tau)$ is the power spectrum, $E_{11}(\omega)$. Thus

$$\phi_{12}(\tau) = \int_{-\infty}^{\infty} f_1(t) f_2(t+\tau) dt = \int_{-\infty}^{\infty} f_1(t-\tau) f_2(t) dt$$

and

$$\phi_{11}(\tau) = \int_{-\infty}^{\infty} f(t) f(t+\tau)\, dt$$

and so the correlation theorem says that

$$\phi_{12}(\tau)\leftrightarrow E_{12}(\omega) \text{ or } \phi_{12}(\tau)\leftrightarrow F_1(\omega).F_2^*(\omega)$$

$$\phi_{11}(\tau)\leftrightarrow E_{11}(\omega) \text{ or } \phi_{11}(\tau)\leftrightarrow [F_1(\omega)]^2$$

The cross correlation function behaves in accordance with the degree of similarity between the two correlated functions. It grows larger when the two functions are similar and diminishes otherwise. This function becomes zero in the case of completely random data. We see that the time domain cross correlation and auto-correlation functions are reduced to multiplication of the amplitude spectrum. Auto-correlation of two wavelets is equal to the convolution of the first wavelet with the time-reverse of the second wavelet.

### 4.6 Fast Fourier transform

It is important to find the discrete forms of the continuous Fourier transform pair in equations (12) and (13). These can be written as (Smith, 1999)

$$G(n) = \sum_{k=0}^{N-1} g(k) e^{-i\frac{2\pi nk}{N}} \tag{16}$$

and

$$g(k) = \frac{1}{N}\sum_{n=0}^{N-1} G(n) e^{i\frac{2\pi nk}{N}} \tag{17}$$

Both G(n), the discrete amplitude spectrum and the digitized time domain signal, g(k) are periodic as they repeat at every N points. The amplitude spectrum is symmetric while the phase spectrum is antisymmetric.

For real applications, each of equations (16) and (17) will have to be decomposed into the real and imaginary parts with trigonometric argument of $\omega_n = \frac{2\pi nk}{N}$. Thus an N point time domain signal is contained in arrays of N real parts and N imaginary parts for each equation.

Calculating the discrete Fourier transform (DFT) equations takes a considerable time even with high speed computers because of the cycles of computations that must be run. The raw computations of DFT can either be done by use of simultaneous equations (very inefficient for practical use) or by correlation method in which signals can be decomposed into orthogonal basis functions using correlation (not too useful a method). The third and the most efficient method for calculating the DFT is by Fast Fourier Transform (FFT). This is an ingenious algorithm that decomposes a DFT with N points into N DFTs each with a single point. The FFT is typically hundreds of times faster than the other methods. In actual practice, correlation is the preferred techniques if the DFT has less than 32 points, otherwise the FFT is used.

Cooley & Tukey (1965) have the credit for bringing the FFT to the scientific world. The FFT, a complicated algorithm is based on the complex DFT; a more sophisticated version of the real DFT and operates by first decomposing an N – point time-domain signal into N time domain signals, each composed of a single point. The second step is to calculate the N frequency spectra corresponding to these N time domain signals. Lastly, the N spectra are synthesized into a single frequency spectrum.

An interlaced decomposition is used each time a signal is broken into two, that is, the signal is separated into its even and odd numbered samples. There will be $Log_2N$ stages required in the decomposition and $Nlog_2N$ multiplications instead of NxN multiplications without the FFT use. For instance, a 16-point signal ($2^4$) requires 4 stages, 512-point signal ($2^7$) requires 7 stages, a 4096-point signal ($2^{12}$) requires 12 stages of signal decomposition and so on. The FFT algorithm works on a data length of $2^M$, where M is a positive integer (≥5). If the data length is not up to the requirement for FFT operation, then "zeros" are sufficiently added. The decomposition is nothing more than a reordering of the samples in the signal. After the decomposition, the FFT algorithm finds the frequency spectra of a 1-point time domain signal (easy!) and then combine the N frequency spectra in the exact reverse order that the time domain decomposition took place.

The computation of the inverse FFT is very similar to the forward FFT because of the identical nature of the two. Estimation of power spectrum of a signal can be done by means of FFT.

## 5. The concept of spectra

Though we have mentioned spectra or spectrum in our previous discussions, we shall formally explain it here. The word spectrum (plural: spectra) is used to describe the variation of certain quantities such as energy or amplitude as a function of some parameter, normally frequency or wavelength. Optical spectrum of white light (colour spectrum) dispersed by a glass prism or some other refractive bodies (such as water) is a good example.

When a signal is expressed as a function of frequency, it is said to have been transformed into a frequency spectrum. Thus, mathematically, a time-domain signal, f(t) can be expressed by F(ω), where ω represents angular frequency (ω = 2πf; f being the linear frequency). The function F(ω) is in general complex and may be represented by

1. The sum of real and imaginary parts: $F(\omega) = a(\omega) + ib(\omega)$
2. The product of real and complex parts: $F(\omega) = |F(\omega)|e^{i\varphi(\omega)}$

Where

$$|F(\omega)| = \sqrt{a^2(\omega) + b^2(\omega)}$$

and

$$\varphi(\omega) = tan^{-1}\frac{b(\omega)}{a(\omega)} + 2\pi n; n = 0, \pm 1, \pm 2, \pm 3, \ldots$$

The modulus $|F(\omega)|$ is normally called the amplitude spectrum and the argument $\varphi(\omega)$ is called the phase spectrum.

### 5.1 Spectral analysis of periodic functions

In modern analysis, the time function may be expressed through certain mathematical transformation into a function of frequency. The discrete Fourier transform views both the time domain and the frequency domain as periodic. This can be confusing and inconvenient since most of the real time signals are not periodic.

The most serious consequence of time domain periodicity is time domain aliasing. Naturally, if we take time domain signal and pass it through DFT, we find the frequency spectrum. If we could immediately pass this frequency spectrum through the inverse DFT to reconstruct the original time domain signal, we are expected to recover this signal, save for spill over from one period into several periods - a problem of circular convolution. Periodicity in the frequency domain behaves in much the same way (as frequency aliasing), but is more complicated.

### 5.2 Spectral analysis of aperiodic functions

Equations (14) and (15) or any of their similar versions give the Fourier transform pair for a periodic function. For non-repetitive (or aperiodic) signal, the period T→∞ and the Fourier transform pair are expressed as

$$f(t) = \frac{1}{2\pi}\int_{-\infty}^{\infty} F(\omega)e^{i\omega t}d\omega \quad (18)$$

$$F(\omega) = \int_{-\infty}^{\infty} f(t)e^{-i\omega t}dt \quad (19)$$

Note again that f(t) is real time domain signal, while F(ω), the amplitude spectrum is a complex function.

### 5.3 Fourier spectrum

A time function, f(t) such as gravity field may be transform into another function, F(ω), where the amplitude of all frequency components present in f(t) and their corresponding phases are expressed as function of frequency. The two transform relations are already given in equations (18) and (19).
The complex function, $F(\omega)$, is called the Fourier spectrum and its modulus and arguments as earlier explained are called amplitude and phase spectra respectively. The cosine transform part of $F(\omega)[ = a(\omega) + ib(\omega)]$, $a(\omega)$ is called the co-spectrum and the sine transform part, $b(\omega)$ is called the quadrature spectrum.

### 5.4 Power spectrum

If $\bar{E}$ is the mean power of a real function, f(t) whose period is T, then (Thompson 1982)

$$\bar{E} = \lim_{T\to\infty}\frac{1}{2T}\int_{-T}^{T}(f(t))^2dt \quad (20)$$

Where $(f(t))^2$ is termed the instantaneous energy and the complete integration in equation (20) is the total (mean) energy of the function.
We have already noted that for two Fourier pairs, $f_1(t)\leftrightarrow F_1(\omega)$ and $f_2(t)\leftrightarrow F_2(\omega)$, then

$$f_1(t).f_2(t)\leftrightarrow\frac{1}{2\pi}F_1^*(\omega) * F_2(\omega)$$

and that

$$\int_{-\infty}^{\infty}(f(t))^2\,dt = \frac{1}{2\pi}\int_{-\infty}^{\infty}|F(\omega)|^2d\omega \text{ [Parseval's theorem].}$$

The power spectrum $|F(\omega)|^2$ and its total energy $E_T$ are then related by

$$E_T = \frac{1}{2\pi}\int_{-\infty}^{\infty}|F(\omega)|^2d\omega = \frac{1}{\pi}\int_{0}^{\infty}|F(\omega)|^2d\omega,$$

where the power spectrum $|F(\omega)|^2$ is a real quantity.

### 5.5 Spectral windows and their uses

When we were discussing the convolution theorem, we noted that we might run into convolution operations in truncating lengthy data (signal) by use of window functions.
Data windowing can be viewed as the truncation of an infinitely long function, f(t). A box-car function, w(t) = 1, $-T<t<T$ and w(t) = 0 elsewhere, has a value 1 over the required length (2T) and zero elsewhere. The function, w(t), a time window can be used to truncate f(t) and the truncated time function, $f_{tr}(t)$ = f(t).w(t). Using the convolution theorem, the Fourier transform, $F_{tr}(\omega)$ of the truncated function, $f_{tr}(t)$ is given by $F_{tr}(\omega) = \frac{1}{2\pi}F(\omega) * W(\omega)$, where F(ω) and W(ω) are the Fourier transforms of f(t) and w(t) respectively. We can compute the

W(ω) of the rectangular pulse, w(t) as $2\frac{\sin \omega T}{\omega}$. Thus $F_{tr}(\omega) = \frac{1}{2\pi}F(\omega) * \frac{2\sin \omega T}{\omega}$. This shows that truncating a signal brings about a spectrum modification expressed by the convolution operation between the two spectra, F(ω) and W(ω). The truncation of a signal, therefore, introduces a smoothing effect whose severity depends on the window length. The shorter the window length, the greater the degree of smoothing and vice-versa. The truncated Fourier transform $F_{tr}(\omega)$ is often called the average or weighted spectrum (Blackman & Tukey, 1959). Since all observational data or signals have finite length, truncation effect can never be avoided.

In order to minimize spectral distortion from the signal truncation, other types of time windows may be applied. In general, a window which tapers off gradually towards both ends of the signal introduces less distortion than a window which has near-vertical sides (like the box-car function). A least distortive time window should have the following properties:

a. The time interval must be as long as possible. This implies that its corresponding Fourier transform or the spectral window has its energy concentrated to its main lobe.
b. The shape must be as smooth as possible and free of sharp corners. The more smooth the time window is, the smaller the side lobes of the corresponding spectral window become (the box-car function is a dirty window!).

At this juncture, we shall mention some popular time windows. These include the box-car (rectangular), Bartlett (triangular), Blackman, Daniell, Hamming, Hanning (raised cosine), Parzan, Welch and tapered (rectangular windows). Excellent treatise on spectral windows can be consulted.

In general, the Fourier transforms of time domain windows have central main lobe and side lobes in each transform and the magnitudes of the side lobes emphasize the differences between them. Ideally, the main lobe width should be narrow, and the side lobe amplitude should be small.

Windows are also extensively used in designing filters and the window parameters (side lobe amplitude, transition width and stopband attenuation) must be used for the design.

### 5.6 Fourier spectrum of observational data

To compute the spectrum of a function f(t) which obeys Dirichlet conditions, Fourier transform is applied to it directly, particularly the use of Fourier integral equations. We can use the basic theorems already presented to evaluate the transforms.

The common types of functions which are usually subjected to Fourier analysis are those obtained by some kind of physical measurements. Functions which represent observational data are normally converted into digital form (if presented as a continuous plot in profile or map forms) so that their spectra can be computed by numerical Fourier transformation.

Observational functions are usually not continuous and not infinite as the theory of Fourier transformation demands. For this reason, observational spectra suffer from two types of distortions: (1) truncation effect (by a window function) and (2) digitization effect. When a signal is digitized, its spectrum becomes periodic and so the original spectrum (scaled by the inverse of sampling period) becomes repetitive with the same frequency as that used in sampling the signal. Coarse digitization results in distorted spectrum. The extent of digitization effect depends on the sampling frequency as well as on the cut-off frequency of the signal (see Section 3).

## 6. Spectrum of random fields

### 6.1 Random functions

A random variable is a real-valued function defined on the events of probability system. A random variable, f(t) emanates from a random or stochastic process: a process developing in time and controlled by probabilistic laws. A random (or stochastic) process is an ensemble or set of functions of some parameter (usually time, t) together with a probability measure by which we can determine that any member, or a group of members has certain statistical properties. Like any other functions, random processes can either be discrete or continuous.
At any point in a medium, a unit mass or a unit charge or a unit magnetic pole experiences a certain force. This force will be a force of attraction in the case of gravitational field. It will be a force of attraction or repulsion when two unit charges or two magnetic monopoles of opposite or same polarity are brought close to each other. Every mass in space is associated with a gravitational force of attraction. This force has both magnitude and direction. For gravitational field, the force of attraction will be between two masses along a line joining the bodies. For electrostatic, magnetostatic and direct current flow fields, the direction of the field will be tangential to any point of observation. These forces produce force fields. These fields, either global or man-made local fields are used to quantitatively estimate some physical properties at every point in a medium.
Most geophysical potential fields, in particular gravity and magnetic fields are caused by an ensemble of sources distributed in some complex manner, which may be best described in a stochastic or random framework. We shall examine some characteristics of these fields.

### 6.2 Geophysical potential fields

The potential field, $\phi(x, y, z)$ in free space (i.e. without sources) satisfies the Laplace equation

$$\frac{\partial^2 \phi}{\partial x^2} + \frac{\partial^2 \phi}{\partial y^2} + \frac{\partial^2 \phi}{\partial z^2} = 0 \tag{21}$$

When sources are present, the potential fields satisfy the so-called Poisson equation

$$\frac{\partial^2 \phi}{\partial x^2} + \frac{\partial^2 \phi}{\partial y^2} + \frac{\partial^2 \phi}{\partial z^2} = -\rho(x, y, z) \tag{22}$$

Where $\rho(x,y,z)$ stands for the density, magnetization or conductivity depending opun whether $\phi$ stands for gravity, magnetic or electric potential respectively. It is important to know that both global or local fields are subject to inverse square law attenuation of the signal strengths. It is at its simple peak in gravity work where the field due to a point mass is inversely proportional to the square of the distance from the mass, and the constant of proportionality (the gravitational constant, G) is invariant. Magnetic fields, though complex, also obey the inverse square law. The fact that their strength is, in principle, modified by the permeability of the medium, is irrelevant in most geophysical work, where measurements are made in either air or water. Magnetic sources are, however, essentially bipolar and the modifications to the simple inverse-square law due to this fact are important. A dipole here consists of equal-strength positive and negative point sources: a very small distance apart. Field strength here decreases as the inverse cube of distance and both strength and direction change with "latitude" (inclination) of the Earth's magnetic field. The intensity of the field at a point on a dipole axis is double the intensity at a point, the same distance away on the dipole "equator", and in the opposite direction.

Electric current flowing from an isolated point electrode embedded in a continuous homogeneous ground provides a physical illustration of the significance of the inverse square law. All the current leaving the electrode must cross any closed surface that surrounds it. Usually the surface is spherical, concentric with the electrode and the same function of the total current will cross each unit area on the surface of the sphere. The current per unit area will be inversely proportional to the total surface (half-space) of $2\pi r^2$. Current flow in the earth, however, is modified drastically by conductivity variation.

The potential fields of either gravity, magnetic or electrical fields are the ones given by either the Laplace or Poisson equations. Some of the useful properties of $\phi$(x, y, z) are (i) given this potential field (scalar) over any plane, we can compute the primary or force field (vector) at almost all points in the space by analytic continuation and (ii) the points where the force field cannot be computed are the so-called singular points. A closed surface enclosing all such singular points also encloses the sources which give rise to the potential field. Thus the singularities of the potential field are confined to the region filled with sources.

All these properties are best described and accentuated in the Fourier domain. We shall therefore express the Fourier transformation of the potential field in two or three dimensions (see Section 4.5). In two dimensions, the Fourier transform pair, $\Phi$ (u, v) and its inverse $\phi$(x, y) are given by

$$\Phi(u,v) = \iint_{-\infty}^{\infty} \phi(x,y) \exp[-i(ux+vy)]\, dxdy \tag{23}$$

and

$$\phi(x,y) = \frac{1}{4\pi^2} \iint_{-\infty}^{\infty} \Phi(u,v) \exp[i(ux+vy)]\, dudv \tag{24}$$

Where here, u and v are coordinates of the Fourier plane. Equation (24) is also known as the Fourier integral representation of $\phi$(x, y). Equation (23) exists only if and only if

$$\iint_{-\infty}^{\infty} |\phi(x,y)| dxdy < \infty:$$

a condition generally not satisfied in most geophysical situations except for an isolated anomaly (Roy, 2008). However, the Fourier transform of a real function in two dimensions possesses the following symmetry:

$$\Phi(u,v) = \Phi^*(-u,-v), \Phi(-u,v) = \Phi^*(u,-v)$$

$$\Phi(0,0) = \iint_{-\infty}^{\infty} \phi(x,y) dxdy \tag{25}$$

If $\phi$(x, y, z) is the potential field on a plane z, satisfying the Laplace equation (equation (21)), its Fourier integral representation is given by (Naidu 1987)

$$\phi(x,y,z) = \frac{1}{4\pi^2} \iint_{-\infty}^{\infty} \Phi(u,v) H(u,v,z) \exp[-i(ux+vy)]\, dudv \tag{26}$$

Where H(u, v, z) is to be determined by requiring that it also satisfies the Laplace equation and is only true if it satisfies the differential equation

$$\frac{d^2H}{dz^2} - (u^2+v^2)H = 0\,,$$

whose solution is H(u, v, z) = $e^{\pm s z}$ for all values of z, where $s = \sqrt{u^2+v^2}$. For z ≥ 0 equation (26) becomes

$$\phi(x,y,z) = \frac{1}{4\pi^2}\iint_{-\infty}^{\infty} \Phi(u,v)e^{-sz}e^{i(ux+vy)}dudv \tag{27}$$

Equation (27) is a useful representation of the potential field and forms the basis for derivation of many commonly encountered results.

### 6.3 A random interface

Potential fields that are caused by stochastic sources are of two types: a random interface separating two homogeneous media (e.g. sedimentary rocks overlying a granitic basement) and a horizontal layer of finite thickness within the density or magnetization varying randomly. We can relate the stochastic character of the potential fields to this interface or layer with the aim of determining some gross features of the source model (e.g. depth to interface). Figure 2 shows a homogeneous random interface separating two media. Let f(x, y) be a homogeneous (stationary) stochastic field. The spectral representation (or Cramer representation) of a homogeneous random field is given by (Yaglom, 1962)

$$f(x,y) = \frac{1}{4\pi^2}\iint_{-\infty}^{\infty} \mathrm{d}F(u,v)\exp[i(ux+vy)] \tag{28}$$

Where

$$\mathrm{dF(u, v) = F(u + du, v + dv) - F(u, v), (du, dv)}\to 0$$

and F(u, v) is the generalized Fourier transform of f(x, y). Some of the properties of dF(u, v) include:

i. $E\left\{\frac{1}{4\pi^2}dF(u,v)\right\} = E[f(x,y)]: u = v = 0$

$$= 0: (u, v) \neq 0$$

ii. $E\{dF(u,v).dF^*(u',v')\} = 0: (u,v) \neq (u',v')$

iii. When the two points overlap [i.e. (u, v) = (u', v')],

then

$$E\left\{\frac{1}{4\pi^2}dF(u,v).\frac{1}{4\pi^2}dF^*(u,v)\right\} = \frac{1}{4\pi^2}S_f(u,v)dudv$$

Where $S_f(u, v)$ is the spectrum (power spectrum) of the stochastic field, f(x, y).
We can further obtain some basic properties of the random potential field in free space. If $\phi(x, y, z)$ is a random potential field in free space, then similar to equation (26),

$$\phi(x,y,z) = \frac{1}{4\pi^2}\iint_{-\infty}^{\infty} \mathrm{d}\Phi(u,v)H(u,v,z)\exp[i(ux+vy)] \tag{29}$$

Where H(u, v, z) is selected so that $\phi(x, y, z)$ satisfies Laplace equation and dΦ(u, v) is a random function. When equation (29) is substituted in the Laplace equation, we obtain

$$\frac{d^2H(u,v,z)}{dz^2} - (u^2+v^2)H = 0$$

whose solution is

$$H(u,v,z) = \exp[\pm\sqrt{u^2+v^2}\,z]$$

and so equation (29) becomes

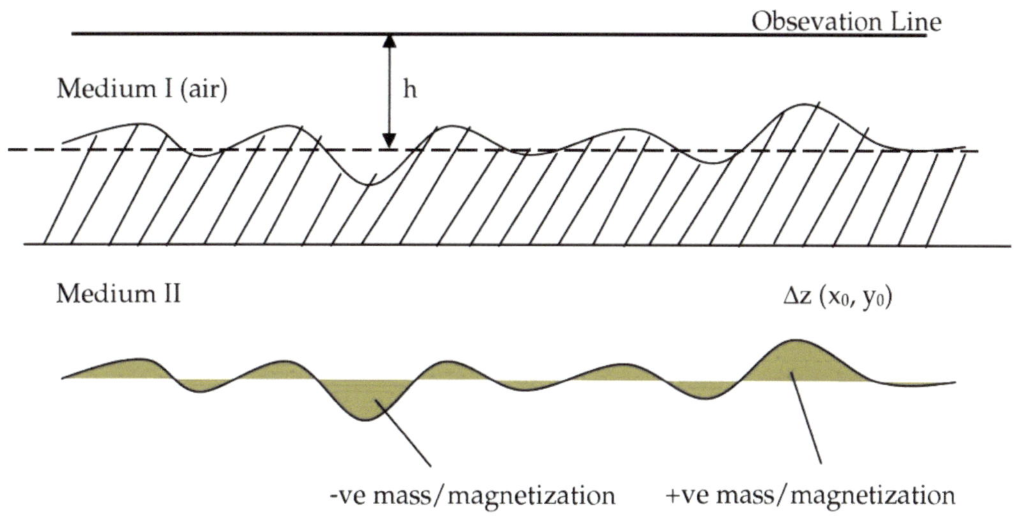


Fig. 2. A homogeneous random interface separating two media of uniform density/magnetization

$$\phi(x,y,z) = \frac{1}{4\pi^2}\iint_{-\infty}^{\infty} \mathrm{d}\Phi(u,v)\exp(-sz)\exp[i(ux+vy)] \quad (30)$$

The spectral representation of the three components (x, y, z) of the potential field is easily obtained by differentiating equation (30) with respect to the coordinate axis

$$\left[f_x(x,y,z) = -\frac{\partial\phi}{\partial x}, f_y(x,y,z) = -\frac{\partial\phi}{\partial y}, etc\right]$$

and the spectrum of the potential field and its components may also be obtained from equation (30).

The cross-spectrums between different components can also be computed from the field equations resulting from the coordinate components obtained from equation (30).

We again look at the random interface (Fig. 2) for the cases of gravity and magnetic fields. Here we see that the potential field is observed on a plane, h unit above the interface. Medium I can be replaced with a vacuum (i.e. where density/magnetization = 0) and the lower medium (Medium II) is characterized by density or magnetization equal to the difference in density or magnetization of Medium I and Medium II. It is easy to see that the observed field will consist of two components: a constant part representing the field due to a semi-finite medium with its upper surface as a horizontal surface, and a random component representing the field due to the random interface.

For the gravity aspect, the vertical gravity field due to a random interface on the observation plane may be expressed as (Telford et al., 1990; Roy, 2008)

$$f_z(x,y,h) = -\iint_{-\infty}^{\infty}\int_0^{\Delta z} \frac{G\Delta\rho(h+z_0)}{[(x-x_0)^2+(y-y_0)^2+(h+z_0)^2]^{3/2}} dx_0 dy_0 dz_0 \quad (31)$$

Where $\Delta\rho$ is density contrast between Medium II and Medium I with the mass element located at point ($x_0$, $y_0$, $z_0$) and $\Delta z(x_0, y_0)$ is a homogeneous random function. Equation (31) can be further expressed using a generalized Fourier transform, $\Delta Z(u, v)$ of the random function, $\Delta z(x, y)$ as (Naidu & Mathew, 1998)

$$f_z(x,y,h) = -\frac{1}{2\pi} G\Delta\rho \iint_{-\infty}^{\infty} \sum_{n=1}^{\infty} d\Delta Z_n(u,v) \frac{(-s)^{n-1}}{n!} \exp(-sh) \exp[i(ux+vy)] \quad (32)$$

Where $\Delta Z_n(u, v)$ is define from the following relation

$$\Delta z^n(x,y) = \frac{1}{4\pi^2} \iint_{-\infty}^{\infty} d\Delta Z_n(u,v) \exp[i(ux+vy)]$$

When $\Delta z(x_0,y_0)<<h$

$$f_z(x,y,h) = -\frac{1}{2\pi} G\Delta\rho \iint_{-\infty}^{\infty} d\Delta Z(u,v) \exp(-sh) \exp[i(ux+vy)] \quad (33)$$

The magnetic field due to a random interface may be obtained either by repeating the mathematical procedure for a magnetic field or by making use of the relationship between the magnetic and gravity potential (Poisson's relation). The magnetic potentials and fields can be estimated from gravitational potential using Poisson's relation expressed as $\phi = \frac{I}{G\rho}\frac{\partial\psi}{\partial\lambda}$, where φ is the magnetic potential, ψ is the gravitational potential, I, the magnetic polarization, ρ, the volume density of the medium and G is the gravitational constant. The horizontal and vertical components of the magnetic field are respectively,

$$H_x = -\frac{\partial\phi}{\partial x} \text{ and } H_z = -\frac{\partial\phi}{\partial z}.$$

By using the latter approach, the magnetic potential, $f_T(x, y, h)$ is expressed as (Naidu & Mathew, 1998)

$$f_T(x,y,h) = -\frac{1}{2\pi} G\Delta\rho \iint_{-\infty}^{\infty} \frac{\Gamma(u,v)}{s} \sum_{n=1}^{\infty} d\Delta Z_n(u,v) \frac{(-s)^{n-1}}{n!} \exp(-sh) \exp[i(ux+vy)] \quad (34)$$

Where $\Gamma(u, v) = (im_x u + im_y v - m_z s)(i\alpha u + i\beta v - \gamma s)$ and $(m_x, m_y, m_z)$ are components of the magnetization. Note that $m_x = I_x\Delta k$, $m_y = I_y\Delta k$ and $m_z = I_z\Delta k$, where $\Delta k = (k_{medII} - k_{medI})$ is the magnetization contrast between the medium II and medium I and (α, β, γ) are the direction cosines of the gradient direction and $(I_x, I_y, I_z)$ are the three components of the inducing magnetic field. When $|\Delta z(x_0, y_0)| \ll h$

$$f_T(x,y,h) = -\frac{1}{2\pi} \iint_{-\infty}^{\infty} \frac{\Gamma(u,v)}{s} \exp(-sh)\, d\Delta Z(u,v) \exp[i(ux+vy)] \quad (35)$$

### 6.4 Discrete sampling of potential fields

Potential fields are continuous functions of space. They have to be sampled for processing on a digital computer. How do we sample them?

A homogeneous random field, f(x) is sampled at $x = n\Delta x$, $n = 0, \pm1, \pm2, \ldots$, where $\Delta x$ is the sampling interval. The accuracy of the sampling is judged by whether by using the samples, we can recover the original function with as small an error as possible (i.e. with the mean square error identically zero). This can be achieved as we have earlier noted that (i) the spectrum of the field be band-limited and (ii) sampling rate ($= 1/\Delta x$) is at least twice the highest frequency present. Band-limitedness of the spectrum of the field means $S_f(u) = 0\ for\ |u| \geq u_0$, where $u_0 = 2\pi/\lambda_0$; $\lambda_0$ being the smallest wavelength corresponding to the highest frequency and condition (ii) implies that $\Delta x = \lambda_0/2$.

A random field in two dimensions can be sampled in more than one way: we may use a square or rectangular grid, polar grid or hexagonal grid. Such a choice of sampling patterns is not available for one dimensional field. The Fourier transform of discrete data turns out to

be one of the basic steps in data processing, but the FFT algorithm is most often used for the computation when the data should have been windowed appropriately.
Spectral windowing, as we have earlier noted comes in because potential data are available over a finite area (as against the infinite length demanded from the mathematics). The finiteness of data is either because no measurements were made outside the area of investigation or the data are found to be homogeneous only over a finite area. Such a situation may be modeled as a product of a homogeneous random process and a window function. Thus the model of the finite data is

$$f_0(m, n) = f(m, n).w_0(m, n)$$

Where $f_0(m, n)$ is the observed field over a finite area and $w_0(m, n)$ is a discrete window function and f(m, n) is the random potential field function. Hence from equation (28), we have

$$f(m,n) = \frac{1}{4\pi^2}\iint_{-\infty}^{\infty} \mathrm{d}F(u,v)\exp[i(um+vn)] \tag{36a}$$

$$w_0(m,n) = \frac{1}{4\pi^2}\iint_{-\infty}^{\infty} \mathrm{W}_0(u,v)\exp[i(um+vn)]dudv \tag{36b}$$

$$f_0(m,n) = f(m,n).w_0(m,n)$$

$$= \frac{1}{4\pi^2}\iint_{-\pi}^{\pi} \mathrm{F}_0(u,v)\exp[i(um+vn)]dudv \tag{36c}$$

Where

$$F_0(u,v) = \frac{1}{4\pi^2}\iint_{-\pi}^{\pi} dF(u',v')W_0(u-u',v-v').$$

The DFT coefficients of $f_0(m, n)$, $0 \leq m \leq M\text{-}1$ and $0 \leq n \leq N\text{-}1$ (area of investigation is a rectangle of size M x N) and so for $0 \leq k \leq M\text{-}1$ and $0 \leq l \leq N\text{-}1$

$$F_0(k,l) = \frac{1}{4\pi^2}\iint_{-\pi}^{\pi} dF(u',v')W_0(\frac{2\pi k}{M}-u',\frac{2\pi l}{N}-v') \tag{37}$$

And so the spectrum of a finite random potential field is (Kay 1989)

$$S_{f_0}(k,l) = E\left\{\frac{1}{MN}|F_0(k,l)|^2\right\}$$

$$= \frac{1}{4\pi^2}\iint_{-\pi}^{\pi} S_f(u',v')\left|W_0(\frac{2\pi k}{M}-u',\frac{2\pi l}{N}-v')\right|^2 du'dv' \tag{38}$$

Equation (38) gives a relationship between the determined spectrum from finite 2D data which are observed and the spectrum of infinite 2D data which are not observed. The factor $|W_0(u,v)|^2$ in the spectrum expression is the spectrum of the window function. The angular variation of the spectrum of the potential field may be distorted unless the window possesses the following properties:

a. Window spectrum most be closer to a delta function. A time domain delta function, $\delta(t)$ has that $\delta(t) \geq 0$ for some finite value of t, $\delta(t - a) = 0$ for $t \neq a$ and $\int_{-\infty}^{\infty}\delta(t)dt = 1$. A window spectrum must be closer to the properties of $\delta(t)$.
b. The leakage of power is minimum which is possible by controlling the height of sidelobes of window spectrum and
c. The spectrum of the window must be isotropic or close to being isotropic.

### 6.5 Angular and radial spectra of the field

Naidu (1969) and Mishra & Naidu (1974) gave the spectrum of a two-dimensional magnetic survey as

$$S_K(u,v) = \frac{1}{K}\sum_{k=-1}^{K}\frac{1}{L_x L_y}\left|X_k(u,v)\right|^2 \tag{39a}$$

Where S (u, v) is the power spectrum of a 2-D aeromagnetic field, X (u, v) is the Fourier transform of the field, $L_x$ and $L_y$ are length dimensions, and u and v are frequency in the x and y directions respectively given as u = $2\pi/L_x$ and v = $2\pi/L_y$. A two-dimensional spectrum can be expressed in a condensed form as one-dimensional spectra; radial and angular (Spector & Grant, 1970; Mishra & Naidu, 1974; Naidu, 1980; Naidu & Mishra, 1980; Naidu & Mathew, 1998). The radial spectrum is defined as

$$R_f(s) = \frac{1}{2\pi}\int_0^{2\pi} S_f(s\cos\theta, s\sin\theta)d\theta \tag{39b}$$

Where again $s = \sqrt{u^2 + v^2}$ is the magnitude of the frequency vector and $\theta = \tan^{-1}(v/u)$ is the direction of the frequency vector in the spatial frequency plane v and u (in radian/km) in the x- and y-directions respectively. The angular spectrum is defined as

$$A(\theta) = \frac{1}{\Delta s}\int_{s_0}^{s_0+\Delta s} S_f(s\cos\theta, s\sin\theta)ds \tag{40}$$

Where, $\Delta s$ is the radial frequency band starting from $s_0$ to $s_0 + \Delta s$, over which the averaging is carried out. It is useful as a rule to look at power spectra in one-dimensional or profile form rather than in two-dimensional or map form. This is because S is a somewhat bumpy function of θ (Fedi et al., 1997) when the width of the model is moderately large, and the bumpiness imparts a certain irregularity to the contours (Spector & Grant, 1970). The power spectra in one-dimension also enables ensemble of magnetic block parameters (average magnetic moment/unit depth, average depths to top, average thickness and average widths) to be factored out completely for effective analysis.

Usually the angular spectrum is normalized with respect to the radial spectrum so as to free this spectrum from any radial variation. In this case (Naidu & Mathew, 1998)

$$A_{norm}(\theta) = \frac{1}{\Delta s}\int_{s_0}^{s_0+\Delta s} \frac{S_f(s\cos\theta, s\sin\theta)}{R_f(s)}ds \tag{41}$$

The computations of spectra in equations (40) and (41) may require the use of template. The radial spectrum is computed by averaging the 2D spectrum over a series of annular regions while the angular spectrum is computed by averaging over angular sectors in the template. Naidu (1980) and Naidu & Mathew (1998) have shown that the angular spectrum of the total field of a uniform magnetized layer having an uncorrelated random magnetization is given by

$$S_{fT}(u,v) = e^{-2hs} I_0^2 s^4 \left[\gamma^2 + (\alpha^2 + \beta^2)\cos^2(\theta - \theta_0)\right]^2 \tag{42}$$

Where h is depth to the magnetic layer, $I_0$ is the direction of the current earth's magnetic field, $\alpha$, $\beta$ and $\gamma$ are directional cosines while $\theta_0$ is the declination of the earth's magnetic field. The non-exponent component of equation (44) is exactly the same as the square of $\Gamma(u, v)$ expressed in equation (34). The angular spectrum, $A_{norm}(\theta)$ may now be expressed as

$$A_{norm}(\theta) = \Omega[\gamma^2 + (\alpha^2 + \beta^2)\cos^2(\theta - \theta_0)]^2$$

Where $\Omega$ is a constant. Thus the angular spectrum is a maximum in the direction of polarization vector. Naidu (1970) had earlier shown that the shape of the angular spectrum is a product of a number of factors such as rock type, strike and polarization vector, but the latter two factors influence the shape of the spectrum significantly. Thus the presence of peaks in the angular spectrum gives an indication of linear features in the map.

On the other hand, the radial spectrum gives a measure of the rate of decay with respect to radial frequency of the spectral power, which may represent a deep-seated phenomenon (Bhattacharyya, 1966; Naidu, 1970, Spector & Grant, 1970).

### 6.6 Estimation of radial and angular spectra of aeromagnetic data

We shall apply the concept of radial and angular spectra to synthetic and real data. For the synthetic data, we have taken the field over a magnetic dipole buried at a depth of 4 km. Other parameters of the dipole include: inclination and declination of the inducing field (and remanent field of the dipole) on the dipole are respectively $6^o$ and $8^o$ (i.e. at low magnetic latitudes), where the field strength of the inducing field is 33510 nT and the magnetization intensity is 0.01 A/m, while its susceptibility is assumed as $9.5 \times 10^{-5}$cgs (0.0012 SI). For space, the anomaly map is not shown. In computing the angular spectrum of this dipole anomaly, the 50x50 data grid was padded with sufficient zeros and cosine tapered to make data matrix amenable for FFT computations. This resulted in a data matrix size of 64x64. The angular spectrum is then calculated for three frequency sub-bands (1-10, 10-20 and 20-30 frequency numbers) for angular interval of $180^o$. The highest frequency in the data is 32.

Figure 3 shows the curves representing the low (1-10), middle (10-20) and high (20-30) frequency numbers. Observation of these curves shows that $8^o$ and $90^o$ spectral peaks appear in the three frequency subbands followed by a peak at $156^o$ in the mid to high frequency sub-bands though with reduced magnitude. Thus we see the emergence in the three frequency sub-bands, of $8^o$: the declination of both the inducing and remanents fields. The other angular feature ($90^o$) at subdued level is due to the tapered window used in the analysis. The other peak ($156^o$) corresponds to the direction of the polarization in the direction of the inducing field at this magnetic latitude.

The real data used for the computation of radial and angular spectra were obtained from aeromagnetic total field intensity of the Middle Benue Trough, Nigeria(MBT) collected from 1974 to 1976. The MBT is the central part of the main Benue Trough of Nigeria. The Benue trough is linked genetically to the oil/gas bearing rocks of the inland Nigerian Niger Delta area. With an upbeat in petroleum efforts in the inland basins, attention is now focused on the Benue trough with more emphasis laid on the structural setting of the basin.

The composite total field intensity data were corrected for the main field using the IGRF 1975 model and this resulted in the residual field map (Fig. 4) used for the present analysis.

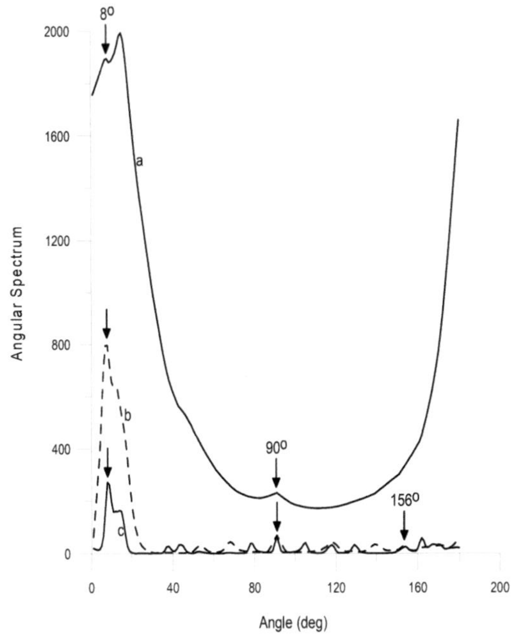


Fig. 3. Angu lar spectrum of the dipole anomaly (properties described in Section 6.6) with labels as follows: (a) low frequency band (1-10), (b) mid frequency band (10-20) and (c) high frequency band (20-30)

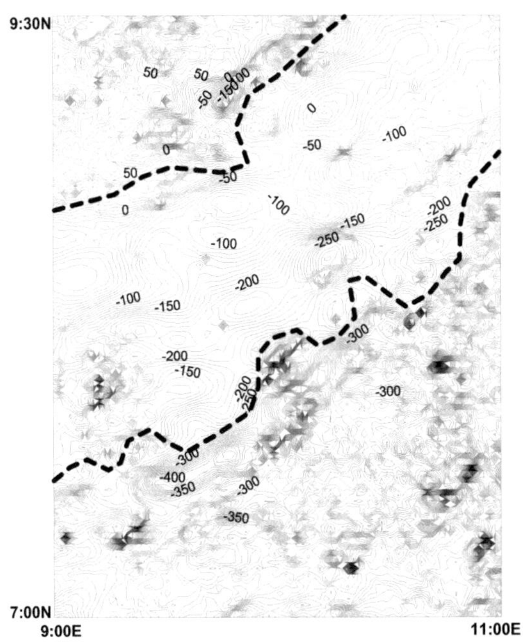


Fig. 4. Total field magnetic anomaly over the Middle Benue Trough, Nigeria. The 1975 IGRF model (epoch date 1st January 1974) field has been removed. Approximate sedimentary-basement boundary is indicated in dashed lines. Contour interval is 10 nT.

The angular and radial spectra of the residual total field magnetic map (Fig. 4) were computed. First, for detailed delineation of angular features, the map was divided into two sub-areas: north/south or upper/lower. The upper part (enclosing nearly 90% of the sediments) has a size of 128x128 and the lower portion has a size of 57x120. It is computationally efficient to use the Fast Fourier Transform (FFT) algorithm to accomplish the analysis of the angular and radial spectra. The upper part of the map (128x128) is already amenable for FFT application since the matrix size is already $2^N$, where N is an integer. The lower part however, is cosine tapered and then padded with sufficient zeros to make data matrix amenable for FFT application. This resulted into another data of 128x128. Next, the averaging was carried out over three frequency bands (1-20, 20-40 and 40-60 frequency numbers) for each data matrix. This division represents the low frequency band (1-20), mid frequency band (20-40) and high frequency band (40-60), noting that the highest frequency number in the data is 64. The angular spectra for the three sub-bands in each sub-area of the map are shown (Figs. 6a and b). Figure 6a shows correspondences between the three spectra indicating peaks at 4°, 36°, 56°, 75°, 102°, 140° and 169°; where the first peak (4°) appears in the three sub-bands, while the other six peaks appear in the mid and high frequency bands. A similar pattern is obtained in the lower part of the map area where also 4° peak appears in the three sub-bands (Fig. 6b) and prominent peaks at 12° and 157° at the mid and high frequency sub-bands amid a subdued peak at 149°.

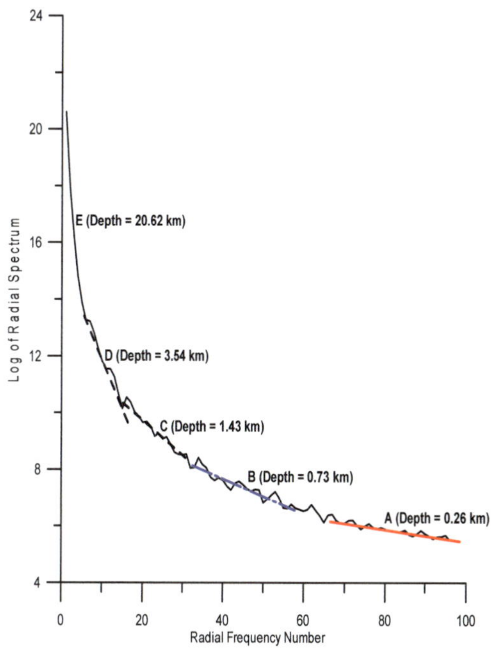


Fig. 5. Radial spectrum plotted against frequency number of the total field magnetic anomaly over the Middle Benue Trough, Nigeria. Five linear segments are marked (A, B, C, D, E). Estimates of depths from slopes are indicated.

The 4° peak appearing in the three sub-bands corresponds to the value of the inclination of the Earth's magnetic field in the area. The peak at 56° corresponds to the general trend of the Benue Trough (N56E). Other angular values have been fully mapped in the area (Benkhelil, 1982, 1989; Likkason, 2005).

To get the overall picture of the depth estimates of the area, the radial spectrum and its log of the aeromagnetic field of the Middle Benue Trough, Nigeria were computed and the latter plotted against frequency number (Fig. 5). Logarithmic spectra are preferred for the analysis than linear spectra because of the additive property of the former, which gives better performance of ensemble average parameters (Spector and Grant 1970) as influences simply add. The plotted radial spectrum seems to support the possibility of five linear segments (marked **A**, **B**, **C**, **D** and **E**). The segment marked **E** represents the spectrumdominated by deep-seated contribution, while the mid-layers marked **D, C** and **B** come from near-surface contribution and other plate sources of intermediate depths. The low gradient portion (marked **A**) represents two contributions: from near-surface structures and magnetic terrain effect. From the slopes of the segments at these points it may be inferred that the depths to corresponding magnetic layers are 20.62 km (portion **E**), 3.54 km (portion **D**), 1.43 km (portion **C**), 0.73 km (portion **B**) and 0.26 km (portion **A**). Flight height above mean sea level (m s l) in the survey area is 0.275 km. We suggest that the deep band at 20.62 km is probabily the lowest boundary of static magnetic sources.

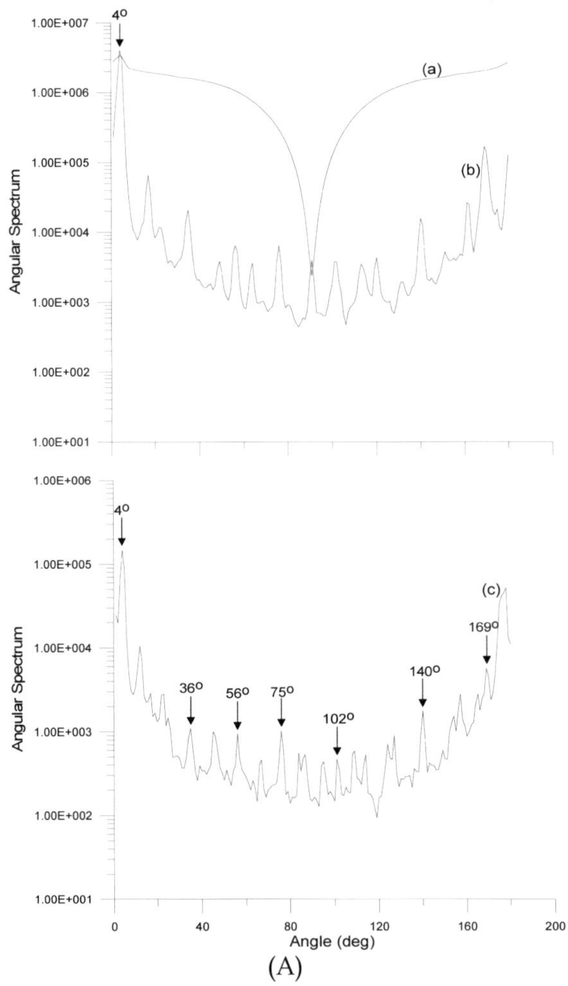


(A)

Fig. 6. (continued)

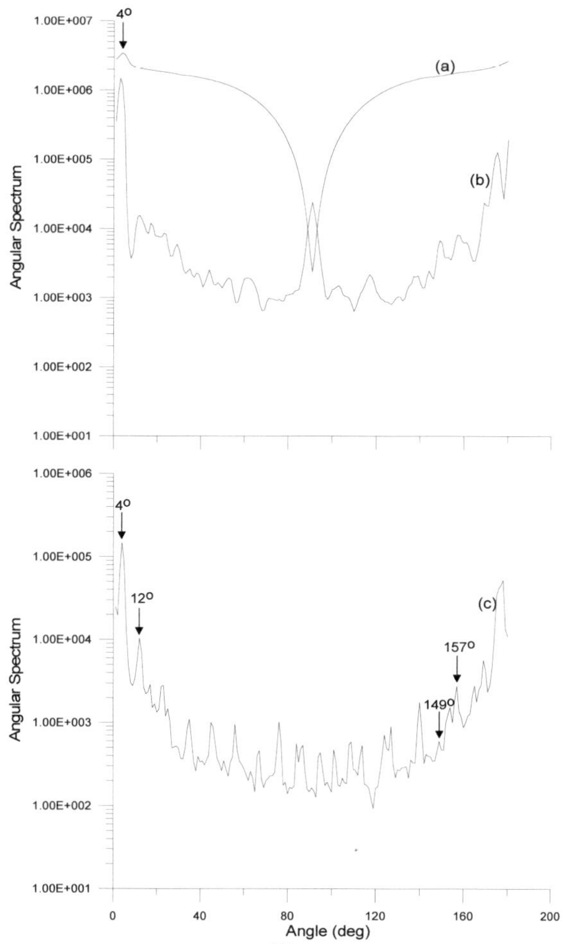


(B)

Fig. 6. Angular spectrum of the total field magnetic anomaly over (A) the upper sub-area and (B) the lower sub-area of the Middle Benue Trough, Nigeria. In each of the curves, the labels represent as follows: (a) the low frequency band (1-20), (b) the mid-frequency band (20-40) and (c) the high frequency band (40-60).

## 7. Conclusion

Spectrum analysis is a basic tool in signal processing and shows how the signal power is distributed as a function of spatial frequencies. These Fourier-based methods have found usefulness in the analysis of geophysical data and what has been presented in this chapter is just tip of the iceberg. It is important as a rule to have a good understanding of the signal and the corrupting noise. This will lead to a successful extraction of the desired signal from the observed map data. Parameter estimation from such processes have to be carefully done as some operations involved in the processes bear on the resultant data. For example, the effect of the window function bears to some extent on the final outcomes of the computations of angular spectra of both the synthetic and real aeromagnetic field data used in the last section. The window effects must be recognized and pointed out.

## 8. References


Bath, M. (1974). *Spectral analysis in geophysics*, Elsevier Amsterdam

Benkhelil, M. J. (1982). Benue Trough and Benue Chain, *Geol. Mag.* 119, 155-168

Benkhelil, M. J. (1989). The origin and evolution of the Cretaceous Benue Trough, Nigeria, *J. Afric. Earth Sci.* 8, 251-282

Bhattacharyya, B. K. (1966). Continuous spectrum of the total magnetic field anomaly due to a rectangular prismatic body, *Geophysics* 31, 97-121

Blackman, R. B. & Tukey, J. W. (1959). *The measurement of power spectrum*, Dover Publications, New York

Cooley, J. W. & Tukey, J. W. (1965). An algorithm for the machine computation of complex Fourier series, *Math. Comput.* 19, 297 - 301

Fedi, M., Quarta, T. & De Santis, A. (1997). Inherent power-law behaviour of magnetic field power spectra from a Spector and Grant ensemble, *Geophysics* 62, 1143-1150

Kay, S. M. (1989), *Modern spectrum analysis*, Prentice Hall, New Jersey

Likkason, O. K. , Ajayi, C. O., Shemang, E. M. & Dike, E. F. C. (2005). Indication of fault expressions from filtered and Werner deconvolution of aeromagnetic data of the Middl;e Benue Trough, Nigeria, *J. Mining and Geology* 41 (2), 205 - 227

Mishra, D. C. & Naidu, P. S. (1974). Two-dimensional power spectral analysis of aeromagnetic fields, *Geophys. Prosp.* 22, 345-353

Naidu, P. S. (1969). Estimation of spectrum and cross spectrum of aeromagnetic field using fast digital Fourier transform (FDFT) techniques, *Geophys. Prosp.* 17, 344-361

Naidu, P. S. (1970). Statistical structure of aeromagnetic field, *Geophysics* 35, 279-292

Naidu, P. S. (1980). Spectrum of potential fields due to randomly distributed sources, *Geophysics* 33, 337-345

Naidu, P. S. (1987). Characterization of potential field signal in frequency domain, *J. Assoc. Exploration Geophys.* (India) 8, 1-16

Naidu, P. S. & Mathew, M. P. (1998). Digital analysis of aeromagnetic maps: detection of a fault, *J. Applied Geophysics* 38,169-179

Naidu, P. S. & Mishra, D. C. (1980). Radial and angular spectrum in geophysical map analysis, In: *Application of Information and Control Systems* (D.G. Lainiotis and N. S. Tzannes, Eds.). Riedel Publishing Co., Dordrecht, 447-454

Papoulis, A. (1962). *The Fourier integral and its application*, McGraw Hill, New York

Roy, K. K. (2008). *Potential theory in applied geophysics*, Springer, New York

Smith, S. W. (1999). *The Scientist and Engineer's guide to digital signal processing* (2nd edition), California Technical Publishing

Spector, A. & Grant, F. S. (1970). Statistical models for interpreting aeromagnetic data, *Geophysics* 35, 293-302

Telford, W. M., Geldart, L. P., Sheriff, R. E. & Keys, D. A. (1990). *Applied geophysics*, Cambridge University Press, London

Thompson, D. J. (1982). Spectrum estimation and harmonic analysis, *Proc. IEEE* 70 (7), 1055 - 1096

Yaglom, A. H. (1962). *Introduction to theory of stationary random functions*, Prentice Hall, New Jersey

# 3

# Quantitative Evaluation of Spatial Interpolation Models Based on a Data-Independent Method

Xuejun Liu, Jiapei Hu and Jinjuan Ma
*Key Laboratory of Virtual Geographic Environment (Nanjing Normal University), Ministry of Education, Nanjing,*
*School of Geography Science, Nanjing Normal University, Nanjing, China*

## 1. Introduction

Spatial interpolation, i.e. the procedure of estimating the value of properties at unsampled sites within areas covered by existing observations (Algarni & Hassan, 2001), appears various models using local/global, exact/approximate and deterministic/geostatistical methods. As being an essential tool for estimating spatial continuous data which plays a significant role in planning, risk assessment and decision making, interpolation methods have been applied to various disciplines concerned with the Earth's surface, such as cartography (Declercq, 1996), geography (Weng, 2002), hydrology (Lin & Chen, 2004), climatology (Attorre et al, 2007), ecology (Stefanoni & Ponce, 2006), agriculture and pedology (Wang et al, 2005; Robinson & Metternicht, 2006), landscape architecture (Fencik & Vajsablova, 2006) and so on.

Since spatial interpolation is based on statistics, there are inevitably a certain assumptions and optimizations. As a result, errors introduced by spatial interpolation and their propagation in analysis models will certainly influence the quality of any decision-making supported by spatial data. This has been one of the hot issues of geographical information science in recent years (David et al, 2004; Shi, W. Z, et al, 2005; Weng, 2006). There are many factors affecting the performance of spatial interpolation methods. The errors are mainly generated from sample data density (Stahl et al., 2006), sample spatial distribution (Collins and Bolstad, 1996), data variance (Schloeder et al., 2001), grid size or resolution (Hengl, 2007), surface types (Zimmerman et al., 1999) and interpolation algorithms (Weng, 2006). However, there are no consistent findings about how these factors affect the performance of the spatial interpolators (Li & Heap, 2011). Therefore, it is difficult to select an appropriate interpolation method for a given input dataset.

With the increasing applications of spatial interpolation methods, there is a growing concern about their accuracies and evaluation measures (Hartkamp et al., 1999). The previous studies have greatly focused on individual evaluation methods of spatial interpolation (Weber & Englund, 1992 & 1994; Erxleben et al, 2002; Chaplot, 2006; Weng, 2006; Erdogan, 2009; Bater & Coops, 2009). It is necessary to explore comprehensive evaluation methods of interpolation accuracy. Two fundamental issues related to assessment measures of interpolation are addressed here as follows.

1. Comparison results: most commonly used methods for evaluation of spatial interpolation models compare the measured data with the interpolated data. However, it is no doubt that measured data are always unsatisfactory. This leads to unknown errors inherent in measured data (Zhou & Liu, 2002). The results may not always keep consistent and even get some controversial conclusions. For example, Laslett et al. (1987), Javis & Stuart (2001) and Erdogan (2009) thought Thin Plate Spline interpolation model can give better interpolated results, while Bater & Coops (2009) argued that Nature Neighbour Interpolation is with more accurate interpolated value. Meanwhile, some researchers (Hosseini et al., 1993; Gotway et al., 1996; Zimmerman et al. 1999; Erxleben et al., 2002; Vicente-Serrano et al., 2003; Attorre et al, 2007; Piazza et al, 2011) found that Kriging is the best one among all the existing interpolation models. Another phenomenon should be mentioned is that the frequency of interpolation methods compared varies considerably among methods and different studies have compared a suite of different methods, which makes it difficult to draw general conclusions.
2. Assessment indices: there are two typical assessment indices, i.e. statistical measures and spatial accuracy measures. The statistical measures such as Root Mean Squared Error (RMSE), Standard Deviation (SD) and Mean Error (ME) are most frequently used (Weber & Englund, 1994; Weng, 2002; Vicente-Serrano et al., 2003; Hu et al., 2004; Weng, 2006; Tewolde, 2010), whereas incapable of describing the spatial pattern of errors. Then the morphological accuracy measures such as accuracy surface and spatial autocorrelation (Weng, 2002; Weng, 2006; Tewolde, 2010) are employed. However, in order to obtain full evaluation of the interpolations, following problems should be further addressed: (1) most of the evaluations are still concentrated on the statistical measures, while the spatial accuracy ones are likely to be ignored relatively; (2) the maintenance of integrity of an interpolated surface has attracted little attention and a suitable quantitative index is still lack; (3) without consideration of the robustness of interpolation algorithms to data errors.

To overcome the above-mentioned problems, the author (2002, 2003 & 2004) developed a quantitative, data-independent method to evaluate algorithms in Digital Terrain Analysis. With this method, six slope/aspect algorithms and five flow routing algorithms were evaluated properly. Here we hope to employ this method to comprehensively evaluate spatial interpolation models and identify a set of accuracy measures.

## 2. Unified interpolation models

So far, more than ten spatial interpolation models have been developed in different fields. Here eight commonly used interpolation algorithms are examined and discussed, e.g. Inverse Distance Weighted (IDW), Kriging, Minimum Curvature (MC), Natural Neighbor Interpolation (NNI), Modified Shepard's Method (MSM), Local Polynomial (LP), Triangulation with Linear Interpolation (TLI) and Thin Plate Spline (TPS). According to the range of interpolation, these interpolations can be classified as global interpolation, block interpolation and point-by-point interpolation. While in view of mathematical mechanism, they can also be grouped into deterministic algorithms and geostatistical algorithms.

Although there are various spatial interpolation algorithms with diverse functions, they share the same essential factors, i.e. on the basis of describing the relationships between data points, and computing the values of unmeasured points through different function combinations of sample points. In another word, the relationships depict the spatial correlations between the known points, while the combined functions are the performance of interpolations in mathematics, both of which constitute the commonality of interpolation functions in mathematics and physics. Therefore, they can be unified as one general interpolation model, just as follows:

$$Z_p = w_1 Z_1 + w_2 Z_2 + ... + w_n Z_n + m = WZ + m \tag{1}$$

where $Z_p$ is the estimated value of an interpolated point $P(x_p, y_p)$, $Z_i$ denotes a sample point with $w_i$ indicating its corresponding weight, m presents a constant, and n is the total number of sample points.

In this united model shown as Formula 1, any interpolation function can be regarded as a linear combination of sample points, with the difference of rules for weight allocation. In other words, the determination of the weight vector W is essential and critical for interpolations. For example, IDW determines its weight according to the distance between sample points directly, while NNI employs Thiessen polygons and Kriging uses semivariable functions instead. As for the moving curved surface fitting interpolation, though, the weight function is not obvious, surface-fit functions are employed to allocate weights, implying the spatial relationships of data points. The united interpolation models of the eight interpolation algorithms discussed in this study have been separately listed in Tab. 1.

It has been proved in Tab. 1 that in spite of various interpolation algorithms and models, they have the same intrinsic interpolation mechanism, and any common interpolation method can be transformed into a united model. From the mathematical mechanism, any spatial interpolation is actually a process of assigning weights to sample points, and

| Interpolation models | Interpolation functions | Weight vector(W) | Constant (m) | Parameter Specification |
|---|---|---|---|---|
| IDW, Inverse Distance Weighted | $Z_p = \sum_{i=1}^{n} w_i Z_i$ | $w_i = \frac{d_i^{-k}}{\sum_{i=1}^{n} d_i^{-k}}$ | 0 | $d_i$: the distance between $P_0$ and $P_i$; $k$: a power parameter |
| Kriging | $Z_p = \sum_{i=1}^{n} w_i [z_i - m] + m$ | $w_i$ | $m(1 - \sum_{i=1}^{n} w_i)$ | |
| MC, Minimum Curvature | $Z_p = \sum_{i=1}^{n} w_i R(d_i) + T(x,y)$ | $w_i$ | $T(x,y)$ | $d_i$: the distance between $P_0$ and $P_i$; $R(d_i)$: the principal curvature function; $T(x,y)$: a 'trend' function |

| Interpolation models | Interpolation functions | Weight vector(W) | Constant (m) | Parameter Specification |
|---|---|---|---|---|
| NNI, Nature Neighbor Interpolation | $Z_p = \sum_{i=1}^{n} \frac{a_i}{a} z_{p_i}$ | $w_i = \frac{a_i}{a}$ | 0 | $a_i$ : area of Thiessen polygon(Pi); $a$ : the total areas of all Thiessen polygons |
| MSM, Modified Shepard's Method | $Z_p = \sum_{i=1}^{n} w_i Q_i$ | $w_i = \frac{d_i^{-k}}{\sum_{i=1}^{n} d_i^{-k}}$ | 0 | $Q_i$ : a quadratic polynomial at the interpolated point Pi; $d_i$ : the distance between $P_0$ and $P_i$; $k$ : a power parameter |
| LP, Local Polynomial | $Z_p = A[P^T P]^{-1} P^T Z$ | $W = A[P^T P]^{-1} P^T$ | 0 | $A$ : position vector of unknown point; $P$ : position vector of known points |
| TLI, Triangulation with Linear Interpolation | $Z_p = AP^{-1}Z$ | $W = AP^{-1}$ | 0 | $A$ : position vector of unknown point; $P$ : position vector of known points |
| TPS, Thin Plate Spline | $Z_p = \sum_{i=1}^{n} w_i(\sigma . d_i)\ln(\sigma . d_i) + w_{n+1}$ | $w_i$ | $w_{n+1}$ | $d_i$ : the distance between $P_0$ and $P_i$; $\sigma$ : the optimal parameter |

* In this study, the power parameter k of IDW function is set as 2, as well as MSM; the quadratic polynomial is applied in LP interpolation.

Table 1. Unified interpolation models of common spatial interpolation algorithms different interpolation models have different patterns of weight allocation. While concerning the meaning of geography, the essence of interpolations lies in the spatial correlations between unmeasured points and sample points, reflected during the course of weight allocation. Both sides of mathematics and geography mentioned here can not only give a hypostatic explanation for the spatial interpolation physical mechanism, but can also provide certain guidance for further analysis and evaluation of spatial interpolation models.

## 3. Methods and procedures

In order to achieve the objectives proposed above, a data-independent experiment has been carried out, which allowed us to quantitatively analyze and evaluate different spatial interpolation models. Fig. 1 shows the flowchart of the whole process employed for our experiment. More specific procedures are illustrated as follows: (1) Constructing a mathematical surface with a known-formula; (2) Discretizing the mathematical surface and then randomly sample N points from those discrete ones; (3) Adding errors with varying levels to the randomly sampling points, so that we can get discrete points with the same distribution but varying error-levels; (4) Making interpolated operations separately on the sampling points without errors and the ones with varying error-levels, using the eight interpolation models mentioned above; (5) Analyzing and evaluating the results acquired from different interpolation algorithms according to different evaluation indices.

It is noted that all of the eight interpolation algorithms applied in this study are fulfilled by Surfer 8.0, a powerful contouring, gridding and 3D surface mapping package. Another aspect should be indicated is about the parameter-setting during interpolation. The parameters here mainly consist of three kinds: (1) a search neighhood including its search radius and the number of sampling points, which should be set for the local interpolation methods such as LP, IDW, MSM and TPS; (2) the maximum residual and the maximum number of cycles when gridding with MC method; (3) variogram models like linear, gaussian and logarithmic models for Kriging interpolator. Except variogram models used in Kriging interpolation, the parameters in the others interpolation methods are control parameters and can be set as default of Surfer 8.0, for they have no effect on weight allocation. While for Kriging, the choice of variogram models has a close connection with weight allocation and may affect the results of interpolation. Through repeated tests and validations, the linear model is selected in this study.

### 3.1 Design of mathematical surfaces

In this study, we took the similar approach as reported by Zhou and Liu (2002, 2003 & 2004) by employing pre-defined standard surfaces for testing and comparing selected algorithms. As a result, the 'true' attribute value of any point on the standard surfaces which are pre-defined by known mathematical formulas can be acquired without errors. Our focus is on the difference between the values calculated by interpolation methods and the 'true' valuesto compare these interpolation algorithms objectively. According to the complexity of the surfaces, three surfaces have been selected for test, namely a simple surface, a more complex surface and a Gauss synthetic surface, which are defined by the equations below:

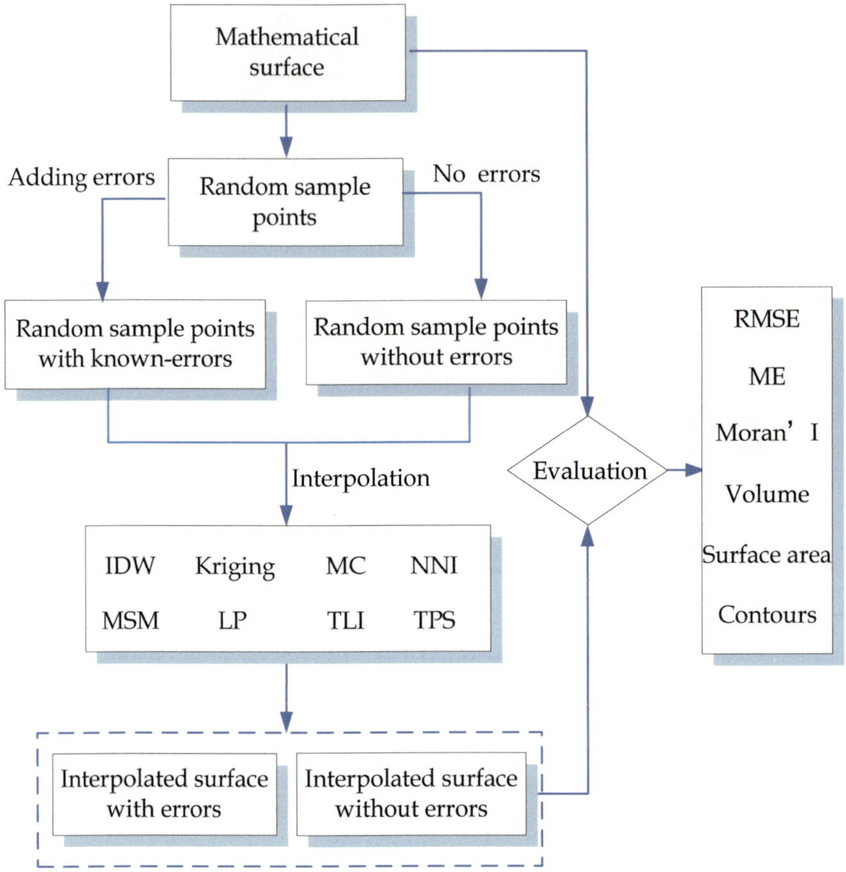


Fig. 1. Flowchart of the scheme to analyze and evaluate the spatial interpolation models

$$\text{Surface1}: \backslash\ f(x,y) = 30 \times \sin(\frac{x}{60}) \times \cos(\frac{y}{100}) + y \times \frac{20}{100} \backslash\ (50 \le x \le 150; 0 \le y \le 100) \quad (2)$$

$$\text{Surface2}: \backslash\ f(x,y) = (\sin(\frac{x}{y}) - \sin(\frac{x \times y}{800})) \times 10 + 100 \backslash\ (-100 \le x \le 0; 10 \le y \le 110) \quad (3)$$

$$\text{Surface3}: \backslash\ f(x,y) = 3(1-x^2)e^{-x^2-(y+1)^2} - 10(0.2x - x^3 - y^5)e^{-x^2-y^2} - \frac{1}{3e^{-(x+1)^2-y^2}}$$

$$(-50 \le x \le 50; -50 \le y \le 50) \quad (4)$$

Then the three selected surfaces are separately scattered into discrete points, from which one thousand points were randomly sampled. All of the three simulated mathematical surfaces are showed in Tab. 2, as well as the distribution of their randomly sample points. After that, add different errors with the same mean 0 but varying Root Mean Square Errors (RMSE), which are in turn 0.5, 1, 1.5, 2, 4, 6, 8, 10, to these sample points, making their errors with the same distribution but different levels of values.

| | Surface 1 (S1) | Surface 2 (S2) | Surface 3 (S3) |
|---|---|---|---|
| Mathematical surfaces | 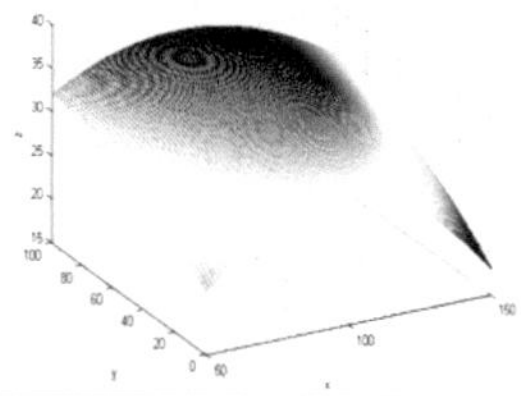 | 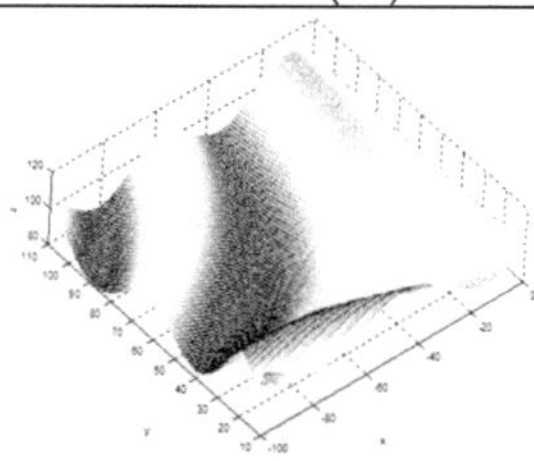 | 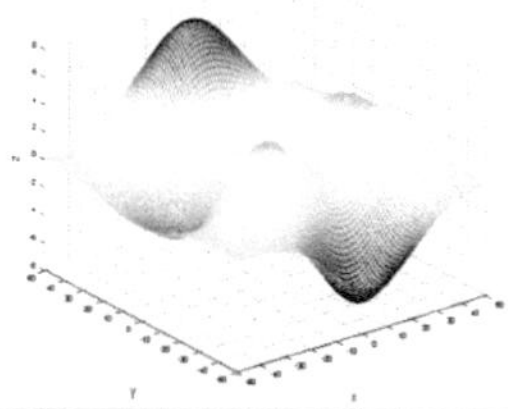 |
| Randomly sample points | 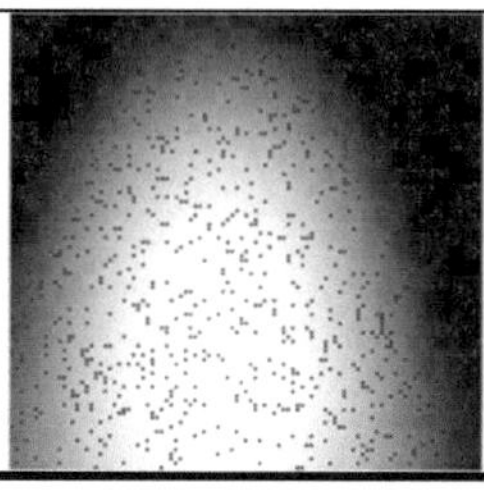 | 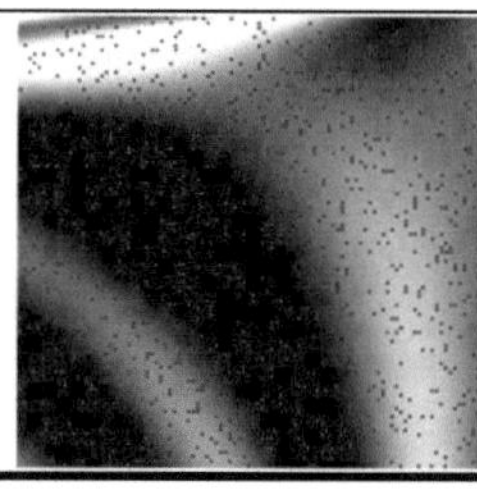 | 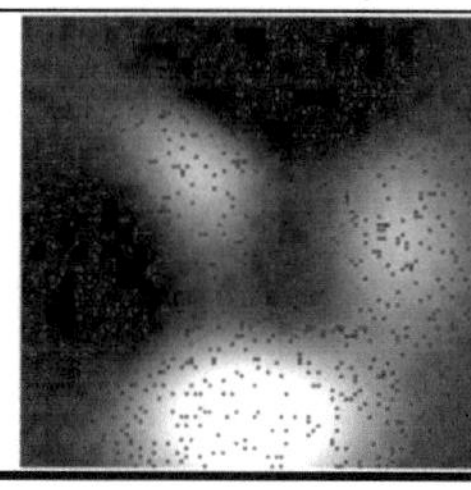 |

Table 2. Mathematical surfaces and distribution of randomly sample points

### 3.2 Design of evaluation indices

The interpolation result can be regarded as an original surface recovered by sample points. It has two implications, i.e. one is to reflect the closeness between the original surface and the recovered one on the value, and the other one is to recover the structural features of the original surface. It means that the interpolated surface should as far as possible keep the characteristics of the original surface both on statistics and structures, which should be considered for the accuracy assessment of the interpolation results as well.

The evaluation indices about statistical features mainly include RMSE (Root Mean Square Error), ME (Mean Error) and spatial autocorrelation. In this study, RMSE and ME are selected to describe the quality of the interpolation functions. For following the first law of geography (Tobler, 1970), the original surface itself has a strong spatial autocorrelation, so as to the whole interpolated surface. As a result, the surface acquired by interpolations should keep the spatial autocorrelation measured by Moran'I here, or else leading to a meaningless result with an almost randomly interpolated surface. What's more, another two spatial indices, volume and surface area are chosen to reflect the maintenance of overall performance after interpolation. The volume stands for the room above a datum plane and under an original surface or an interpolated surface whose area is measured by surface area.

Structural characteristic is the other important evaluation method. It can be regarded as the skeleton of a surface, determining its geometric shape and basic trend, on which the interpolated surface should be in accord with the original one. For the integrity of the structural characteristic, so far there is lack of a suitable quantitative index. In this study, a method of contour-matching has been applied to compare and analyze different interpolations qualitatively, by means of overlaying the contours generated from an interpolated surface and the original one. If these overlaid contours match generally without great deviation or distortion, it can be induced that the structural characteristic of the surface has been kept well after being interpolated, or the structural characteristic will be lost leading to a fault result.

## 4. Results and discussion

### 4.1 RMSE and ME

As the RMSE statistics of the interpolated results from the three surfaces shown in Tab. 3, it is not difficult to identify that all of the three interpolated surfaces present similar variation tendency as a whole. The RMSEs of the interpolated surfaces keep pace with the increasing errors of the original surface, leading to a decreasing interpolated accuracy.

| | S1 | | | S2 | | | S3 | | |
|---|---|---|---|---|---|---|---|---|---|
| | 0 | 1 | 10 | 0 | 1 | 10 | 0 | 1 | 10 |
| IDW | 0.13 | 0.46 | 1.38 | 1.14 | 1.19 | 1.72 | 0.25 | 0.50 | 1.30 |
| Kriging | 0.02 | 0.65 | 2.13 | 0.11 | 0.65 | 1.99 | 0.01 | 0.67 | 1.98 |
| MC | 0.07 | 0.83 | 2.79 | 0.29 | 0.90 | 2.53 | 0.06 | 0.88 | 2.51 |
| NNI | 0.03 | 0.62 | 2.02 | 0.27 | 0.67 | 1.91 | 0.05 | 0.64 | 1.88 |
| MSM | 0.02 | 1.07 | 3.62 | 0.29 | 1.07 | 3.22 | 0.01 | 1.11 | 3.28 |
| LP | 0.22 | 0.25 | 0.43 | 2.51 | 2.49 | 2.50 | 0.56 | 0.58 | 0.69 |
| TLI | 0.03 | 0.70 | 2.31 | 0.26 | 0.76 | 2.17 | 0.05 | 0.72 | 2.14 |
| TPS | 0.02 | 0.84 | 2.86 | 0.07 | 0.87 | 2.58 | 0.01 | 0.90 | 2.56 |

* 1) 0, 1, 10 are the RMSE added to the original sample points; 2) For the interpolation methods of NNI and TLI cannot deal with the boundary problem well, therefore their boundary values which are replaced with the maximum have been excluded when calculated in statistics.

Table 3. RMSE statistics of the interpolated results from the three surfaces

As shown in Tab.3, Fig. 2 and Fig. 3, when sample points have no errors, the RMSEs of the interpolated results for different methods have an decreasing sequence as LP > IDW, MC > NNI, TLI > Kriging, MSM, TPS. However, the interpolated results vary with the augment of data errors. When the RMSE of sample points increases to 10, the RMSE of the surface interpolated by MSM achieves the maximum, with the minimum gained by LP and the sequence of RMSE for different interpolations changes to MSM > TPS, MC > TLI, Kriging, NNI > IDW > LP. The results show that if the original data has a better quality, the methods of TPS, MSM and Kriging can get a high precision for the interpolated results, while the quality of the original data becomes poorly, the result of LP turns to be relatively reliable.

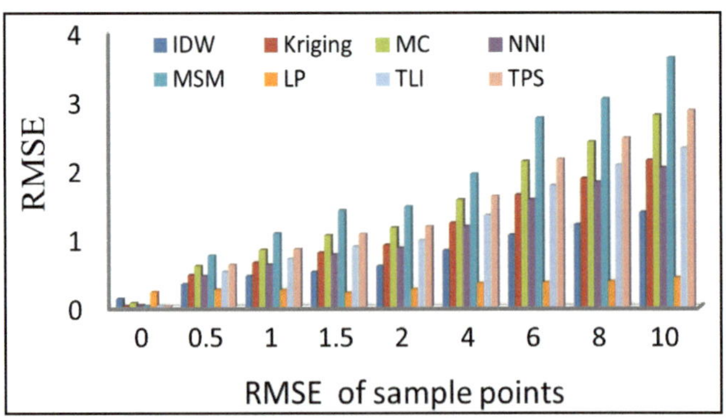


Fig. 2. RMSE statistics of interpolated results from S1

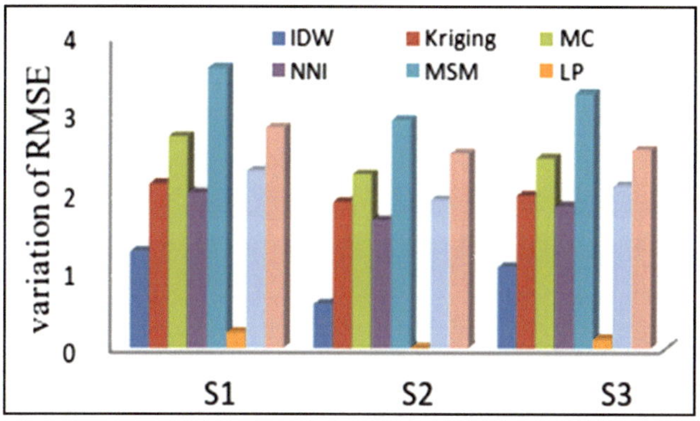


Fig. 3. Changes in RMSEs of the three surfaces before and after adding errors

Actually, it is not difficult to explain the results. When sample points have no errors or small errors, these sample points themselves can portray the characteristics of the original surface in a relatively accurate degree. Using semi-variogram, the geostatistical method of Kriging recovers the spatial correlation of the original surface exactly, while TPS and MSM are means of finding a proper way to allocate weights to sample points according to the distance between known points and interpolated points. With the increasing of sample data errors, the surface generated by sample points starts to deviate from the original surface, meaning the sample surface can no longer describe the characteristics of the original surface completely. No matter Kriging or TPS, the surfaces they want to depict or recover are just sample surfaces. For LP, although not all of the sample points are strictly passed through, this interpolated method can make a certain restraint on the original data errors, showing a role of peak-clipping and valley-filling for the interpolation. Furthermore, the restraining effect can also reflect the variation tendency of the surfaces created by sample points, bringing about a higher interpolated precision. The interpolated results for the three surfaces with different complexity levels have been showed by Fig. 3. On the whole, the changing tendencies of these three surfaces present a roughly consistent pace, that is to say the largest change of RMSE before and after adding errors belongs to MSM with the minimum belonging to LP, and an ascending order between the two extremes are as follows: TPS, MC, TLI, NNI and IDW. It has been further proved that the interpolated method of LP is less vulnerable to data errors appearing a superior resistance to errors, while MSM is extremely sensitive to data errors showing a worst error-resistance.

### 4.2 Moran'I index

The Moran'I statistics of S1 with no data-errors and higher data-errors acquired from different interpolation methods have been compared in Fig. 4 and Fig. 5. For the data without errors, the Moran'I of Kriging 0.9980 reaches the topmost showing the best spatial correlation, then followed by LP, TPS and IDW with a better spatial correlation, and the lowest Moran'I belongs to MC whose spatial correlation is the worst (refer to Fig. 4). However, with the increasing of the sample data errors, the Moran'I of LP, 0.9977, changes to the highest with the optimal spatial correlation and by contrast, MSM turns to the lowest as shown by Fig. 5. The other two surfaces present a similar variation regularity or tendency for the entire, although there are a few differences among individual interpolation methods. In order to further interpret the impact caused by the increasing data-errors on spatial

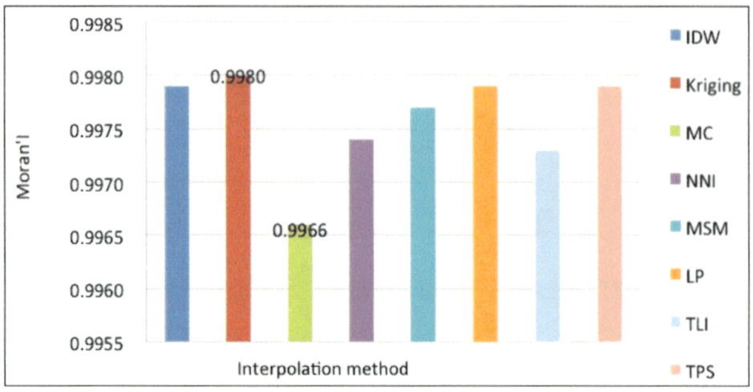


Fig. 4. Moran'I statistics of S1 (RMSE = 0)

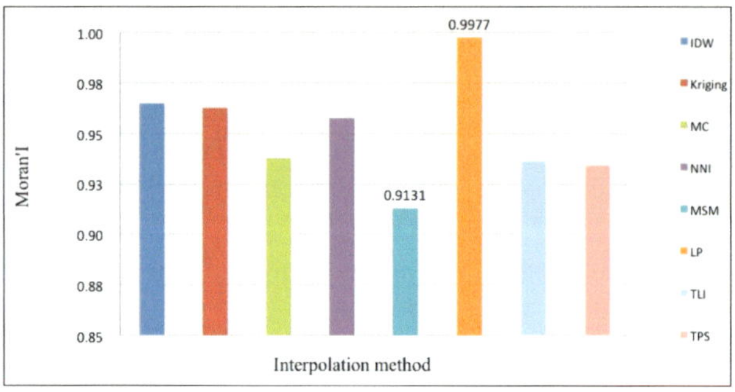


Fig. 5. Moran'I statistics of S1 (RMSE = 10)

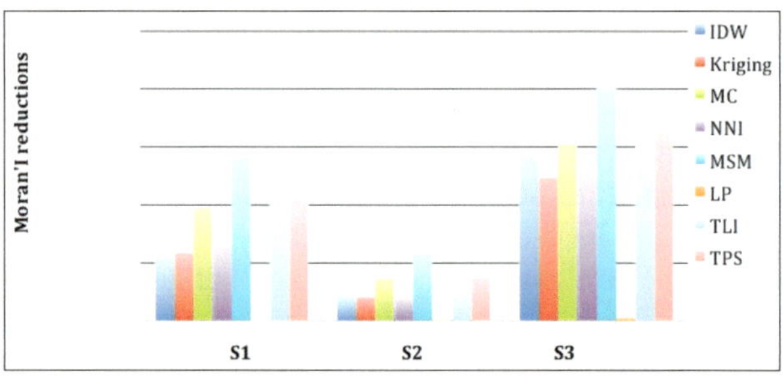


Fig. 6. Comparisons of Moran'I reductions among three surfaces

correlation, Fig. 6 reveals the overall variations of the three surfaces. Though Moran'I reductions of the three surfaces have a few differences, they share the same changing tendency. No matter what surface it is, the Moran'I reductions caused by interpolation LP is the smallest with its spatial correlation kept best, while MSM loses most. And the increasing sequence between LP and MSM is listed as follows: IDW, Kriging, NNI < MC, TLI, TPS, which is nearly in accordance with the statistical results of RMSE and ME given in Section 4.1.

## 4.3 Volume and surface area

To further analyze the maintenance of overall performance after interpolation, the absolute differences between the 'true' volume and volumes calculated by surfaces interpolated by different models have been compared, except NNI and TLI for their boundary effect. Still taking the first surface for example, the absolute volume difference showed in Fig. 7 stands

for the difference of the 'true' volume and the volume between an interpolated surface and the datum plane whose elevation is 0. All of the results are calculated by Surfer 8.0.
As shown in Fig. 7, when the original data has no error, the absolute volume differences of MSM, TPS and Kriging are smaller, and by comparison, LP, MC and IDW are relatively larger, with the minimum belonging to MSM and the maximum belonging to LP. However, their relationships make changes after adding a certain errors, similar as the variation of RMSE in Section 4.1. Aside from LP, the absolute volume differences of other interpolation methods are increased with mounting errors, keeping a consistent sequence of LP < IDW < Kriging < TPS < MC < MSM. Beyond doubt, the above analysis results are approximately accordant with the results of RMSE, ME and Moran'I. Moreover, judging from the variation tendency, MSM changes greatest with LP changing least, which has demonstrated the powerful robustness of LP to data errors again. Similar conclusions as volume index can be got from Fig. 8, which presents the absolute differences of the 'true' surface area and different interpolated surfaces areas.

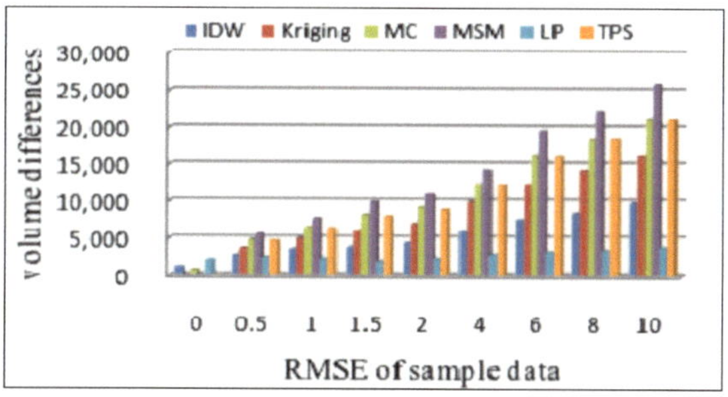


Fig. 7. Absolute volume differences of S1

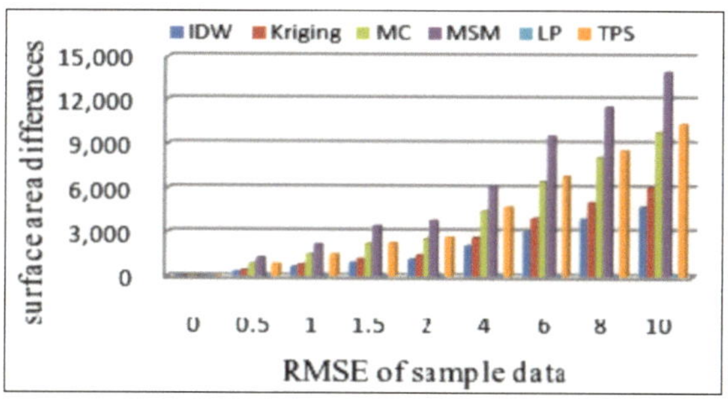


Fig. 8. Absolute surface area differences of S1

### 4.4 Contour matching

Comparison between the contours of the original surface and those of the interpolated surfaces will be discussed in this section. Fig .9 takes the second surface for instance to present the comparisons when the original data is without errors. As shown in Fig. 9, the contours generated by Kriging, MSM and TPS perform preferably smooth, matching with contours of the original surface well. For NNI and TLI, the inside shapes of the contours maintain well with smooth lines, however, some abnormalities like figure losses appear on the boundary, further verifying their boundary effect mentioned above. Though the shapes of contours produced by LP keep well too, it is easy to notice that their positions shift on the whole. On the contrary, the contours of IDW and MC display evident deformation and distortion, especially an obvious bull's eye effect appears for IDW. By contrast, contours of MSM which shares a similar interpolation theory as IDW maintain a better shape without the bull's eye effect, for its improvement in the weight function. As a result, it has been proved again that weight allocation and its corresponding spatial relationship between interpolated points and known points are the ultimate causes for the results of different interpolation methods.

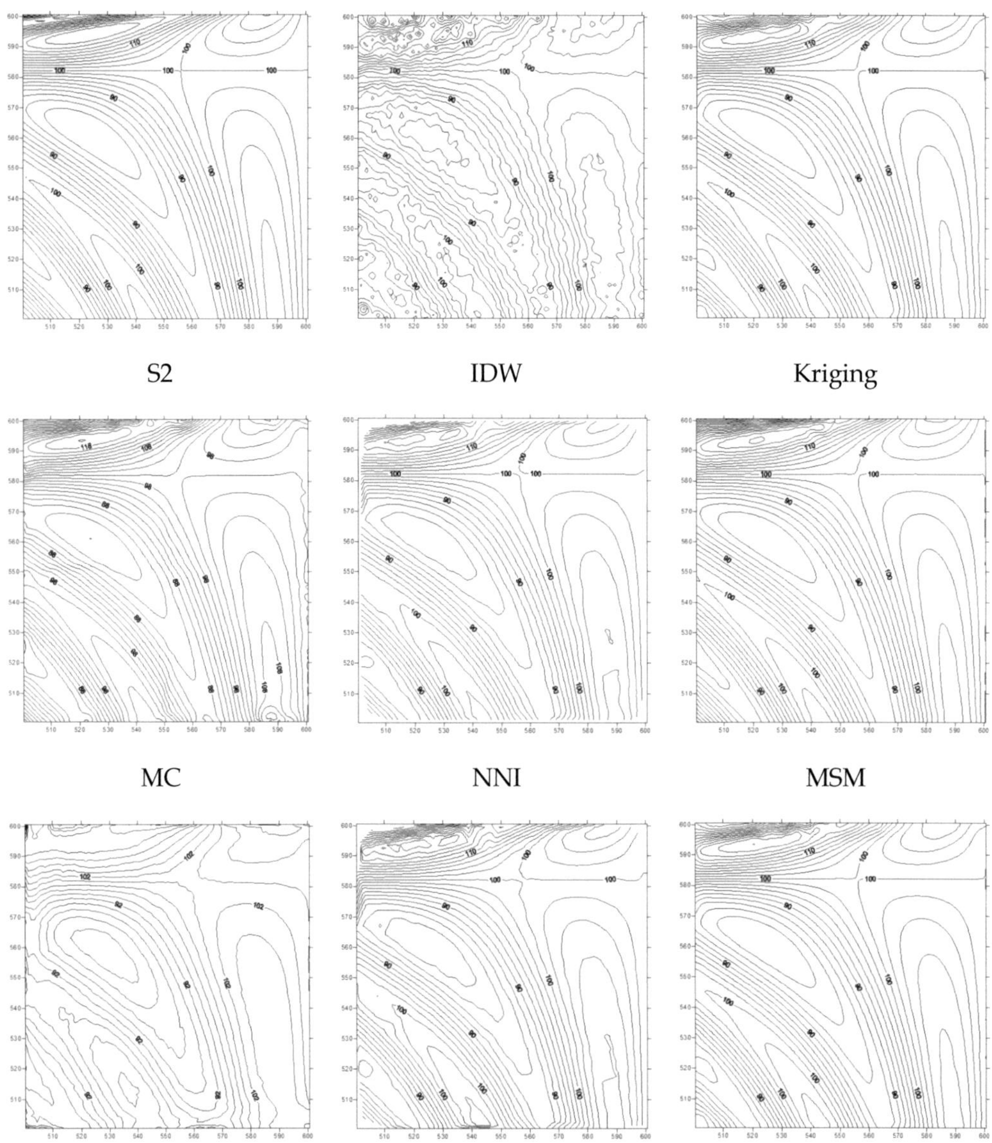


Fig. 9. Contour comparison among different interpolated surfaces from non-error original data (cell size = 1m)

When RMSE of sample points rises to 8, the contours produced by the same eight interpolation methods have been showed in Fig. 10, most of them deforming or distorting drastically except LP. More specifically, the shape or distribution of the deformed contours can be divided into two cases: as for IDW, Kriging, MC, MSM and TPS, the bull's eye effect appears in the regions with high errors, and for TLI and NNI, their contours display as roughly fold-lines with an uneven intensity. Compared with other methods, the contours of LP, though, are not that smooth as the original ones, their shape and distribution are both kept relatively intact, showing a powerful robustness to errors.

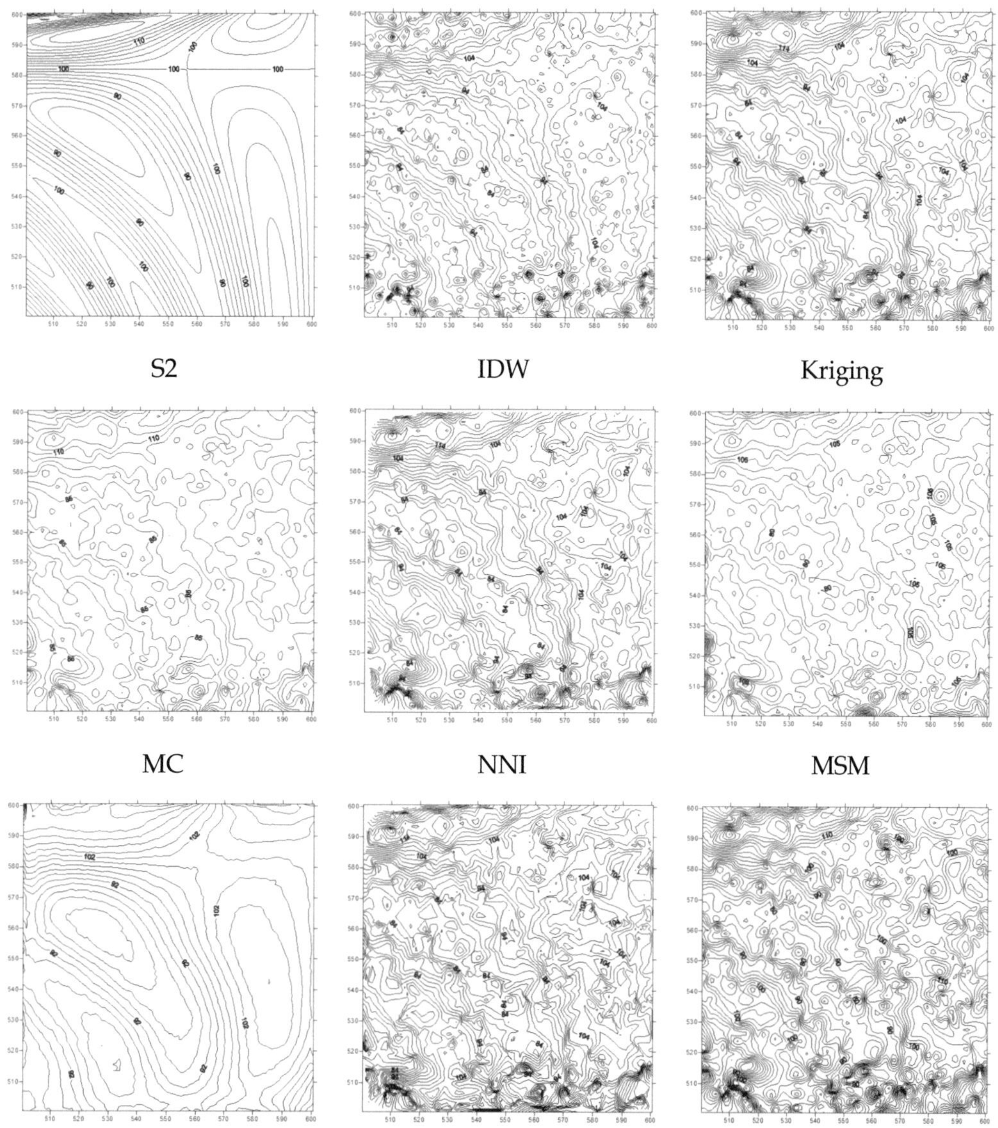


Fig. 10. Contour comparison among different interpolated surfaces when RMSE of original data is 8 (cell size = 1m)

### 4.5 Comprehensive evaluation

Combing the various evaluation indices discussed above, Tab. 4 gives a comprehensive evaluation for various interpolation models. The levels of interpolation accuracy are defined as: lowest, lower, high, higher and highest, while the levels of robustness to errors are set as: weakest, weaker, strong, stronger and strongest. When the original data has no errors, the interpolation accuracy of TPS is the highest, followed by MSM, and MC is the lowest. After higher errors being added to the original data, the interpolation accuracy of TPS changes from the highest to the lowest, while the precision of LP alters from lower to the highest. As a result, the strongest robustness to errors is LP, and the weakest is MSM by contrast. As for MC, regardless of the original data with errors or not, its interpolation accuracy always keeps lower.

| Models | Accuracy with non-error data | Accuracy with error data | Robustness |
|---|---|---|---|
| IDW | lower | higher | stronger |
| Kriging | high | higher | strong |
| MC | lowest | lower | weaker |
| NNI | high | high | stronger |
| MSM | higher | lowest | weakest |
| LP | lower | highest | strongest |
| TLI | high | lower | weaker |
| TPS | highest | lower | weaker |

Table 4. Comprehensive Evaluation for interpolation models

## 5. Conclusions

From the mechanism of spatial interpolation, weight allocation and its corresponding spatial relationship between interpolated points and known points, this article proposes an evaluation and analysis approach of spatial interpolation in GIS based on data-independent method, with the construction of mathematical surfaces without errors to objectively reflect the precision of different interpolation algorithms and with the addition of varying degree-errors to examine their robustness to errors. Based on our study, following conclusions can be given: (1) when the quality of original data is relatively well, TPS and Kriging can acquire more reliable results; (2) when the quality of original data becomes worse, for its resistance to data errors, LP can maintain a preferable interpolated precision, showing a powerful robustness to errors; (3) the validity of weight function and its corresponding spatial relationship are the kernel for design and analysis of weight function; (4) a kind of data

smooth process or data precision improvement method can be an effective way to advance the interpolated accuracy.
In order to further quantitatively depict the differences of morphological characteristics between original surfaces and interpolated surfaces, further studies will be focused on developing a visually quantitative index, such as an area enclosed between homologous contours (two level contours separately derived from an original surface and an interpolated surface). The real-world tests will also be conducted to compare with the findings by the theoretical analysis and a set of high-accuracy data should be needed for test.

## 6. Acknowledgements

This study is supported by the National High Technology Research and Development Program of China (No. 2011AA120304) the National Natural Science Foundation of China (No. 40971230), the Doctoral Fund of Ministry of Education of China (No. 20093207110009) and the Project Funded by the Priority Academic Program Development of Jiangsu Higher Education Institutions.

## 7. References

Algarni D & Hassan I. (2001). Comparison of thin plate spline, polynomial C-function and Shepard's interpolation techniques with GPS derived DEM. International Journal of Applied Earth Observation and Geoinformation 3(2): 155-161.

Chaplot V, Darboux F, Bourennane H, Leguédois S, Silvera N & Phachomphon K. (2006). Accuracy of interpolation techniques for the derivation of digital elevation models in relation to landform types and data density. Geomorphology 77: 126-141.

Collins, F.C. & Bolstad, P.V. (1996). A comparison of spatial interpolation techniques in temperature estimation. Proceedings, Third International Conference/ Workshop on Integrating GIS and Environmental Modeling, Santa Fe, NM. National Center for Geographic Information and Analysis, Santa Barbara, Santa Barbara, CA.

Christopher W. Bater & Nicholas C. Coops. (2009). Evaluating error associated with lidar-derived DEM interpolation. Computers & Geosciences 35: 289-300.

Declercq FAN. (1996). Interpolation methods for scattered sample data: accuracy, spatial patterns, processing time. Cartography and Geographical Information Systems 23(3): 128-44.

David W. Wong, Lester Yuan & Susan A. Perlin. (2004). Comparison of spatial interpolation methods for the estimation of air quality data. Journal of Exposure Analysis and Environmental Epidemiology 14: 404-415.

Di Piazza, F. Lo Conti, L.V. Noto, F. Viola & G. La Loggia (2011). Comparative analysis of different techniques for spatial interpolation of rainfall data to create a serially complete monthly time series of precipitation for Sicily, Italy. International Journal of Applied Earth Observation and Geoinformation 13: 396 - 408.

Erxleben, J., Elder, K. & Davis, R. (2002). Comparison of spatial interpolation methods for estimating snow distribution in the Colorado Rocky Mountains. Hydrological Processes 16: 3627 - 3649.

Fabio Attorre et al. (2007). Comparison of interpolation methods for mapping climatic and bioclimatic variables at regional scale. INTERNATIONAL JOURNAL OF CLIMATOLOGY. 27: 1825 - 1843.

Fencik R & Vajsablova M. (2006). Parameters of interpolation methods of creation of digital model of landscape. Ninth AGILE Conference on Geographic Information Science. April 20-22: 374-381.

Gotway, C.A., Ferguson, R.B., Hergert, G.W. & Peterson, T.A. (1996). Comparison of kriging and inverse-distance methods for mapping parameters. Soil Science Society of American Journal 60: 1237 - 1247.

Hernandez-Stefanoni, J. L. & Ponce-Hernandez, R. (2006). Mapping the spatial variability of plant diversity in a tropical forest: comparison of spatial interpolation methods. Environmental Monitoring and Assessment 117: 307 - 334.

Hosseini, E., Gallichand, J. & Caron, J. (1993). Comparison of several interpolators for smoothing hydraulic conductivity data in South West Iran. American Society of Agricultural Engineers 36: 1687 - 1693.

Hengl, T. (2007). A Practical Guide to Geostatistical Mapping of Environmental Variables. JRC Scientific and Technichal Reports. Office for Official Publication of the European Communities, Luxembourg.

Hartkamp A.D., De Beurs K., Stein A. & White J.W. (1999). Interpolation Techniques for Climate Variables. CIMMYT, Mexico, D.F.

Hu K., Li B., Lu Y. & Zhang F. (2004). Comparison of various spatial interpolation methods for non-stationary regional soil mercury content. Environmental Science 25: 132 - 137.

Javis, C.H. & Stuart, N. (2001). A comparison among strategies for interpolating maximum and minimum daily air temperature. Part II: the interaction between number of guiding variables and the type of interpolation method. Journal of Applied Meteorology 40: 1075 - 1084.

Jin Li & Andrew D. Heap. (2011). A review of comparative studies of spatial interpolation methods in environmental sciences: Performance and impact factors. Ecological Informatics 6: 228 - 241.

Lin G. & Chen L. (2004). A spatial interpolation method based on radial basis function networks incorporating a semivariogram model. Journal of Hydrology 288:288 - 298.

Laslett, G.M., McBratney, A.B., Pahl, P.J. & Hutchinson, M.F. (1987). Comparison of several spatial prediction methods for soil pH. Journal of Soil Science 38: 325 - 341.

Mussie G. Tewolde, Teshome A. Beza, Ana Cristina Costa & Marco Painho. (2010). Comparison of Different Interpolation Techniques to Map Temperature in the Southern Region of Eritrea. 13th AGILE International Conference on Geographic Information Science.

Qihao Weng. (2002). Quantifying Uncertainty of Digital Elevation Models Derived from Topographic Maps. Symposium on Geospatial Theory, Processing and Applications.

Shi, W. Z, et al. (2005). Estimating the Propagation Error of DEM from Higher-order Interpolation Algorithms. International Journal of Remote Sensing 14: 3069-3084.

Stahl, K., Moore, R.D., Floyer, J.A., Asplin, M.G. & McKendry, I.G. (2006). Comparison ofapproaches for spatial interpolation of daily air temperature in a large region with complex topography and highly variable station density. Agricultural and Forest Meteorology 139: 224 - 236.

Schloeder, C.A., Zimmerman, N.E. & Jacobs, M.J. (2001). Comparison of methods for interpolating soil properties using limited data. Soil Science Society of American Journal 65: 470 - 479.

Saffet Erdogan. (2009). A comparision of interpolation methods for producing digital elevation models at the field scale. Earth Surf. Process. Landforms 34: 366-376.

T.P. Robinson & G. Metternicht. (2006). Testing the performance of spatial interpolation techniques for mapping soil properties. Computers and Electronics in Agriculture 50: 97-108.

Tobler W. R. (1970). A computer movie simulating urban growth in the Detroit region. Economic Geography 46: 234-240.

Vicente-Serrano, S.M., Saz-Sánchez, M.A. & Cuadrat, J.M. (2003). Comparative analysis of interpolation methods in the middle Ebro Valley (Spain): application to annual precipitation and temperature. Climate Research 24: 161 - 180.

Wang H., Liu G. & Gong P. (2005). Use of cokriging to improve estimates of soil salt solute spatial distribution in the Yellow River delta. Acta Geographica Sinica 60: 511 - 518.

Weng Q. (2006). An evaluation of spatial interpolation accuracy of elevation data. In Progress in Spatial Data Handling, Riedl A, Kainz W, Elmes GA (eds). Springer-Verlag: Berlin, 805-824.

Weber, D. & Englund, E. (1992). Evaluation and comparison of spatial interpolators. Mathematical Geology 24: 381 - 391.

Weber, D. & Englund, E., (1994). Evaluation and comparison of spatial interpolators II. Mathematical Geology 26: 589 - 603.

Zimmerman D., Pavlik C., Ruggles A. & Armstrong M.P. (1999). An experimental comparison of ordinary and universal kriging and inverse distance weighting. Mathematical Geology 31: 375 - 390.

Zhou, Q. & Liu, X. (2002). Error assessment of grid-based flow routing algorithms used in hydrological models. International Journal of Geographical Information Science 16 (8): 819-842.

Zhou, Q. & Liu, X. (2003). The accuracy assessment on algorithms that derive slope and aspect from DEM. In: Shi, W., Goodchild, M.F., Fisher, P.F. (Eds.), Proceedings of the Second International Symposium on Spatial Data Quality. 19 - 20 March, Hong Kong, pp. 275 - 285.

Zhou, Q. & Liu, X. (2004). Analysis of errors of derived slope and aspect related to DEM data properties. Computers & Geosciences 30: 369–378.

4

# Integrated Geochemical and Geophysical Approach to Mineral Prospecting – A Case Study on the Basement Complex of Ilesa Area, Nigeria

Emmanuel Abiodun Ariyibi
*Earth and Space Physics Research Laboratory,*
*Department of Physics,*
*Obafemi Awolowo University (OAU), le – Ife,*
*Nigeria*

## 1. Introduction

The crust of the earth is composed of solid rocks. When the rocks are closely examined, they are found to be composed of discrete grains of different sizes, shapes, and colours. These grains are minerals, which are the building blocks of all rocks (Mazzullo, 1996).

The formation of soils from rocks generally involves the combination of mechanical and chemical weathering resulting from surface processes. Climatic conditions under which the weathering is affected determine which of the two forms of weathering becomes more pronounced than the other. In arid climates where there is little or no water and where there are appreciable diurnal variations in temperature, chemical weathering is considerably subordinated to mechanical weathering and the rocks simply become broken into increasingly small grains and pieces in which the individual minerals that constitute the rock are easily recognized. If, on the other hand, the climate is warm and humid with appreciable rainfall, chemical weathering becomes markedly pronounced and the individual minerals that form the rock are each subjected to rather intense chemical and comparatively modest mechanical weathering with the formation of different products which are all constituents of soils (Adewunmi,1984). In most Basement Complex rocks, weathered products, reflect certain characteristics (geochemical and mineralogical) of the parent rock.

Previous studies by Ako et al. (1979), Ako (1980), Ajayi (1988) and Elueze (1977) on the Ilesa area have been suggestive ofsulphide mineralization but the scope of coverage have been limited to the Amphibolite and the area around the Iwara fault. The present work is regional in scope and seeks to uniquely use an integrated geochemical and geophysical approach around the Ilesa area which is within the Schist belt of southwestern Nigeria and consists of Schist Undifferentiated, Gneiss and Migmatite Undifferentiated, Pegmatite, Schist Epidiorite Complex, Quartzite and Quartz Schist, Granite Gneiss, Ampibolite, Schist Pegmatised and Granulite and Gneiss aimed at delineating the area for mineral exploration. The geochemical data from 61 sampling locations were subjected to multivariate analysis and interpreted to delineate geochemical anomalous zone. The geophysical investigation of the anomalous zone that follows employed the Very – Low Frequency Electromagnetic (VLF – EM), Electrical resistivityand magnetic methods.

## 2. Geologic setting

The Ilesa area, lies within the Southwest Nigerian Basement Complex (Schist Belt) which is of Precambrian age (De Swardt, 1953). De Swardt (1947) and Russ (1957) suggested that the Nigerian Basement Complex is Polycyclic. This was confirmed by Hurley (1966, 1970) who used radiometric method to determine the age of the rocks. The Nigerian basement is believed to have had structural complexity as a result of folding, igneous and metamorphic activities with five major rock units recognized within the Basement Complex by Rahaman (1976). These are:

The migmatite-gneiss-quartzite complex

i. Slightly migmatized to unmigmatized paraschists and meta-igneous rocks which consists of pellitic schists, quartzites, amphibolites, talcose rocks, metaconglomerate, marble and calc-silicate rocks.
ii. Charnockitic rocks
iii. Older Granites and
iv. Unmetamorphosed acid-basic intrusives and hypabyssal rocks

Geochronological data on all these rock groups were summarized by Rahaman (1988). The geological map of Ilesa area is as shown in Figure 1.

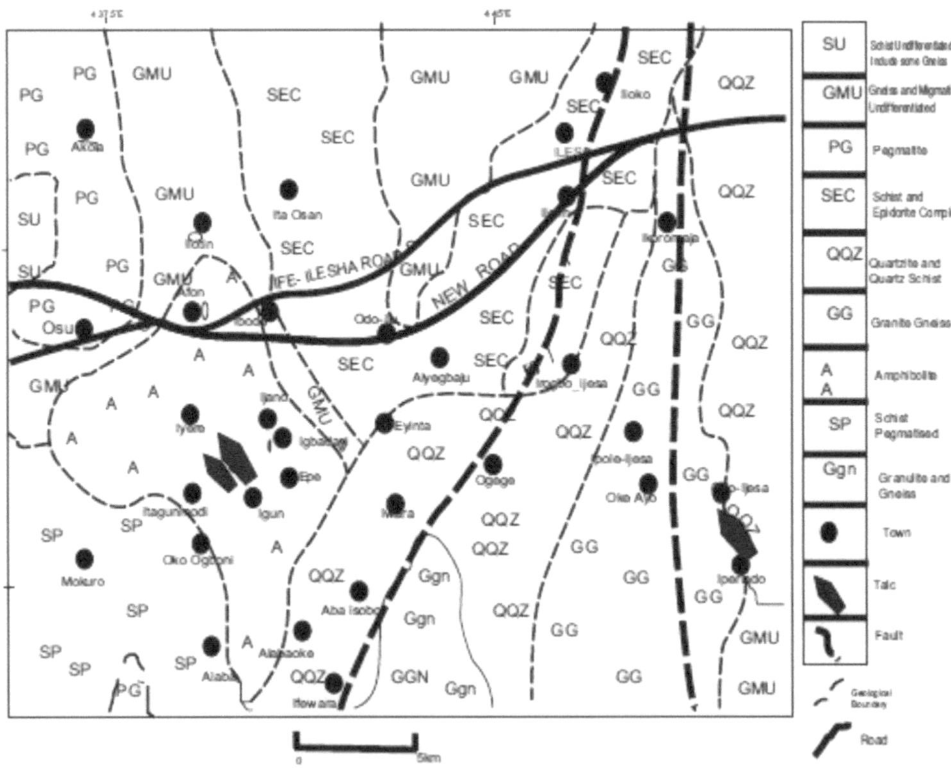


Fig. 1. Geological map of the study area (After Adebayo, 2008 and Adelusil, 2005)

The Schist Belts group of which Ilesa area is part have been variously termed" Newer Metasediments" (Oyawoye, 1964), "Younger metasediments" (McCurry, 1976); "Schist belts" (Ajibade, 1976) and" Slightly migmatized to non-migmatized metasedimentary and meta-igneous rocks" (Rahaman, 1988). The Nigerian Schist belts occur as prominent N-S trending features in the Nigerian Basement Complex. Lithologically, the schist belt consists of metamorphosed pelitic to semi-pelitic rocks, quartzites, calc-silicates rocks, metaconglomerates and pebbly schists; amphibolites and metavolcanic rocks. On the basis of lithology, metamorphism, structure, geochemical characteristics such as the tholeitic affinities of the amphibolites, and economic potentials, the schist belts show similarities to typical Archaean greenstone belts of the world (Wright and McCurry, 1970; Hubbard, 1975; Elueze, 1977; Klemm et al, 1979, 1984).
However, certain differences exist between these schist belts and typical greenstone belts. Though the schist belts of Nigeria have the sedimentary and ultramafic rock groups as defined by Anhaeusser et al, (1969), the so-called, "greenstone group" comprising serpentinites, crystalline limestones, rhyolites, banded ironstones of chemical origin and massive carbonated "greenstones" are either absent or less conspicuous. Also the ratio of metasediments to metavolcanics is much higher in the Schists belts of Nigeria than in the typical Archaean greenstone belts (Wright and McCurry, 1970). Mineralization is also not strongly developed in the schist's belts as in well known greenstone belts of the Canadian, Indian, Australian, and South African Precambrian shield areas.

## 3. Location, geomorphology and relief of study area

The area of study is located in the southern part of Ilesa and is geographically enclosed within Latitude $7^0$ 30`N to $7^0$36` N and Longitude $4^0$ 38`E to $4^0$ 50`E. It covers an area of about 1200 square kilometers. There are a good number of major and minor roads that link up the towns and villages in the study area. There are also main paths and footpaths linking the communities. These make the entire study area fairly accessible. The landscape of the area of study can be generally described as undulating, rising gently to steeply, but in some areas is punctuated by hilly ridges. The ridges are formed by quartz-schist or quartzite that rises abruptly from the enveloping basins and trend in the North to South direction. On the other hand, hills which are probably products of fragments of coarse-grained batholitic granite or resistance gneisses have a positive relief and are covered up by vegetation. Dissected topography also develops over the easily eroded basic rocks which according to De Swardt (1953) reflect the erosion cycles that separately occur in the area.

## 4. Research methodology

Preliminary work in the survey area took the form of a reconnaissance geological mapping followed by the statistical analysis of geochemical data to delineate areas with geochemical anomalies indicative of possible mineralization (Ariyibi et. al. 2010). This was aimed at studying the geology and selecting geophysical traverses approximately normal to the strike of the geochemical anomaly. In some cases however, the dense vegetation necessitated the cutting of traverses, although most of the traverses used were along existing roads and footpaths. The field data acquisition involved ground magnetic, electrical resistivity and VLF – EM survey methods. Essentially, the magnetic and VLF - EM measurements were carried out simultaneously on the chosen traverses located on the delineated area using topographical and geochemical anomaly maps.

The magnetometer used for this work is the GEM - 8 Proton magneto-meter which measures the Earth`s total field. The potable VLF-EM equipment used for this work was the GEONICS-EM16 (with GBR station) and on frequency of 16 kHz and it measures the real and imaginary part of the signal. The ABEM Terrameter was used for the electrical resistivity measurement.

## 5. Geoscientific methods used, data analysis and results

### 5.1 General

The geoscientific methods that can be used in mineral prospecting include, magnetic, gravity, electrical, electromagnetic, radiometric, geothermal and seismic methods. However, the choice of method (s) would depend basically upon their resolution with respect to problems encountered or conditions sought after within a given locality. More often, consideration is given to the cost, time, portability and reliability of instruments used in small scale surveys (Adewusi, 1988). The methods considered in this work include the geochemical, electrical resistivity, VLF - EM and magnetic methods.

#### 5.1.1 The geochemical data, analysis and results

Geochemical methods of exploration should be viewed as an integral component of the variety of weapons available to the modern prospector. As the goal of every exploration method is, of course the same - to find clues that will help in locating hidden ore, the geochemical prospecting for minerals, as defined by common usage, includes any method of mineral exploration based on systematic measurement of one or more chemical properties of a naturally occurring material (Suh, 1993). The chemical property measured is most commonly the trace content of some element or group of elements; the naturally occurring material may be rock, soil, gossan, glacial debris, vegetation, stream sediment or water. The purpose of the measurements is the discovery of abnormal chemical patterns, or geochemical anomalies, related to mineralization.

Sampling and analysis of residual soil is by far the most widely used of all the geochemical methods. The popularity of residual-soil surveying as an exploration method is a simple reflection of the reliability of soil anomalies as ore guides (Gill, 1997). Practical experiences in many climates and in many types of geological environments has shown that where the parent rock is mineralized, some kind of chemical pattern can be found in the residual soil that results from the weathering of that rock. Where residual soil anomalies are not found over known ore in the bedrock, further examination usuallyshows either that the material sampled was not truly residual or that an unsuitable horizon or size fractionof the soil was sampled, or possibly that an inadequate extraction method was used. In other words, when properly used, the method is exceptionally reliable in comparison with most other exploration methods.

By definition, an anomaly is a deviation from the norm. A geochemical anomaly, more specifically, is a departure from the geochemical patterns that are normal for a given area or geochemical landscape. Strictly speaking, an ore deposit being a relatively rare or abnormal phenomenon is itself a geochemical anomaly. Similarly, the geochemical patterns that are related either to the genesis or to the erosion of an ore deposit are anomalies (Hawkes and Webb,1962). Anomalies that are related to ore and that can be used as guides in exploration are termed significant anomalies. Anomalies that are superficially similar to significant anomalies but are unrelated to ore are known as non-significant anomalies.

As with all geochemical surveys, the first step in approaching an operational problem is to conduct an orientation survey. Such a survey normally consists of a series of preliminary experiments aimed at determining the existence and characteristics of anomalies associated with mineralization. This information may then be used in selecting adequate prospecting techniques and in determining the factors and criteria that have a bearing on interpretation of the geochemical data.

Although the orientation study will provide the necessary technical information upon which to base operational procedures, the final choice of methods to be used must also take into account other factors, such as cost of operation, availability of personnel, and the market value of the expected ore discoveries. The nature of the overburden, whether it is residual or is of glacial, alluvial, or wind-borne origin, is the first question that must be answered by the orientation survey. Sometimes it is surprisingly difficult to discriminate between residual and transported soil. The safest method therefore, is to make critical and careful examination of complete sections of the overburden at the start of every new field survey. If road-cut exposures are not available, the soil should be examined by pitting or auguring. Previous orientation studies carried out by Olorunfemi (1977) and Adewunmi (1984) in parts of southern Ilesa established that the C horizon is the preferredhorizon for sampling. Details of laboratory procedures and analysis of the samples were reported by Ariyibi et al. (2010)

**5.1.1.1 Statistical results**

The multivariate technique, has proven to be viable and credible when applied on geochemical data as reported by Grunfeld (2003). The Principal Component Analysis (PCA) which isa multivariate technique, describep observable random variables $x_1$, $x_2$, ….,$x_p$ in terms of the joint variation of a fewer number, k (<p) variables. The purpose of PCA is to determine factors (i. e. principal components) in order to explain as much of the total variation in the data as possible with as few of these factors as possible. This will uncover their qualitative and quantitative distinctions.

Table1 shows the descriptive statistics of the data. The data are not widely dispersed from the average when values of standard deviation are compared with the raw data. The measured values of Fe in the samples are quite large and so account for the large value of standard deviation (5. 0903) as seen in the table. The covariance matrix is shown in Table 2 and the corresponding correlation coefficients are shown in Table 3 and this was used to obtain the coefficients of the principal component using MATLAB as shown in Table 4 from standardized variable.

Observation elements are on the rows of Table 4. For example, Pb is denoted by $X_1$ and Fe by $X_2$ and so on. $U_1$, ………,$U_8$ are the principal components. The loadings or coefficients of the principal components are on the vertical columns. The magnitude of loadings greater than or equal to 0. 5 is to be considered for interpretation (Dillon and Goldstein,1984) as this will give the element with higher association ratio. On the last two rows of Table 4 is the eigenvalues of covariance matrix of the data and the Hotelling's $T^2$ statistic which gives a measure of the multivariate distance of each observation from the centre of the data set. A plot of the variability (in %) and the Principal component is as shown in Figure 2.

The three principal components with elements having loadings of 0. 5 and above (for which also the variability is greater than 10%) are: $U_1$, $U_2$and $U_3$ and these combined, account for 85. 34 % variability in the data. In $U_1$ are the elements : Fe and Mn in association (i. e. Fe-Mn)

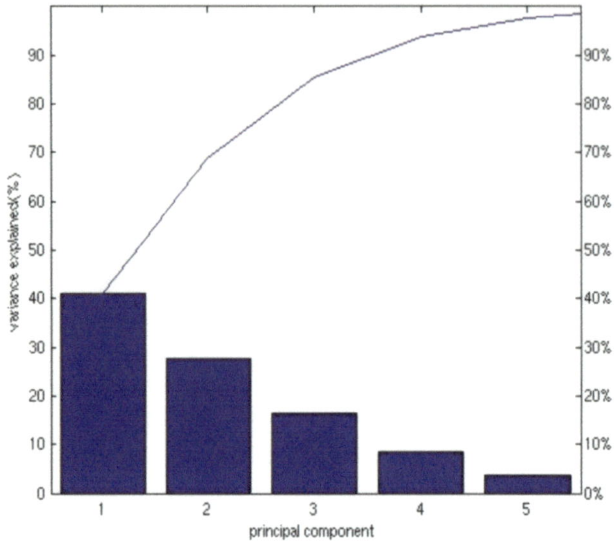


Fig. 2. A plot of the variability of the Principal Component, with five components accounting for 97. 44%.

accounting for 41. 11%. In $U_2$ the elements,Pb and Cr (i. e. Pb – Cr) accounting for 27. 65%and $U_3$ the elementsCd and Zn in association (i. e. Cd-Zn) with 16. 58 %.

| Element | Threshold (ppm) | Range (ppm) | Geometric Mean (ppm) | Std. dev. |
|---|---|---|---|---|
| Pb | 4. 50 | 3. 00 – 8. 00 | 1. 10 | 0. 0512 |
| Fe | 400. 00 | 50. 00 – 1750. 00 | 346. 00 | 5. 0903 |
| Ni | 5. 00 | 1. 50 – 7. 00 | 4. 00 | 0. 2121 |
| Cd | 0. 80 | 0. 03 – 1. 70 | 0. 50 | 0. 3043 |
| Cr | 0. 40 | 0. 01 – 1. 00 | 0. 30 | 0. 3512 |
| Cu | 0. 04 | 0. 01 – 0. 90 | 0. 03 | 0. 0090 |
| Zn | 0. 20 | 0. 01 – 0. 40 | 0. 14 | 0. 0113 |
| Mn | 10. 00 | 0. 20 – 26. 0 | 7. 30 | 1. 2015 |

Table 1. Descriptive statistics of the elements

| Element | Pb | Fe | Ni | Cd | Cr | Cu | Zn | Mn |
|---|---|---|---|---|---|---|---|---|
| Pb | 6. 41 | 438. 90 | -0. 02 | 0. 08 | -0. 05 | 0. 01 | 0. 08 | 4. 73 |
| Fe | 438. 90 | 281721. 80 | 339. 80 | 50. 00 | 3. 26 | 0. 14 | 2. 32 | 3602. 40 |
| Ni | -0. 02 | 339. 8 | 1. 81 | 0. 21 | 0. 06 | 0. 01 | 0. 02 | 4. 26 |
| Cd | 0. 08 | 50. 00 | 0. 21 | 0. 18 | 0. 02 | -0. 01 | 0. 02 | 0. 81 |
| Cr | -0. 05 | 3. 26 | 0. 06 | 0. 02 | 0. 05 | 0. 01 | 0. 01 | 0. 15 |
| Cu | 0. 01 | 0. 14 | 0. 01 | -0. 01 | 0. 01 | 0. 01 | 0. 01 | 0. 02 |
| Zn | 0. 08 | 2. 32 | 0. 02 | 0. 02 | 0. 01 | 0. 01 | 0. 01 | 0. 03 |
| Mn | 4. 73 | 3602. 4 | 4. 26 | 0. 81 | 0. 02 | 0. 02 | 0. 03 | 52. 41 |

Table 2. Covariance Matrix of the elements

| Element | Pb | Fe | Ni | Cd | Cr | Cu | Zn | Mn |
|---|---|---|---|---|---|---|---|---|
| Pb | 1. 00 | | | | | | | |
| Fe | 0. 35 | 1. 00 | | | | | | |
| Ni | 0. 01 | 0. 48 | 1. 00 | | | | | |
| Cd | 0. 12 | 0. 22 | 0. 37 | 1. 00 | | | | |
| Cr | -0. 05 | 0. 03 | 0. 22 | 0. 17 | 1. 00 | | | |
| Cu | 0. 15 | 0. 01 | 0. 01 | -0. 03 | 0. 08 | 1. 00 | | |
| Zn | 0. 33 | 0. 04 | 0. 14 | 0. 37 | 0. 11 | 0. 40 | 1. 00 | |
| Mn | 0. 29 | 0. 94 | 0. 44 | 0. 27 | 0. 09 | 0. 09 | 0. 03 | 1. 00 |

Table 3. Correlation coefficients of the elements

| | % | $U_1$ (41. 11) | $U_2$ (27. . 65) | $U_3$ (16. 58) | $U_4$ (8. 31) | $U_5$ (3. 79) | $U_6$ (2. 51) | $U_7$ (0. 04%) | $U_8$ (0%) |
|---|---|---|---|---|---|---|---|---|---|
| Pb | $X_1$ | -0. 0022 | 0. 5905 | -0. 1851 | -0. 4640 | 0. 3429 | -0. 1434 | -0. 2725 | 0. 4351 |
| Fe | $X_2$ | -0. 5046 | 0. 2445 | 0. 0376 | 0. 0729 | -0. 1184 | 0. 3095 | 0. 6589 | 0. 3674 |
| Ni | $X_3$ | -0. 3897 | -0. 3036 | -0. 1220 | 0. 5128 | 0. 5782 | -0. 0440 | -0. 2619 | 0. 2707 |
| Cd | $X_4$ | -0. 1052 | -0. 3599 | -0. 6629 | -0. 1409 | -0. 4200 | -0. 3702 | 0. 0425 | 0. 2912 |
| Cr | $X_5$ | 0. 0880 | -0. 5187 | 0. 4400 | -0. 3967 | -0. 0550 | 0. 2898 | -0. 1527 | 0. 5118 |
| Cu | $X_6$ | 0. 4073 | 0. 2412 | 0. 2444 | 0. 5520 | -0. 3130 | -0. 2636 | -0. 0585 | 0. 4905 |
| Zn | $X_7$ | 0. 4034 | 0. 0405 | -0. 5018 | 0. 1695 | 0. 0080 | 0. 7336 | -0. 0911 | 0. 0929 |
| Mn | $X_8$ | -0. 5001 | 0. 2024 | 0. 0696 | 0. 0757 | -0. 565 | 0. 2304 | -0. 6214 | -0. 0809 |
| Eig. of Cov. Matrix | | 3. 2890 | 2. 2120 | 1. 3262 | 0. 6653 | 0. 3033 | 0. 2008 | 0. 0034 | 0. 0000 |
| $T^2$ | | 6. 1250 | 6. 1250 | 6. 1250 | 6. 1250 | 6. 1250 | 6. 1250 | 6. 1250 | 6. 1250 |

Table 4. Coefficients of the Principal component transformation (Numbers in Bracket are the proportion of total variance in %) of the elements

The principal components $U_4$, $U_5$, $U_6$, $U_7$ and $U_8$ have smaller variability and have respective eigenvalues of 0. 6653, 0. 3033, 0. 2008, 0. 0034 and 0. 0000. These when compared to the first three components are very small which is an indication of their relative decrease in significance in the data. So any association suggested by them cannot be realistic.

Results of a previous regional geochemical survey carried out by Ajayi and Suh (1993) using the factor analysis statistical techniquerevealed the existence of : Zn - Co - Cd, Ni - Cr, Fe - Mn and Cu - Pb as relevant associations mainly in the Amphibolites rocks. The results from the present study compare favourably well to these earlier results obtained from factor analysis but the association region extends to the quartz-schist rocks as can be seen in Figures 3, 4 and 5. Figure 3 shows the association ratio of Fe - Mn. The range of values is shown by the coloured circles. Values are classified as low (in purple circles) with concentration less than 20ppm, moderate (in yellow circles) with concentration range of 21 - 40ppmand high (in blue circles) with concentration greater than 41ppm. It can be seen that higher values are clustered at the central part of the figure which is on the Amphibolite, Schist and Epidiorite Complex and also on the Quartzite/Quartz Schist. At other parts, the values are distributed between moderate and lower values, though, with some isolated higher values probably due to rock intrusions. For examples, the higher values near Iperindo are surrounded by lower and moderate values.

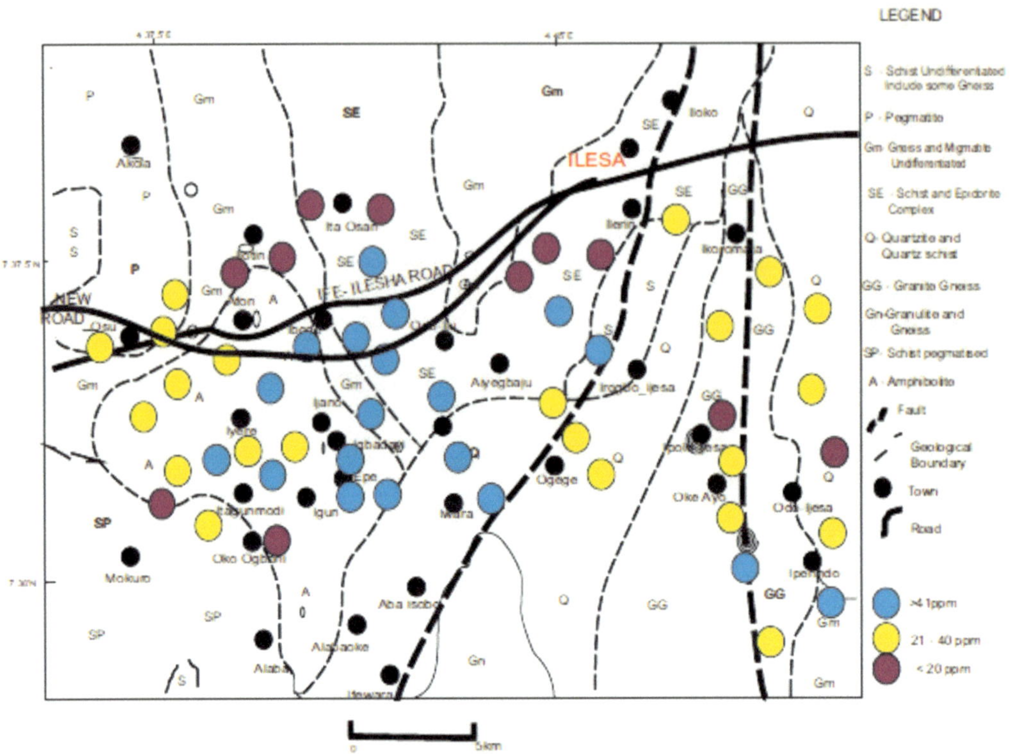


Fig. 3. Map showing the association ratio of Fe – Mn in the study area.

The association ratio of Pb - Cr are as shown in Figure 4. The higher values with concentration greater than 21ppm are more widely distributed than the Fe – Mn association ratio. However, most of the higher values are still concentrated on the Amphibolite, Quartzite/Quartz Schist and Schist Epidiorite Complex, though, with some moderate and lower values in between them. The association ratio of Cd – Zn is shown in Figure 5. The values are seen to be more widely distributed for most part of the figure. This shows the spread of the association over the basement rocks. Cd is known universally to associate with Zn. It actually reflects the strong lithologic influence related to the mafic minerals with which Zn is associated. It can thus be seen, that for the plotted concentration ratios, higher values of the metallic association are distributed largely towards the centre of the study area. This partly corresponds to the suggested mineralized areas by previous geological and geophysical studies of Ako (1980) and Ajayi (1988). Similar studies (such as the present one) were carried out on the Elura Deposit in the Cobar Mining district of Central New South Wales, Australia. The study identified high grade Zn – Pb - Agsulphide mineralization and this was later followed by extensive geophysical surveys to map the siliceous, pyriticand the pyrrhotitic ores (Palacky, 1988). A combined map of Figures 3, 4 and 5 is as shown in Figure 6. This gives the combined map for the geochemical anomaly over the study area. The Fe – Mn ratio is represented in red circles, the Pb – Cr ratio is represented in green circles while the Cd – Zn ratio is represented in blue circles. The anomaly is seen to be widely distributed over the delineated area. Also shown on the map are the geophysical locations and traverses to investigate the anomaly.

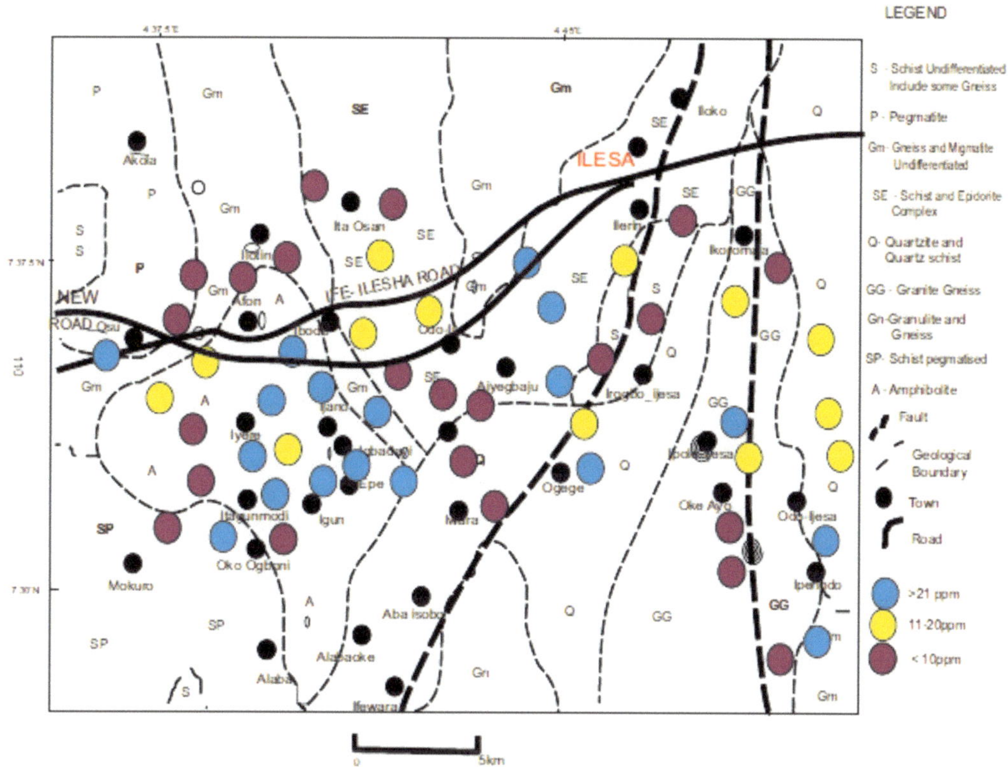


Fig. 4. Map showing the association ratio of Pb - Cr in the study area.

### 5.2 The electrical resistivity data, analysis and results

Electrical methods are generally referred to as "resistivity surveys". Metallic minerals are relatively good conductors of electricity. In contrast, common rock forming minerals are generally poor conductors. This fact is the basis for geophysical exploration methods which measure conductivity to evaluate the metal content of rocks. The methods also provide some limited information about the geometry of the subsurface metallic mineralization. Surface electrical methods are limited to shallow depths (<200m), but the electrical properties of rocks can be measured at much greater depths by using electrical borehole instruments sent down deep drill holes. The location of the seven (7) VES data points are as shown in Figure 6. They are located so as to provide subsurface geological information including, layer resistivity, thickness and depth to the bedrock which will be useful in the geophysical modeling that will follow in the subsequent section. The Wenner configuration was used with electrode separation (a) varying from 1 - 96m. VES curves were interpreted using the partial curve matching technique followed with computer iteration procedure on the WinGLink computer software version 1. 62. 08. On this software, computed curves are compared with observed field curves. Where a good fit (i. e. 95 % correlation and above) was obtained between the two curves, the interpretation results was considered satisfactory; otherwise the geoelectric parameters were modified as appropriate and the procedure repeated until a satisfactory fit was obtained. The result from the computer iteration is as summarized with their geoelectric parameters in Table 5.

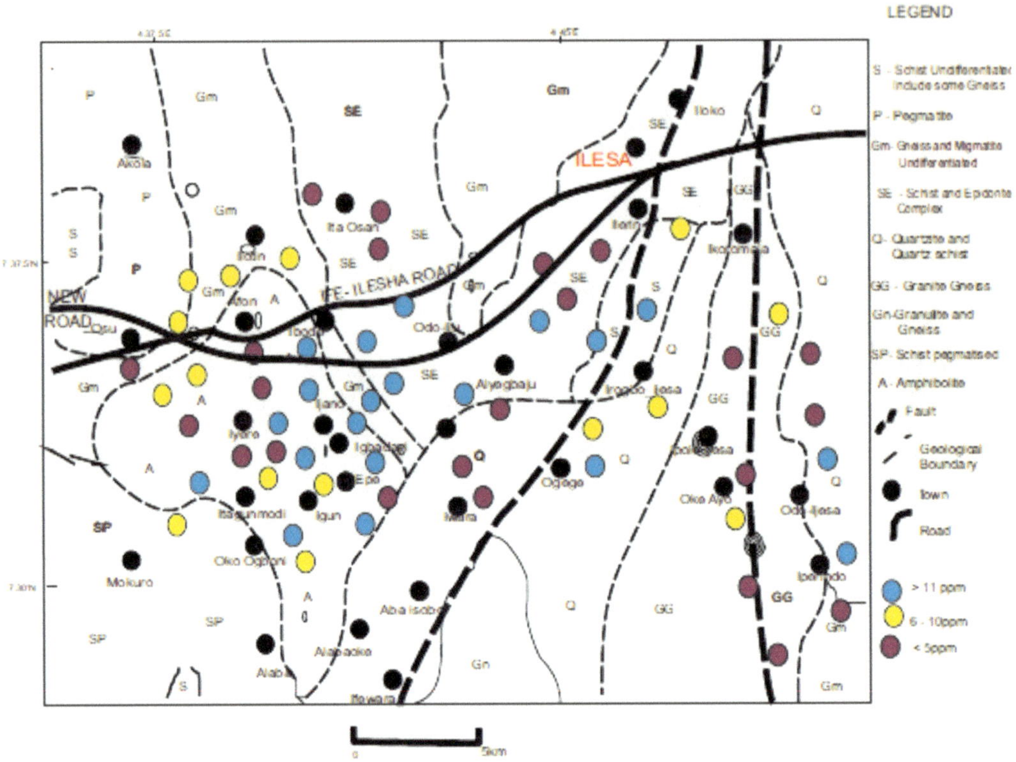


Fig. 5. Map showing the association ratio of Cd - Zn in the study area.

## 5.3 VLF– EM data, analysis and results

The Very Low Frequency (VLF) electromagnetic method uses powerful remote radio transmitters set up in different parts of the world for military communications (Klein and Lajoie, 1980). In radio communications terminology, VLF means very low frequency, of about 15 to 25 kHz. Relative to frequencies generally used in geophysical exploration, these are actually very high frequencies. The radiated field from a remote VLF transmitter, propagating over a uniform or horizontally layered earth and measured on the earth's surface, consists of a vertical electric field component and a horizontal magnetic field component each perpendicular to the direction of propagation.

These radio transmitters are very powerful and induce electric currents in conductive bodies thousands of kilometers away. Under normal conditions, the fields produced are relatively uniform in the far field at a large distance (hundreds of kilometers) from the transmitters. The induced currents produce secondary magnetic fields that can be detected at the surface through deviation of the normal radiated field. The VLF method uses relatively simple instruments and can be a useful reconnaissance tool. Potential targets include tabular conductors in a resistive host rock such as faults in limestone or igneous terrain. The depth of exploration is limited to about 60% to 70% of the skin depth of the surrounding rock or soil. Therefore, the high frequency of the VLF transmitters means that in more conductive environments, the exploration depth is quite shallow; for example, the depth of exploration might be 10 to 12 m in 25-Ωm material (Milsom, 1989). Additionally, the presence of conductive overburden seriously suppresses response from basement conductors, andrelatively small variations in overburden conductivity or thickness can themselves generate significant VLF anomalies. For this reason, VLF is more effective in areas where the host rock is resistive and the overburden is thin.

In the VLF method, two orthogonal components of the magnetic field were measured, and normally the tilt angle, α, and ellipticity, *e*, of the vertical magnetic polarization ellipse are derived. Real (in-phase) and imaginary (quadrature) are used in the Karous - Hjelt Fraser filter (KHFFILT) programme. These components are based on the tilt angle, (α) and ellipticity *(e)* as:

$$Re = \tan(\alpha)\, 100\% \quad \text{and} \quad Im = e\, 100\%. \tag{1}$$

| VES No | Layer Resistivity (Ω -m) | Layer thickness (m) | Curve type | Remark |
|---|---|---|---|---|
| 1 | 231<br>368<br>892<br>233 | 3. 77<br>6. 01<br>13. 50<br>- | AK | AMP* |
| 2 | 127<br>439<br>104<br>324 | 0. 59<br>2. 23<br>13. 28<br>- | KH | AMP |
| 3 | 795<br>202<br>455<br>3065 | 1. 03<br>11. 45<br>7. 68<br>- | HA | AMP |
| 4 | 407<br>257<br>49<br>320 | 0. 60<br>2. 11<br>4. 87<br>- | QH | GMU |
| 5 | 350<br>177<br>445<br>238 | 1. 13<br>1. 80<br>36. 43<br>- | HK | GMU |
| 6 | 550<br>1195<br>1804<br>186 | 0. 50<br>6. 28<br>25. 03<br>- | AK | SEC |
| 7 | 330<br>455<br>2212<br>319 | 1. 26<br>6. 16<br>22. 82 | AK | QQS |

*AMP = AmphiboliteGMU = Gneiss and Migmatite Undifferentiated
QQS = Quartzite/Quartz Schist SEC= Schist and Epidiorite Complex

Table 5. Geoelectric layer parameters of the study area

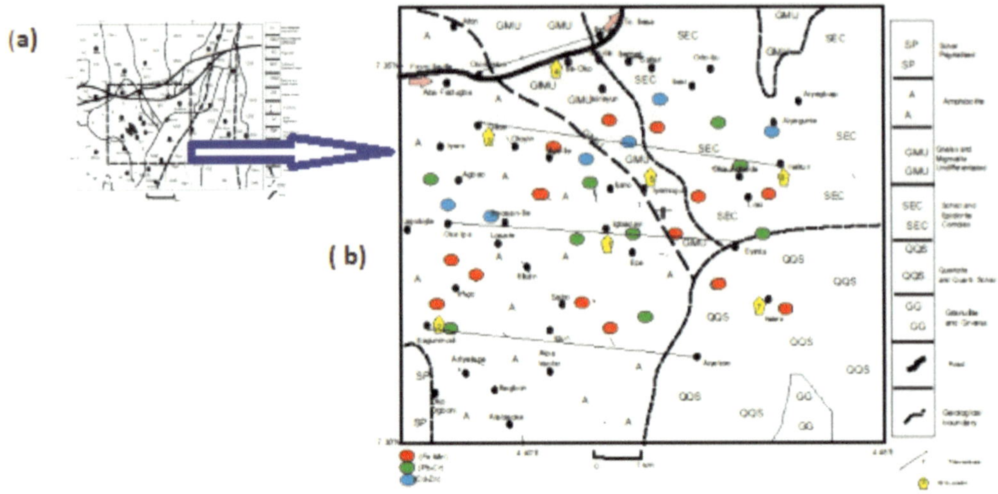

Fig. 6. The geochemical anomalous zone is enclosed by the rectangle with dashed lines on the geological map (a) and the geophysical traverses and locations on the zone (b).

The real and imaginary data values were first plotted using the Microsoft Excel. Next was the plot using the KHFFILT programme (Pirttijarvi, 2004) to obtain the Fraser - filtered plot and the Karous - Hjelt filtered pseudo-section.

The Fraser and Karous - Hjelt filtering are the two methods widely used in processing VLF -EM data (Fraser, 1969: Karous and Hjelt, 1983). The Fraser filter transforms the zero-crossing points into positive peaks which indicate conductive structures. The Karous - Hjelt filter is also used to obtain relative current density pseudo-sections, in which lower values of relative current density correspond to higher values of resistivity (Benson et al., 1997). The areas of high current density (represented in red colour) flow correspond to positive values and low current density (in blue colour) flow to negative values on the accompanying colour scale (for example see Figure 8).

### 5.3.1 The four west – east VLF (EM) profiles

The filtered response for the four West - East profiles are presented as in Figure 7 based on their respective locations in Figure 6 to correlate and describe the fractures across the area with the geochemical anomaly. The magnitude of the filtered response is varied along the four profiles due to the nature of the conductivity of the underlying materials. It is clear that the response along Olorombo to Ibode (7a) differs from those along Gada to Iwikun, Okeipa to Eyinta and that of Itagunmodi to Aiyetoro which are similar (Figures (7 b– d)).

The positive peaks labeled F1 is seen to occur across the three profiles in Figures 7 b,7c and 7d. It occurs at about station 1000m along each profile on the Amphibolite. This shows that the linear feature, interpreted as mineralized fracture is consistent in occurrence in the Amphibolite. The positive peaks labeled F2 and F3 are observed to cut across the three profiles also on the Amphibolite as shown in Figures 7 c and 7 d. The linear feature labeled F4 is observed to be consistent in occurrence in Figures 7c and 7d and actually lie near the boundaries of the amphibolites and gneiss/migmatite Undifferentiated rocks (Figure 7c) and amphibolites and quartzite/quartz schist rocks (Figure 7d). The linear feature labeled F4 is not visible in Figure 7b due largely to the nature of the material hosted by the gneiss/migmatite Undifferentiated and the schist/epidiorite complex rocks along this profile.

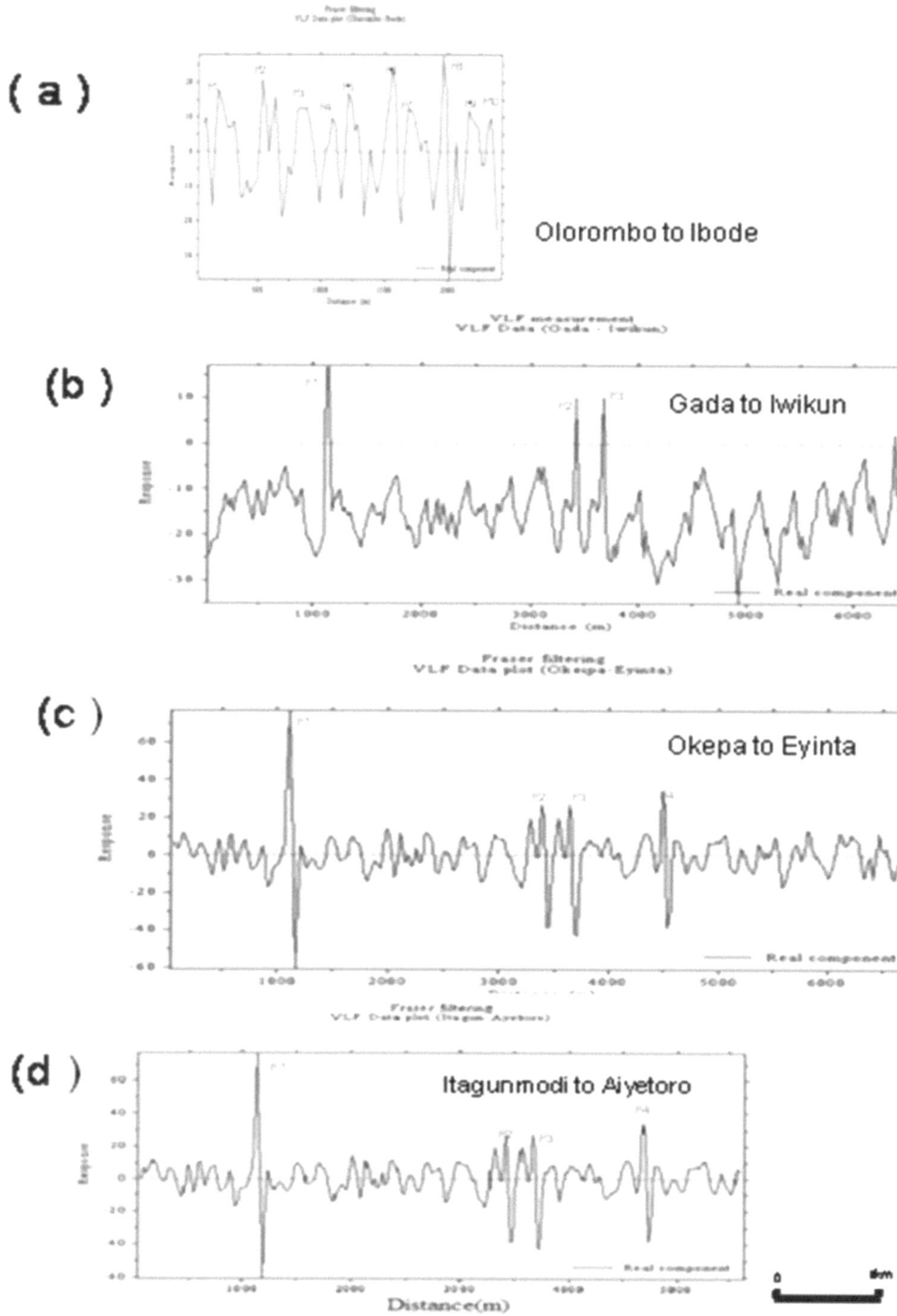


Fig. 7. The VLF – EM filtered real profiles for the four West – East traverses

Figure 8 is the pseudosections for the four West – East profiles. The figure shows the character of the labeled linear features earlier discussed. The linear feature labeled F1 is observed to be dipping conductors (as are observed on the Amphiboliteat about station 1000m) in all the three traverses of Gada to Iwikun, Okeipa to Eyinta and Itagunmodi to Aiyetoro. These are seen to have values thatrange between 15% and 70%. The occurrence of these dipping fractures from near the surface to depth of 100m and approximately along the N – S direction suggest that it is one and the same linear feature which cuts across the study area. The other fractures (F2 and F3) also on the amphibolites which are almost vertical and at between stations 3000m and 4000m occur at a depth range of 20 -70m are also one and the same linear features which cut across along N – S direction in the Amphibolite. Vertical conductive structures are also observed in the schist/epidiorite Complex and inthe quartzite/quartz schist rocks. The occurrence of these linear features in the study area has implications for mineralization, geotechnical and groundwater studies (Palacky,1988). Minerals that are structurally controlled such as lateritic nickel, gold, talc and clay deposits can be prospected along the identified linear features.

### 5.4 The magnetic data, analysis and results

Magnetism has been studied for a very long time in human history. Early Greek philosophers knew about the attraction of iron to a magnet. The first magnets consisted of a naturally occurring rock called lodestone, a variety of massive magnetite (almost pure iron oxide). Magnetite is the only naturally occurring mineral with distinctly obvious magnetic properties. Only a few other minerals have any detectable magnetism. However, extremely sensitive magnetometers can detect trace magnetism in many different minerals. Iron, because of its atomic structure, has the greatest tendency to become magnetized. Other elements, such as cobalt and nickel, have fewer tendencies to become magnetic. Any mineral or rock which contains any of these elements is likely be more magnetic.

The Earth possesses a magnetic field caused primarily by sources in the core. The form of the field is roughly the same as would be caused by a dipole or bar magnet located near the Earth's center and aligned sub-parallel to the geographic axis. The intensity of the Earth's field is customarily expressed in S. I. units as nanotesla (nT) or in an older unit, gamma (γ):1 γ = 1 nT = $10^{-3}$ μT. Except for local perturbations, the intensity of the Earth's field can vary between about 25 and 80 μT. Many rocks and minerals are weakly magnetic or are magnetized by induction in the Earth's field, and cause spatial perturbations or "anomalies" in the Earth's main field. Man-made objects containing iron or steel are often highly magnetized and locally can cause large anomalies up to several thousands of nT. Magnetic methods are generally used to map the location and size of objects that have magnetic properties.

In order to produce a magnetic anomaly map of a region, the data have to be corrected to take into account the effect of latitude and, to a lesser extent, longitude (Reynolds, 1997). As the Earth’s magnetic field strength varies from 25000nT at the magnetic equator to 69000nT at the poles, the increase in magnitude with latitude needs to be taken into account. Survey data at any given location can be corrected by subtracting the theoretical field value $F_{th}$, obtained from the International Geomagnetic Reference Field, from the measured value, $F_{obs}$. Regional latitudinal (φ) and longitudinal (θ) gradients can be determined for areas concerned and tied to a base value, $F_o$. Gradients northwards ($\delta F/\delta\varphi$) and westwards

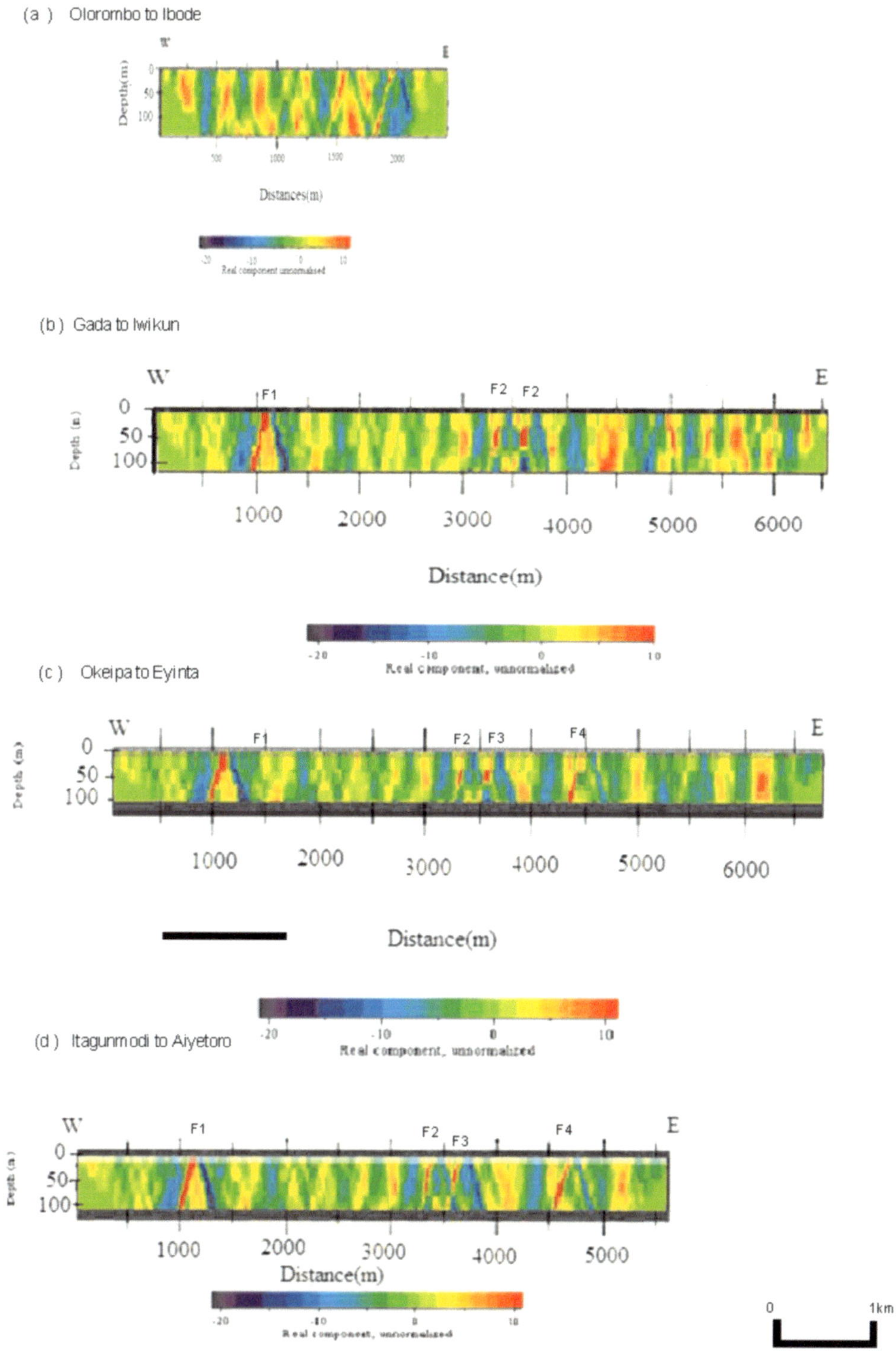


Fig. 8. The VLF – EM pseudosection for the four West – East traverses

(δF/δθ) are expressed in nT/km. Consequently the anomalous value of the total field (δF) can be calculated from

$$\delta F = F_{obs} - (F_o - {}^{\delta F}\!/_{\delta\phi} + {}^{\delta F}\!/_{\delta\theta}) \quad (nT) \tag{2}$$

Geophysical (Magnetic) traverses over the delineated area are as shown in Figure 6b. In this survey, the total magnetic field was measured. The data reduction involved removing the regional field, reduction to the pole and vertical continuation before the plot of profile along the traverse. Removing the regional field helps to emphasize the magnetic anomaly of interest and this was done by subtracting the known regional field from the measured value. The reduction to the pole helps to change the actual inclination to the vertical. It was performed by convolving the magnetic field with a filter whose wave number response is the product of a polarization-orientation factor and the field -orientation factor (Baranov, 1957; Gunn, 1975; Spector and Grant, 1985). Also the field upward continuation attenuates anomalies caused by local, near - surface sources relative to anomalies caused by deeper more profound sources.

### 5.4.1 The magnetic models

Information available from the magnetic profiles along the four West - East traverses, VLF - EM, VES data and the geochemical anomaly are used in the modeling of magnetic data to confirm the existence of the linear features, basement depth (and topography) and the basement tectonic framework which are revealed in the area. The magnetic profiles were modeled using the 2. 5D modeling algorithm of the WingLink software programme (version 1. 62. 08). The profiles which are along the West - East include: Olorombo to Ibode, Gada to Iwikun, Okeipa to Eyinta and Itagunmodi to Aiyetoro. The modelling of these profiles shows very reasonable fit between the observed and the calculated magnetic profiles.

Figure 9A shows the observed and the calculated anomaly for the magnetic profile along Olorombo to Ibode and the corresponding geologic section in Figure 9B. Three model bodies are involved in the computation and these include the Amphibolite (s = 280 SI unit), Gneiss/Migmatite Undifferentiated (s = 300 SI unit) and Schist Epidiorite Complex (s = 100 SI unit) together with the overburden (s= 5 SI unit). The contact between the rock types may represent structural or lithological contacts especially in the area of basement outcrops (El-Shayeb, Personal communication). The model lithological contacts when correlated with the known geology shows that the contact between the Amphibolite and the Gneiss/Migmatite Undifferentiated partly correlated but the contact between the Gneiss/Migmatite Undifferentiated and the Schist Epidiorite does not correlate with the known geology. An overburden thickness of 2m is observed on theGneiss/Migmatite Undifferentiated rock at station 750m while the thickness is 1. 5m in the Schist Epidiorite Complex rock. The overburden is deepest (14m) in the Gneiss/Migmatite Undifferentiated rock at station 1350m.

Figure 10A shows the observed and the calculated anomaly for the magneticprofile along Gada to Iwikun and the corresponding geologic section in Figure 10B which is on the central part of the delineated area. Three model bodies are involved in the computation and these include the Amphibolite (s = 250 SI unit), Gneiss/Migmatite Undifferentiated (s = 130 SI unit) and Schist Epidiorite Complex (s = 150 SI unit) together with the overburden (s= 10 SI unit) and a Quartz vein (s=35 SI unit). The model has three bodies representing the three

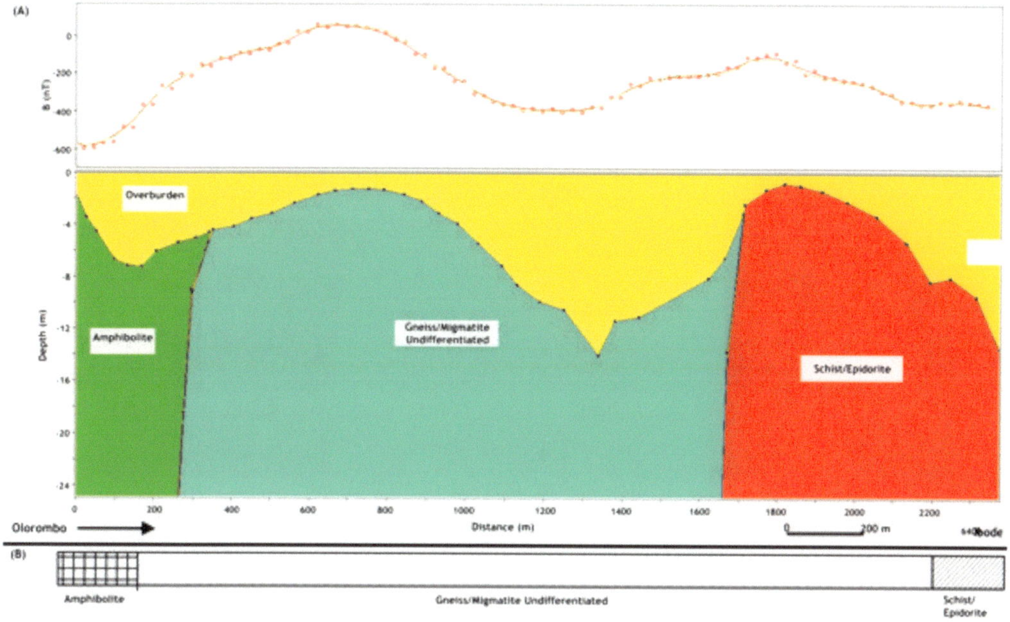


Fig. 9. The observed and calculated magnetic profile along Olorombo to Ibode

basement blocks. The contact between bodies represents structural or lithological contacts. The model lithological contacts when correlated with the known geology show afair correlation. The model result suggests a westward shift in the geological boundary. The model also revealed the existence of a Quartz vein (s = 35 SI Unit) sandwiched between the Amphibolite and the Gneiss/Migmatite Undifferentiated rocks at stations between 2400 - 2600m along the profile. The model reveals the outcropping of the Schist Epidiorite Complex at stations 4100m and 5500m. The overburden is deepest (48m) in the Amphibolite rock at station 400m.

Figure 11A shows the observed and the calculated anomaly for the magneticprofile along Okepa to Eyinta and the corresponding geologic section in Figure 11B which is on the central part of the delineated area. Three model bodies are involved in the computation and these include the Amphibolite (s = 200 SI unit), Gneiss/Migmatite Undifferentiated (s = 100 SI unit) and Schist Epidiorite Complex (s = 300 SI unit) together with the overburden (s= 5 SI unit) and a dyke body (s=40 SI unit). The contact between bodies represents structural or lithological contacts. The model lithological contacts do not correlate with the known geology probably due to the "masking" effect of the overburden. The model reveals the existence of a dyke body (s = 400 SI unit) in the Quartzite/Quartz Schist rock just before station 6400m. Observed also are the five fractures in the Amphibolite at stations 200, 1400, 1600, 2300 and 3000m along the profile. The modeled basement blocks magnetic susceptibility contrast ranges between 100 - 300 SI units and reflects the variation in the composition of the basement rocks across the profile.

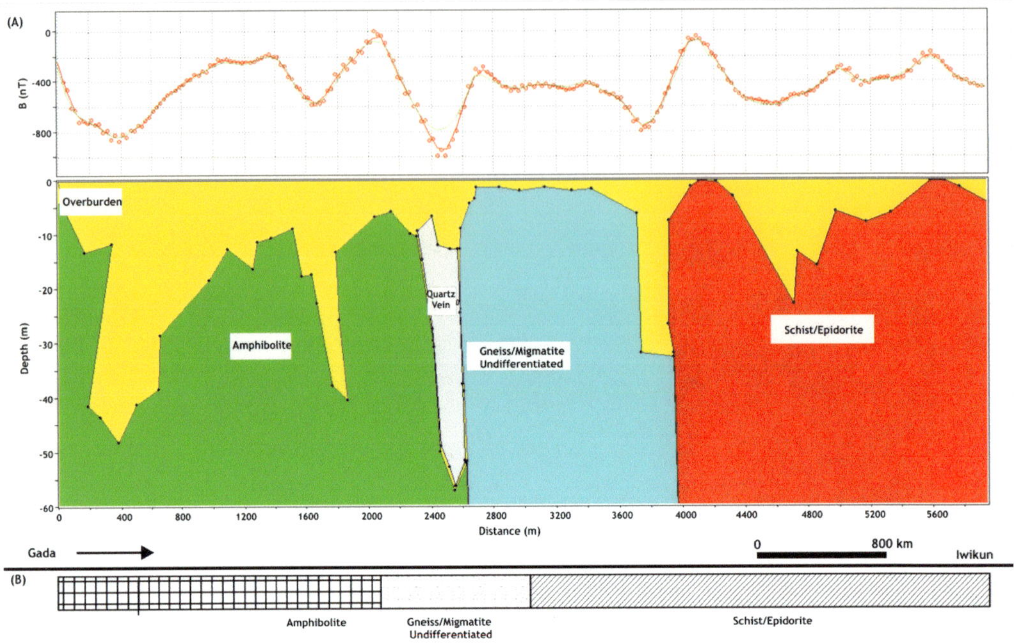


Fig. 10. The observed and calculated magnetic profile along Gada to Iwukun

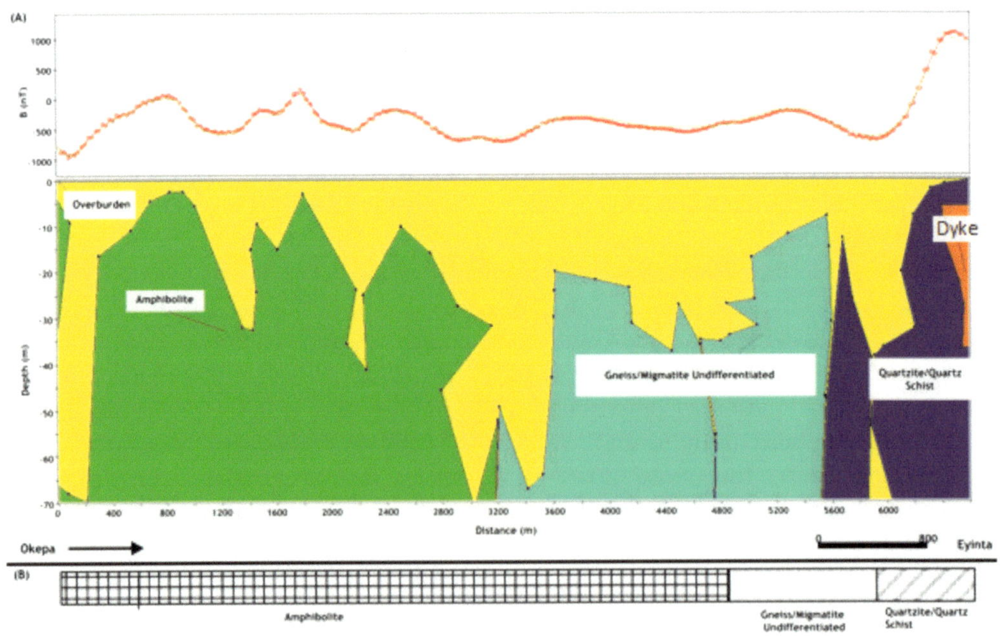


Fig. 11. The observed and calculated magnetic profile along Okepa to Eyinta

Figure 12A shows the observed and the calculated anomaly for the magneticprofile along Itagunmodi to Aiyetoro and the corresponding geologic section in Figure 12B which is on the southern part of the delineated area. Two model bodies are involved in the computation and these include the Amphibolite (s = 280 SI unit), Quartzite/Quartz Schist (s = 400 SI unit) together with the overburden (s= 10 SI unit). The contact between bodies represents structural or lithological contacts. The model lithological contact when correlated with the known geology shows a good correlation. Observed also are many fractures/faults in the Amphibolite along the profile. The modeled basement blocks magnetic susceptibility contrast ranges between 280 - 400 SI units.

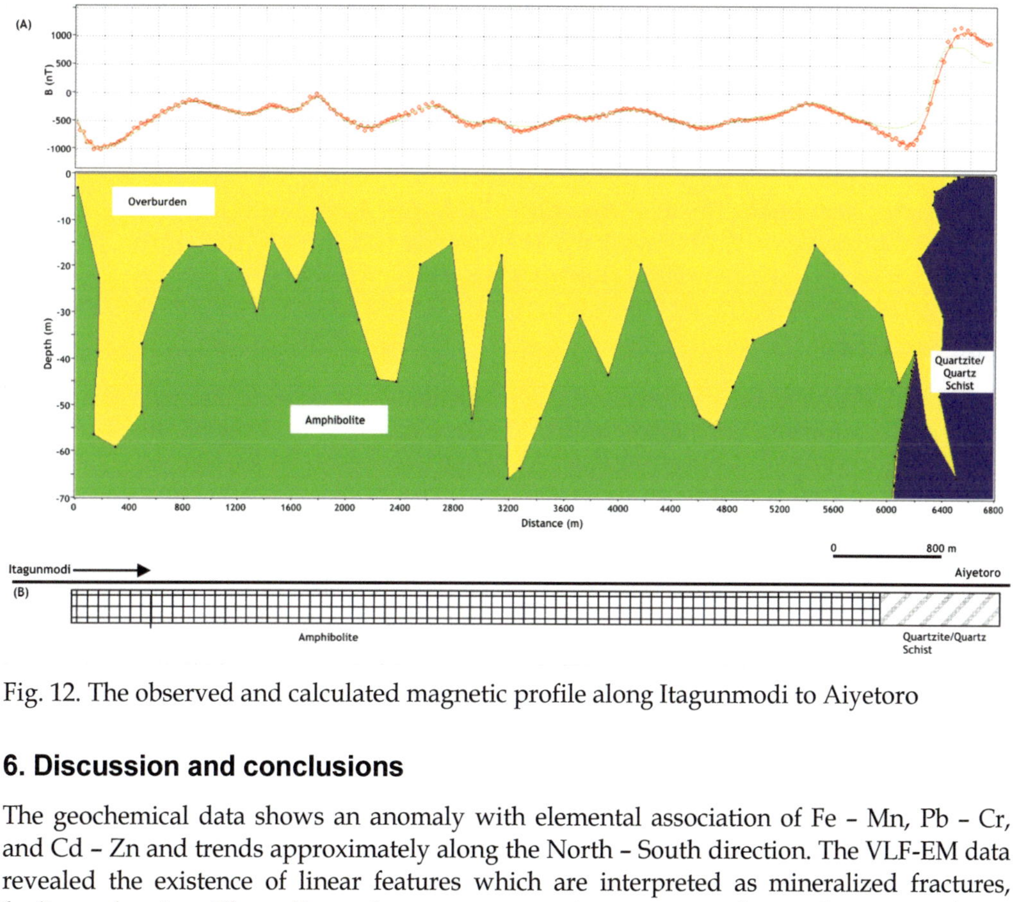


Fig. 12. The observed and calculated magnetic profile along Itagunmodi to Aiyetoro

## 6. Discussion and conclusions

The geochemical data shows an anomaly with elemental association of Fe - Mn, Pb - Cr, and Cd - Zn and trends approximately along the North - South direction. The VLF-EM data revealed the existence of linear features which are interpreted as mineralized fractures, faults and veins. These linear features are consistent across the study area and run approximately in the N - S direction. These structures exist from near the surface to a depth of up to 100m. The magnetic data also mapped the linear features, magnetized bodies (interpreted as dykes) and the geological contacts. The coincidence of electromagnetic (VLF) and magnetic anomalies and their correlation with the geochemical anomaly, especially on the amphibolites, is an indication of the occurrence of sulphide deposits rich in Zn-Pb-Cralong the identified linear features . The geophysical methods engaged have helped to map the structural complexity such as evidence of faulting, mineralized fractures, joints and dykes in the study area. In addition, it is now evident that mineralization is not limited to within the amphibolites, but also exist in the other rock types. There is an improved delineation of the rock contacts which has hitherto been difficult to map as a result ofdearth of outcrops and the presence of overburden.

Based on the combined geological, geochemical and geophysical data which are available (and the modeled results) and interpreted over the delineated area, a new geological map is proposed for the area. Figure 13 shows the disposition of the prominent linear features, dykes and the inferred mineralized zone found in the area. The fractures are seen to trend approximately in the N - S direction. The frequency of fracturing is a function of rock elastic properties, structure and tectonic history. The rock with high fracture frequency is known to be highly brittle and prone to fracturing. From this study the occurrence of fractures on the Amphibolite is high and so also that on the Quartzite/Quartz Schist and Gneiss/Migmatite Undifferentiated rocks are moderately high. The Quartzites/Quartz Schist isalso strongly magneticprobably due to the occurrence of iron oxides as described by Ajayi (1988).

## 7. Acknowledgments

The chapter contribution to the Geoscience Text was done while the author was on Associateship visit to ICTP, Trieste, Italy. The financial and material support by ICTP that ensured the successful completion of the work is greatly acknowledged.

## 8. References

Adebayo, A. B. (2006). Physico-chemical characteristics of groundwater in two mining areas of contrasting lithologies in the Ife-Ilesa Schist Belt. Unpubl. B. Sc. Thesis, Department of Geology, Obafemi Awolowo University, Ile- Ife, 80p.

Adelusi, A. O. (2005). Multi-method geophysical investigation for groundwater study in southeastern part of Ilesa area, Osun state Southwestern Nigeria. Ph. D Thesis, Department of Applied Geophysics, Federal University of Technology, Akure. 289p

Adewunmi, A. O. (1984). Geochemical investigation of soil, stream sediment and surface water samples in ilesa Area and their probable relationships to thesurrounding rock types. M. Sc. Thesis, Department of Geology, Obafemi Awolowo University, Ile-Ife . 185p.

Adewusi, G. A. (1988). Geophysical study of the Iwara fault and associated mineralization. M. Sc. Thesis, Department of Geology, Obafemi Awolowo University, Ile-Ife . 247p

Ajayi, T. R. (1988). Integrated exploration and statistical studies of geochemical data in Ife - Ilesa Goldfield. Ph. D Thesis. Obafemi Awolowo University, Ile - Ife, Nigeria, 459pp.

Ajibade, A. C. (1976). Provisional classification and correlation of the Schist Belts ofNorthwestern Nigeria (In)

C. A. Kogbe (ed) Geology of Nigeria, Elizabethan Publishing Co. Lagos, pp85 -90.

Ako, B. D. Ajayi, T. R. and Alabi, A. O. (1978). A geoelectrical study of the Ifewara area. *Journal of Mining and Geology* 15 (2) p84-89

Ako, B. D. (1979). Geophysical prospecting for groundwater in part ofsouthwestern Nigeria. PhD Thesis, University of Ife, Ile-Ife. 371p

Ako, B. D. (1980). A contribution to mineral exploration in the Precambrian belt of part of southwestern Nigeria. *Journal of Mining and Geology* 17 (2) p 129-138.

Anhaeusser, C. R., Mason, R., Viljoen, M. J. and Viljoen, R. P. (1969). A reappraisal of some aspects of Precambrian shield Geology. Bull. Geol. Soc. Amer. Vol. 80, 2175-2200

Ariyibi, E. A., Folami S. L., Ako B. D., Ajayi, T. R. and Adelusi, A. O. (2010). Application of Principal Component Analysis on geochemical data: A case study in the basement complex of southern Ilesa area, Nigeria. Arabian *Journal of Geosciences*. DOI 10. 1007/s12517-010-0175-5

Baranov, V., (1957). A new method for interpretation of aeromagnetic maps: Error analysis for remote reference magnetotellurics: *Geophysics*, 22 359-367

Benson, A. K., Payne, K. L. and Stubben, M. A. (1997). Mapping groundwater contamination using dc resistivity and VLF geophysical method-A casestudy. *Geophysics* 62 (1), 80 - 86.

Boyd, D. 1967. The contribution of airborne magnetic surveys to geological mappingIn: Mining and groundwater geophysics. Published by The Geological Survey of Canada Economic Geology Report No. 26. 213 - 227p.

DeSwardt, A. M. J. (1947). The Ife-Ilesa goldfield. (Interim report no. 2). Geol. Surv. Nigeria, Annu. Rep., pp 14-19.

De Swardt, A. J. (1953). The geology of the country around Ilesa. Geological Surveyof Nigeria Bulletin 23 ; 55p.

Dillon W. R., and Goldstein M. (1984). Multivariate Analysis Methods andApplications. John Wiley and Sons, Inc. New York. 587p

El-Shayeb, H. M., El-Meliegy, M. A., Meleik, M. L. and Abdel-Raheim, R. M. (Personal Communication). Magnetic Interpretation of Esh-Mallaha area, northern eastern desert, Egypt 19p

Elueze, A. A. (1977). Geological and geochemical studies in the Ife-Ilesa Schist belt in relation to gold mineralization. Unpubl. M. Phil. Thesis, Univ. of Ibadan, Nigeria

Fraser, D. C., (1969). Contouring of VLF-EM data. *Geophysics* 34, p958–967.

Gill, R. (1997). Modernanalytical geochemistry. Addison Wesley Longman Limited. 210p.

Grunfeld K. (2003). Interactive visualization applied to multivariate geochemical data: A case study. *Journal Physics* IV France 107 p577-580.

Gunn, F. (1975). Evaluation of terrain effects in ground magnetic surveys. *Geophysics*, 32 (2): 582-589

Hammer, S. (1939). Terrain corrections for gravimeter stations. *Geophysics* 4, 184 - 194

Hawkes H. E. and J. S. Webb. (1962). Geochemistry in mineral Exploration. Addison Wesley Longman Limited. 110p.

Hubbard, F. H. (1975). Precambrian crustal development in western Nigeria, indications from Iwo region. *Geological Society of America Bulletin* Vol. 86. pp548 - 554.

Hurley, S. (1966). Electromagnetic reflections in salt deposits. *Journal of Geophysics*, 32: 633 - 637.

Hurley, S. (1970). Radar propagation in rock salt. *Geophysical Prospecting*, 18 (2):312-328

Karous, M., Hjelt, S. E., (1983). Linear filtering of VLF dip-angle measurements. *Geophysical Prospecting* 31, 782–794.

Klein, J., and Lajoie, J. (1980). Electromagnetic Prospecting for minerals. Practical geophysics for the Exploration geologists. Northwest Mining Association, Spokane, WA, pp 239 -290

Klemm, D. D., Schneider, W. and Wagner, B., (1979). Geological and geochemical investigations of the metasedimentary belts of the Ife-Ilesa area of Southwestern

Nigeria. Presented at the Conference on African geology, University of Ibadan July 1978.
Klemm, D. D., Schneider, W. and Wagner, B., (1984). Precambrian metavolcano - sedimentary sequence east of Ife and Ilesa, S. W. Nigeria - A Nigerian Schist belt? *Journal of African Earth Sciences*, 2 (2) : 161 - 176.
Mazzullo, J. (1996). Investigations into Physical geology, A laboratory manual. Saunders College Publishing 361p.
McCurry, P. (1976). The geology of the Precambrian to lower Paleozoic rocks of
Northern Nigeria - A review, in C. A. Kogbe (ed) Geology of Nigeria. Elizabethan pub. Co., Lagos, pp 15- 40.
Milsom, J. (1989). Field Geophysics: Milton Keynes: Open university press. 172p
Olorunfemi, B. N. (1977). A geochemical soil survey in the Ife-Ifewara- Ogudu Area of Oyo State, Nigeria. M. Sc. Thesis, Department of Geology, University of Ife99pp.
Oyawoye, M. O. (1964). Geology of the Nigerian Basement Complex - A survey of our present knowledge of them. Journal of Mining Geology and Metal vol. 1 no 2pp87 - 103.
Palacky, G. J., (1988). Resistivity characteristics of geologic targets. Electromagnetic Methods in Applied Geophysics, vol. 1. SEG, Tulsa, OK, pp. 106–121.
Pirttijärvi, M., (2004). Karous–Hjelt and Fraser filtering of VL measurements. Manual of the KHFFILT Program. 26p
Rahaman, M. A. (1976). Review of the basement Geology of southwesternNigeria. Geology of Nigeria,
Kogbe, C. A.; (Ed). Elizabethan Publ. Co., Lagos, Nigeria. 41-58.
Rahaman, M. A. (1988). Recent advances in the study of the Basement Complex of Nigeria. Precambrian Geology of Nigeria. Pp 11 - 39
Reynolds, J. M. (1997). An introduction to applied and environmental geophysics. John Wiley and sons. 796p.
Russ, P. (1957). Airborne electromagnetics in review. *Geophysics*, 22 691-713
Spector, A., and F. S. Grant. (1985). Statistical models for interpreting magnetic data, *Geophysics*, 35, 293- 302,
Suh, C. E. (1993). Primary metal dispersion patterns for gold exploration in the Amphibolites of Ife - Ilesa schist belt. M. Sc. Thesis. Department of Geology, Obafemi Awolowo University, Ile -Ife. 145p
Wing Link Integrated Geosystem software version 1. 62. 08 - 20030519. Microsoft Corp. 23p
Wright, J. B. and McCurry, P. (1970). A reappraisal of some aspects of the Precambrian shield geology. A discussion. *Bulletin of Geological Society of America*, 81: 3491 - 3492.

# 5

# Regularity Analysis of Airborne Natural Gamma Ray Data Measured in the Hoggar Area (Algeria)

Saïd Gaci[1], Naïma Zaourar[1], Louis Briqueu[2] and Mohamed Hamoudi[1]
*[1]University of Sciences and Technology Houari Boumediene, Algiers,*
*[2]Laboratoire Géosciences- University Montpellier 2- CNRS, Montpellier,*
*[1]Algeria*
*[2]France*

## 1. Introduction

The airborne Gamma Ray (GR) measurements have been used since decades in geophysical research. The airborne measurement of gamma radiation emitted by naturally occurring elements finds applications in: geological mapping (Graham and Bonham-Carter, 1993; Jaques *et al.*, 1997; Doll *et al.*, 2000, Aydin *et al.*, 2006; Sulekha Rao *et al.*, 2009), regolith and soil mapping (Cook *et al.*, 1996; Wilford *et al.*, 1997; Bierwirth and Welsh, 2000), mineral exploration (Brown *et al.*, 2000), and hydrocarbon research (Matolín and Stráník, 2006).
Potassium (K), Uranium (U) and Thorium (Th) are the three most abundant, naturally occurring radioactive elements. The K element is the main component of mineral deposits, while Uranium and Thorium are present in trace amounts, as mobile and immobile elements, respectively. The concentration of these different radioelements varies between different rock types, thus the information provided by a gamma-ray spectrometer can be exploited for needs of the rocks cartography. The obtained maps allow to localize radioelement anomalies corresponding to zones disrupted by a mineralizing system.
The approach presented in this chapter deepens the results derived from the conventional study. It consists on a mono(two)-dimensional fractal analysis of natural radioactivity measurements recorded over the Hoggar area (Algeria).
The natural radioactivity measurements, like other geophysical signals, contain a deterministic and a stochastic components. The former part holds information related to the regional aspect, while the latter reflects the local heterogeneities. As the raw spectrometric data need to be processed before any exploitation, the stochastic component can be altered and some information about heterogeneities is lost.
Here, we show first the fractal behavior of the analyzed GR measurements. In addition, it is demonstrated that this behavior is not affected by all the pre-processing operations (spectrometric corrections and 2D-interpolations). The corrections are not then necessary. Since the analyzed data exhibit a fractal exponent varying with the spatial position, they are modeled as paths of multifractional Brownian motions (mBms) (Peltier and Lévy-Véhel, 1995).

The local Hölder exponent (or local regularity) maps obtained from the GR data recorded in the K, Th and U channels, using a multiple filter technique that we generalize to a 2D-case, exhibit almost an identical image. Besides, they allow to locate the faults affecting the studied zone.

## 2. Regional geology

The Hoggar is a large shield area covering approximately 550,000 km$^2$. It includes an important surface of the Tergui shield, prolonged in South-east, in Mali, by the solid mass of Iforas and in the East, in Niger, by the solid mass of Aïr (Fig. 1).
The Hoggar belongs to the Trans-Saharan pan-African chain (Cahen *et al.*, 1984, Liégeois *et al.*, 1994). It is crossed by two major submeridian faults, located at longitudes 4°50' and 8°30', which delimit three longitudinal compartments (Eastern, Central and Western), with different structural and lithological characteristics. This geological configuration resulted by an extreme E-W compression, during the pan-African (600 My), of the Touareg shield by two rigid plates: the Western African craton and the Eastern African craton (Bertrand and Caby, 1978; Black *et al.*, 1979).

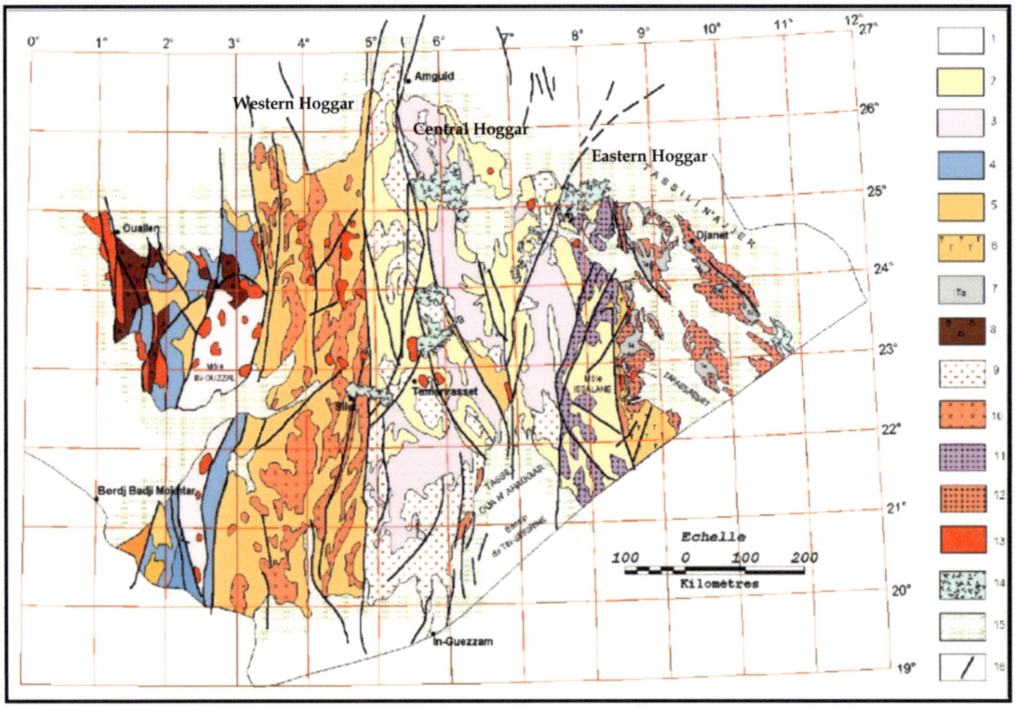


**1** - Archaean granulites; **2** - Gneiss and metasediments, series of Arechchoum (Pr1); **3** - Gneiss with facies amphibole, series of Aleskod (Pr2); **4** - Indif. gneiss (Pr3); **5** - Pharusian Greywackes; **6** - Arkoses and conglomerates, series of Tiririne (Pr4); **7** - Volcano-sediments of Tafassasset (Pr4); **8** - Molasses (purple series) of Cambrian; **9** – Pan-African syn-orogenic granites; **10** - Pan-African Granites; **11** - Pan-African post-orogenic granites; **12** - Granites of Eastern Hoggar; **13** - Late pan-African Granites; **14** - Basalts and recent volcanism; **15** - Paleozoic cover; **16** - Fault.

Fig. 1. A simplified geological map of the Hoggar (Caby *et al.*, 1981, modified)

## 3. Overview on the analyzed GR measurements

The analyzed GR measurements are recorded during a magneto-spectrometric survey accomplished, between 1971 and 1974 over the Hoggar, for the purpose of the mining research and the regional geological mapping.
The technical characteristics of the survey are:

- Two types of planes :
  - Douglas DC-3.
  - Aero Commander.
- Navigation System: Doppler type A DRA-12
- Magnetic Compass of type Sperry CL 2, with a resolution of 1°.
- Radar altimeter with an accuracy of 30 feet (type Honeywell Minneapolis).
- Camera with a continuous 35 mm-film
- Acquisition system of data (type Lancer) for the recording of the numerical data on magnetic tapes of 1/2".
- Two types of graphic recorders: with 2 and 6 channels for the graphic monitoring of the magnetic and spectrometric profiles respectively.
- Two types of magnetometers:
  - Magnetometer with optical pumping with the Cesium (model VARIAN) of resolution of 0.02 NT (nano Tesla).
  - Magnetometer Flow-gate of a resolution of 0.5 NT.
- NaI(Tl) spectrometer with four (04) channels: Total Count (TC), Uranium (U), Thorium (Th) and Potassium (K).

The parameters of airborne spectrometric survey carried out over the Hoggar area are:

- The average of the flight height is fixed at 500 feet (approximately 150 m).
- The direction of the profiles: perpendicular to the geological structures.
- The distance between lines varies from 2 to 5 kilometers according to the areas, but on average it is about two kilometers.
- The distance between the observation points is approximately 46.2 m (152 feet).

## 4. Corrections of the airborne natural activity measurements

The measurements acquired during an airborne spectrometric survey can not be exploited in a raw state, but need to be corrected mainly from aircraft background, stripping (or Compton) effect and height effect (IAEA, 2003).

Background corrections

There are three components of the background correction:

- The instrument background (called "aircraft background" in airborne gamma spectrometry),
- The cosmic background arisen from the reaction of primary cosmic radiation with atoms and molecules in the upper atmosphere.
- The effect of atmospheric radon. In portable or car-borne gamma ray surveys, the background component is usually small relative to the signal from the ground.

The observed count rates in the four channels: Total Count (TC), Potassium (K), Uranium (U) and Thorium (Th), are corrected for the background effects using the following formulae:

$$\begin{aligned} TC_{corr} &= TC_{obs} - BC_{TC} \\ K_{corr} &= K_{obs} - BC_K \\ U_{corr} &= U_{obs} - BC_U \\ Th_{corr} &= Th_{obs} - BC_{Th} \end{aligned} \tag{1}$$

where all these values are expressed in counts per second (cps).
$TC_{corr}$, $K_{corr}$, $U_{corr}$ and $Th_{corr}$ : Values of the corrected count rates in the four channels,
$TC_{obs}$, $K_{obs}$, $U_{obs}$ and $Th_{obs}$ : Values of the observed count rates in the four channels,
$BC_{TC}$, $BC_K$, $BC_U$ and $BC_{Th}$ : Values of the background correction in the four channels. In this case, the estimated values are (Groune, 2009): $BC_{TC}$ = 250 cps, $BC_K$ = 72 cps, $BC_U$ = 17 cps and $BC_{Th}$ = 5 cps.

Stripping correction

This correction, also known as the channel interaction correction, consists of removing ('strips') count rates from each of the K, U and Th for gamma rays not originating from the radioelement or decay series being monitored. For example, Th series gamma rays appear in both the U and K channels, and U series gamma rays appear in the K channel. The corrections are given by:

$$\begin{aligned} U_{corr} &= U_{obs} - \alpha Th_{obs} \\ K_{corr} &= K_{obs} - \beta Th_{obs} - \gamma U_{obs} \end{aligned} \tag{2}$$

$K_{corr}$ and $U_{corr}$: Values of the corrected count rates in the K and U channels respectively.
$K_{obs}$, $U_{obs}$ and $Th_{obs}$ : Values of the observed count rates in the K, U and Th channels respectively.
$\alpha$, $\beta$ and $\gamma$ : Stripping coefficients. The used coefficients in this application are (Aeroservice, 1975): $\alpha = 0.45$, $\beta = 0.59$ and $\gamma = 0.94$.

Height correction

This correction is applied only on airborne gamma spectrometric measurements. The gamma radiation decreases exponentially with the elevation. Since the height of the aircraft changes continuously, the airborne Gamma Ray spectrometric data need to be corrected to a nominal survey height above the ground.

$$\begin{aligned} TC_{corr} &= TC_{obs}\exp\left[\mu_{TC}(h-h_0)\right] \\ K_{corr} &= K_{obs}\exp\left[\mu_K(h-h_0)\right] \\ U_{corr} &= U_{obs}\exp\left[\mu_U(h-h_0)\right] \\ Th_{corr} &= Th_{obs}\exp\left[\mu_{Th}(h-h_0)\right] \end{aligned} \tag{3}$$

$TC_{corr}$, $K_{corr}$, $U_{corr}$ and $Th_{corr}$ : Values of the corrected count rates in the four channels,
$TC_{obs}$, $K_{obs}$, $U_{obs}$ and $Th_{obs}$ : Values of the observed count rates in the four channels,
$h$ : Real survey height ,
$h_0$ : Nominal survey height ($h_0$=150 m),
$\mu_{TC}$, $\mu_K$, $\mu_U$ and $\mu_{Th}$ : Linear attenuation coefficients in the four channels. The estimated values of these coefficients are (Groune, 2009) : $\mu_K$ = 6.8617 $10^{-3}$ $m^{-1}$, $\mu_U$ = 6.3726 $10^{-3}$ $m^{-1}$ , and $\mu_{Th}$ = 5.2247 $10^{-3}$ $m^{-1}$ . The $\mu_{TC}$ is calculated as the approximate average of the three coefficients: $\mu_{TC}$ = 6.56 $10^{-3}$ $m^{-1}$ .

## 5. Impact of the pre-processings on the fractal properties of the airborne gamma ray measurements

Once all the corrections are applied, the corrected measurements grid is regrided using two-dimensional interpolation algorithms to get a regular sampled grid which is processed by a local regularity analysis.

The set of operations (corrections and interpolations) affects the stochastic component of the raw airborne spectrometric measurements, which holds information about heterogeneities. Therefore the fractal properties of the raw data may be changed.

In the first stage, we have obtained the ''corrected'' and the ''corrected and interpolated'' data grids from the ''raw'' grid data corresponding to the measurements of the three channels: K, Th and U. The 2D-interpolation algorithms used in this study are: the triangle-based linear, the triangle-based cubic and the nearest neighbor interpolation algorithms. Since the results obtained by the different interpolation methods are close, only those related to the triangle-based linear algorithm are presented.

First, five vertical profiles are extracted from the three considered grids (''raw'', ''corrected'' and ''corrected and interpolated '' grids) from the measurements of the three channels (Fig.2). The Fourier amplitude spectrum and the local Hölder exponent $H(x)$ are computed for each data profile.

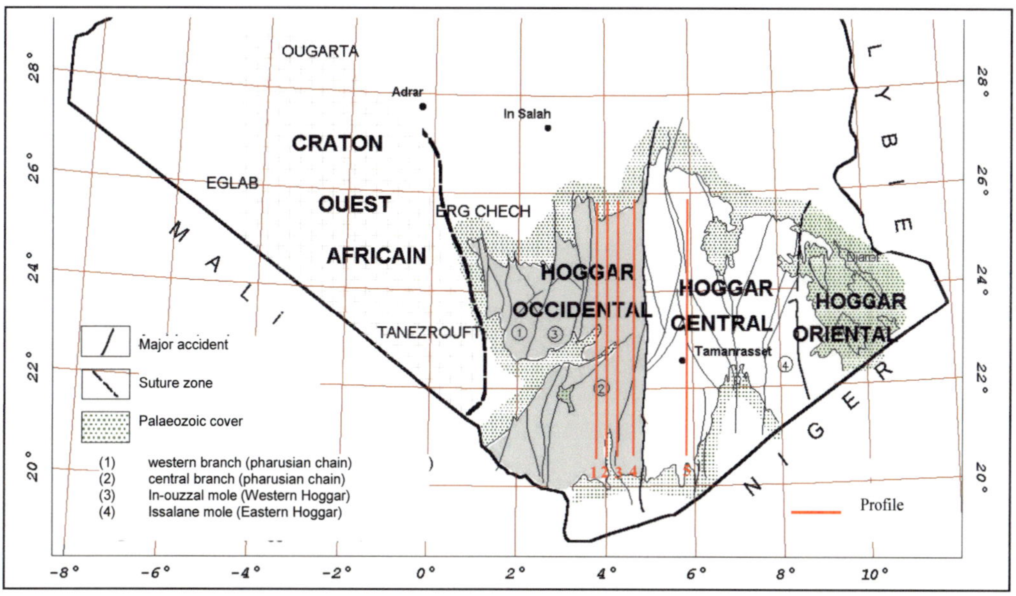


Fig. 2. Position of the five profiles extracted from the GR measurements (in red). (The geological map of the Hoggar, from Caby *et al.*, 1981).

Regarding the computation of $H(x)$, we need a sequence $S_{k,n}(i)$ defined by the local growth of the increment process:

$$S_{k,n}(i) = \frac{m}{n-1} \sum_{j \in [i-k/2, i+k/2]} \left| X(j+1) - X(j) \right|, \; 1<k<n \tag{4}$$

where $n$ is the signal $X$ length, $k$ is a fixed window size, and $m$ is the largest integer not exceeding $n/k$.
The local Hölder function $H(x)$ at point

$$x = \frac{i}{n-1} \tag{5}$$

is given by (Peltier and Lévy-Véhél, 1994, 1995; Muniandy *et al.*, 2001 ; Li *et al.*, 2007, 2008; Gaci *et al.*, 2010):

$$\hat{H}(i) = -\frac{\log\left[\sqrt{\pi/2}\; S_{k,n}(i)\right]}{\log\;(n-1)} \tag{6}$$

From figure 3, it can be seen that all the calculated amplitude spectra, represented in a log-log plan, decay algebraically, the analyzed data exhibit then a fractal behavior. Moreover, the latter is described by a Hölder exponent varying with the latitude of the measure. Hence the data can be considered as paths of multifractional Brownian motions (mBms) (Peltier and Lévy-Véhel, 1995; Gaci *et al.*, 2011).
A significant result deserves to be noted is the fact that the spectra obtained from the ''raw'', ''corrected'' and ''corrected and interpolated'' measurements display a similar form. That is the applied operations (corrections and interpolations) do not affect the fractal aspect of the raw data.

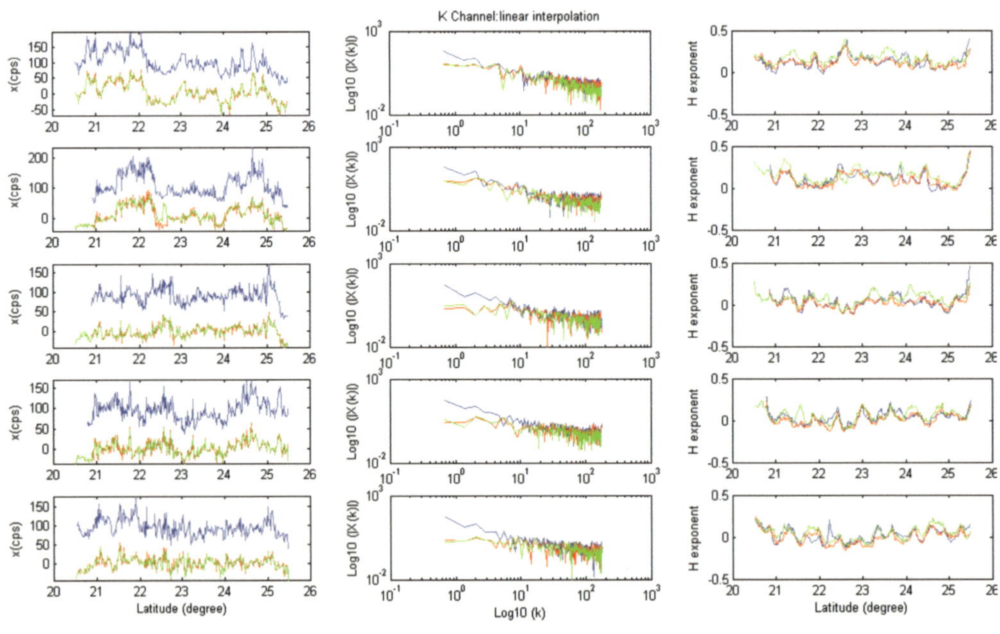


a)

Fig. 3. (Continued)

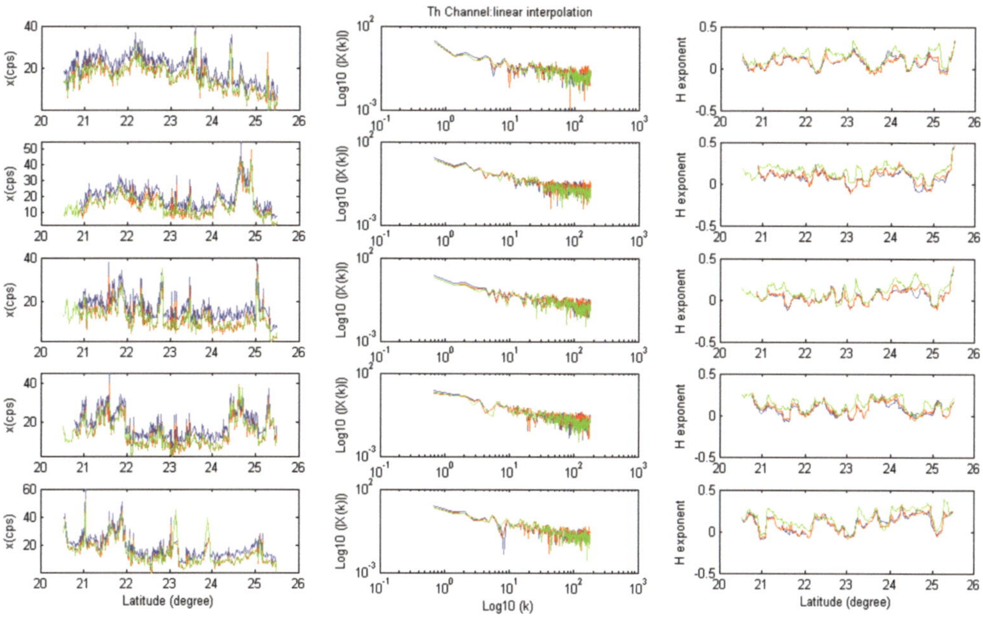


b)

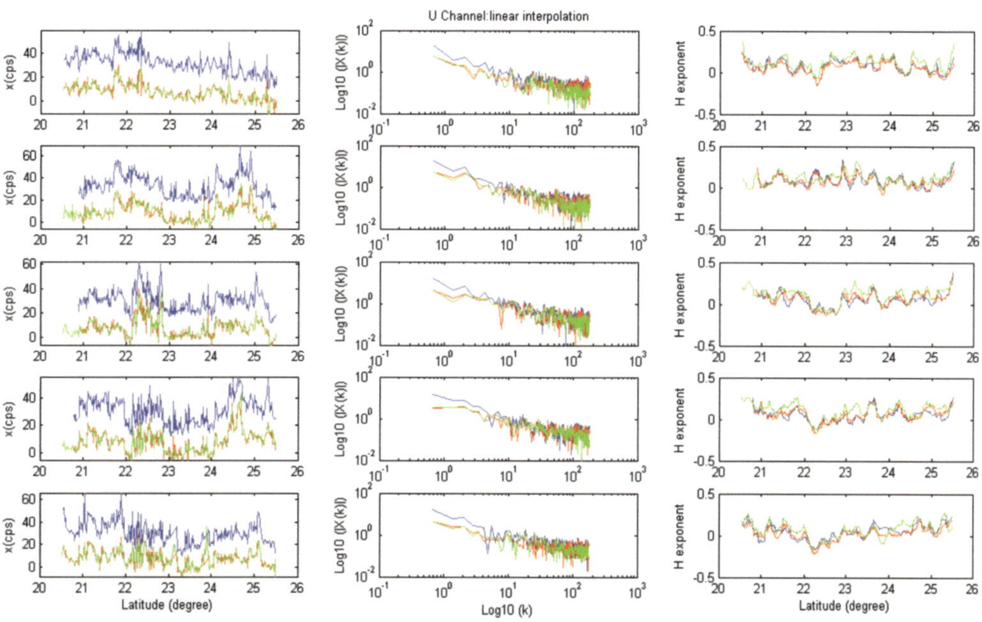


c)

On the left, the measurements profile, in the middle the module of the amplitude spectrum of the measurements profile versus the wavenumber (rad/degree) in the log-log scale, and on the right, the local Hölder function.

The raw data (blue), the corrected data (red) and the corrected and interpolated data (green).

Fig. 3. Investigation of the impact of pre-processings on the fractal properties of the five profiles of the airborne GR data recorded in the channel: (a) K, (b) Th, (c) U.

Moreover, the estimated Hölder functions obtained from the three types of measurements present very close values. Again, we confirm that the fractal properties of the raw data are not modified by both pre-processing operations. The implementation of the different 2D-interpolation algorithms illustrates that the choice of the interpolation algorithm has a very slight effect on the estimated $H$ value. An important result to be noted: the spectrometric corrections are not necessary for a fractal analysis which can be carried out directly on the raw measurements. By doing so, the stochastic component of the measurements is kept intact.

## 6. Local regularity analysis of airborne spectrometric data

In this section, we establish local two-dimensional regularity maps, from the interpolated raw GR data measured in the three channels: K, Th and U, using a wavelet-based algorithm via the two-dimensional Multiple Filter Technique (2D MFT). We obtained the latter technique by generalizing the mono-dimensional version (Dziewonski *et al.*, 1969; Li, 1997) to the 2D-case (Gaci, 2011).

### 6.1 Spectrometric data interpolation

Considering the limitations of the computer's processing capacity, we consider the GR measurements recorded, in the K, Th and U channels, over the zone whose geographical coordinates are defined by: longitude: 3° 13′ 58″- 6°59′ 26″ E, and latitude: 20° 27′ 35″-25° 06′ 37″ N.

The 2D-interpolation of the raw spectrometric data is performed owing to the kriging algorithm. The interpolated GR grids data related to the K, Th and U channels are illustrated respectively by figures 4, 5 and 6.

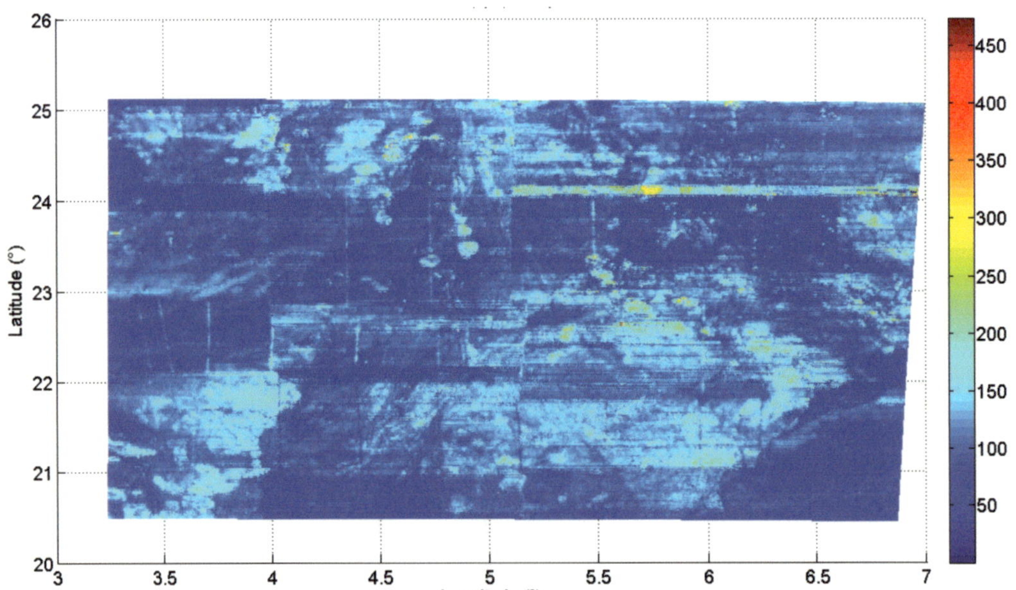


Fig. 4. Interpolated Gamma Ray measurements (in cps) related to the K channel .

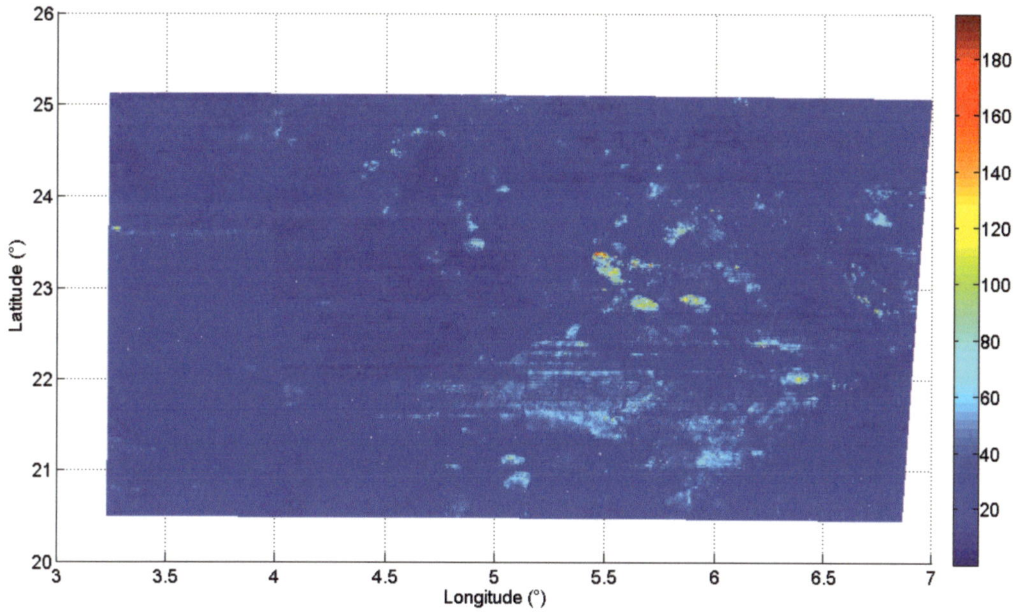


Fig. 5. Interpolated Gamma Ray measurements (in cps) related to the Th channel.

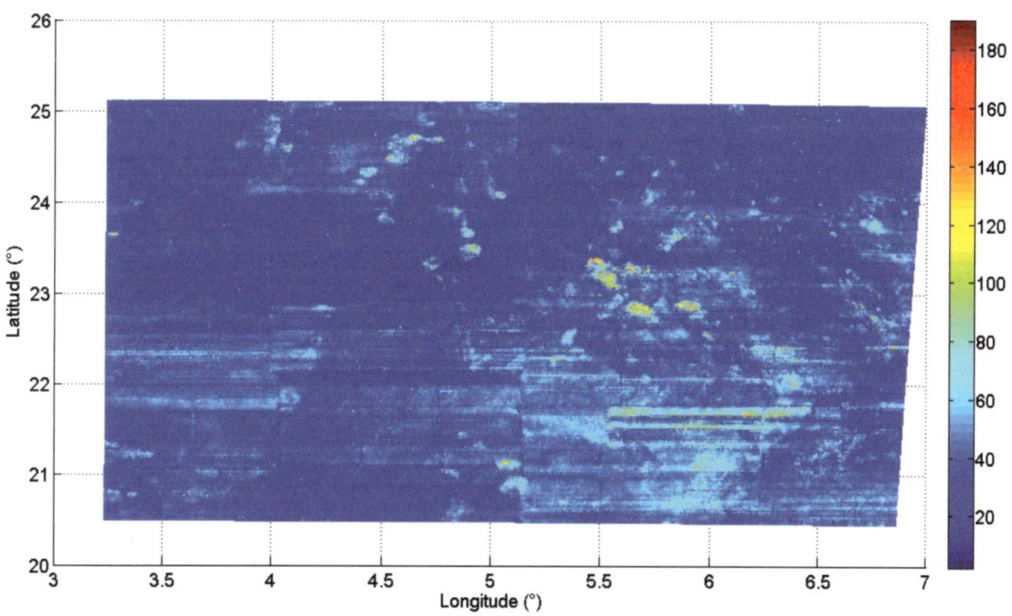


Fig. 6. Interpolated Gamma Ray measurements (in cps) related to the U channel.

### 6.2 Establishment of local regularity maps from interpolated spectrometric data

Using a wavelet-based algorithm, we estimate Hölder exponent maps, from the interpolated GR measurements recorded in the three channels (K, Th and U).

Recall that the two-dimensional continuous wavelet transform (2D- CWT) is given by a convolution product of a signal $s(x,y)$ and an analyzing wavelet $g(x,y)$ (Chui, 1992; Holschneider, 1995):

$$S(a, b_x, b_y) = \frac{1}{\sqrt{a}} \int_{-\infty}^{\infty} s(x,y) \bar{g}\left(\frac{x-b_x}{a}, \frac{y-b_y}{a}\right) dx\, dy$$

where "$a$" is the scale parameter, "$b_x$" and "$b_y$" are the respective translations according to X-axis and Y-axis (the symbol " $-$ " denotes the complex conjugate).
Alternatively, it can be computed via the Fast Fourier Transform:

$$S(a,b_x,b_y) = FFT^{-1}\left(\hat{s}(\xi,\nu).\ \sqrt{a}\ \bar{\hat{g}}(a\xi,a\nu)\right)$$

Here, we compute the wavelet coefficients via FFT using the two-dimensional multiple filter technique (2D MFT). The latter technique is obtained by generalizing the one-dimensional version (1D MFT), suggested by Dziewonski et *al.* (1969) and improved by Li (1997), to the two-dimensional case. It consists of filtering a two-dimensional signal using a Gaussian filter $G(k,\xi_n,\nu_m)$ given by (Gaci, 2011):

$$\begin{aligned} G(k,\xi_n,\nu_m) &= G_1(k,\xi_n)G_2(k,\nu_m) \\ &= e^{-\alpha\left(\frac{k-\xi_n}{\xi_n}\right)^2} e^{-\alpha\left(\frac{k-\nu_n}{\nu_n}\right)^2} \end{aligned} \tag{7}$$

Where $\xi_n$ and $\nu_m$ are variable center angular frequencies (or wavenumbers) of the respective filters $G_1(k,\xi_n)$ and $G_2(k,\nu_m)$. The bandwidths $\Delta k_1$ and $\Delta k_2$ of both filters are calculated as:

$$\begin{aligned} \Delta k_1 &= \xi_2 - \xi_1 = \beta.Ln(\xi_n) \\ \Delta k_2 &= \nu_2 - \nu_1 = \beta.Ln(\nu_m) \end{aligned} \tag{8}$$

Where $\beta$ is a constant, ($\xi_1$ , $\xi_2$) and ($\nu_1$ , $\nu_2$) are respectively the – 3 dB points of the Gaussian filters $G_1$ and $G_2$ , respectively.
A fractal surface $s(x,y)$ verifies the self-affinity property (Mandelbrot, 1977, 1982; Feder, 1988) :

$$s(\lambda x,\lambda y) \cong \lambda^H.s(x,y),\ \forall \lambda > 0 \tag{9}$$

Where $H$ is the Hurst exponent (or the self-affinity parameter). The symbol $\cong$ means the equality of all its finite-dimensional probability distributions.
For sufficiently large values of $k$, the scalogram, defined as the square of the amplitude spectrum: $P(k,x,y) = |S(k,x,y)|^2$, can be expressed as:

$$P(k,x,y) = P'(x,y).k^{-\beta(x,y)} \propto k^{-\beta(x,y)} \tag{10}$$

Where

$$\beta(x,y) = 2H(x,y) + 1 \tag{11}$$

is the local spectral exponent which is related to the local Hurst (or Hölder) exponent, $H(x,y)$. The spectral exponent $\beta(x,y)$ in each point $(x,y)$ is computed as the slope of the scalogram versus the wavenumber in the log-log plan, the $H(x,y)$ value is then derived using the equation (11).

The implementation of the wavelet-based algorithm, using the generalized multiple filter technique, allows to establish regularity maps from the interpolated GR measurements recorded in the three channels (K, Th and U) (Fig. 7). In order to interpret the resulting maps in terms of geology, a geological map of the studied zone is considered.

The results show that the *H* maps, derived from the measurements of all the channels, exhibit almost an identical image of the local regularity. By reporting the faults affecting the studied zone on the obtained regularity maps, we remark that the faults locations correspond to local minima of *H* values. The main accident (the 4°50' fault) is noticeable on almost all the regularity maps. However, the regularity maps present local minima of *H* values in some places, probably due to less important faults which have to be checked on updated detailed geological maps.

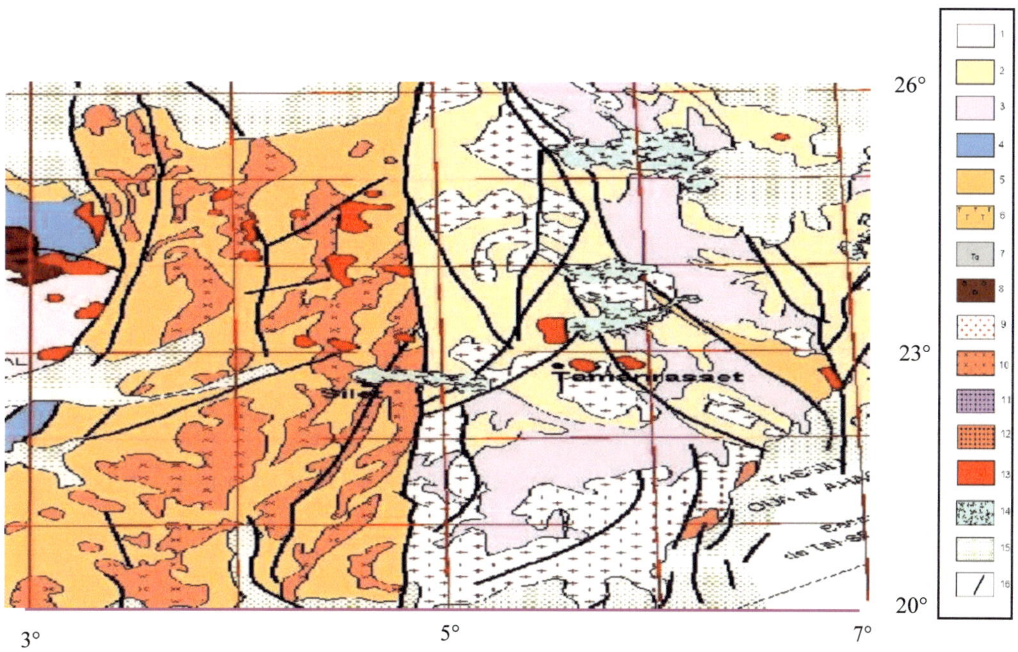


(a) A geological map of the studied zone

**1** - Archaean granulites; **2** - Gneiss and metasediments, series of Arechchoum (Pr1); **3** - Gneiss with facies amphibole, series of Aleskod (Pr2); **4** - Indif. gneiss (Pr3); **5** - Pharusian Greywackes; **6** - Arkoses and conglomerates, series of Tiririne (Pr4); **7** - Volcano-sediments of Tafassasset (Pr4); **8** - Molasses (purple series) of Cambrian; **9** – Pan-African syn-orogenic granites; **10** - Pan-African Granites; **11** - Pan-African post-orogenic granites; **12** - Granites of Eastern Hoggar; **13** - Late pan-African Granites; **14** - Basalts and recent volcanism; **15** - Paleozoic cover; **16** - Fault.

Fig. 7. (Continued)

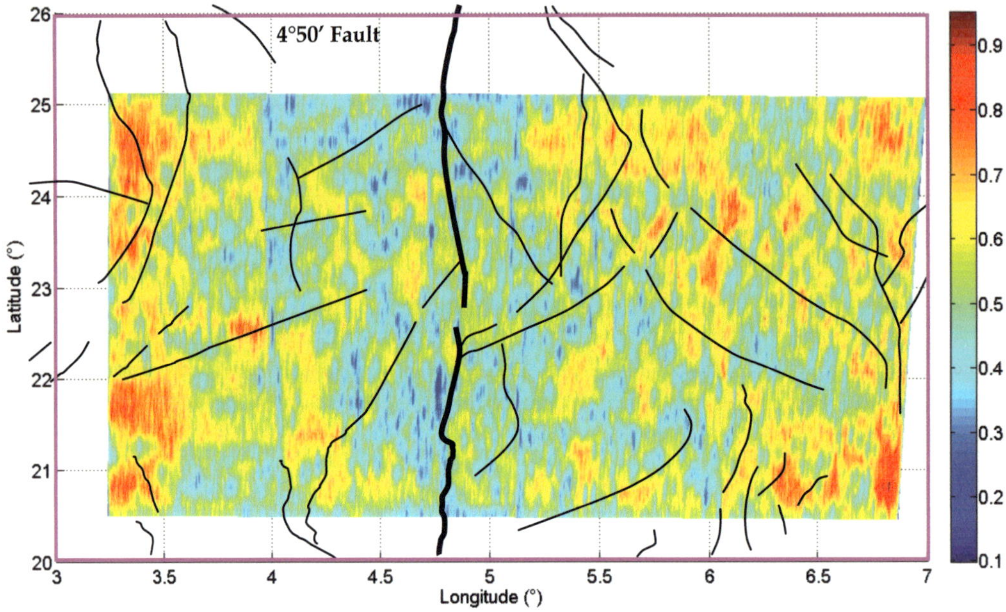


(b) Regularity map obtained from GR measured in the K channel

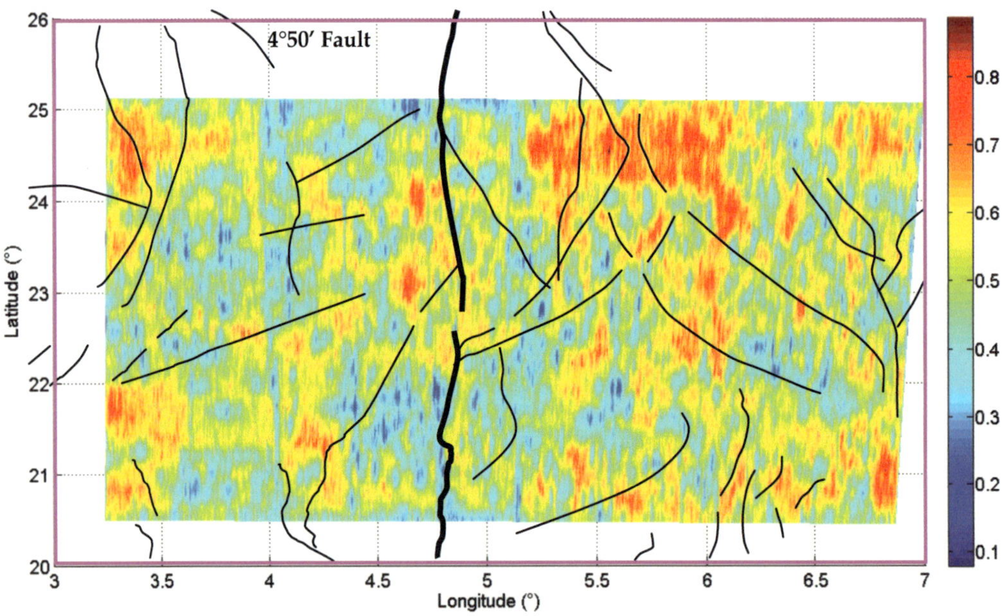


(c) Regularity map obtained from GR measured in the Th channel

Fig. 7. (Continued)

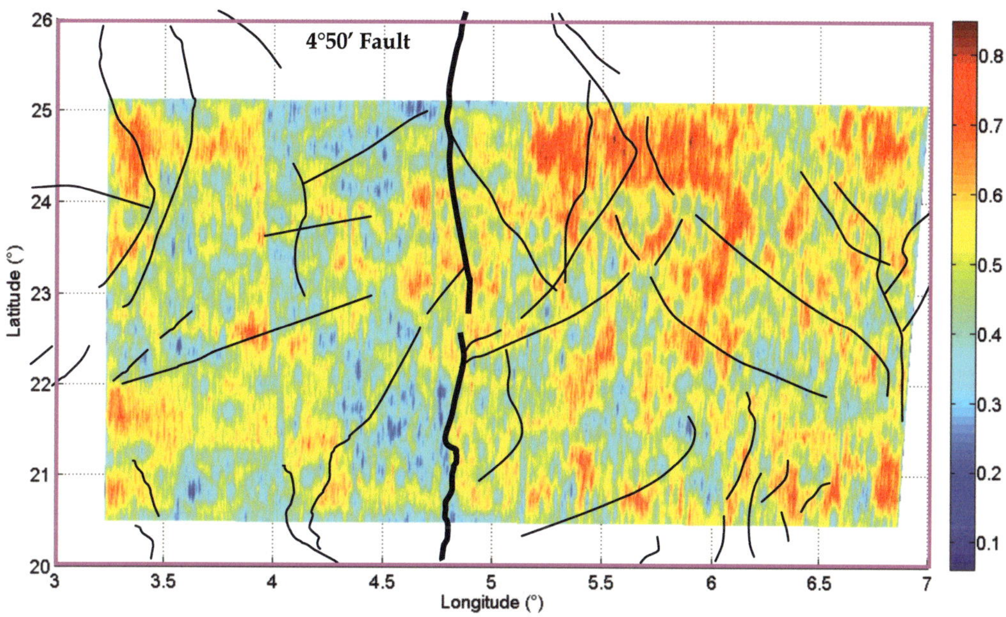


(d) Regularity map obtained from GR measured in the U channel

Fig. 7. Comparison of regularity maps obtained from GR measured in (b)K, (c)Th and (d)U channel and the geological map of the studied zone (a).
The faults affecting the studied area are projected on the *H* maps.

Now, we try to establish a correspondence between the obtained regularity maps and the geological map of the area. Since the obtained regularity maps are similar, we choose that estimated from the measurements recorded in the Th channel. Then, on the considered geological map and *H* map, we delimit in dotted lines the geological formations; the same color corresponds to the same geological facies (Fig. 8). These two maps show that a considered lithology is not characterized by the same value of the *H* coefficient. These obtained preliminary results reveal that the *H* value can not be used as an attribute to characterize lithology, while it could be used for the recognition and the establishment of the network faults.

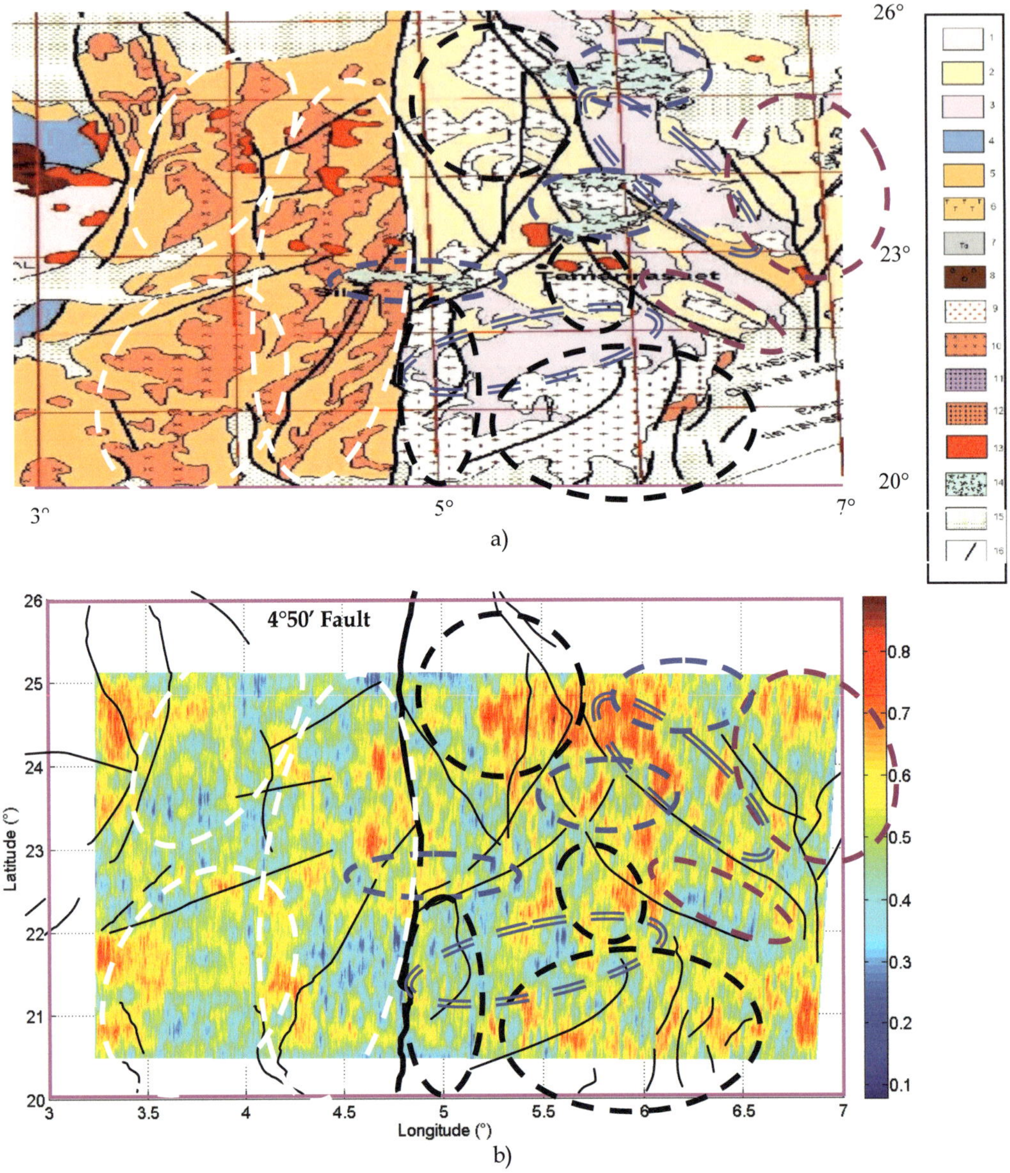


**1** - Archaean granulites; **2** - Gneiss and metasediments, series of Arechchoum (Pr1); **3** - Gneiss with facies amphibole, series of Aleskod (Pr2); **4** - Indif. gneiss (Pr3); **5** - Pharusian Greywackes; **6** - Arkoses and conglomerates, series of Tiririne (Pr4); **7** - Volcano-sediments of Tafassasset (Pr4); **8** - Molasses (purple series) of Cambrian; **9** – Pan-African syn-orogenic granites; **10** - Pan-African Granites; **11** - Pan-African post-orogenic granites; **12** - Granites of Eastern Hoggar; **13** - Late pan-African Granites; **14** - Basalts and recent volcanism; **15** - Paleozoic cover; **16** - Fault.

Fig. 8. Correlation of the regularity map (b) obtained from the GR measurements recorded in the Th channel with the geological map of the studied zone (a). The ellipses in dotted lines delimit the geological formations: black (pan-African syn-orogenic granites), white (pan-African granites), simple blue line (basalts and recent volcanism), doubled blue line (gneiss with amphibole facies), brown (gneiss and metasediments).

## 7. Conclusion

This study presents a regularity analysis undertaken on the airborne spectrometric natural radioactivity measured, in three channels: K, Th and U, over the Hoggar area (Algeria). It reveals that the investigated data exhibit fractal properties depending on the spatial measurement location, thus can be modeled using multifractional Brownian motions. As the spectrometric corrections do not affect these properties, the regularity analysis can be carried out directly on the interpolated raw measurements.

The Hölder exponent maps, obtained from the Gamma Ray measurements recorded in the three channels, show a similar local regularity. Besides, a strong correlation is derived between the $H$ exponent values and the faults locations. Indeed, a fault corresponds to local minima $H$ values, the $H$ exponent value can then be used to identify the faults. However, it does not allow to characterize the lithological facies.

## 8. Acknowledgements

This work was supported by the Algerian -French program CMEP—PHC Tassili N°09 MDU 787.

## 9. References

Aeroservice Corporation (1975) Aero-magneto-spectrometric survey of Algeria. Final report, 3 volumes, Houston, Philadelphia.

Aydin, I.; Selman Aydoğan, M. ; Oksum, E. & Koçak, A. (2006) An attempt to use aerial gamma-ray spectrometry results in petrochemical assessments of the volcanic and plutonic associations of Central Anatolia (Turkey). Geophys. J. Int. Vol. 167, pp. 1044–1052.

Bertrand, J.M.L. & Caby, R. (1978) Geodynamic evolution of the pan-african orogenic belt: A new interpretation of the Hoggar shield (Algerian Sahara), Geol. Rundschman, Vol. 67, pp. 357-388.

Bierwirth, P.N., & Welsh, W.D. (2000) Delineation of Recharge Beds in the Great Artesian Basin Using Airborne Gamma Radiometrics and Satellite Remote Sensing, Bureau of Rural Sciences, Kingston, Australia.

Black, R.; Caby, R. & Moussine-Pouchkine, A. (1979) Evidence for late Precambrian plate tectonics in west Africa, Nature, Vol. 278, pp. 223-227.

Brown, W.M. ; Gedeon, T.D. ; Groves, D.I. & Barnes, R.G. (2000) Artificial neural networks: a new method for mineral prospectivity mapping. Australian Journal of Earth Sciences, Vol. 47, 757–770.

Caby, R.; Bertrand, J.M.L. & Black, R. (1981) Pan-African closure and continental collision in the Hoggar-Iforas segment, central Sahara. in Kroner A (ed) Precambrian Plate Tectonics, Elsevier, Amst. pp. 407-434.

Cahen, L., Snelling, N.J. ; Delhal, J. & Vail J.R. (1984) The geochronology and evolution of Africa, Clarendon Press, Oxford., 512 pp.

Chui, C.K. (1992) An introduction to wavelets. Academic Press.

Cook, S.E.; Corner, R.J.; Groves, P.R. & Grealish, G.J. (1996) Use of airborne gamma radiometric data for soil mapping, Australian Journal of Soil Research. Vol. 34, No.1, pp. 183–194.

Doll, W.E.; Nyquist, J.E.; Beard, L.P. & Gamey,T.J. (2000) Airborne geophysical surveying for hazardous waste site characterization on the Oak Ridge Reservation, Tennessee, Geophysics. Vol. 65, pp. 1372–1387.

Dziewonski, A.; Bloch, S. & Landisman, M. (1969) A technique for the analysis of transient seismic signals: *Bull. Seismol. Soc. Am.*, Vol. 59, pp. 427–444.

Feder, J. (1988). Fractals, p. 283, Plenum (Ed.), New York.

Gaci, S.; Zaourar, N. ; Hamoudi, M. & Holschneider, M. (2010). Local regularity analysis of strata heterogeneities from sonic logs. *Nonlin. Processes Geophys.*, Vol. 17, pp. 455-466, http: www.nonlin-processes-geophys.net/17/455/2010/doi:10.5194/npg-17-455-2010

Gaci, S. (2011). Multifractional analysis of geophysical signals (*in French*). PhD thesis. Univ. of Sciences and Technology Houari Boumdiene (Algeria).

Gaci, S.; Zaourar, N.; Briqueu, L. & Djeddi, M., (2011). Fractal characterization of natural radioactivity measurements in the Hoggar region (Algeria), EGU Proceedings, Vienna, Austria.

Graham, D.F. & Bonham-Carter, G.F. (1993) Airborne radiometric data: A tool for reconnaissance geological mapping using a GIS, Photogrammetric Engineering and Remote Sensing, Vol. 59, pp. 1243–1249.

Groune D. (2009) Magneto-spectrometric analysis of the airborne geophysical data of the large Pharusian ditch (Western Hoggar). Msc thesis, University of Boumerdes. Algeria (in French).

Holschneider, M. (1995). Wavelets: an Analysis Tool. Clarendon. Oxford, England.

International Atomic Energy Agency (IAEA) (2003) Guidelines for radioelement mapping using gamma ray spectrometry data. Vienna, Austria. 179 pp.

Jaques, A.L.; Wellman, P.; Whitaker, A. & Wyborn, D. (1997) High-resolution geophysics in modern geological mapping, AGSO Journal of Australian Geology and Geophysics. Vol. 17, No. 2, pp. 159–173.

Li, X-P (1997). Decomposition of vibroseis data by the multiple filter technique. *Geophysics*, Vol. 62, No. 3, pp. 980–991.

Li, M.; Lim, S.C. ; Hu, B-J. & Feng, H. (2007). Towards describing multi-fractality of traffic using local Hurst function. *Lecture Notes in Computer Science*, Vol. 4488, pp. 1012-1020.

Li, M.; Lim, S.C. & Zhao, W. (2008). Investigating multi-fractality of network traffic using local Hurst function, Advanced Studies in Theoretical Physics, Vol. 2, No. 10, pp. 479–490.

Liégeois, J. P.; Black, R.; Navez, J. & Latouche, L. (1994) Early and late Pan-African orogenies in the Aïr assembly of terranes (Tuareg shield, Niger), Precambrian Research, Vol. 67, No. 1-2, pp.59-88.

Mandelbrot, B.B. (1977). Fractals : Form, Chance and Dimensions. Freeman, San Francisco.

Mandelbrot, B.B. (1982) The Fractal Geometry of Nature. Freeman, San Francisco.

Matolín, M. & Stráník, Z. (2006) Radioactivity of sedimentary rocks over the Ždánice hydrocarbon field. Geophys. J. Int., Vol. 167, pp. 1491–1500.

Muniandy, S.V.; Lim, S.C. & Murugan, R. (2001). Inhomogeneous scaling behaviors in Malaysian foreign currency exchange rates, Physica A, Vol. 301, No. 1–4, pp. 407–428, 2001.

Peltier, R.F. & Lévy-Véhel, J. (1994). A New Method for Estimating the Parameter of Fractional Brownian motion, Technical report, INRIA RR 2396.

Peltier, R.F. & Lévy-Véhel, J. (1995). Multifractional Brownian Motion: Definition and preliminary results, Technical report, INRIA RR 2645.

Sulekha Rao, N. ; Sengupta, D. ; Guin, R. & Saha S. K. (2009) Natural radioactivity measurements in beach sand along southern coast of Orissa, eastern India. Environ Earth Sci. Vol. 59, pp. 593–601.

Wilford, J.R.; Bierwirth, P.N. & Craig, M.A. (1997) Application of airborne gamma-ray spectrometry in soil/regolith mapping and applied geomorphology, Journal of Australian Geology and Geophysics, Vol. 17, No. 2, pp. 201–216.

6

# Two-Dimensional Multifractional Brownian Motion- Based Investigation of Heterogeneities from a Core Image

Saïd Gaci and Naïma Zaourar
*University of Sciences and Technology Houari Boumediene, Algiers, Algeria*

## 1. Introduction

A core sample is a cylindrical section obtained by driving a hollow tube into the undisturbed medium and withdrawing it with its content. In practice, the sample is pushed more or less unbroken into the tube. Once removed from the tube in the laboratory, it is analyzed by different techniques and equipment depending on the desired type of data. The hole made for the core sample is called the "core hole". A variety of core samplers exist to sample different media under diverse conditions. For instance, sediments or rocks are sampled with a hollow steel tube called a core drill.

A scientific coring has been used in the first time for sampling the ocean floor. Then, it is soon exploited to analyze lakes, ice, mud, soil and wood. Cores provide precious information about the evolution of climate, species and sedimentary composition during geologic history.

In petroleum engineering, core analysis presents a way of measuring well conditions downhole by studying samples of reservoir rocks. It gives the most accurate estimations of porosity, permeability, fluid saturation and grain density. These measurements help to understand the conditions of the well and its potential productivity.

In addition to the basic petrophysical properties estimated from the core, a special core analysis can be undertaken in order to determine permeability, wettability, capillary pressure, and electrical properties. Petrographic studies and sieve analysis can also be carried out in such analysis.

In recent years, numerical analysis has been widely used for the investigation of images, since it yields results more objective and reliable than those obtained by conventional methods based on human observations. Fractal analysis has been introduced to examine images texture (Bourissou *et al.*, 1994; Lévy-Véhel and Mignot, 1994; Liu and Li, 1997; Lévy-Véhél, 1995, 1997, 1998; Pesquet-Popescu and Lévy-Véhel, 2002; Malladi *et al.*, 2003; Tahiri *et al.*, 2005).

In this study, we suggest to go beyond the conventional core analysis, and to perform a new approach to extract the maximum features from a core image using a fractal analysis. The conventional fractal model used previously in image processing, the two-dimensional fractional Brownian motion (2D- fBm), presents a constant Hölder function $H$, thus does not allow to explore the spatial evolution of the local regularity. To do so, we suggest to

consider a generalized fractal model, the two-dimensional multifractional Brownian motion (2D-mBm), which presents a regularity varying in space.
The mBm model, initially proposed by Peltier and Lévy-Véhél (1995), and Benassi *et al.* (1997), is used in many disciplines: images processing (Bicego and Trudda, 2010), traffic phenomena (Li *et al.*, 2007), geophysics (Wanliss, 2005; Wanliss and Cersosimo, 2006; Cersosimo and Wanliss, 2007; Gaci *et al.*, 2010; Gaci and Zaourar, 2010, 2011). For the estimation of the mBm processes' local regularity, we propose three algorithms based on the two-dimensional continuous wavelet transform (2D- CWT). The wavelet coefficients are calculated by Fast Fourier Transform (FFT) using the Morlet wavelet and the Mexican hat for the first and the second algorithms, respectively. However for the third algorithm, the coefficients estimation is carried out using the multiple filter technique (2D MFT) that we generalized to two dimensions (Gaci, 2011), from the one-dimensional case (1D MFT) (Li, 1997; Gaci *et al.*, 2011).
This chapter is organized as follows. First, we give a brief theory on 2D-mBm model and the wavelet-based estimators of the local regularity. The potential of the suggested algorithms is then demonstrated on synthetic 2D-mBm paths. The results showed that the 2D MFT algorithm yields the best Hölder exponent estimates. Next, the suggested regularity analysis is extended to digitalized image data of a core extracted from an Algerian borehole. It is shown that the data exhibit a fractal behavior. In addition, the derived regularity maps, obtained with the 2D MFT algorithm, show a strong correlation with the core heterogeneities.

## 2. (Multi)fractional Brownian motion

### 2.1 Fractional Brownian motion

Fractional Brownian motion (fBm) is one of the most popular stochastic fractal models for studying rough signals. It was introduced by Kolmogorov (1940) and studied by Mandelbrot and Van Ness (1968).
A fBm, denoted by $B_H(t)$, is the zero-mean Gaussian process with stationary increments. It is parameterized by a constant Hurst parameter $H$. The fBm is $H$-self affine, *i.e.*:

$$B_H(\lambda t) \cong \lambda^H B_H(t), \ \forall \lambda > 0 \tag{1}$$

Where $\cong$ means the equality of all its finite-dimensional probability distributions.

The bidimensional isotropic fractional Brownian motion, or Lévy Brownian fractional field, with Hurst parameter $H$ is a centered Gaussian field $B_H$ with an autocorrelation function (Kamont, 1996):

$$E\left( B(\vec{x})B(\vec{y}) \right) \propto \|\vec{x}\|^{2H} + \|\vec{y}\|^{2H} - \|\vec{x}-\vec{y}\|^{2H}, \text{ with } 0 < H < 1 \tag{2}$$

where $\vec{x}, \vec{y} \in R^2$ and $\|.\|$ is the usual Euclidian norm.

For $H$=1/2, the fractional Brownian motion is reduced to a Wiener process.

The regularity of the 2D-fBm is measured by the pointwise Hölder exponent $\alpha_{B_H}(\vec{x})$. Indeed, it is shown that almost surely: $\alpha_{B_H}(\vec{x}) = H$. Therefore, the higher the $H$ value, the smoother the 2D-fBm paths.

### 2.2 Multifractional Brownian motion

Multifractional Brownian motion (mBm) was introduced by Peltier and Lévy-Véhél (1995), and Benassi *et al.* (1997) by allowing $H$ to vary over time. Even if no longer stationary nor self-similar compared to the fBm, the mBm presents the advantage to be very flexible since the function $H(t)$ can model phenomena whose sample paths display a time changing regularity.

For a continuous function $H(\vec{x}): R^2 \to R$, the isotropic multifractional Brownian field $W_{H(\vec{x})}$ is a centered Gaussian field with a covariance function

$$E\left(W_{H(\vec{x})}(\vec{x})\,W_{H(\vec{y})}(\vec{y})\right) \propto \|\vec{x}\|^{H(\vec{x})+H(\vec{y})} + \|\vec{y}\|^{H(\vec{x})+H(\vec{y})} - \|\vec{x}-\vec{y}\|^{H(\vec{x})+H(\vec{y})} \tag{3}$$

Identically to the 2D-fBm, the local regularity of the 2D-mBm paths is measured by means of the pointwise Hölder exponent. For a differentiable function $H$, the relation $\alpha_{B_H}(\vec{x}) = H(\vec{x})$ is demonstrated, almost surely, for all $\vec{x} \in R^2$.

## 3. Regularity analysis using the wavelet transform

### 3.1 Two- dimensional continuous wavelet transform

The two-dimensional continuous wavelet transform (2D- CWT) of a signal $s(x,y)$ is given by (Chui, 1992; Holschneider, 1995):

$$S(a, b_x, b_y) = \frac{1}{\sqrt{a}} \int_{-\infty}^{\infty} s(x,y)\,\bar{g}\left(\frac{x-b_x}{a}, \frac{y-b_y}{a}\right) dx\,dy \tag{4}$$

where $g(x,y)$ is the analyzing wavelet, "$a$" is the scale parameter, "$b_x$" and "$b_y$" are the respective translations according to X-axis and Y-axis. The symbol "–" denotes the complex conjugate.

Let $s(x,y)$ be a self-affine fractal surface. It satisfies then the relation:

$$s(\lambda x, \lambda y) \cong \lambda^H . s(x,y) \tag{5}$$

with the Hurst exponent $H$ and a positive factor $\lambda$. If $s$ is a stochastic process, the two sides of the relation follow the same law.

Let us define the function $s_{x_0,y_0}(x,y)$ in each point $(x_0,y_0)$ by: $s_{x_0,y_0}(x,y) = s(x+x_0, y+y_0) - s(x,y)$. This function also satisfies the self-affine property described by (Mandelbrot, 1977, 1982; Feder, 1988; Vicsek 1989; Edgard, 1990) :

$$s_{x_0,y_0}(\lambda x, \lambda y) \cong \lambda^H\, s_{x_0,y_0}(x,y) \tag{6}$$

This property is reflected by the 2D-CWT provided that the analyzing wavelet decreases swiftly enough to zero and has enough vanishing moments (Holschneider, 1995):

$$S(\lambda\, a, x_0 + \lambda b_x, y_0 + \lambda b_y) \cong \lambda^{H(x_0,y_0)+\frac{1}{2}}\, S(a, x_0 + b_x, y_0 + b_y)\,,\ \lambda > 0 \tag{7}$$

By taking the scale "*a*" inversely proportional to the wavenumber *k*: $a \propto 1/k$, the wavelet coefficients will be expressed in ($k$, $b_x$, $b_y$) plane.
The scalogram can be defined as the square of the amplitude spectrum: $P(k,x,y) = |S(k,x,y)|^2$. For large values of *k*, it can be expressed as:

$$P(k,x,y) = P'(x,y).k^{-\beta(x,y)} \propto k^{-\beta(x,y)} \tag{8}$$

Where

$$\beta(x,y) = 2H(x,y) + 1 \tag{9}$$

is the local spectral exponent which is related to the local Hurst (or Hölder) exponent, $H(x,y)$. The spectral exponent $\beta(x,y)$ in each point $(x,y)$ is computed as the slope of the scalogram versus the wavenumber in the log-log plan, the $H(x,y)$ value is then derived using the equation (9).

### 3.2 Used analyzing wavelets

The analyzing wavelets used in this application are the Morlet wavelet and the Mexican hat (Fig. 1). This choice is motivated by their adequate properties for the regularity analysis.

- The Morlet wavelet:

$$g(x,y) = e^{-i(\omega_{0x}x + \omega_{0y}y)}.e^{-\frac{1}{2}(x^2+y^2)}$$

$$\hat{g}(\xi,\nu) = e^{-\frac{1}{2}\left[(\xi-\omega_{0x})^2 + (\nu-\omega_{0y})^2\right]}; \quad \text{with } \left(\omega_{0x}^2 + \omega_{0y}^2\right)^{1/2} \geq 5 \tag{10}$$

- The Mexican hat:

$$g(x,y) = (2 - x^2 - y^2)\,e^{-(x^2+y^2)/2} \; ; \hat{g}(\xi,\nu) = (\xi^2 + \nu^2).e^{-(\xi^2+\nu^2)/2} \tag{11}$$

### 3.3 Wavelet-based estimators of the local regularity

As explained earlier, the computation of the local Hölder exponents *H*(*x*,*y*) requires to calculate the two-dimensional wavelet continuous transform. Here, we suggest three algorithms for the implementation of the 2D- CWT, which are:

#### 3.3.1 FFT-based algorithms

These algorithms are based on the property that wavelet coefficients, expressed by the equation (4), can be performed via the Fourier transform using the Morlet wavelet and the Mexican hat:

$$S(a,b_x,b_y) = FFT^{-1}\left(\hat{s}(\xi,\nu).\,\sqrt{a}\,\bar{\hat{g}}(a\xi,a\nu)\right) \tag{12}$$

These algorithms are accurate but slow, and the signal length must be a power of 2.

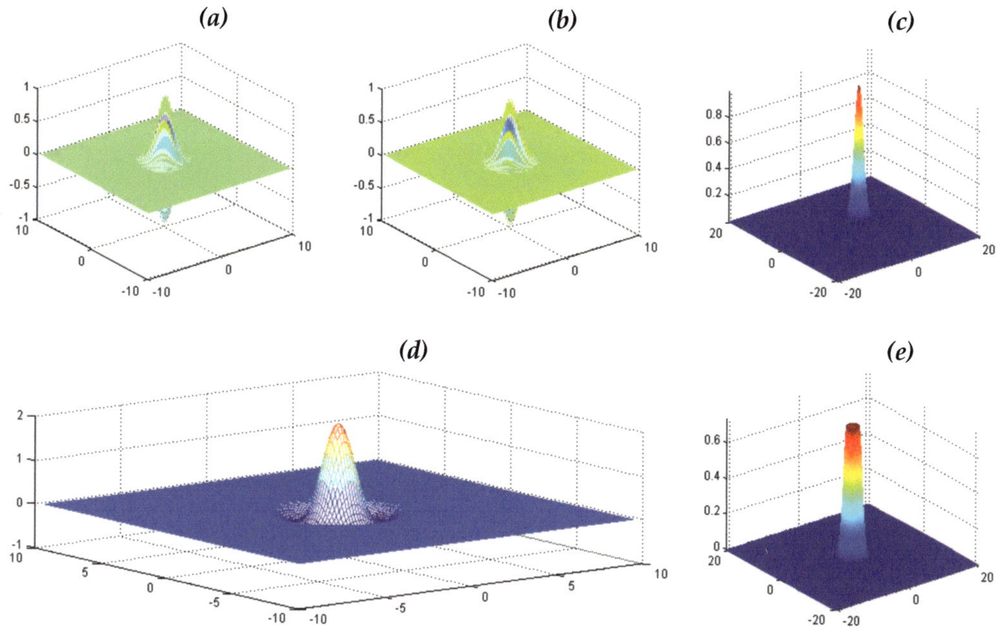


Fig. 1. Used bidimensional analyzing wavelets:
Morlet wavelet *($\omega_{0x}=\omega_{0y}=5.$)*.
Spatial representation: (a) real part (b) imaginary part
Fourier transform representation (c)
Mexican hat.
Spatial representation (d)
Fourier transform representation (e)

### 3.3.2 Generalized two-dimensional multiple filter technique

The 1D MFT was initially suggested by Dziewonski et *al.* (1969). It consists on carrying out a decomposition using a Gaussian filter:

$$G(k,k_n)=e^{-\alpha\left(\frac{k-k_n}{k_n}\right)^2} \tag{13}$$

where $k_n$ is a variable center angular frequency (or wavenumber) of the filter $G(k,k_n)$, and $\alpha$ is a shaping parameter of the filter.
In order to overcome the poor "time" and "low-frequency" domains resolution toward the low frequencies, Li (1997) suggests a varying quality factor $Q$ and a varying bandwidth $\Delta k$:

$$\Delta k = k_2 - k_1 = \beta.Ln(k_n) \tag{14}$$

Where $\beta$ is a constant, $k_1$ and $k_2$ are the – 3 dB points of the Gaussian filter.

Here, we propose to extend the 1D MFT enhanced by Li (1997) to 2 dimensions. The idea consists on decomposing the two-dimensional signal using a Gaussian filter $G(k,\xi_n,\nu_m)$ defined as:

$$G(k,\xi_n,\nu_m) = G_1(k,\xi_n)G_2(k,\nu_m) = e^{-\alpha\left(\frac{k-\xi_n}{\xi_n}\right)^2} e^{-\alpha\left(\frac{k-\nu_n}{\nu_n}\right)^2} \quad (15)$$

Where $\xi_n$ and $\nu_m$ are variable center angular frequencies (or wavenumbers) of the respective filters $G_1(k,\xi_n)$ and $G_2(k,\nu_m)$. The bandwidths $\Delta k_1$ and $\Delta k_2$ of both filters are calculated as above (Eq. 14).

## 4. Application to simulated 2D-mBm paths

In this section, the suggested estimators of the local regularity are tested on synthetic 2D-mBm paths whose lengths are 256 x 256, generated using the kriging method (Barrière, 2007). Three types of Hölder function $H$ are chosen:

$$\text{bilinear}: H_1(x_1,x_2) = 0.8 - 0.6x_1x_2$$

$$\text{logistic}: H_2(x_1,x_2) = 0.7 - \frac{0.4}{1+\exp(-20(x_2-0.5))}$$

$$\text{periodic}: H_3(x_1,x_2) = 0.5 + 0.3\sin(2\pi x_1)\cos\left(\frac{3}{2}\pi x_2\right)$$

The regularity functions and the simulated 2D-mBm paths corresponding to the three theoretical $H$ functions are presented in Figure 2. The larger $H$ value, the smoother the modeled surface.

Using the three algorithms, we have estimated $H$ maps. For the first wavelet-based algorithm, we use the Morlet wavelet with $\omega_{0x}=\omega_{0y}=8.9443$, while for 2D MFT, the selected parameters for the two-dimensional Gaussian filters are:

- The shaping factor $\alpha=40$,
- The minimal central wavenumber of the filter $\xi_{min}=\nu_{min}=2\pi.10^{-3}$ rad/m,
- The maximal central wavenumber of the filter $\xi_{max}=\nu_{max}=2\pi.10^{0.5}$ rad/m,
- The number of the central wavenumbers of the filter $N=100$,
- The sampling rate is selected as 0.1524m.

The $H$ maps obtained by the three estimators, presented in Figure 3, show that the regularity estimated by the first wavelet-based algorithm using the Morlet wavelet are better than that calculated by the second algorithm with the Mexican hat. In addition, the suggested 2D MFT provides the best estimations of the regularity maps with the least errors. For this reason, we retain only this estimator in the following. It can be also remarked that all the used algorithms yield large absolute values of the estimation error in the limits of the analyzed 2D-mBms.

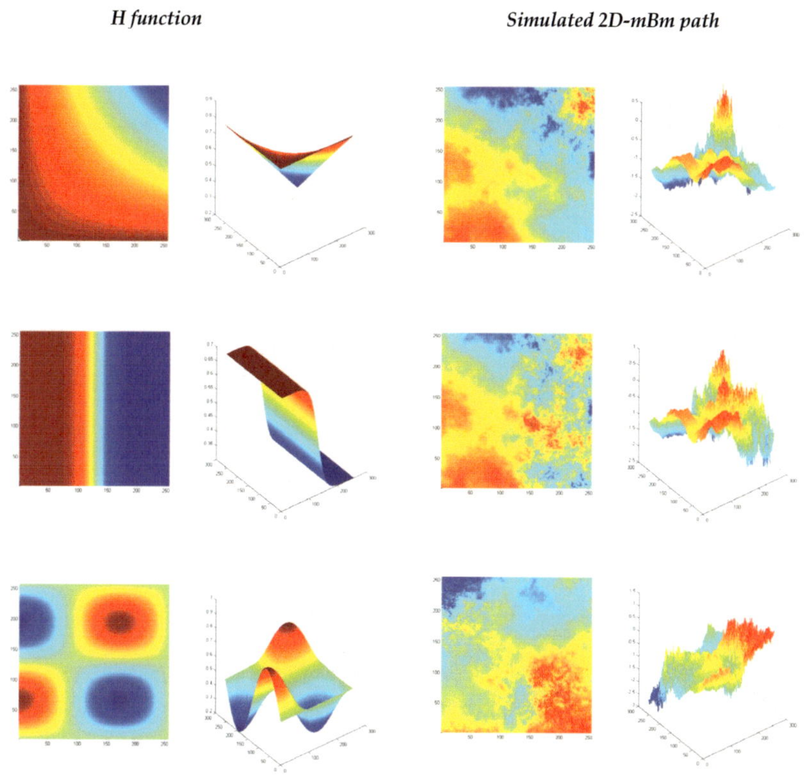


Fig. 2. A 3D-view representation and a XY-plan projection of the theoretical *H* function and the corresponding simulated 2D-mBm path, for the three types of *H* functions. Bilinear (top), logistic (middle) and periodic (bottom).

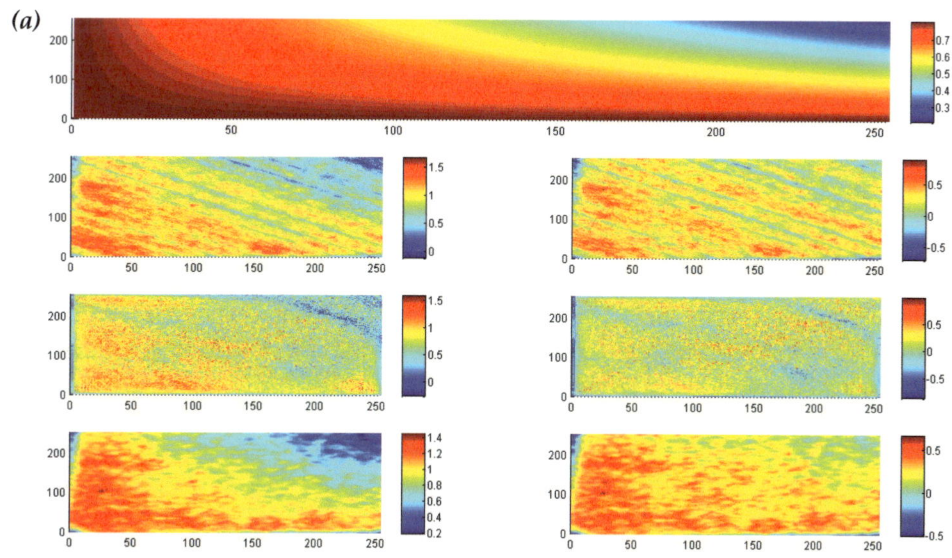


Fig. 3. (Continued)

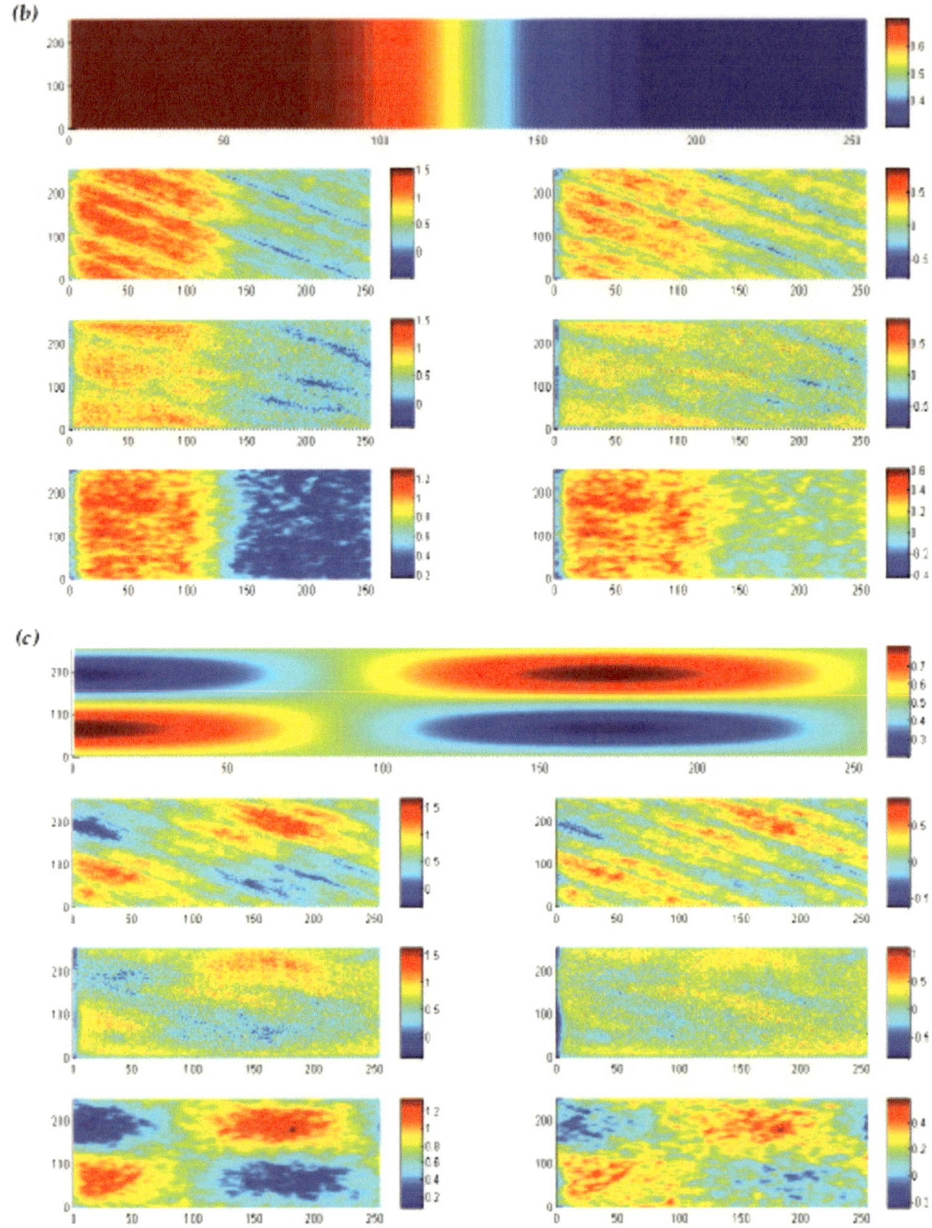


Fig. 3. Regularity functions obtained by the three algorithms from the 2D-mBm paths, represented in Fig.2, simulated with three types of *H* function: (a) bilinear (b) logistic (c) periodic

Line 1: theoretical *H* function; Lines 2 & 3: regularity functions estimated using FFT-based algorithms with, respectively, the Morlet wavelet and the Mexican hat; Line 4: regularity function estimated using 2D MFT ($\alpha$=40).

The columns from left to right represent: (1) the estimated *H* function , (2) the estimation error , calculated as the difference between the estimated *H* function and the theoretical *H* function.

## 5. Application to digitalized core image data

Here, the local regularity analysis is performed on digitalized core image data. The analyzed core is extracted from a well drilled in an Algerian basin. It is chosen since it represents the main geological features of the studied region (Fig.4).

The core presents medium to fine quartzitic sandstone, clay and quartz cemented cross stratifications underlined by mud films. The formation is affected by a main fracture F1 with a high angle dip (≈75°) filled with quartz. It is also noted the presence of another fracture F2 sub-parallel to F1 but less important.

Fig. 4. Core image.

The processing of the core image requires first its digitalization. The core image is digitalized and codified in gray levels with integer values ranged between 0 and 255. The obtained digitalized core image, illustrated by figure 5, corresponds to a matrix of 3642 x 996 with a sampling rate $\Delta x = \Delta y = 0.0121$ cm.

First, the fractal behavior of the digitalized core image data is inspected on five horizontal and five vertical profiles as shown in Figure 5.

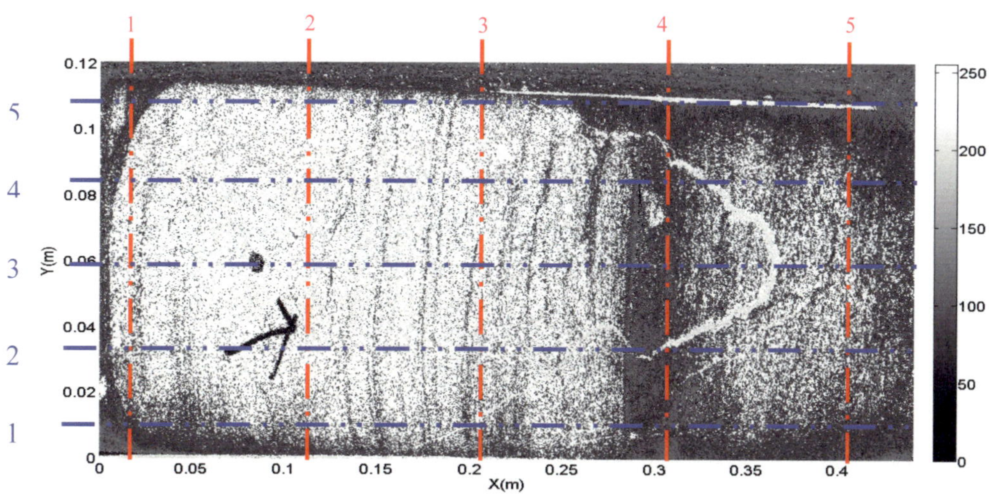


Fig. 5. Positions of the horizontal and the vertical profiles on the digitalized core image. The horizontal profiles numbered from 1 to 5 (in blue) correspond to the respective positions $y$ = 0.0120 m, 0.0361 m, 0.0603 m, 0.0845 m and 0.1087 m, while the vertical profiles numbered from 1 to 5 (in red) correspond to the respective positions $x$ = 0.0241 m, 0.1208 m, 0.2175 m, 0.3142 m and 0.4109 m.

For each profile, the Fourier amplitude spectrum is computed and represented in a double logarithmic scale. Then, we estimate the local Hölder function $H(x)$ using an algorithm based on the local growth of the increment process $S_{k,n}(i)$ (Peltier and Lévy-Véhél, 1994, 1995; Muniandy *et al.*, 2001; Li *et al.*, 2008; Gaci *et al.*, 2010):

$$S_{k,n}(i) = \frac{m}{n-1} \sum_{j \in [i-k/2, i+k/2]} \left| X(j+1) - X(j) \right|, \ 1<k<n \tag{16}$$

where $n$ is the signal $X$ length, $k$ is a fixed window size, and $m$ is the largest integer not exceeding $n/k$.
The local Hölder function $H(x)$ at point

$$x = \frac{i}{n-1} \tag{17}$$

is given by

$$\hat{H}(i) = -\frac{\log\left[\sqrt{\pi/2}\ S_{k,n}(i)\right]}{\log\ (n-1)} \tag{18}$$

The obtained results corresponding to the horizontal profiles and the vertical profiles are respectively exposed in figures 6a and 6b.
It can be noted that all the resulted amplitude spectra exhibit an algebraic decay; that illustrates the fractal properties of the digitalized data. Besides, the analyzed profiles present a varying regularity with the position according to X- and Y-axis. They can be then regarded as paths of multifractional Brownian motions (mBms) (Peltier and Lévy-Véhel, 1995). The variation of $H$ exponent value is related to the local lithological changes of the core composition.
The next step consists on establishing regularity maps from the digitalized data using the 2D MFT algorithm. The implementation of the latter algorithm requires the "reconditioning" of the data so that the matrix dimensions corresponding to the digitalized data are a power of 2. For the purpose of processing the digitalized data, and considering the limitations of the available computer's capacities, we have splited the obtained matrix (3642 x 996) into two overlapping sub-matrixes whose size is 2048 x 1024. The sub-matrixes are padded by zeros so that their dimensions following $Y$-axis, initially equal to 996, reach 1024.
The parameters selected for the 2D MFT are as follow:

- The minimal center wavenumbers of the filter : $\xi_{min}=\nu_{min}=2\pi.10^{-1}$ rad/m;
- The maximal center wavenumbers of the filter:

$$\xi_{max}=\nu_{max} = 2\pi/(2\ \Delta x) \approx 25964 \text{ rad/m};$$

The other parameters ($\alpha$ and $N$) are similar to those used in the previous section.
The final regularity map is constructed from the $H$ sub-maps related to the two sub-matrixes (Fig. 7). The $H$ values in the overlapping zone are calculated as the average of the $H$ values corresponding to the $H$ values in the sub-maps.

(a) Horizontal profiles

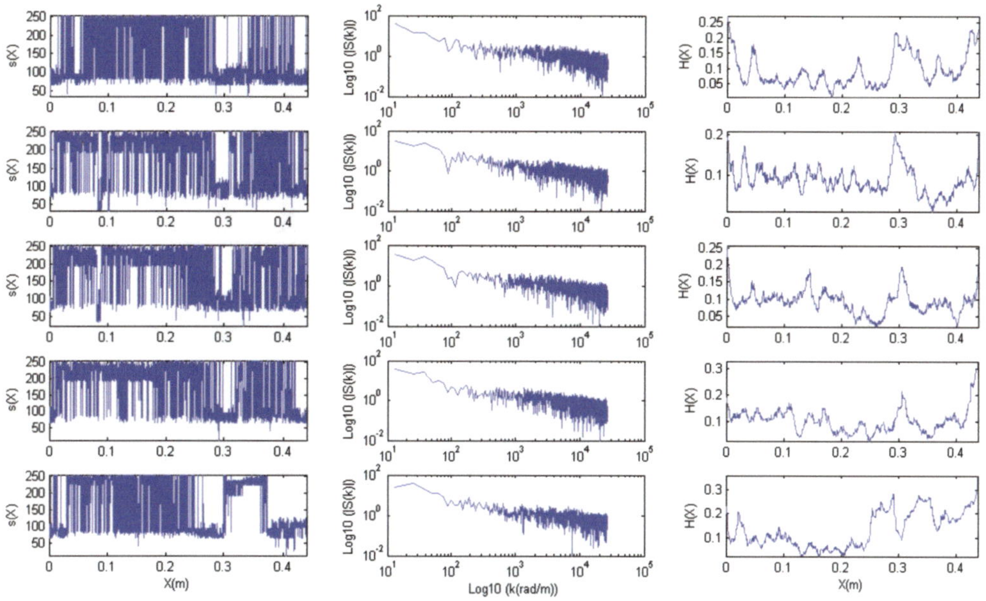


(b) Vertical profiles

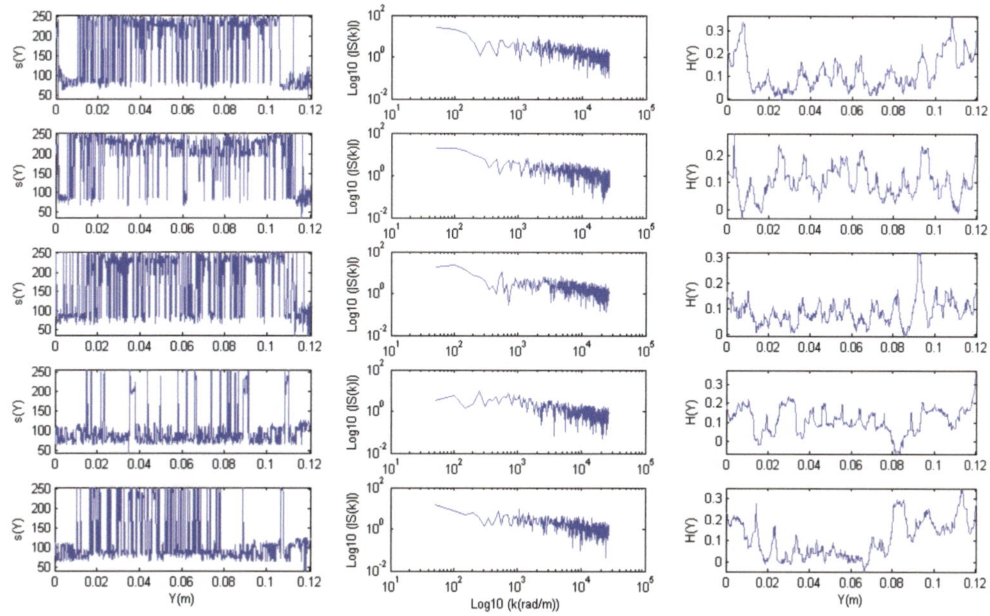


Fig. 6. Investigation of the fractal properties of the five horizontal (a) and vertical (b) profiles extracted from the digitalized core image. The five lines in (a) (resp. (b)), from top to bottom, correspond to the respective horizontal (resp. vertical) profiles numbered from 1 to 5. Left: the profile of the digitalized core image data, middle: the amplitude spectrum module of the data with respect to the wavenumber in the log-log scale, right: the local $H$ exponent.

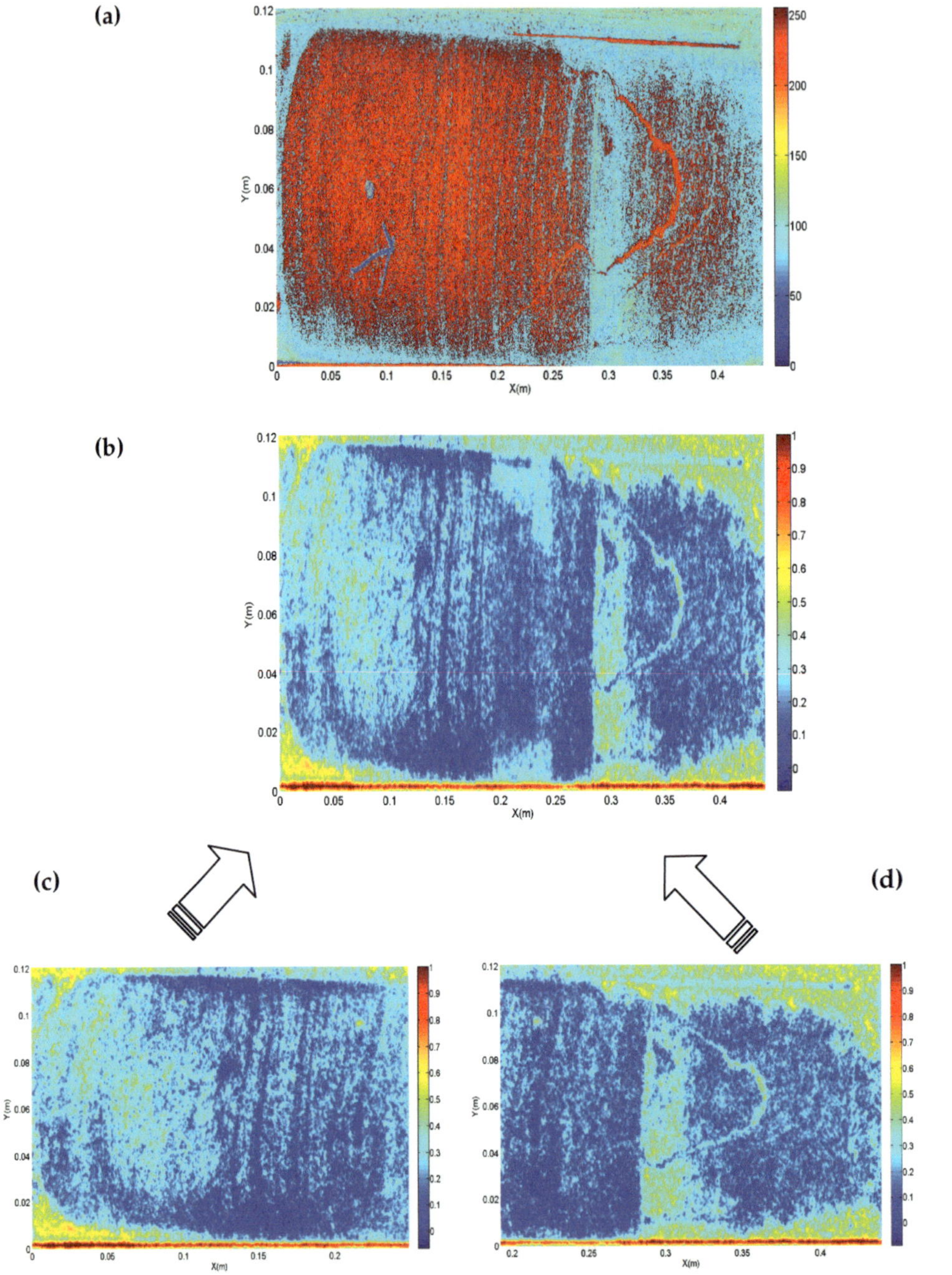


Fig. 7. A regularity map (b) obtained by 2D MFT from the digitalized core image data (a). The regularity map (b), corresponding to the data of the whole core, is obtained from the regularity maps (c) and (d), related to the two ''sub- zones'' of the core image.

From Figure 7, it can be seen that the analyzed data present a varying regularity in the XY plan. It is again confirmed that the digitalized core image data can be modeled as a trajectory of a 2D-mBm. The obtained $H$ maps highlight well the main fault F1, the break and the mud films. However, the minor fault F2 is locally noticeable. We note that these lithological changes are marked by local maxima of $H$ values which are higher than those characterizing the surrounding medium.

Now, we aim to establish a correspondence between the digitalized data values, which are the gray levels values representing the geological facies, and the $H$ value via a statistical analysis. In order to avoid the abnormally high values of $H$ due to the limits effects, we consider the digitalized data corresponding to a central zone extracted from the core image (X: 10.8671- 43.5047cm; Y: 0.9550- 10.2627 cm). Thereafter, six classes are determined by fitting the results yielded by the application of the k-means method on the selected data. The six classes of the gray levels resulted from this classification are: class 1 = [0, 62 [, class 2 = [62,80 [, class 3 = [80,160 [, class 4 = [160, 210 [, class 5 = [210, 235 [ and class 6 = [235, 255 ].

From figure 8, it can be seen that the histograms of $H$ values calculated by 2D MFT follow a normal distribution. For each class, the statistical parameters (mean and standard-deviation) are estimated from the histograms of the gray level values, and the corresponding $H$ values (Table 1). It is worth noting that for the six classes, the statistical parameters of $H$ values, estimated from the histograms, present very close values. These results show that the Hölder exponent value can not characterize a geological facies represented by the gray level, whereas its local variation reflects local lithological changes as explained earlier.

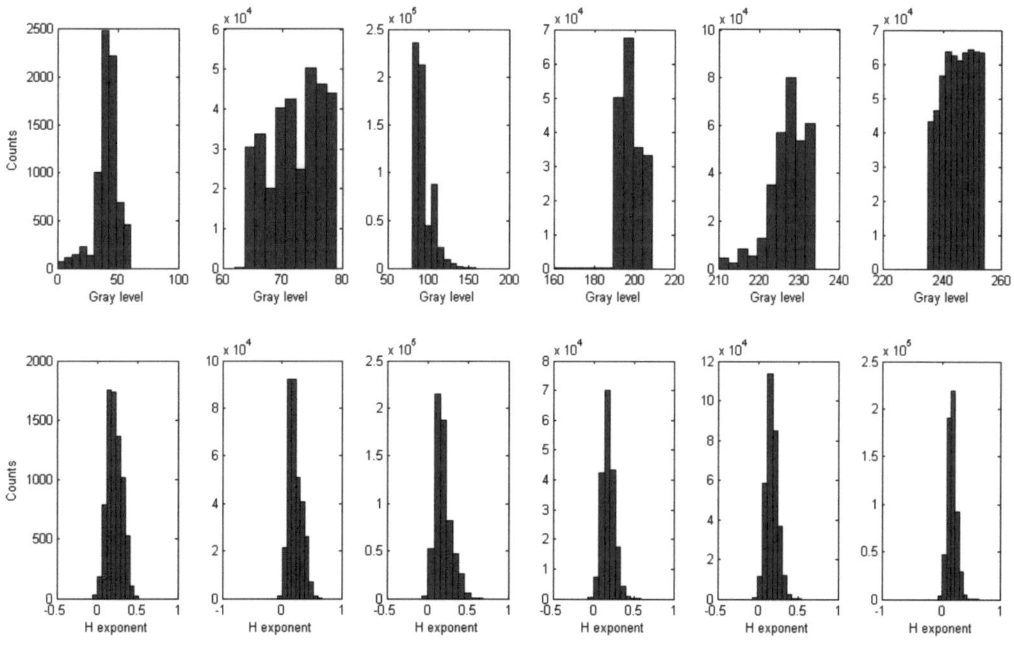


Fig. 8. Histograms of the digitalized data values extracted from the core image, and the corresponding $H$ values estimated by 2D MFT, for the six classes. The six columns from left to right correspond respectively to the classes 1 to 6.

| | | Class 1 [0, 62] | Class 2 [62,80] | Class 3 [80,160] | Class 4 [160, 210] | Class 5 [210, 235] | Class 6 [235, 255] |
|---|---|---|---|---|---|---|---|
| Digitalized data | Mean | 40,811 | 72,093 | 93,054 | 198,517 | 227,367 | 245,087 |
| | Standard-deviation | 9,761 | 4,432 | 11,085 | 4,945 | 4,626 | 5,558 |
| *H* value | Mean | 0,218 | 0,222 | 0,192 | 0,181 | 0,172 | 0,163 |
| | Standard-deviation | 0,090 | 0,107 | 0,095 | 0,071 | 0,070 | 0,070 |

Table 1. Statistical parameters estimated from the histograms (gray level and corresponding *H* values) related to the six classes.

## 6. Conclusion

In this study, the 2D-mBm has been successfully used for a local Hölder regularity-based modeling of the core image. We have presented three methods for estimating the local regularity. The first and the second ones are FFT-based algorithms using, respectively, the Morlet wavelet and the Mexican hat, while the third method is obtained by extending the one-dimensional multiple filter technique to 2 dimensions (2D MFT). The application of these methods on synthetic 2D-mBm paths showed that the 2D MFT yields the best estimations of the *H* functions.

The analysis of profiles extracted from the digitalized core image data reveals a fractal behavior. Furthermore, the regularity maps obtained by 2D MFT from the digitalized data can characterize heterogeneities from the analyzed core. Although a Hölder exponent value does not describe a specific geological facies, its local variation reflects the lithological changes (faults, breaks, stratifications, etc.). The presented analysis must be undertaken on a large number of cores in order to establish a relation between a geological facies, the corresponding gray level and *H* values.

## 7. Acknowledgements

I would like to thank Mr. Tenkhi for his comments and suggestions.

## 8. References

Barrière, O., 2007. Synthèse et estimation de mouvements browniens multifractionnaires et autres processus a régularité prescrite. Définition du processus auto-régulé multifractionnaire et applications (*in french*). PhD thesis. Univ. of Nantes (France).

Benassi, A.; Jaffard, S. & Roux, D. (1997). Elliptic Gaussian random processes. *Rev. Mat. Iberoamericana*, Vol. 13, No.1, pp. 19–90.

Bicego, M. & Trudda, A. (2010). 2D shape classification using multifractional brownian motion. *Lecture Notes in Computer Science*, Vol. 5342, pp. 906-916.

Bourissou, A.; Pham, K. & Lévy-Véhel, J. (1994). A Multifractal Approach for Terrain Characterization and Classification on SAR Images; *International Geoscience and Remote Sensing Symposium (IGARSS)*, Vol. 3, pp. 1609–1611. doi : 10.1109/IGARSS.1994.399514. August 8-12, 1994.

Cersosimo, D.O. & Wanliss, J.A. (2007). Initial studies of high latitude magnetic field data during different magnetospheric conditions. *Earth Planets Space*, Vol. 59, No.1, pp. 39-43.

Chui, C.K. (1992) An introduction to wavelets. Academic Press.

Dziewonski, A.; Bloch, S. & Landisman, M. (1969) A technique for the analysis of transient seismic signals: *Bull. Seismol. Soc. Am.*, Vol. 59, pp. 427–444.

Edgard, G.A. (1990) Measures, Topology anf Fractal Geometry. Springer Verlag, Berlin.

Feder, J. (1988). Fractals, p. 283, Plenum (Ed.), New York.

Gaci, S.; Zaourar, N.; Hamoudi, M. & Holschneider, M. (2010). Local regularity analysis of strata heterogeneities from sonic logs. *Nonlin. Processes Geophys.*, Vol. 17, pp. 455-466, http: www.nonlin-processes-geophys.net/17/455/2010/doi:10.5194/npg-17-455-2010

Gaci, S. & Zaourar, N. (2010). A new approach for the investigation of the local regularity of borehole wire-line logs. *J. hydrocarb. mines environ. res.*, Vol. 1, No.1, pp. 6-13.

Gaci, S. (2011). Multifractional analysis of geophysical signals (*in French*). PhD thesis. Univ. of Sciences and Technology Houari Boumdiene (Algeria).

Gaci, S. & Zaourar, N. (2011). Heterogeneities characterization from velocity logs using multifractional Brownian motion. *Arab J. Geosci.* Vol. 4, No 3-4 pp, 535–541. doi: 10.1007/s12517-010-0167-5.

Gaci, S.; Zaourar, N.; Briqueu, L. & Holschneider, M. (2011). Regularity analysis applied to sonic logs data: a case study from KTB borehole site. *Arab J Geosci.* Vol. 4, No. 1-2, pp. 221-227. doi 10.1007/s12517-010-0129-y

Holschneider, M. (1995). Wavelets: an Analysis Tool. Clarendon. Oxford, England.

Kamont, A., 1996. On the Fractional Anisotropic Wiener Fields. Journal of Probability and Mathematical Statistics, Vol. 16, No.1, pp.85-98.

Kolmogorov, A.N. (1940) Wienersche spiralen und einige andere interessante kurven im hilbertschen raume. *Doklady*, Vol. 26, pp. 115–118.

Lévy-Véhel, J. & Mignot, P. (1994). Multifractal segmentation of images; *Fractals*, Vol. 2, No. 3, pp. 371–378.

Lévy-Véhel, J. (1995). Fractal approaches in signal processing, *Fractals,* Vol. 3, No. 4, pp. 755-775. Symposium in Honor of Benoit Mandelbrot (Curaçao, 1995).

Lévy-Véhel, J. (1997). Introduction to the multifractal analysis of images. Springer Verlag.

Lévy-Véhel, J. (1998). Fractals Images Encoding and Analysis, Springer Verlag.

Li, X-P (1997). Decomposition of vibroseis data by the multiple filter technique. *Geophysics,* Vol. 62, No. 3, pp. 980–991.

Li, M.; Lim, S.C.; Hu, B-J. & Feng, H. (2007). Towards describing multi-fractality of traffic using local Hurst function. *Lecture Notes in Computer Science*, Vol. 4488, pp. 1012-1020.

Li, M.; Lim, S.C. & Zhao, W. (2008). Investigating multi-fractality of network traffic using local Hurst function, Advanced Studies in Theoretical Physics, Vol. 2, No. 10, pp. 479–490.

Liu, Y. & Li, Y. (1997). New approaches of multifractale image analysis; Proceedings of International Conference on information, communications and signal processing, Vol. 2, pp. 970–974.

Malladi, R.K.; Kasilingam, D. & Costa, A.H. (2003). Speckle filtering of SAR images using Hölder regularity analysis of the sparse code; *IEEE Int. Geosci. Remote Sens. Symp.*, Vol. 6, pp. 3998–4000.

Mandelbrot, B.B. & Van Ness, J.W. (1968). Fractional brownian motion, fractional noises and applications. *Siam Review*, Vol. 10, No. 4, pp. 422–437.

Mandelbrot, B.B. (1977). Fractals : Form, Chance and Dimensions. Freeman, San Francisco.

Mandelbrot, B.B. (1982) The Fractal Geometry of Nature. Freeman, San Francisco.

Muniandy, S.V.; Lim, S.C. & Murugan, R. (2001). Inhomogeneous scaling behaviors in Malaysian foreign currency exchange rates, Physica A, Vol. 301, No. 1–4, pp. 407–428.

Peltier, R.F. & Lévy-Véhel, J. (1994). A New Method for Estimating the Parameter of Fractional Brownian motion, Technical report, INRIA RR 2396.

Peltier, R.F. & Lévy-Véhel, J. (1995). Multifractional Brownian Motion: Definition and preliminary results, Technical report, INRIA RR 2645.

Pesquet-Popescu, B. & Lévy-Véhel, J. (2002). Stochastic Fractal Models for Image Processing. *IEEE Signal Processing Magazine*. Vol. 19, No. 5, pp. 48-62.

Tahiri, A.M.; Farssi, S.M. & Touzani, A. (2005). Textures in images classification using a multifractal approach. *IEEE SITIS*, pp. 56-61.

Vicsek, T. (1989). Fractal Growth Phenomena. World Scientific, Singapour.

Wanliss, J.A. (2005). Fractal properties of SYM-H during quiet and active times. *Journal of Geophysical Research,* Vol. 110, No. A03202, pp 12. doi : 10.1029 /2004JA010544.

Wanliss, J.A. & Cersosimo, D.O. (2006). Scaling properties of high latitude magnetic field data during different magnetospheric conditions. Proceedings 8th International Conference Substorms, Banff, Canada, 325-329.

7

# 3D Seismic Sedimentology of Nearshore Subaqueous Fans – A Case Study from Dongying Depression, Eastern China

Yang Fengli[1,*], Zhao Wenfang[1], Sun Zhuan[1], Cheng Haisheng[2] and Peng Yunxin[3]

[1]*School of Ocean and Earth Science, Tongji University, Shanghai,*
[2]*Jiangsu Oilfield, SINOPEC, Yangzhou, Jiangsu,*
[3]*Chengdu University of Technology, Chengdu, Sichuan,*
*China*

## 1. Introduction

The nearshore subaqueous fan, also known as the steep bank sublacustrine fan (Zhao, 2000), or submarine fan (Catuneanu et al., 2002; Richard & Bowman, 1998; Takahiro & Makoto, 2002), is the fan-shaped sedimentary accumulation of sand-conglomerate body located in the footwall of major fault in rift basins (Zhang, J.L & Shen, 1991; Zhang, M. & Tian, 1999), and commonly composed by three sub- facies including a root sub-fan, a mid sub-fan and a marginal sub-fan. Its formation and development are controlled by basin boundary conditions, paleotopography, tectonic evolution, the nature of the provenance, paleoclimate, paleocurrent and other factors (Lu, 2008; Xie et al., 2004; Yan et al., 2005). In the Bohai Bay Basin, eastern China, a nearshore subaqueous fan system of Paleogene age is widely developed in the lower Es4 Formation in Paleogene in the northern Dongying Depression (Fig.1). Analyses found that the mid sub-fan is the main part of the fan, characterized by the pebbly sandstones, conglomerates and block sandstones in the braided channel microfacies, intra-channel microfacies and leafy sandbody microfacies (Gao et al., 2008; Song, 2004; Yan et al., 2005). Oil and gas exploration in the Dongying Depression has demonstrated that the sand-conglomerate developed in the mid sub-fan is the effective oil and gas reservoir. Due to a deep burial (>3500m), multi staged sub-fan development, and small seismic impedance differences, however, to describe and predict the distribution of the effective sand-conglomerate reservoir in the nearshore subaqueous fan is difficult. Studies have also found that conventional 3D seismic data with industry-standard and seismic acoustic impedance inversion data from ac+den loggings could not distinguish the effective sand-conglomerate reservoir from sand-conglomerate sedimentary body (Song, 2004).

Seismic sedimentology is the use of seismic data to study sedimentary rocks and processes by which they form (Zeng et al., 2001, 2004). Since its first introduction in 1998 (Zeng et al., 1998), the concept has been applied in the identification of paleorivers, the sedimentary facies and sedimentary environment evolutions of carbonate platform and the slope fans with many good results (Carter, 2003; Chen.& Meng, 2004; Crumeyrolle et al., 2007; Darmad et al., 2007; Gee & Gawthorpe, 2007; Handford & Baria, 2007; Ling et al., 2005; Lin et al.,

2007; Liu, B.G. & Liu, L.H., 2008; Nordfjord et al., 2005; Posamentier & Killa, 2003; Prather, 2003; Sullivan et al., 2007; Schwab et al., 2007; Wang et al , 2004; Wu et al, 2005; Zeng et al., 2003, 2004, 2007; Zhang et al., 2007), but rarely used to study nearshore subaqueous fans. In this paper, we took the nearshore subaqueous fan in the Dongying Depression as a case, and used the pseudo-acoustic 3D seismic inversion method on characteristic logs to reconstruct 3D seismic sedimentological structures of the nearshore subaqueous fans including the distribution of the effective sand-conglomerate reservoirs and the temporospatial evolution of individual nearshore subaqueous fan system.

Over the years, six exploratory wells were drilled into the lower Es4 Formation in the northern Dongying Depression and four of them encountered commercial oil and gas. The logging data from all six wells yield good coverage with 0.125 m or 0.25 m sampling spacing. An industry- standard 3D seismic data of 600 $km^2$ acquired in 2005 was processed using high-fidelity prestack time migration technique with 25m × 25m track spacing, 1ms sampling interval, 25HZ dominant frequency and 10-60Hz effective frequency bandwidth in the target formation.

## 2. Characteristics of the nearshore subaqueous fan in the northern Dongying Depression

The Dongying Depression is a typical sub-structural unit in the Bohai Bay Basin, Eastern China, surrounded by a series of uplifts, including the Luxi Massif in the south, Chenjiazhuang Uplift in the north, Qingtuozi Uplift in the east, Binxian Uplift and Qingcheng Uplift in the west (Fig.1). As a rift basin, the Dongying Depression is

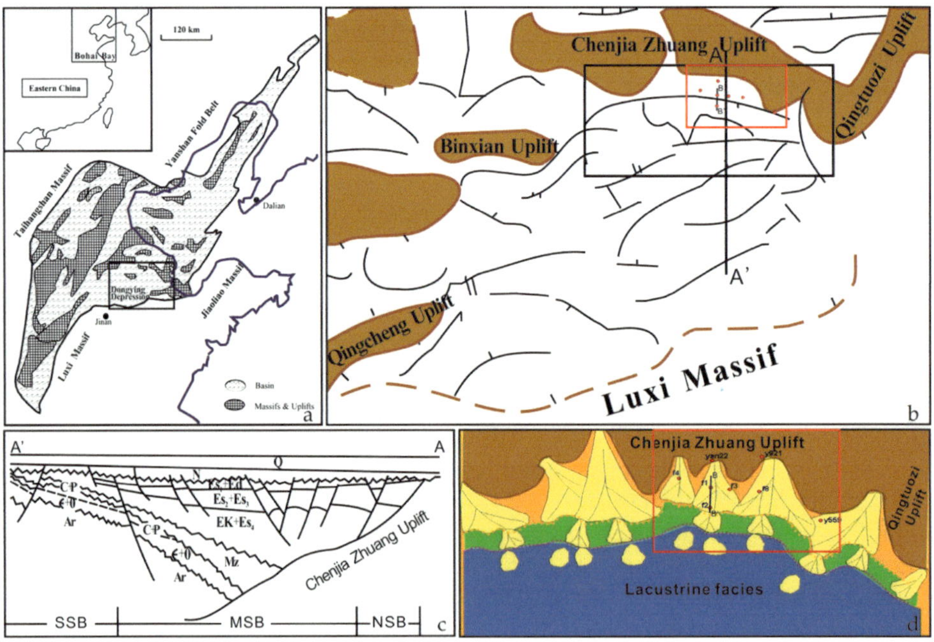


Fig. 1. Location and distribution of sedimentary facies of the lower Es4 Formation in the northern Dongying Depression, Bohai Bay Basin, Eastern China. 3D seismic area is marked by Red box.

characterized by a structural style of half-graben with a northern faulting and southward overlaping. The Depression can be divided in to three structural belts: a northern steep belt (NSB), a middle sag belt (MSB) and a southern slope belt (SSB) (Fig.1c). During the lower Es4 Formation in the early Paleogene, under the controlling of the northern Chenjiazhuang boundary extensional fault, many nearshore subaqueous fans developed in the footwall of the Chenjiazhuang major fault along the northern steep belt and extending into the deep and semi-deep lacustrine facies of the middle sag belt (Fig.1d) with sources mainly from the northern Chenjiazhuang Uplift (Gao et al., 2008; Xie et al., 2004; Yan et al., 2005) (Fig. 1). The burial depth of these subaqueous fans now reaches more than 3500m.

Facies analyses shows that the Dongying nearshore subaqueous fan consists of three sub-facies including a root sub-fan, a mid sub -fan and a marginal sub -fan (Gao et al., 2008; Song, 2004; Yan et al., 2005). The root sub-fan is composed by one or more major channel

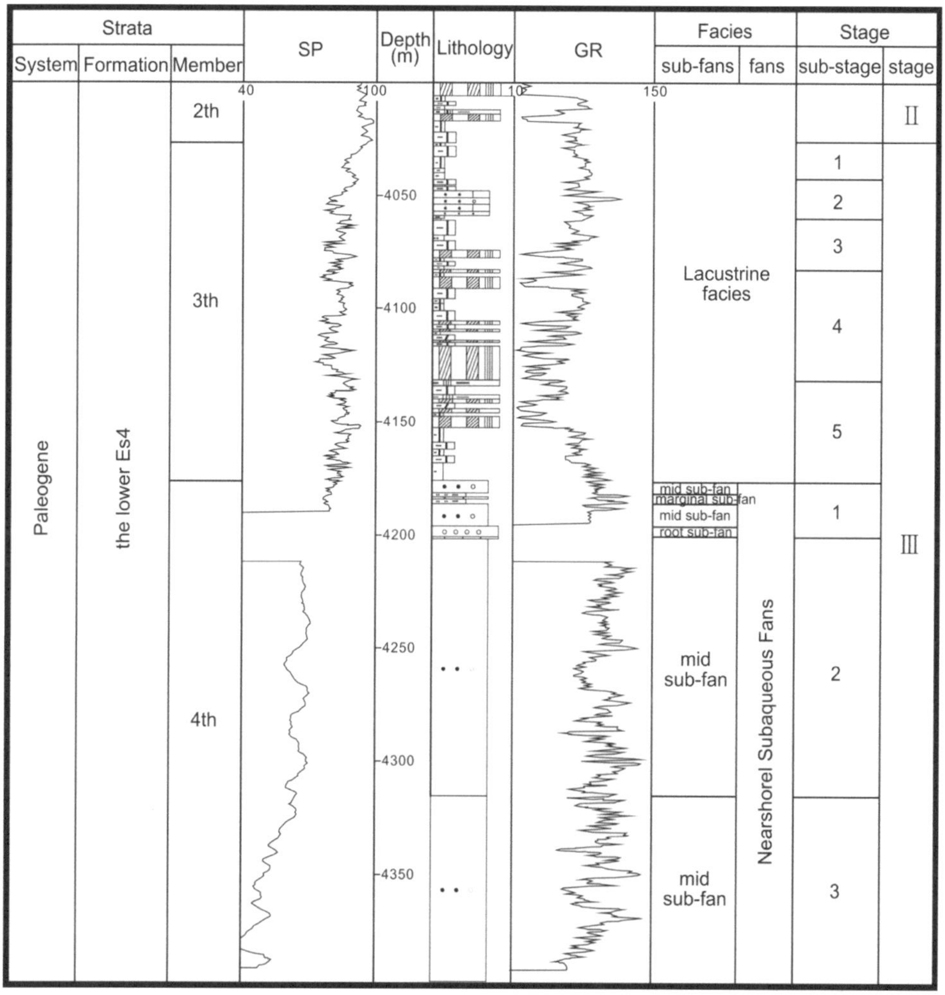


Fig. 2. The single-well facies analysis of well f8 shows that the 4th member of the lower Es4 Formation is a nearshore subaqueous fans with three sub-facies (modified from Gao et al., 2008; Song, 2004; Yan et al., 2005).

sediments and the main lithology includes gray matrix-supported conglomerates, sandy conglomerates and black shales. The mid sub-fan, which is the main part of the nearshore subaqueous fan and forms the effective reservoir in the study area, is characterized by braided channels with braided channel microfacies, intra-channel microfacies and leafy sandbody microfacies. The main lithology of the mid sub-fan includes pebbly sandstones, conglomerate and block sandstones, with thickness varying between 1 and 55 m. The marginal sub-fan consists of siltstones, muddy siltstones and mudstone interbedding rocks (Fig.2). According to well stratigraphic cyclicities and 3D seismic reflection features, the lower Es4 Formation can be divided further into 5 members, standing for 5 individual nearshore subaqueous fans with several sub-facies (Fig. 2).

In general, 3D seismic reflection profiles of the nearshore subaqueous fan are characterized by wedge-shaped, mound-shaped or lenticular-shaped systems, and sub-fans can be further identified. On the synthetic seismograms record calibration, the root sub-fan is characterized by weak reflection, non-reflection or chaotic reflection, the mid sub-fan is characterized by weak to moderate intensity amplitudes, sub-parallel, weak continuous reflection, and the marginal sub-fan is characterized by continuous medium frequency, moderate to low intensity amplitudes. The deep lacustrine facies in the sag belt is characterized by either weak reflection or non- reflection. However, to identify the sub-facies of the nearshore subaqueous fan using the 3D seismic section is difficult (Fig.3).

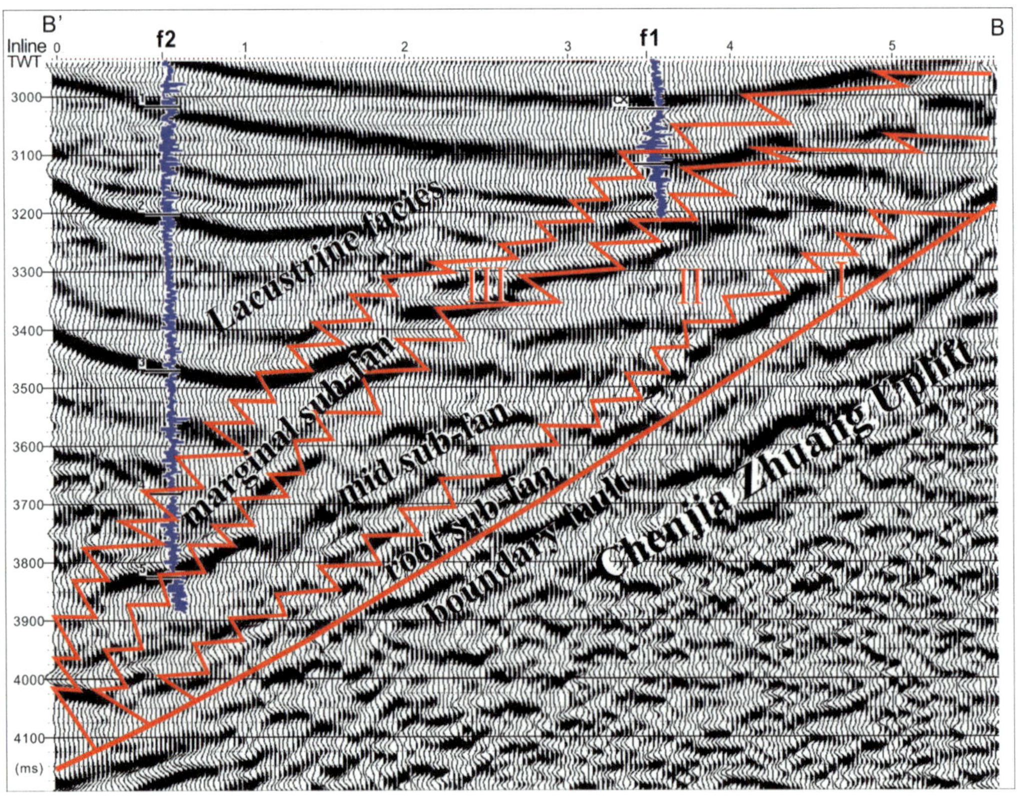


Fig. 3. 3D seismic reflection characteristics of the nearshore subaqueous fans along line B'B (see Line location in Fig.1)

# 3. The pseudo-acoustic 3D seismic inversion based on Logs reconstruction

## 3.1 Methodology

The pseudo-acoustic 3D seismic inversion method different from the conventional 3D seismic impedance inversion method not only in working through logs reconstruction, inversion, interpolation and extrapolation, but also adding or replacing characteristic curves to the density logs or, more commonly, velocity logs in order to achieve the ability to identify the reservoir from the surrounding rock in the case of small impedance difference (Shen & Yang, 2006; Zhang et al., 2005). The potential reservoir may show no direct relationship with the seismic reflection but can be distinguished from different lithological changes.

The velocity and the time-depth relationship after logs reconstruction may change so deviations between seismic reflection horizon and synthetic seismogram calibration's horizon should be established to reflect these changes (Luo et al., 2006). The pseudo-acoustic seismic inversion results based on logs reconstruction may be not accurately reveals the corresponding lithological changes of the target layers. To solve this problem, the zero Mean-Based logs reconstruction techniques, which keeps the original time-depth relationship unchanged, can be applied. The principle is to set the characteristic curves or logs involved in seismic inversion to a mean of 0, that is $\Sigma Ai = 0$ (Ai is characteristic curve sample values of target layers). Adding or subtracting the normalized curve and acoustic characteristics curve, then properly magnifying the normalized characteristics curve in order to highlight lithology information. This process can be expressed as:

pseudo-acoustic curve = acoustic logs ± characteristic curves × K,

while K stands for the curve amplification factor.

As the characteristic curves keep the information of the target layer, and the velocity curves of the upper and lower target layers are kept unchanged, so the original time-depth relationship will remaine unchange (Luo et al., 2006).

The implementation process of this method includes the following: ①selection of the characteristic curves; ②standardization of the characteristic curves; ③normalization and reconstruction of the pseudo-acoustic curve; ④seismic wavelet extraction and the establishment of the initial model; ⑤ pseudo-acoustic seismic inversion.

## 3.2 The pseudo-acoustic 3D seismic inversion based on logs reconstruction

1. Selection of the characteristic curves

To select the characteristic curves of the target layers, quantitative and semi-quantitative correlations through statistical analysis are established between different lithologyies (such as the conglomerate in the fan-root, sand–conglomerate in the mid-fan, mudstone, gypsum-salt rock in the marginal-fan), effective reservoir (such as gas sand–conglomerate in the mid-fan), logs (such as acoustic time (ac), natural gamma (gr), neutron porosity (cnl), spontaneous potential (sp), and logging parameters that correspond to different lithology types in different fans.

The results show that single logs parameter cannot identify the different lithologies in different fans, but combinations of any two of logging parameters (ac, gr or sp) can effectively indentify them to some extent. Further analysis also show that any two logs parameter's combinations between ac, gr, and sp could distinguish the effective and ineffective sand –conglomerate reservoir with a thicknesses greater than 6 m (Fig.4). Therefore, we can use any two logs combination between ac, gr, and sp as the characteristic curves.

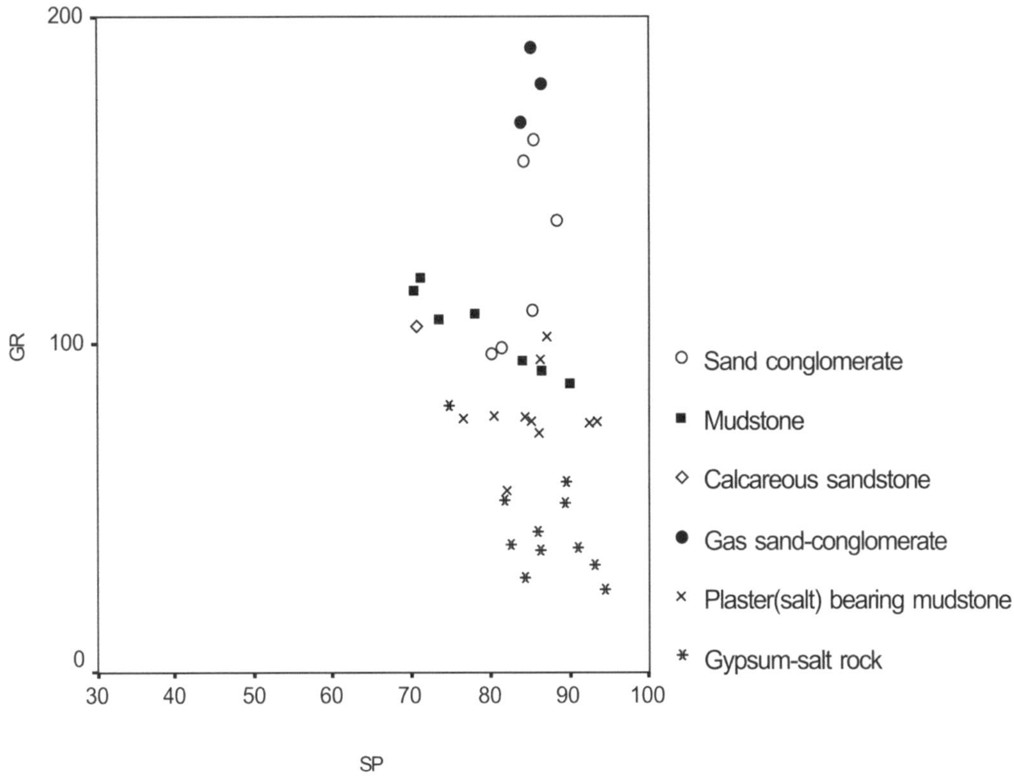


Fig. 4. Statistical analysis between the different lithology, logging parameters of gr and sp and effective reservoir with a thicknesses of > 6 m in wells f1,f2,f3,f8.

2. The standardization of the characteristic curve

In order to eliminate the systematic error caused by different measuring apparatuses and time, the characteristic curves need to be standardized by depth correction, environment correction, mudstone baseline correction, outliers removal, wave filtering and so on.

3. Normalization and the creation of the pseudo-acoustic curve

In order to avoid the systematic error caused by differences in dimension and value range, the characteristic curves need to be normalized before creating the pseudo-acoustic curve. Firstly, the natural gamma (gr) and spontaneous potential (sp) logs will be normalized by regulating the numerical range to the [0 ,1] ,and conducting the [0 ,100] amplification process before summing them for a GS (gr+sp) curve. Then, the asonic logging curve (ac) is processed for the treatment filter values that exceed 100 in order to remain the low-frequency information and eliminate high-frequency information of ac. Finally, the pseudo-acoustic curve GS is obtained by adding the characteristic curve (GS) to the filtered ac. It is clear that the pseudo-acoustic curve GS contain not only the high frequency information of both gr and sp, but also the low frequency information of ac, thus the ability to identify lithologies and strata is greatly improved.

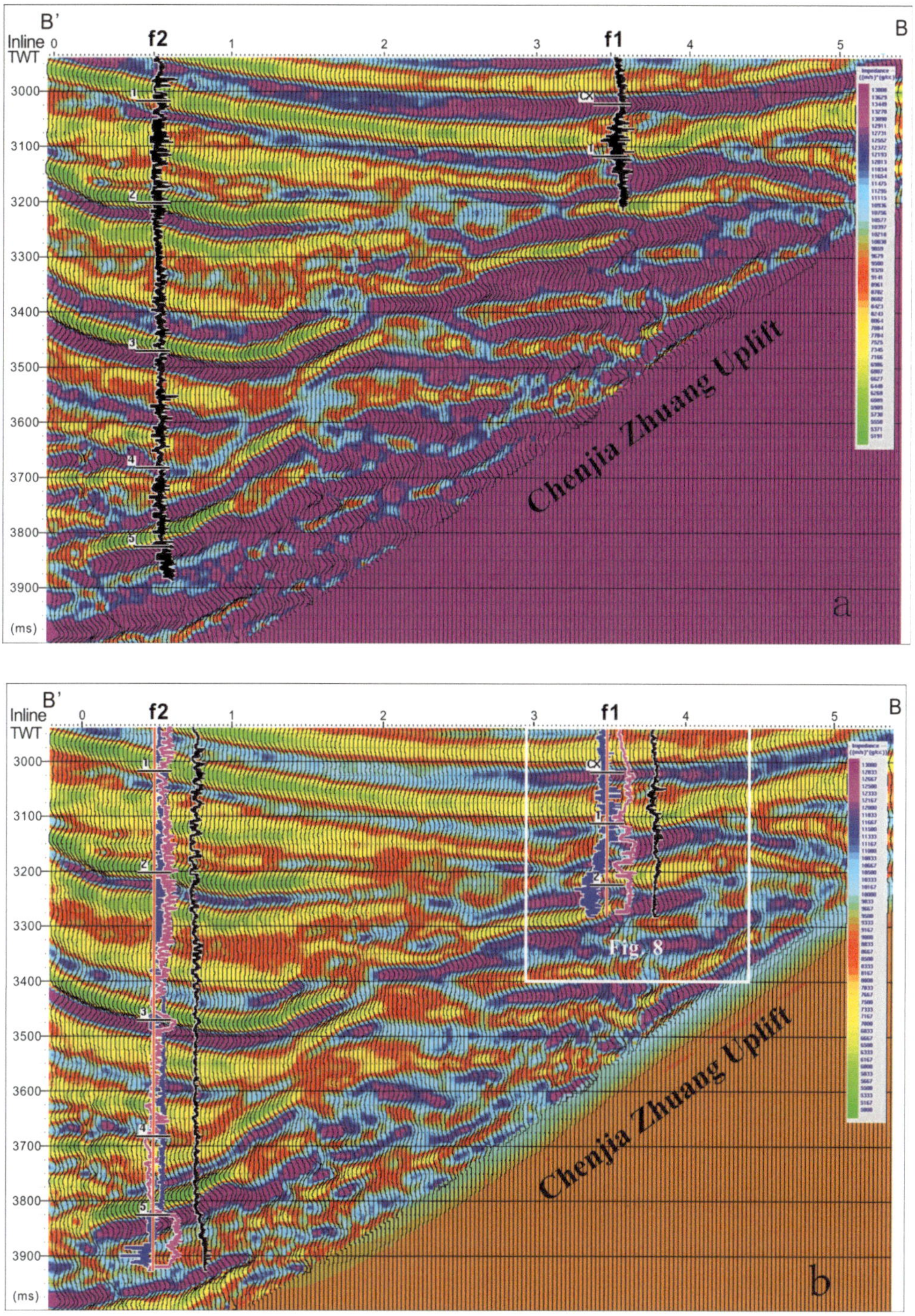


Fig. 5. Comparison between results using different 3D seismic inversion parameters.
a. Conventional 3D seismic inversion data using ac+den loggings; b. Pseudo-acoustic (GS) 3D seismic inversion data based on gr + sp Logs reconstruction

4. Seismic wavelet extraction and initial model creation

Establishing a reasonable initial geological model is the key for getting a good pseudo-acoustic seismic inversion. In fact it is a process of deciphering interpolation and extrapolation of well data under the constraints of the geological concept; the quality of the seismic inversion results are largely dependant on the initial model, which is decided by previous geological knowledge. In order to acquire a good model of impedance inversion, we not only replace the sonic logging curve (ac) by the GS logging curve and by extract Ricker wavelet from the target layer, but also combine the available well information based on the synthetic seismograms calibration and test runs repeatedly.

5. Pseudo-acoustic 3D seismic inversion

On the Strata5.2 inversion software platform, the GS, the GS pseudo-acoustic 3D seismic inversion data are obtained by calculation after importing the GS. The results show that 3D seismic inversion data based on gr+sp logs reconstruction is better than the conventional 3D seismic inversion using ac+den loggings to distinguish the internal structure of the nearshore subagueous fans (Figs. 5, 6)

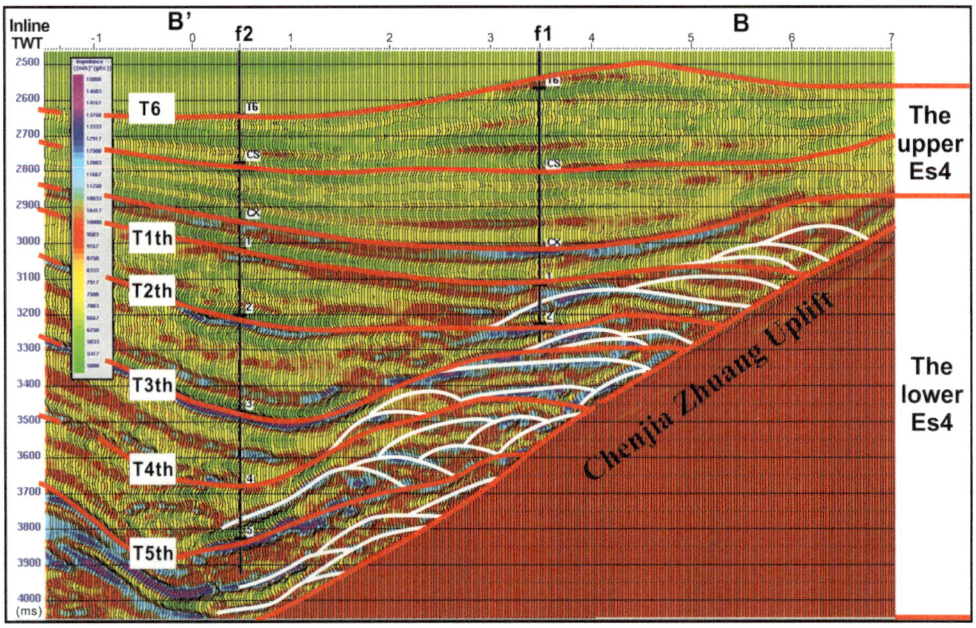


Fig. 6. 3D seismic reflection characteristics of the internal structure in the nearshore subaqueous fans based on GS 3D seismic inversion data along line B'B (see Line L location in Fig.1)

## 4. 3D seismic sedimentology analysis of nearshore subaqueous fans

### 4.1 Evolution characteristics of seismic palaeogeomorphology of nearshore subaqueous fans

By using the GS pseudo-acoustic 3D seismic inversion data coupled with calibration of the synthetic seismograms, the internal sub-facies in each member of the lower Es4 Formation can be identified and the temporospatial evolution of the nearshore subaqueous fans can be extrapolated (Fig.6). The analysis finds that each member of the lower Es4 generally consists of 2-5 sub-facies (Fig.6). The time for high frequent sub-facies development is during deposition of the 4th member of the lower Es4 Formation, which includes at least 5 sub-facies. Fig.7 shows the instantaneous frequency level slices of sub-layers' bottom boundary of the lower Es4 Formation as characterized by a low frequency in the main channel in the fan-root, a middle-low frequency in the mid-fan and a high frequency in the marginal-fan. These results clearly reveal the paleogeographic characteristics and different temporospatial evolution stages of sub-facies in the nearshore subaqueous fan system.

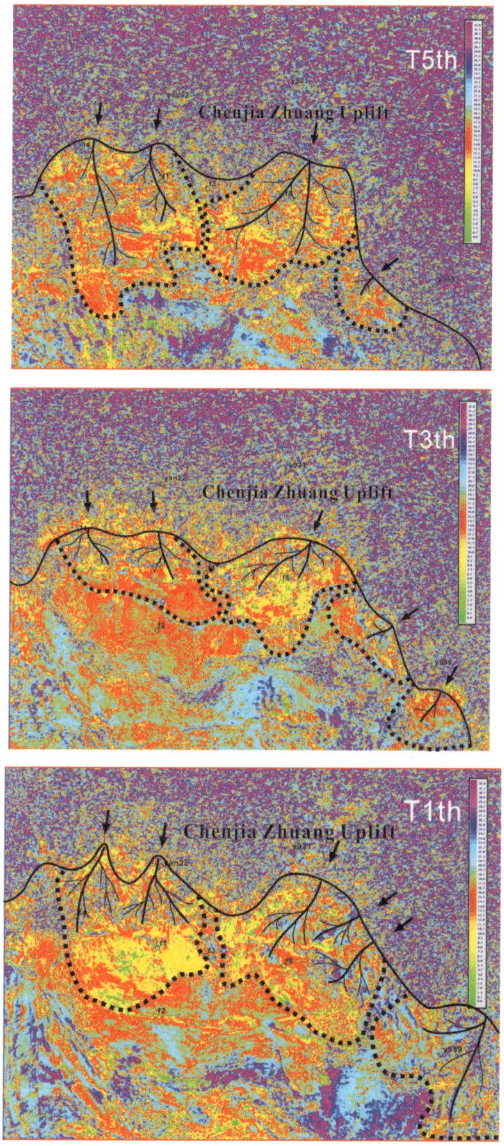


Fig. 7. The instantaneous frequency level slices of sub-layers' bottom boundary of the lower Es4 Formation reflect the paleogeographic characteristics and space-time evolution of different sub-layers

### 4.2 The distribution characteristics of effective reservoir in nearshore subaqueous fan

The synthetic seismogram calibration results show significantly higher dimension values of 12000-15500 in the GS pseudo-acoustic 3D seismic inversion for the effective sand-conglomerate reservoir but lower dimension values<12,000 for the ineffective reservoir in the lower $Es_4$ Formation (Fig.8). Accordingly, quantifying the thickness and the distribution of the effective sand-conglomerate reservoir in the lower $Es_4$ Formation can be relatively easy (Fig.8).

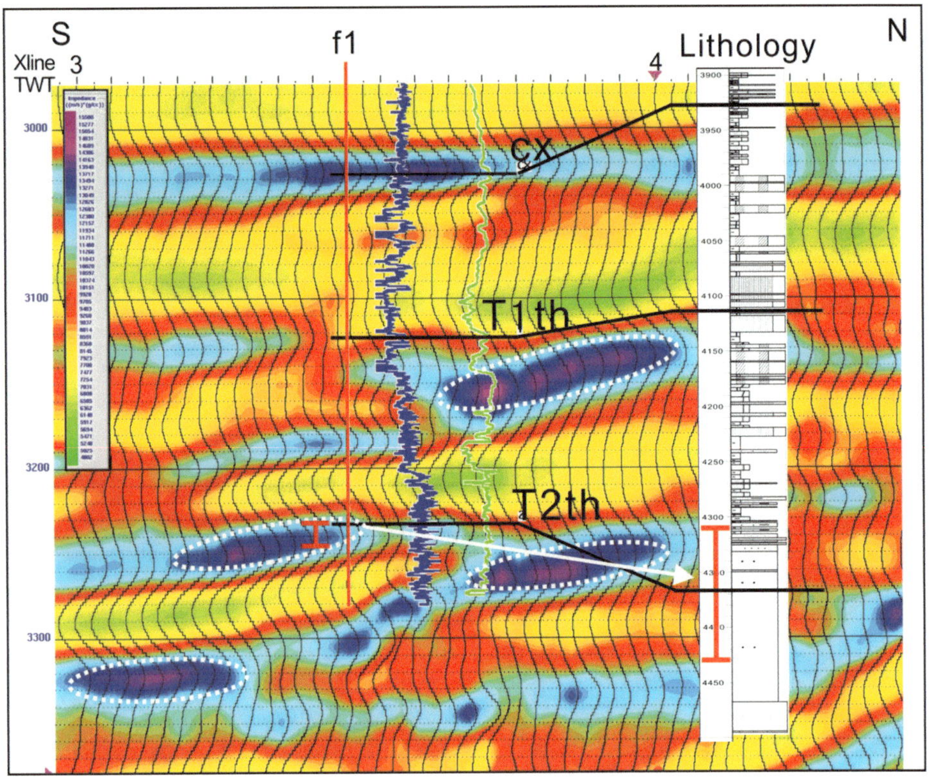


Fig. 8. The effective reservoirs range of values in the GS 3D seismic inversion data for the blue zone (see location in Fig.5b)

## 5. Conclusions

1. Nearshore subaqueous fans of Paleogene age are well developed in the lower Es4 Formation of the Dongying Depression, in the Bohai Bay Basin, eastern China. Research and oil and gas exploration in the northern Dongying Depression have demonstrated that the sand-conglomerate in the mid sub-fan is not only the main part of the nearshore subaqueous fan, but also the effective oil and gas reservoir in the region.
2. Statistical analyses on different lithology and effective reservoir and logging parameters show that the acoustic (ac), natural gamma (gr), spontaneous potential (sp) can be used as characteristic curves for seismic inversion calculation. Any combinations of two logs can distinguish the effective from the ineffective sand -conglomerate reservoirs with a thicknesses greater than 6 m.

3. Compared with the conventional 3D seismic inversion, the pseudo-acoustic 3D seismic inversion based on characteristic logs reconstruction greatly improves the ability to identify internal seismic sub-facies. Several internal sub-facies in each member of the nearshore subaqueous fan in the lower Es4 Formation have been identified.
4. The pseudo-acoustic 3D seismic inversion technique based on logs reconstruction reveals the 3D seismic sedimentological characteristics of nearshore subaqueous fans including the internal sub-facies structure and various temporospatial evolution stages in different sub-facies. The distribution of the effective sand-conglomerate reservoirs can be better quantified by using this method than the conventional 3D seismic impedance inversion.

## 6. References

Carter, D. (2003). 3-D seismic geomorphology: Insights into fluvial reservoir deposition and performance, Widuri field, Java Sea. *AAPG Bulletin,* Vol..87, No.6, (June 2003), pp.909–934, ISSN 0149-1423

Catuneanu, O.; Hancox, P. & Cairncross B. et al.. (2002). Foredeep sunmarine fans and forebulge deltas : orogenic off-loading in the underfilled Karoo Basin. *Journal of African Earth Sciences,* Vol.35, No.4, (November 2002), pp.489-502, ISSN 0899-5362

Chen, S.T. & Meng, X.L. (2004). A method for prediction of reservoirs of thin interbeded layers (in Chinese). *Geophysical Prospecting for Petroleum,* Vol.43, No.1, (January 2004), pp.33-36, ISSN 1000-1441

Chen, P. (2006). Sand reservoir prediction of steep slope zone in Biyang Sag (in Chinese with English abstract). *Petroleum Exploration and Development,* Vol.33, No.2, (April 2004), pp.198-200, ISSN 1000-0747

Darmad, Y.; Willis, B. & Dorobek, S. (2007). Three-dimensional seismic architecture of fluvial sequences on the low-gradient sunda shelf offshore Indonesia. *Journal of Sedimentary Research,* Vol.77, No.3, (March 2007), pp.225–238, ISSN 1527-1404

Gao, X.C.; Zhong, J.H. & Lei, M. et al. (2008). Sedimentary characteristics and controlling factors of the deep glutenite fans in the northern steep slope. Dongying sag-taking Well Fengshen-1 area as example (in Chinese). *Petroleum Geology and Engineering,* Vol.22, No.1, (January 2008), pp.5-8, ISSN 1673-8217

Gee, M. & Gawthorpe, R. (2007). Early evolution of submarine channels offshore Angola revealed by three-dimensional seismic data, In: *Seismic Geomorphology: Applications to Hydrocarbon Exploration and Production.Davies, Geological Society,* R.J. Davies.; H.W. Posamentier.; L.J. Wood. & J.A. Cartwright, (Ed.), pp.223-235, Special Publications, ISBN 978-1-86239-223-6, London

Handford, C. & Baria, L. (2007). Geometry and seismic geomorphology of carbonate shoreface clinoforms Jurassic Smackover Formation, north Louisiana, In: *Seismic Geomorphology: Applications to Hydrocarbon Exploration and Production, Geological Society,* R.J. Davies.; H.W. Posamentier.; L.J. Wood. & J.A. Cartwright, (Ed.), pp.171-185, Special Publications, ISSN 978-1-86239-223-6, London

Li, T.H.; Liang, H.L. & Hu, Y.J. et al. (2005). A research of nearshore subaqueous fans in Guanjiapu beach area and its exploration meaning (in Chinese). *Oil Geophysical Prospecting,* Vol.40, No.5, (October 2005), pp.561-564, ISSN 1000-7210

Lin, J.X.; Shi, Z.J. & Ling, Y. et al. (2007). A study of depositional enviroment evolution of the target zone through basic seismic attributes analysis (in Chinese with English abstract). *Journal of Chengdu University of Technology(Science & Technology Edition),* Vol.34, No.2, (April 2007), pp.174-179, ISSN 1671-9727

Ling, Y.; Sun, D.S. & Gao, J. (2005). Interpretation of depositional bodies in the parasequence sets from 3-D seismic data (in Chinese). *Geophysical Prospecting for Petroleum,* Vol.44, No.6, (November 2005), pp.568-577, ISSN 1000-1441

Liu, B.G. &Liu, L.H. (2008). Application of applied seismic sedimentology in sedimentary facies analysis (in Chinese). *Geophysical Prospecting for Petroleum,* Vol.47, No.3, (May 2008), pp.266-271, ISSN 1000-1441

Lu, Z.Y. (2008). Influence of the Paleogene structural styles on deposition and reservoir in Chezhen Sag, Bohai Bay Basin (in Chinese with English abstract). *Journal of Palaeogeography,* Vol.10, No.3, (June 2008), pp.277-285, ISSN 1671-1505

Luo, Q.S.; Zhao, M. & Zhang X.J. (2006). Application of Zero Mean-Based Curve Reconstruction to Seismic Inversion (in Chinese with English abstract). *Xinjiang petroleum geology,* Vol.27, No.4, (August 2006), pp.478-480, ISSN 1011-3873

Nordfjord, S.; Goff, J; Jr, J. & Sommerfield, C. (2005). Seismic geomorphology of buried channel systems on the New Jersey outer shelf: assessing past environmental conditions. *Marine Geology,* Vol.214, No.4, (February 2005), pp.339-364., ISSN 0025-3227

Posamentier, H. & Kolla, V. (2003). Seismic geomorphology and stratigraphy of depositional elements in deep-water settings. *Journal of Sedimentary Reserch,* Vol.73, No.3, (May 2003), pp. 367–388, ISSN 1527-1404

Prather, B. (2003). Controls on reservoir distribution, architecture and stratigraphic trapping in slope settings. *Marine and Petroleum Geology,* Vol..20, No.6-8, (June-September 2003), pp.529-545, ISSN 0264-8172

Richard, M. & Bowman, M. (1998). Submarine fans and related depositional II: variability in reservoir architecture and wireline log character. *Marine and Petroleum Geology,* Vol.15, No.8, (December 1998), pp.821-839, ISSN 0264-8172

Schwab, A.; Tremblay, S. & Hurst, A. (2007). Seismic expression of turbidity- current and bottom-current processes on the Northern Mauritanian continental slope, In: *Seismic Geomorphology: Applications to Hydrocarbon Exploration and Production, Geological Society,* R.J. Davies.; H.W. Posamentier.; L.J. Wood. & J.A. Cartwright, (Ed.), pp.237-252, Special Publications, ISBN 978-1-86239-223-6, London

Shen, X.C. & Yang, J.F. (2006). The Application of the Reconstructed Characteristic Curve of Reservoir in the Inversion (in Chinese with English abstract). *West China Petroleum Geosciences,* Vol..2, No.4, (December 2006), pp.436-439, ISSN 5021-5850

Song, N. (2004). The study of stratigraphic sequence with gravel rock in northern steep of the Dongying sag stratigraphic research study (in Chinese with English abstract). *Ph.D. Thesis,* nanjing university, Nanjing

Song, R.C.; Zhang, S.N. & Dong, S.Y. et al. (2007). Characteristics and Controlling Factors Analyze of Nearshore Subaqueous Fans in Langgu Depression (in Chinese with English abstract). *Journal of Earth Sciences and Environment,* Vol.2, No.29, (June 2007), pp.145-158, ISSN 1672-6561

Sullivan, E.; Marfurt, K.; Blumentritt, C. & Ammerman, M. (2007). Seismic geomorphology of Paleozoic collapse features in the Fort Worth Basin (USA), In: *Applications to Hydrocarbon Exploration and Production, Geological Society,* R.J. Davies.; H.W.

Posamentier.; L.J. Wood. & J.A. Cartwright, (Ed.), pp. 171-185, Special Publications, ISBN 978-1-86239-223-6, London

Takahiro, S. & Makoto, I. (2002). Deposition of sheet-lke turbidite packets and migration of channel- overbank systems on a sandy submarine fan: an example from the Late Miocene-Early Pliocene Forearc Basin, Bosopeninsula Japan. *Sedimentary Geology,* Vol.149, No.4, (June 2002), pp.265-277, ISSN 0037-0738

Tian, J.C. & Fu, D.J. (2001). A research on the reservoir property of the sand-conglomerate in the nearshore submerged fan taking the hird member of the Shahejie Formation in Well C913 and C916 area on the northern zone of the Zhanhua Depression as example (in Chinese with English abstract).*Journal of Chengdu University of Technology,* Vol.28, No.4, (October 2010), pp.762-766, ISSN 1005-9539

Wang, L.W.; Liang, C.X. & Zou, C.N. et al. (2004). Application of comprehensive seismic data interpretation in reservoir prediction in the south part of Songliao basin (in Chinese with English abstract). *Progress in Exploration Geophysics,* Vol.27, No.1, (February 2004), pp.58-62, ISSN 1671-8585

Wu, Y.Y.; Song, Y. & Jia, C.Z. et al. (2005). Sedimentary features in a sequence stratigraphic framework in the north area of Qaidam Basin (in Chinese with English abstract). *Earth Science Frontiers,* Vol.12, No.3, (September 2005), pp.195-203. ISSN 1005-2321

Xie, R.J.; Qi, J.F. & Yang, Q. (2004). Structural Characteristics and Its Control to Deposition in the North of Dongying Depression (in Chinese). *Journal of Jianghan Petroleum Institute,* Vol.26, No.1, (March 2004), pp.17-19, ISSN 1000-9752

Yan, J.H.; Chen, S.Y. & Jiang, Z.X. (2005). Sedimentary characteristics of nearshore subaqueous fans in steep slope of Dongying depression(in Chinese with English abstract). *Journal of the University of Petroleum,* Vol.29, No.1, (February 2005), pp. 12-21, ISSN 1000-5870

Zhang, J.L. & Shen, F. (1991). Characteristics of nearshore subaqueous fan reservoir in damoguaihe formation, Wuerxun Depression (in Chinese with English abstract). *Acta Petrolei Sinica,* Vol.12, No.3, (July 1991), pp.25-35, ISSN 0253-2697

Zhao, J.Q.; Ji, Y.L. & Xia, B. et al. (2005). High-Resolution Sequence Research on Nearshore Subaqueous Fan System (in Chinese with English abstract). *Acta Sedimentologica Sinica,* Vol.23, No.3, (Septemb2005) pp.490-497, ISSN 1000-0550

Zhang, M. & Tian, J.C. (1999). The nomenclature, sedimentary characteristics and reservoir potential of nearshore subaqueous fans (in Chinese with English abstract). *Sedimentary Facies and Palaeogeography,* Vol,19, No,4, (April 1999),pp.42-52, ISSN 1001-7824

Zhang, X.F.; Dong, Y.C. & Shen, G.Q. (2005). Application of log rebuilding technique in constrain inversion (in Chinese with English abstract). *Petroleum Exploration and Development,* Vol.32, No.3, (June 2005) pp.70-72, ISSN 1000-0747

Zhang, Y.H.; Yang, D.Q. & Sun, Y.H. et al. (2007). Application of high-resolution sequence stratigraphy in subtle oil pool prediction (in Chinese). *Geophysical Prospecting for Petroleum,* Vol.46, No.4, (July 2007), pp.378-383, ISSN 1000-1441

Zeng, H.L.; Backusz, M.; Barrow, K. & Tyler, N. (1998). Stratal slicing, Part I: Realistic 3-D seismic model. *Geophysics,* Vol..63, No.2, (April 1998), pp.502-513, ISSN 0016-8003

Zeng, H.L. & Ambrose, W. (2001). Seismic sedimentology and regional depositional systems in Mioceno Norte. Lake Maracaibo, Venezuela. *The Leading Edge,* Vol..20, No.9, (November 2001), pp.1260-1269, ISSN 1070-485x

Zeng, H.L. & Kerans, C. (2003). Seismic frequency control on carbonate seismic stratigraphy: A case study of the Kingdom Abo sequence, west Texas.*AAPG Bulletin,* Vol..87, No.2 (February 2003), pp.273-293, ISSN 0149-1423

Zeng, H.L. & Hentz, T. (2004). High-frequency sequence stratigraphy from seismic sedimentology: Applied to Miocene, Vermilion Block 50, Tiger schoal area,offshore Louisiana. *AAPG Bulletin,* Vol..88, No.2, (February 2004 ), pp.153-174, ISSN 0149-1423

Zeng, H.L.; Loucks, R. & Frank B. (2007). Mapping sediment- dispersal patterns and associated systems tracts in fourth- and fifth-order sequences using seismic sedimentology: Example from Corpus Christi Bay, Texas. *AAPG Bulletin,* Vol.91, No.7, (July 2007), pp.981–1003, ISSN 0149-1423

Zhao, C.L. (2000). *Sedimentation-reservoir geologic essays* (in Chinese), Petroleum Industry Press, ISBN 7-5021-3006-3, Beijing, China

Zhang, R.H.; Yu, S.Y. & Wu J.H. (1997). The effect of sediments supply condiction on sequence stratigraphy analysis in continental lacustrine basins (in Chinese with English abstract). *Earth Science-journal of China University of Geosciences,* Vol. 22, No.2, (March 1997), pp.139-144, ISSN 1000-2383

# 8

# Mapping and Analyzing the Volcano-Petrology and Tectono-Seismicity Characteristics Along the Syrian Rift – NW the Arabian Plate

Ahmad Bilal
*Damascus University,*
*Syria*

## 1. Introduction

The European and African continents are crossed by several N-S-trending rifts, all together major structural features at world scale. They include, from North to South, firstly the Oslo Permian rift (Norway), continued by the Neogene fracture system of Central-Southern Germany (Eifel, Rhine Graben), then the rift system of French massif Central and Rhone valley, ending finally with the great African rift, the major structure of this continent.
These major crustal fractures, extending down in the underlying mantle, have been active at different times, while always keeping the same approximate N-S direction. Periods of major activity are marked by extensive volcanism, with a distinct tendency to show younger ages southwards: Permian in Norway, Neogene in Germany, Neogene to subactual in France, actual (present-day) in Africa. These ages correspond mainly to the initial stage of rift-forming, whereas more ancient accidents (e.g. Norway) could repeatedly play again, at each phase of crustal extension.
In direct continuity with the Dead Sea Fault, the Syrian rift links the rigid Arabian plate to the mobile ophiolite belt of Cyprus and Southern Turkey (Juteau 1974, Parrot 1977). It plays a very important role in the regional geodynamic structure. Its exact position, as well as the related fracture system, has been documented from the analysis of complete aerial photo coverage of the whole Syrian territory (Bilal and Ammar 2004).
Many partial works on the different aspects of this area: tectonics, geodynamics, volcanism, crustal and mantle rocks, and seismicity have been done. But a global synthetic on these aspects are given in this research, using new data in field and laboratory. The results either of my team at Damascus university, or either those of the scientific cooperation projects, from 1998 till now, with the teams of colleagues from the French universities: professors Jean Chorowicz, and Albert Jambon, from Pierre and Marie Curie university; professor Phillipe Huchon, from Ecole Normal Superior of Paris; professor Jacques Touret, from Ecole des Mines of Paris, and professor Jean Ives Cottin from the university Jean Monnet of Saint-Etienne. In addition to international works indicated in the references list. While the global work, at the macro- scale, has been achieved, it still more works to do at the micro –scale: the detailed composition variations of the volcanic rocks, and their geologic process indication; the liaison between the different tectonic unities, and theirs liaison with the regional geotectonic; and the micro- seismic zonation in the country.

## 2. Geodynamic setting

The Arabian plate has a roughly polygonal shape, inserted between the major African plate (including Nubian and Somalian ),to the East, and Eurasiatic and Indian plates, to the North. It is delimitated by the Red Sea in the South-West, the Aden gulf in the South, and the Zagros and Taurus chains in the North and North-East, respectively.
Geophysical investigations confirms the typical continental nature of this plate, with an average crust thickness of 40 Km, which changes, at the level of the Red Sea , to less than 15 Km., on a distance of about 250Km.(Al Damegh et al.2005).
The Arabian plate shows three types of active borders (Fig.1):

Fig. 1. Geodynamic framework of the Arab plate (Barrier et al. 2004).

- Convergent :it include the collision zone of Bitlis-Zagros (Fig.1,A,B) ,(Sosson et al 2005,Molinaro et al 2005,Agard et al.2006),the subduction Makran-Oman zone between

Aurasia-Arabia, as well as the Anatolian fault at the north-west end of the Arabian plate(Cetin et al.2003).

- Divergent: the oceanic rifts (Arabia-Nubia), and the Aden gulf (Arabia-Somalia), (Fig.1, C, D),(Bosworth et al.2005)
- Transform: the senester faults of the Levant (Arabia-Nubia), to the West and the Dexter faults of Owen (Arabia-India),to the East (Barrier et al.2004).

The territory of syria corresponds to the NW corner of the Arabian plate. It is bordered by the Zagros Taurus collision zone, to the North, and the oceanic expansion zone, to the South. In the Western part of Syria, the rift structure, which corresponds to the northern part of the Dead Sea Fault Zone(DSFZ), is named the Levant fault, in continuity with the Red Sea rift zone.

## 3. Interplate volcanism along the Syrian rift

The Syrian rift is marked by an active interplate volcanism, occurring from Jurassic to present (Ponikarov 1967, Laws and Wilson 1997, Giannerini et al.1998). Volcanoes bring to the surface a number of mantle xenoliths, which provide essential information on the nature and composition of the underlying lithospheric mantle (Stein and Hofman 1992, Stein et al 1993, Sharkov et al.1993, Bilal and Touret 2001, Bilal and Sheleh 2004). Most important data are summarized below:

### 3.1 Volcanism

The occurrence of volcanic activity in its geotectonic context shows that this activity covers an important part of the surface of the Arabian plate: in Syria; in Jordan; and in Saudi Arabia (Fig.2). This volcanism covers the Mesozoic and Cenozoic times, but the major eruption is recent. It is distributed over three distinct regions (Mor 1993): (1) the Harrat Ash Shaam plateau; (2) the region from the Homs basalts to the Karasu valley; (3) the Arabian platform and the Southern part of the Bitlis belt (e.g.karacadage volcano).

The Harrat Ash Shaam basalt eruption occurred in three episodes: at 26-22 Ma.;18-13Ma.;and 7to <0,5Ma.(Mor 1993, Ilani et al 2001).The Homs basalts are dated at 6,5-2,0Ma.(Mouty et al.1992,Sharkovet al.1994,1998,Butler et al.1997, Butler and Spencer 1999).In the Ghab basin area and east of it ,the age ranges from 2.0 to 1,1 Ma (Heiman et al 1998). In the Karasu valley and vicinity, the age vary from 1, 6 to 0, 05 Ma. (Capan et al.1987,Heiman et al.1998, Rojay et al 2001,Yurtmen et al 2002).

Summarizing, volcanism in Syria, started during Lias with magmatism associated to the ophiolites in the north of the territory (in the Baer et Bassit region),at the same time of volcanism in the southern of Turkey (Antalya et Hatay ),or in the Mamonia complex in Cyprus (Robertson et al.1991).This volcanism is related to subvertical tension fractures caused by transcurent movement along the Syrian part (the Syrian rift)of the Dead Sea Fault Zone(DSFZ).It can be hypothetized that these fractures induced adiabatic partial melting in the lithosphere (Polat et al.1997,Adiyaman and Chorowwicz 2002,Chorowicz et al.2005).

The volcanic emission extends over about 10% of the whole surface area of Syria (Fig. 3).Volcanism is related to the movement of the Arabian plate towards the Eurasian plate, at a velocity of $18\pm2 mma^{-1}$, in a NNW direction (McClusky et al2000). Eruptions, flooding cover significant areas, where the Cenozoic basaltic lavas may be up to 500 M thick and are covered by Tertiary and Quaternary sediments (Al Mishwat and Nasir 2004).

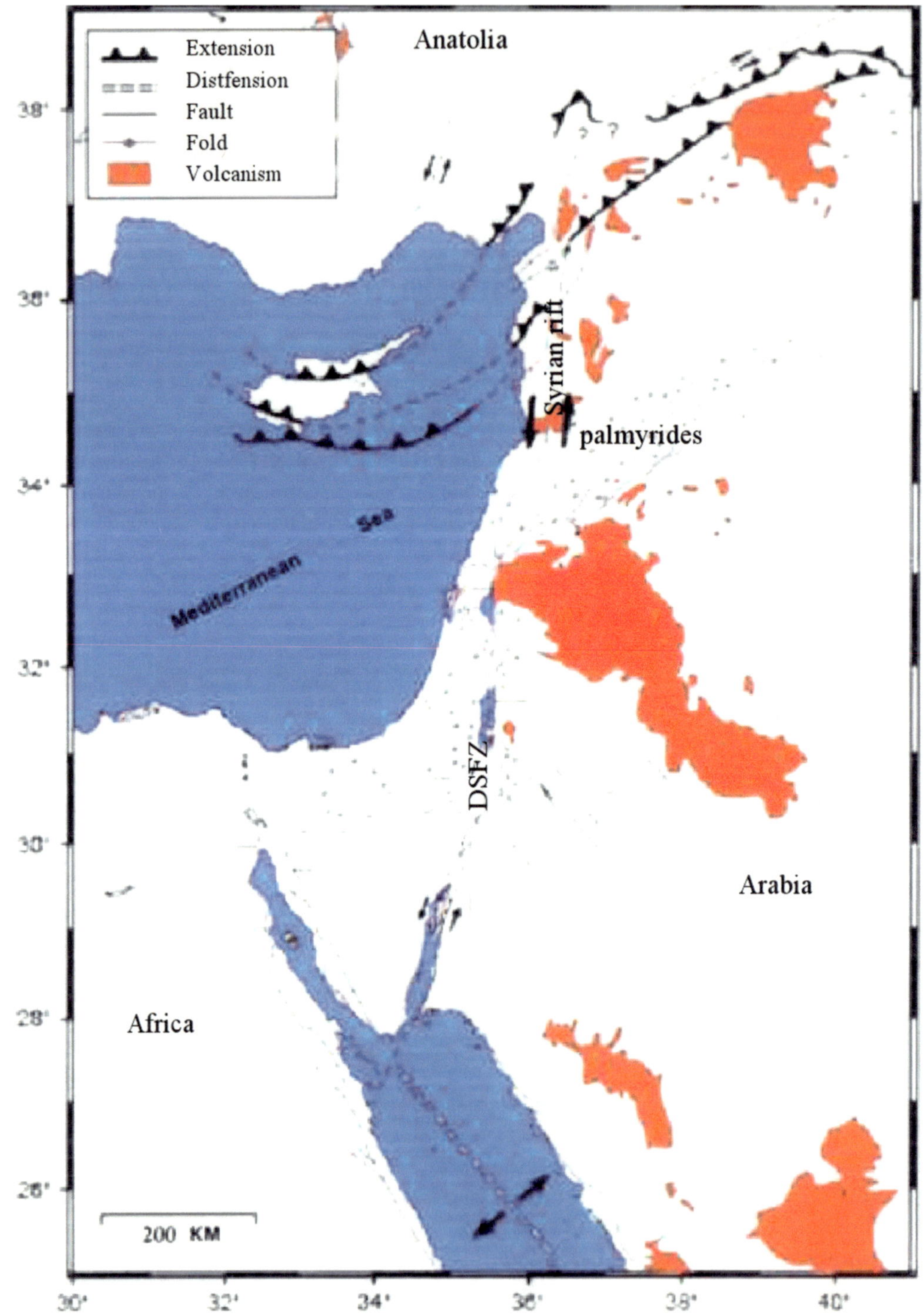


Fig. 2. The Arabian interplatplate volcanism occurrence in its tectonics context. (Adiyaman and Chorowicz, 2002 ).

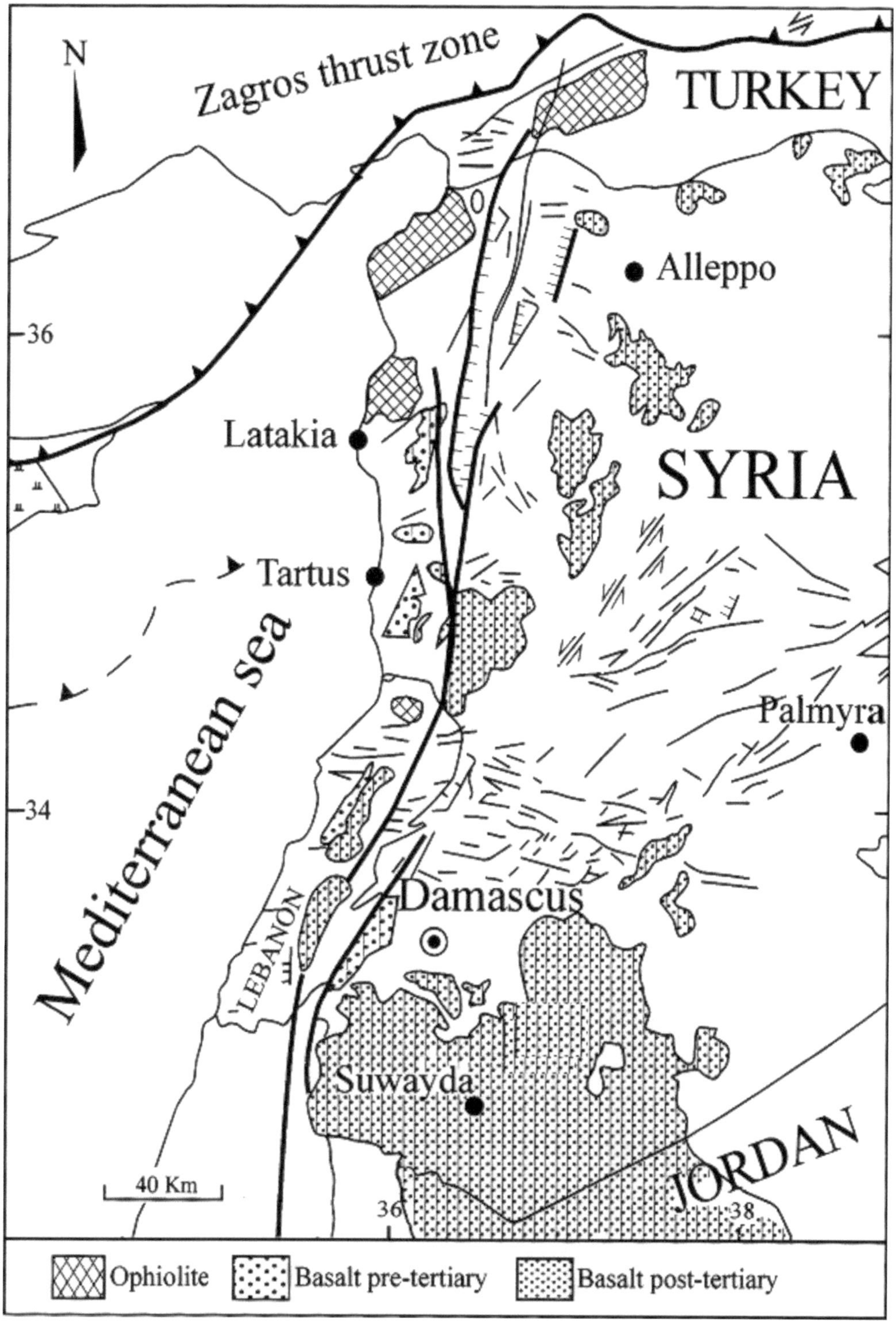


Fig. 3. The volcano-tectonics map of Syria.

The volcanism is divided into two periods: upper Jurassic-lower Cretaceous, a period corresponding to a phase of extension of the Arabian plate margin, it corresponds to the Bhannes-Tayasir area of Syria sequence, constitutes isolated lavas or covered by the Neogene eruption, especially in the south of the country (west of Damascus), and in the center (Nabi Mata region) (Dubertret 1962,Ponikarov 1967,Laws and Wilson 1997).More recent Neogene-Quaternary eruptions are related to the formation of the Red Sea (24-16Ma), and Dead Sea rifts (8-0,4 Ma) (Ponikarov 1967, Bohanon et al.1989,Camp et Roobol 1989,Baker et al 1997,Chorowicz et al 2005). A number of eruptions have been identified from 17 Ma to present. The last volcanic eruption took place in the South of the country, about 10000 years ago, as the end of the last eruption (<1Ma) (Dubertret 1933, Baker et al 1997).

Erupted lavas are in general very basic.rock compositions , cover the basalt-, picrobasalt-, and basanites fields on the diagram of Le Bas et al.(1986), (Fig.4),corresponding to a low differentiated magma(Ismail et al.2008). Major and trace elements data show overall similarities between recent and ancient ones, with however a more distinct alkaline trend and stronger variations of LILE-elements for recent lavas. These data involve small volume melt fractions.

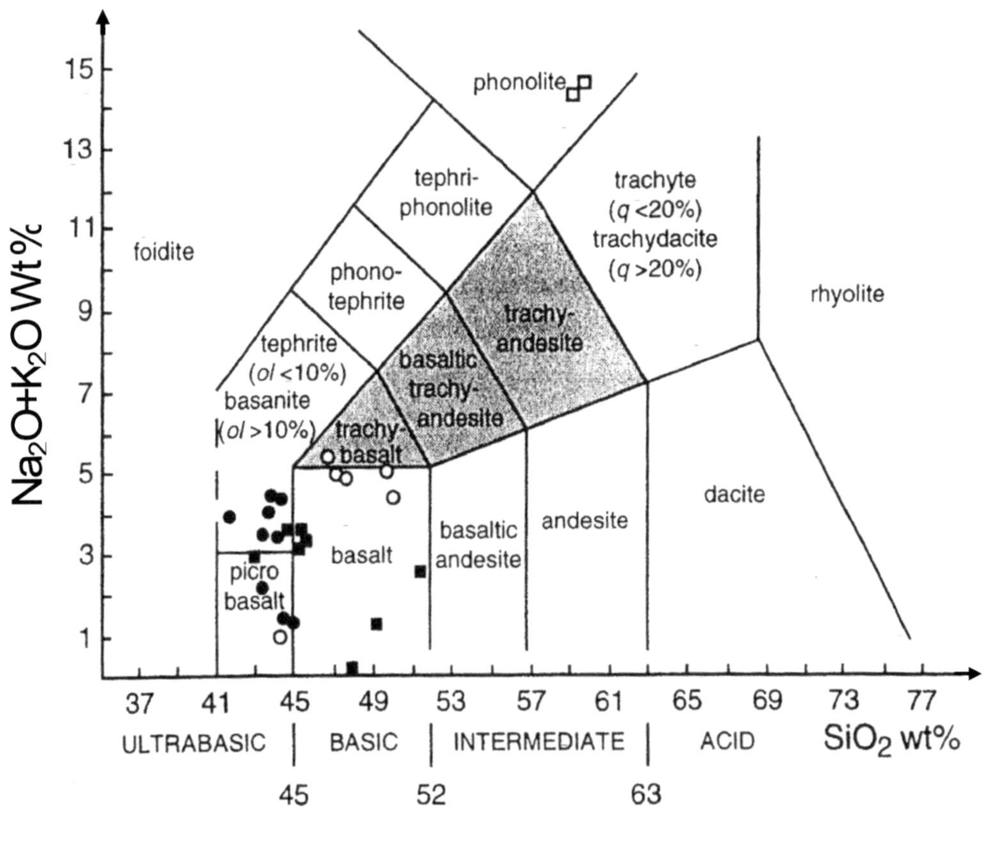


Fig. 4. Total alkalis-silica diagram of Le Bas et al (1986) for the basaltic rocks along the Syrian rift (Bilal and Touret 2001).

### 3.2 Composition of the lithospheric mantle

A number of volcanoes along the rift contain a number of ultrasiques xenoliths, notably lherzolites, harzburgites and pyroxenites(Bilal and Touret 2001).Major rock -forming minerals are Olivine (Ol), Ortho- and Clinopyroxene (Opx and Cpx),with as common accessories spinel and amphibole. Microstructure varies from coarse-grained, coarse-grained-tabular to rare porphyroclastic (Ismail et al.2008).Most mantle peridotites are very well preserved with however a small variable possibility of local melting by the enclosing basalt around intergrain boundaries (Fig.5, A, B, C).

Most abundant rock types are harzburgites (Ol+Opx),which from their mineral composition and geochemistry can be divided into three groups (Ismail et al 2008): Group I, issued from a residual, depleted mantle, Groups II and III which correspond to a refertilized mantle, caused by the percolation of undifferentiated basaltic melt or ephemeral carbonate magmas through the residual lithosphere. Both groups correspond to a different degree of melting of the mantle peridotite (large for Group II, small for Group III).They are characterized by undispread mantle metasomatism with a carbonatite signature (Frezzotti et al 2002, Gregoir et al 2000), as notably indicated by the composition of clinopyroxene in some pyroxenites (Bilal and Sheleh 2004).

Rare garnet-bearing varieties have also been observed in the middle and south domains (Mheilbeh,Tel Thenoun) including few grenatites. These correspond most probably to lower crustal granulites, even if the occurrence of some high-pressure basaltic derivates cannot be excluded (Bilal and Touret 2001). The possible occurrence of xenoliths corresponding to lower crustal granulites is further indicated by the occurrence of sapphirine in some garnet and/or spinel-bearing websterites (Opx and Cpx-bearing pyroxenites, Fig. 5 D ,E) (Bilal 2009b, Bilal et al.2011). These basalts result from a complex polybaric melting process, first starting in the garnet peridotite stability field, then proceeding within the field of spinel peridotite( Bilal and Touret 2001,Bilal et al.2011).

### 3.3 Fluid inclusions

A great of pure CO2-bearing fluid inclusions have been found in olivine and pyroxenes from xenoliths, and in phenocrysts from enclosing basalts (Bilal and Touret 2001).This type of inclusions occur in virtually all mantle xenoliths in basalts worldwide(Roedder 1984),but in the present case some features confirm the occurrence of mantle metasomatism seen above in group II and III.The CO2 density in inclusions is very variable ,most commonly around or lower the critical point(about 0,4 g/Cm3).Fluid pressure at trapping conditions ,for a reference temperature of about 1000C,correspond to a depth of about 5 Km, namely the last magma chamber prior to eruption .But some primary inclusions contain fluid of much higher density recording deeper episode of the rock evolution. Highest fluid densities (up to 1, 15 g/cm$^2$) are found in pyroxenites, notably in clinopyroxene. Fig 5 (F , G, H, I) show primary inclusions, of tubular shape, aligned along orthopyroxene or plagioclase exsolution lamellae within the clinopyroxene host .It is belived that these fluids are formed by a reaction illustrating the mantle metasomatism carbonatite connection:

Olivin+Carbonate (from the Carbonatite)→Clinopyroxene (with plagioclase and orthopyroxene in solid solution)+Co2.

P-T conditions of mineral equilibration in the xenoliths can are deduced from the pyroxene mineral assemblage (pyroxene thermometry) for the temperature (Wells 1977, Bertrand and Mercier 1986,Brey and Kohler 1990, Kohler and Brey 1990),and from the maximum fluid

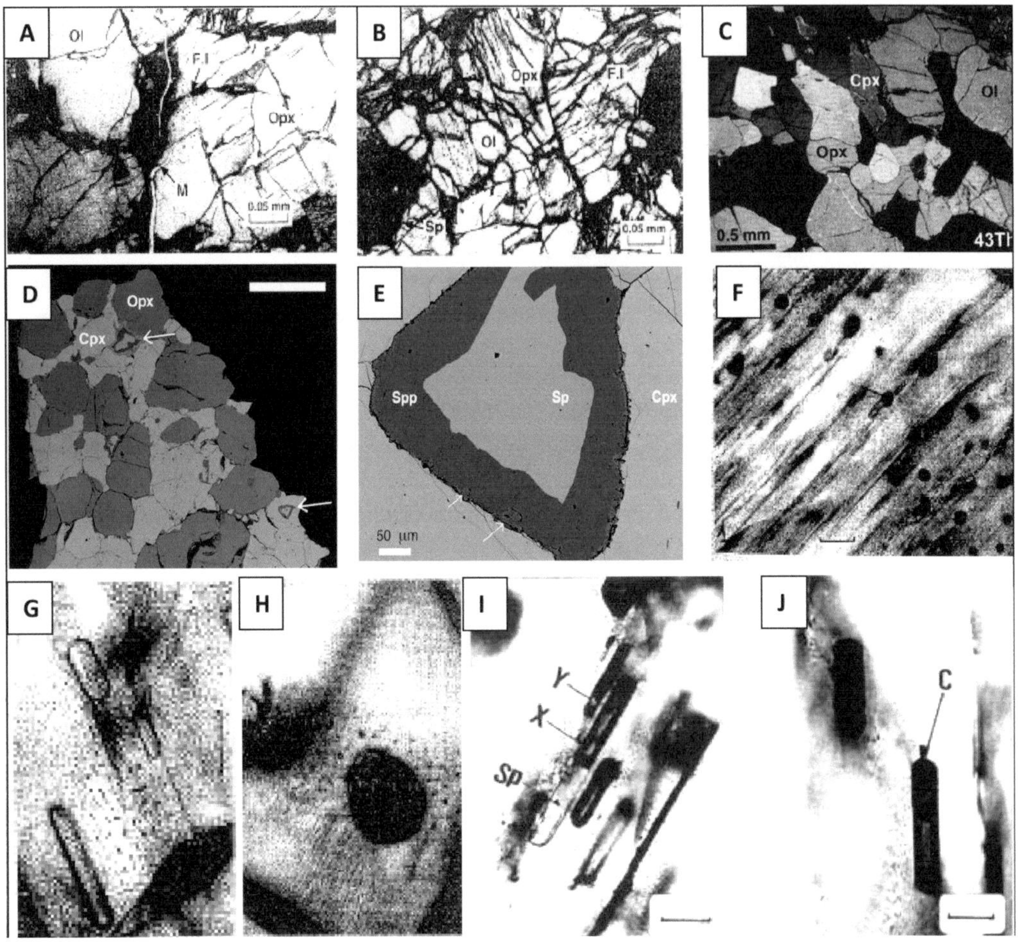


Fig. 5. Photomicrographs of petrographic features of mantle xenoliths , and fluid carbonic inclusions.

A: Coarse-grained microstructure harzburgit (sample37Th ,Tel Thenoun)-large Olivine (Ol) and Clinopyroxene(OPX) crystals shown a dark melt zone(M) at the intergranular boundary containing white veinlet filled with secondary minerals(opal, carbonate),and trail of secondary ,low density ,carbonic inclusions(FI), issued from M, and disposed along the OPX cleavage plan. B: Coarse-grained-tubular microstructure spinel lherzolite (sample 34Th,Tel Thenoun)-large pyroxenite crystals, subordinate olivine(Ol),and large spinel crystals surrounded by a dark reaction zone.Primary CO2 inclusions(FI) are disposed in the core of some clinopyroxene crystals. C:Coarse-grained microstructure in harzburgite (sample 43Th,Tel Thenoun).D:composite BSE image section(sample Th12,Tel Thenoun).The submillimitric coronitic feature correspond to spinel (light grey core) rimmed by sapphirine(dark grey).Scale bar 2mm.E: More clear of coronitic spinel of figure D,where the sapphirine corona(dark grey) is continuous.A thin rim of symplectite is hardly visible between sapphirine and clinopyroxene.F:Carbonic (pure CO2) fluid inclusions(black circles) in exsolution features in the butterfly-wing structure, in pyroxenite(sample 135Th,Tel Thenoun).Scale bar50 micron m. G:Tubular carbonic inclusions in pyroxene (sample135Th,Tel Thenoun),Scale 20 micron m.H:Details of the butterfly-wing structure around inclusion in olivine(sample 11Th,Tel Thenoun).Scale20 micronm. I, J: Carbonic inclusions in clinopyroxene .X and Y supercritical CO2 (homogenization temperature =15 C),(sample 135Th,Tel Thenoun).bar 20 micronm.

density in primary carbonic inclusions, for the pressure (Bilal et Touret 2001,Bilal et Sheleh 2004).Using theses parameters the figure (6),show the obtained Results, which correspond to about 1100-1300C for the temperature, 10-13 Kb for the pressure.
In conclusion it is suggested that the volcanic activity along the Syrian rift is due to the presence under the Arabic plate of a mantle plume(Stein and Katz 1989,Stein and Hofman 1992 ,Stein et al 1993), active since Cretaceous times, locally refertilizing a residual (oceanic-type) lithospheric mantle (Bilal and Touret 2001,Bilal and Sheleh 2004).

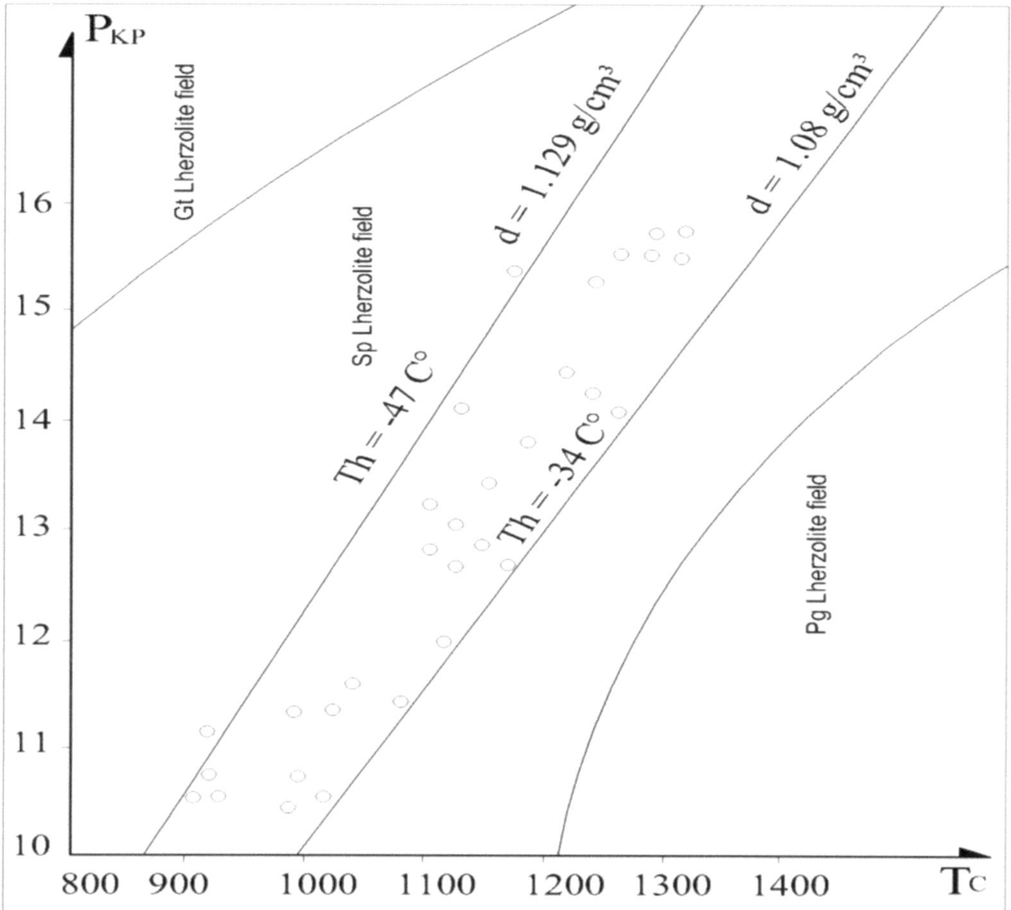


Fig. 6. Tthermparometric conditions (P, T), of Syrian mantle xenoliths (Bilal and Touret 2001, Bilal and Sheleh 2004).Th: homogenization temperature of CO2 fluid inclusions; d: density of CO2 liquid.

## 4. Tectono-seismicity characteristics

The tectono- seismicity characteristics are deducted from the satellite imagery, the field survey, the seismic observatories data, the odometric measurements, and the analyzing of ancient and recent earthquakes.

### 4.1 Tectonics

The Syria platform is constituted from several structural units, which have been formed at different periods since the Permo-Trias. The map of figure (7), shows these principles units: The Palmyrides in the center; the graben of Euphrates and the Djebel Abdelaziz in the East; the Afrin region and the Aleppo plateau in the North-West; and the Coastal chain at the west of the Levant fault ( Al Abdalla 2008).

The West of Syria is occupied by an important structure, locally named the Syrian rift, and worldly known the Levant fault, corresponding to the northern part of the Dead Sea fault zone (DSFZ), which forms the boundary between the African and Arabian plates. The tectonic activity along its northern part, the Yammouneh fault and northward continuation is still a subject of controversy. Between those who maintained that this segment is poorly active(Girdler 1990,Butler et al1997,1998,Butler and Spencer 1999),and others who concluded ,from the earthquakes study ,that this segment would then be active until it

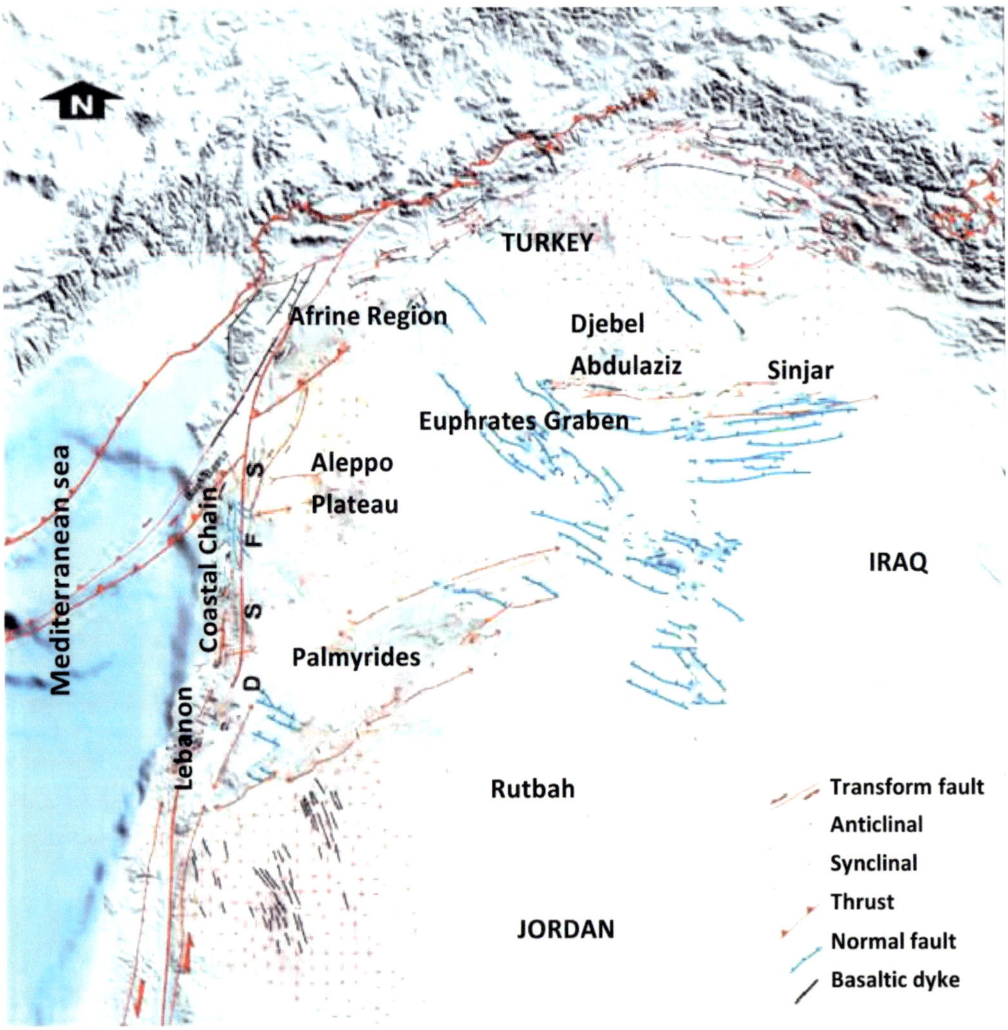


Fig. 7. Principles structural units of Syria, northwest of the Arabian plate (Al Abdalla 2008)

reaches the Africa-Arabia-Eurasia triple junction near Maras(Khair et al 2000,Meghraoui et al.2003), confirmed by recent field studies (Chorowicz et al.2005,Bilal 2009a,b).
Thus, the Syrian rift is an active seismic zone oriented North-South.It could be identified either by a satellite image or in the field (Fig.8). It cross the middle east through more 1200Km., from the Aqaba gulf in the South, to the Turkey in the North (Bilal and Touret 2001), successively through Lebanon, Syria and Turkey.

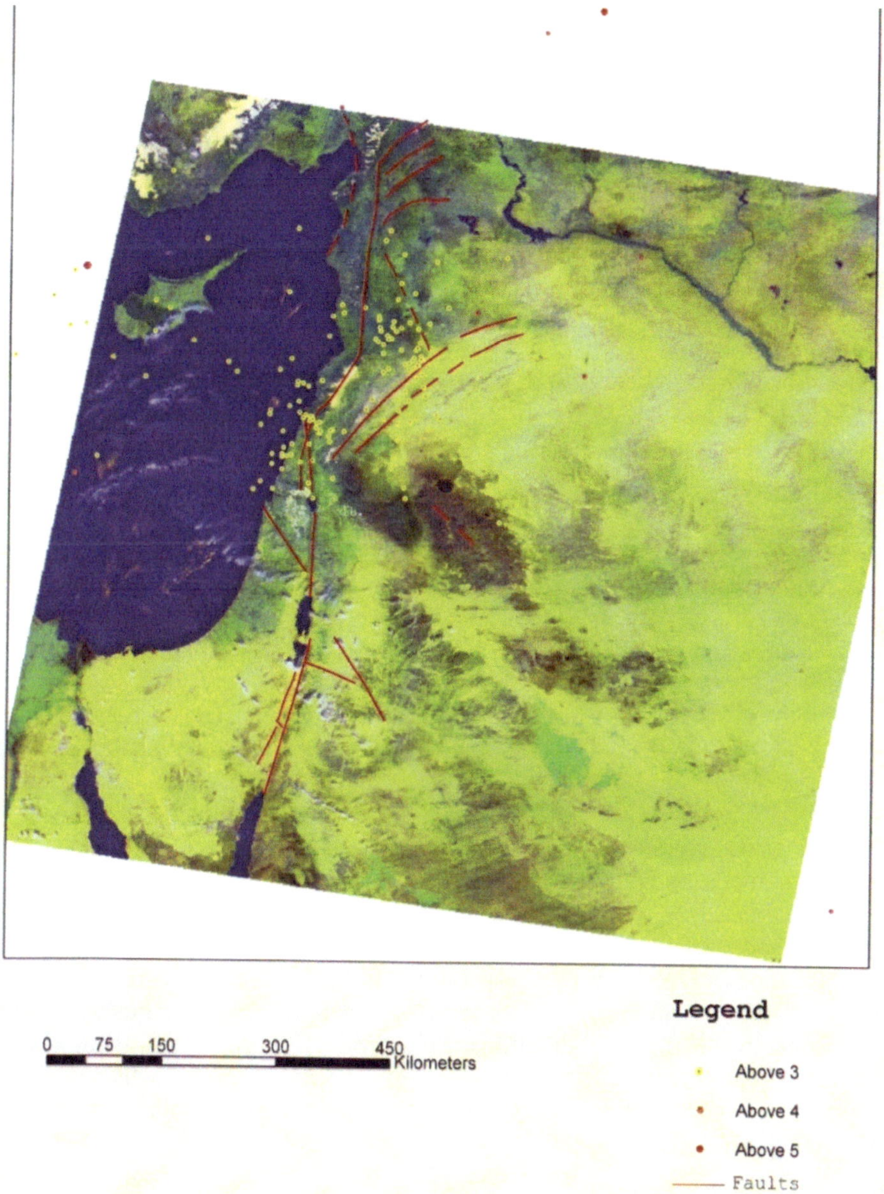


Fig. 8. Structure of the Syrian rift from the satellite image, Landsat 2005, and epicenters of the seisms in 2005 - 2006 after the Syrian seismic centre

In Syria, the Levant fault continues towards the North, between Damascus and Tartous, by the Yammouneh fault. This last structure marks an important transition to the Palmyra chain, making a sort of hinge between the northern and southern parts of the rift. The rift continues to the north through the El Ghab basin, and it disappears in the Taurus zone, in Turkey, in the Maras's triple junction point which relays Africa-Arabia-Eurasia. Many arguments, from structural analysis and field observation, in addition to satellite imagery data, and geomorphologic analysis, point to a recent, up to present day tectonic activity along this structure, e.g. mylonites and fine-grained shear zones, filling of pull-apart basins, deformed small active ravines and formation of scarps.

The Syrian rift corresponds to a transform fault, with lateral displacements decreasing from more than100 Km, to the South, to less than 30 Km, to the North, at the level of the Yammouneh fault (Walley 1988).The secondary faults and fractures deduced by the satellite images (Bilal1994 a,b) ,and detected in the field (Lovelock 1984,Sawaf et al. 1993),support this results, but more studies are needed to more explore others factors.

Many faults, as shown by the structural analysis and field observation, in addition to satellite imagery data and geomorphologic analysis, attest that have recently been or are still active along the Syrian rift. This is notably indicated by several phenomena: 1) the local transformation of basaltic rocks into mylonites and fine-grained shear zones. Carbonate basement rocks may also be deformed; they are more mylonitic because more easily fragmented than basalts. 2) The occurrence of pull-apart basins filled with quaternary sediments. 3) The deformation of small active ravines, with the formation of scarps (Chorowicz et al.2005, Bilal 2009a, b).

### 4.2 Seismology

Syria, the northern part of the Dead Sea Fault Zone (DSFZ), has a long record of active seismicity (Taher 1979, Al Tarazi 1999). Field observations, physical effects on ancient building structures and movement analysis show that tectonic is still active at present time (Khair et al.2000, Meghraoui et al2003, Chorowicz et al 2005).

Most major seisms in Syria occur in two regions: Either within or close to the rift zone, along a North - South direction, or SW - NE oriented, along the Damascus Palmyra mountain chain. This last domain does not contain any volcanic activity. Earthquakes in this region can only be caused by superficial deformation of the sedimentary cover.

Many of seisms take place within the litohsphere, in response to active fault displacement (Lay and Wallace1995,Yeats et al.1997) ,but others could also originate when the crust is subducted into the mantle, or along rifts in response to an ascending hot spot (plume) (Yeats et al.1997,Bilal and Sheleh 2004).This is precisely what may happen in Syria. The study of volcanic xenoliths has identified a hot spot under the Arabic plate (Stein and Hofman 1992, Sharkov et al.1993), starting during Cretaceous and ascending continuously until present time (Bilal and Touret 2002, Bilal and Sheleh 2004). This part aims to analyzing the seismicity distribution in the Syrian territory, using tectonic activity, laboratory measurements, and historic and recent earthquakes records. It will also attempt to compare these data with volcanic parameters.

#### 4.2.1 Seismic parameters

The movement rate-displacement along a given fault could be estimated in the region of Homs, where important basaltic eruptions took place 6 Ma ago (Sharkov et al.1994, Butler et al.1998). The overall displacement along the fault since the time of eruption, in other words the total length of pull-apart, could be estimated from the displacement of the famous "Krak des Chevaliers" in respect to the main Shine volcano (Fig.9). This gives 16 km, significantly less than other estimates (20 Km, Chorowicz et al 2005). This would give an average displacement rate of 2,7 - 3,3 mm/year, in line with estimates along the Wadi Araba fault, in the northern part of the Levant fault (4,6 +/- 2 mm/year, with a decreasing value of 2,3 mm/year for the last 12 ka , Le Beon 2008). In Syria, the displacement rate in the region of Homs corresponds to a maximum value. The movement rate, estimated by different methods, decreases significantly in the others regions. It is: 1-1.5 mm/yr in the region of Palmyra, less than 1 mm/yr in the Eastern and North Eastern parts of the country (Bilal 2009 b, Tab.1).

| Region | Movement rate / mm.Yr. | Method |
|---|---|---|
| El Ghab, the Syria rift | 2, 7-3, 3 | Basaltic displacement |
| Palmyra chain | 1, 5 | Monuments cracks |
| Eastern Syria, Russafeh | 1-1, 5 | Monuments cracks |
| Afrin region | 1 | Field measurements |
| Northern east Syria | < 1 | Field measurement |

Table 1. Movement rate in different regions estimated by different methods

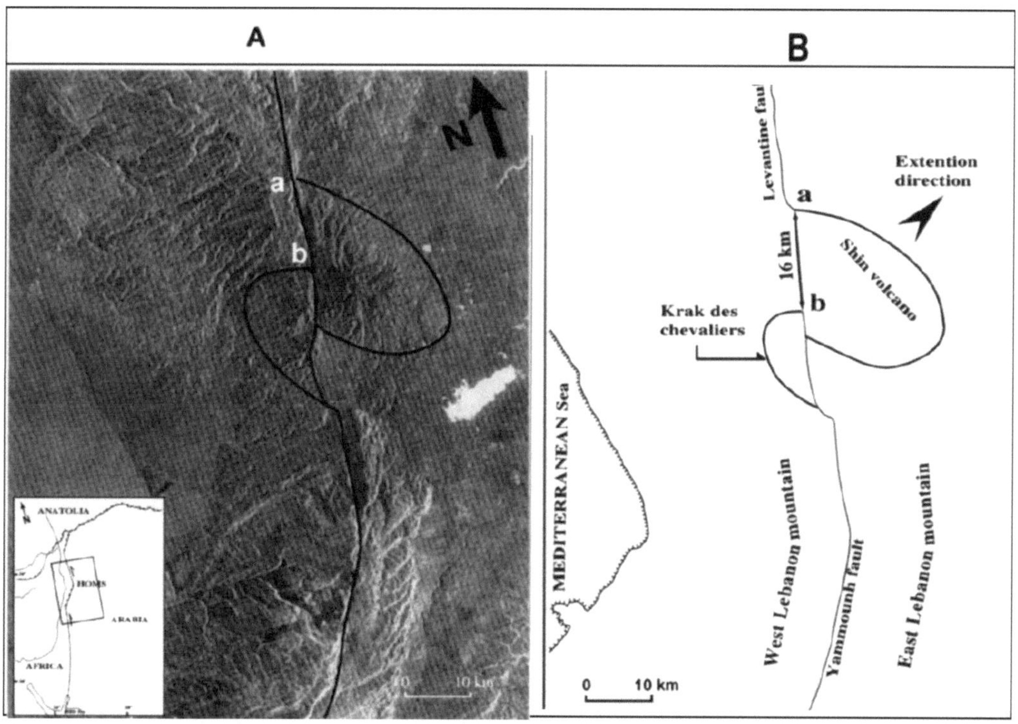


Fig. 9. Movement rate estimate, from field observation and satellite imagery analysis (Landsat TM 1995). ab: displacement, estimated at 16Km during 6 My, the age of the Homs basalt.

These results are confirmed by physical effects on the building structures in the region (field measurement on houses, measurement by Bilal and Ammar 2004, unpublished data). They can be taken as an average value for the movement rate repartition in the whole territory.
For an earthquake of a given intensity, defined by the value of the Magnitude (in Richter), the action on a builded structure, which result in the greatest number of casualties, depends from two sets of parameters: the characteristics of the structure itself and the nature of the ground on which it is built. Several equations are proposed to relate all these variables (Bojoroque and DeRoeck 2007, Ozkan 1998). The Syrian code (2004), used in this work is based on the following equations:

$$Z = V / IKCSW \quad (1)$$

$$C = 1/T \quad (2)$$

$$T \sim 0,1N \quad (3)$$

Where V is the horizontal shear force, I correspond to the type and geometry of the structure (bridges, tunnels, towers, dams, etc.), K the inelasticity coefficient of the structure, C the dynamic coefficient, linked to the nodes propagation period (T), and the number of stages (N), S a coefficient relative to the soil, and finally, W the total weight.
This seismic acceleration coefficient (Z) is a critical parameter. Estimated in cm /s2, it describes the reaction of an object -structure in a limited zone -to an earthquake of a given intensity. Its value changes from region to other after the upon seismic parameters, and so in the same region according to the estimated parameters values, from 0 to a variable value, reaching 1, 5 cm /s2 (Bojoroque and DeRoeck 2007).
For the dams Z is estimated, in Japan, at 0, 1- 0, 12 for weak earthquake zones, and 0, 15 for strong zones. It is taken between 0, 05 and 0, 20 in Turkey, and between 0, 03 and 0, 24 in India (Bilal 2009). A coefficient of 0, 1 indicates that a building is designed so that 0, 1 of its weight can be applied horizontally during an earthquake
In Syria Z ranges between 0 and 0,25, depending on the region.It is 0,25 in Al Ghab region, 0,12 in Palmyra, and 0,05 – 0,1 in Deir Zour, in the East.

#### 4.2.2 Seismicity-time analysis

The repetition of an earthquake (frequency return period of an earthquake in the same locality), namely the seismic cycle is controversial (King 2004, Maderiaga 2004). However if an earthquake is unique for a given locality, the distruction of earthquake activity with time is of major importance.
Therefore, the historical record of ancient earthquakes has been investigated. Only were used these verified by different sources (Taher 1979, Al Tarazi 1999). Available historical data covers a wide period with variable magnitude: Ancient time between 750 and 1800,with magnitude estimated at 7.5-6.5 (Meghraoui 2003), it becomes 6 - 5 between 1800 - 2000 (UNISCO 1983,Stiro1992,Sbeinati and Darawcheh 1992,Al Tarazi1999),less than 5 for the period of 1960 to 2000 and between 4,9-4 at present (USGS 1999). The results are represented by the histogram of figure (10).They show that the seismic intensity tends to decreases with time, in agreement with recent estimates on the movement rate (Le Beon 2008).

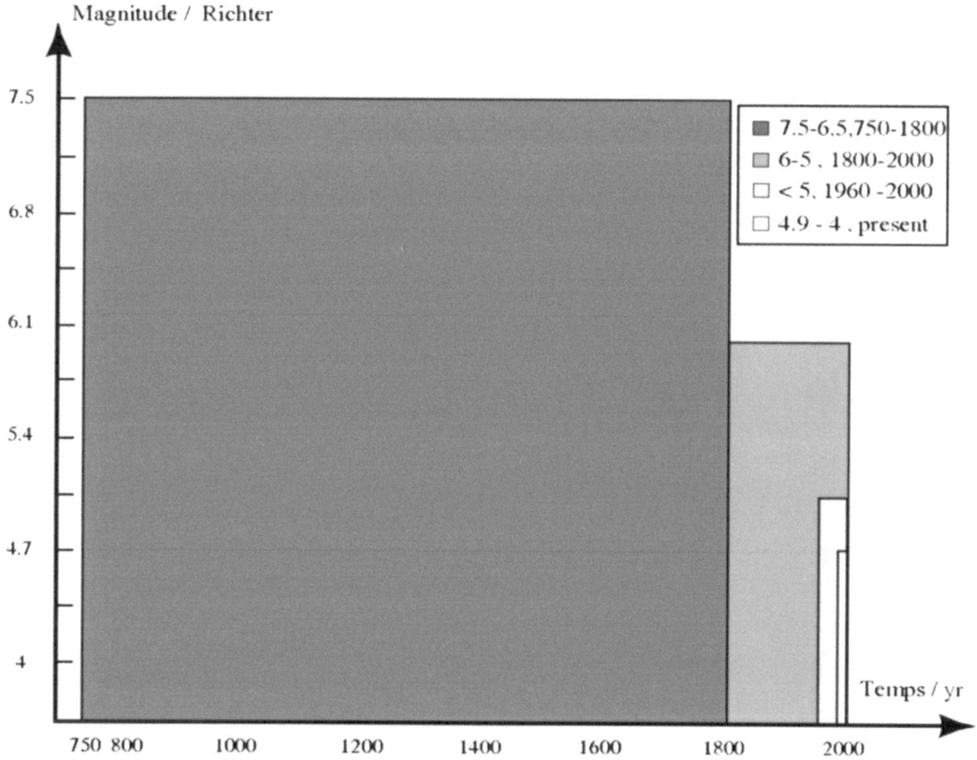


Fig. 10. Histogram of the seismicity evolution in time, after their magnitude and age.

For recent seisms, we have collected good data, especially from the physical effects on the structures and the station seismic net (Fig.11). They are distributed in the whole of the country and in the off shore.

#### 4.2.3 Seismic hazard zoning map of Syria and Arab plate

In order to evaluate the seismic hazards on the whole territory of Syria, and to examine their effect on building structures, odometric experiments on the more representative soil types occur in the country, have been performed in laboratory. Using scale of magnitude intensity of 7, 7.5, 8, and 8.5 respectively, the elasticity modulus (E = power/surface = Kilo Newton / $m^2$), and the Poisson ratio (P=executed displacement / original displacement = deformation %), have been determined. They allow the determination of the soil rigidity and the behavior of building during a seism (interaction soil-seism for a given building). The experiments have been measured on more than 80 litho- logical samples. Measurements were analyzed using the SAP Software (computer and structures Inc. USA, Bilal and Mahmoud 1997). Results are given in terms of relative unity response to seismic hazard, namely relative damage, the result of seism-soil effect on building, estimated from low to strong. It is maximal along the rift (moderate), and decreases gradually towards the East, East South and East North (low) in line with the rate movement, and seismic acceleration coefficient evolution (Fig.12B).

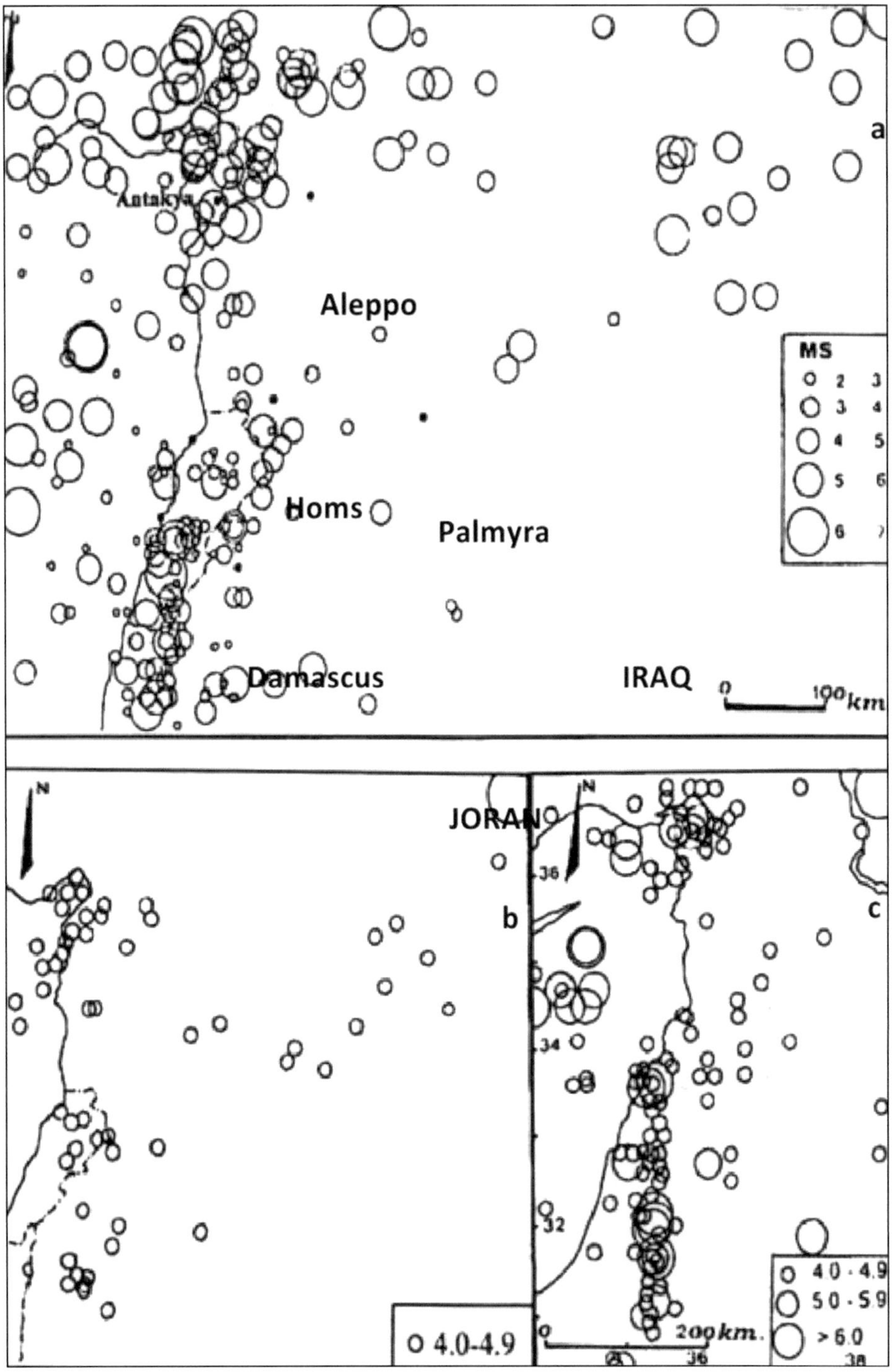


Fig. 11. Occurrence of the recent seismicity in Syria: a-Sbeinati and Darawcheh for the XX century (1992),b-USGS for the period of 1961-1983,(in Stiro 1992);c-USGS for the last years(Monitor,vol.5,1,1995).

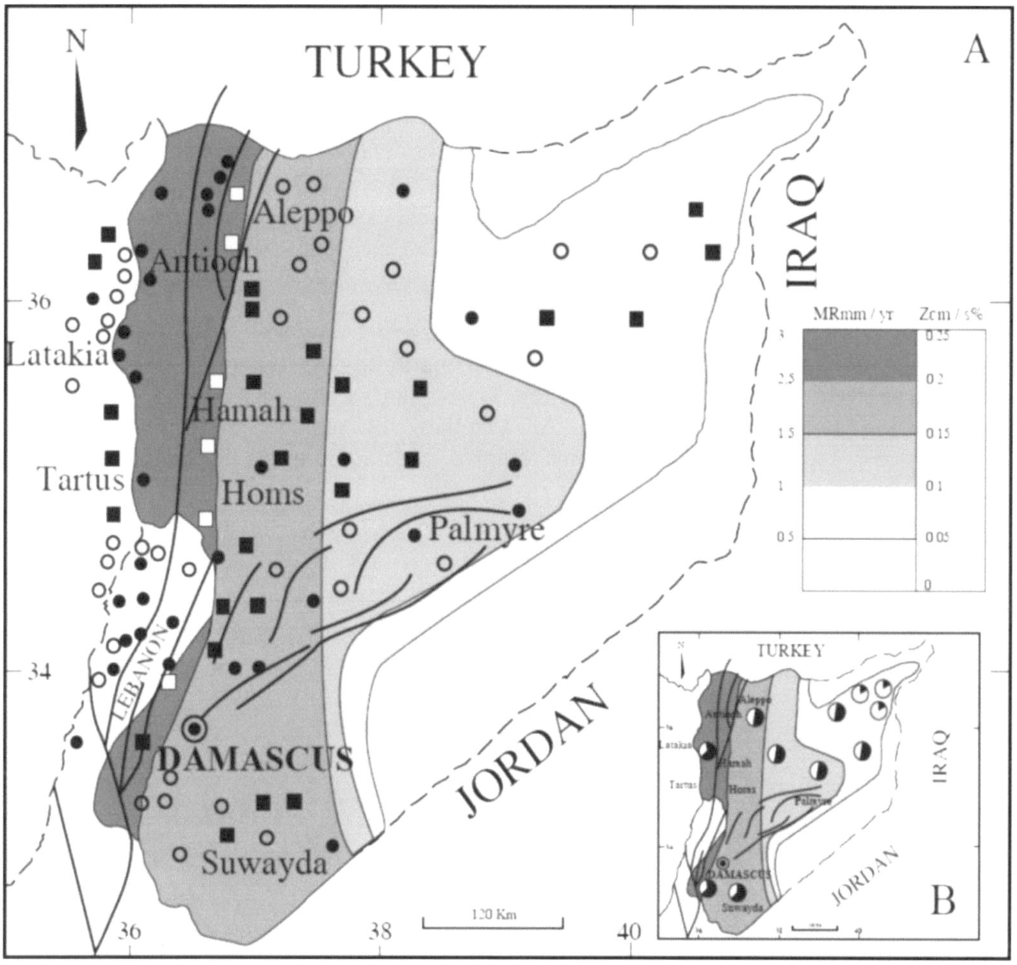


Fig. 12. Seismic map of Syria, showing the five seismic zones, and distribution of the earthquakes which their magnitude more than 4; white square: M> 6 ; black circle: 5< M < 6; black square: M< 5 ;white; circle: M=4,9 - 4(A), and odometric result where the black part indicate the relative damage(B).

According to the movement rate, estimated in the field between <1mm/yr and 2, 7 -3, 3 mm/yr, the Z value estimated at 0-0.25 cm/S2, the relative seismic intensity measured in laboratory (Bilal and Mahmoud 1997), the analysis of recent seismicity documented by the seismic network, and finally historical record, we have identified a number of seismic zones and corresponding seismic intensity (Tab.2).

These zones are represented on a seismic map of Syria (Fig.12A), established here for the first time. This map divides Syria into five zones, each of which corresponds to a given seismic intensity value. Zone one has the highest seismic intensity risk, with most potential damage for the constructions, while zone five has the lowest one.

On the same figure (12A) , the earthquake epicenters with a magnitude higher than 4 are projected. They occur in the whole territory but mostly in zones 1 and 2, associated with volcanism, or along the Damascus-Palmyra mountain chain. Only some earthquakes had a

| MR mm/yr. | Z cm/S2 | Seismic zone | Seismic intensity |
|---|---|---|---|
| 0 | 0 | 0 | none |
| <1 | 0.05 | 0-1 | very low |
| 1 | 01 | 1 | -2A |
| 1.5 | 0.15 | 2A | middle-low |
| 2 | 0.2 | 2B | middle |
| 2.5 | 0.25 | 2C | moderate |

Table 2. Calculated seismic parameters, corresponding seismic zones and their intensity.

magnitude higher than 7, 5- 6, 5, even if this remains controversial. Most recorded earthquakes in the twentieth century have a magnitude less than 6. Out of hundred earthquakes in the region and off shore, 25% have a magnitude between 6-5, and most of them have a magnitude less than 5. All together, the whole Syria shows moderate seismicity, compared to the north, Taurus-Zagros fault zone, or to the south, under the Indian Ocean expansion.

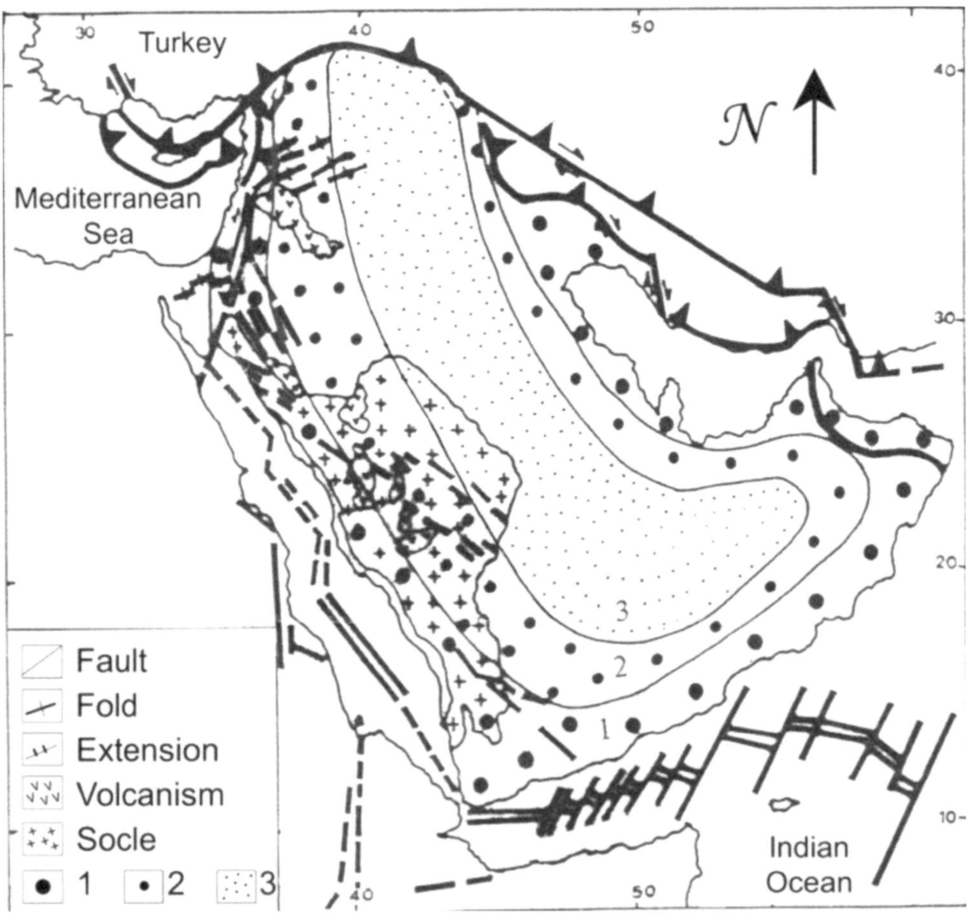


Fig. 13. An extrapolated seismic zoning map of the Arab plate . Zone 1, the highest seismic intensity zon;zone 2,the intermediary one and zone 3,the lowest one.

Using data obtained by deferments sources ( UNESCO 1983,Barrier et al 2004,Al Damegh et al 2005,Le Beon 2008,Bilal 2009b), and the seismo-tectonic parameters, an extrapolated seismic zoning map for the Arab plate is established(Fig.13).It distinguishes between three seismic zones: zone 1 ,the highest seismic zone intensity with major damage risk; zone 3,zone of low seismic intensity with lowest potential risk ,and zone 3,the intermediary zone with intermediary potential risk. These results need to more verification by a qualified team.

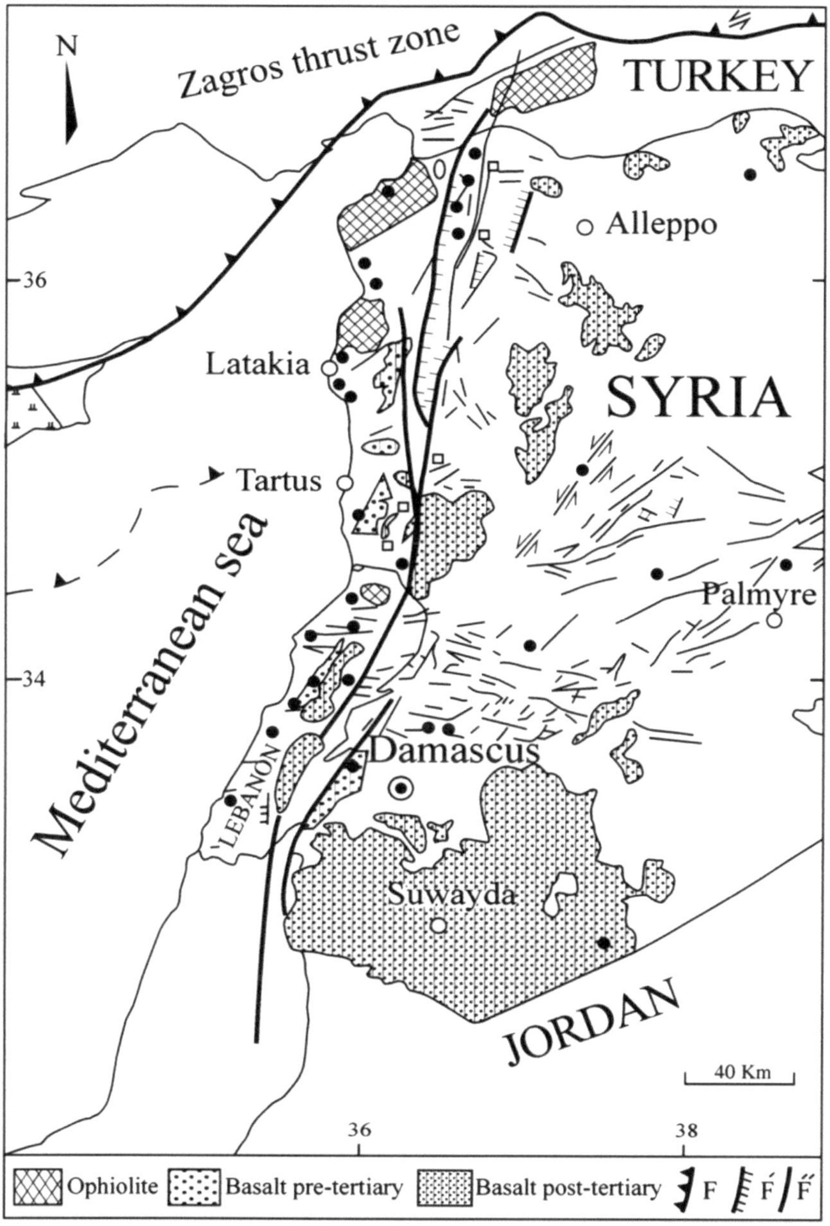


Fig. 14. Regional tectonic map of Syria showing the rift structure, volcanism and the distribution of earthquakes epicenters (in black circles), with the highest magnitude (>5 -7, 5).

## 5. Conclusion

The Syrian rift is a world structure, and constitute the north part of the Dead Sea Fault Zone (DSFZ).Structural analysis using variable techniques attest that many faults have recently or are still active to occuresent along the Syrian rift. This is notably indicated by two phenomena: volcanism and seismicity.
The composition of the underlying lithospheric mantle points to a complex history involving polybaric partial melting at various degrees, starting in the garnet - and proceeding in the spinel stability field. Some clinopyroxenes at least record mantle metasomatism, caused by ephemeral carbonate magmas or percolating basalting melts issued from a mantle plume under the Arabic plate.
Most major seisms in Syria occur in two zones (Fig.14): with the rift zone, in a North – South direction, but not exactly along the fractures. Most epicenters occur westward along the coast or in the sea. The other zone is SW – NE oriented, along the Damascus Palmyra mountain chain. It does not seem to be related to any volcanic activity, but corresponds namely to superficial deformation of the sedimentary cover.
For volcanic-related seismicity, Petrological data from volcanic xenoliths have identified the existence of a hot spot (plume), under the Syrian rift. In the earliest period of volcanic activity (Cretaceous), this plume started at the level of mantel garnet peridotite, leading to a marked explosive volcanism. It may be hypothesized that this type of volcanism did correspond to major seismicity. In more recent time, the plume head tends to rise, while at the same time migrating towards the West. This was accompanied by a more effusive type of volcanism, associated to the moderate seismicity, presently shown. The last eruption (10.000 y) occurred in the large volcanic massif at the South (Djebel Al Arab). With one exception, no major seism relates to this last eruption. This recent massif, by far the larger in Syria, seems to distantiate from the rift zone, at the difference, notably, of older Cretaceous volcanism.
At the scale of the human observation, the seismicity does not seem to be directly related to present-day volcanic activity.Either reminiscence of ancient volcanism, or consequence of superficial deformation. Both phenomena tend to fade out with time, in line with the decrease of major seismic intensity which has occurred during the last millennium.

## 6. References

Adiyaman, O. and Chorowicz, J. (2002)-Late Cenozoic tectonics and volcanism in the northwestern corner of the Arabian plate:a consequence of the strik-slip Dead Sea Fault Zone and lateral escape of Anatolia.*Journal of Volcanologyand Geothermal Research*, 117, 327-345.

Al Abdalla, A. (2008)- *Evolution tectonique de la plat-form Arabe* en *Syrie depuis le Mesozoique*. These Doctorat, Universite Pierre et Marie Curie.302p.

Al Damegh, K; Sandvol, and Barazangi, M. (2005) - Crustal structure of the Arabian Plate: New constraints from the analysis of teleseismic receiver functions. *Earth and Planeta Sciences Letters*, 231, 9, p. 177-196. Contrib. Mineral.Petrol. 50: 79-92.

Al Mishwat, AT. and Nasir, SJ. (2004)- Composition of the lower crust of the Arabian plate: a xenolith perspective. *Lithos*, 72: 45-48.;

Al-Tarazi, E. (1999)- Regional seismic hazard study for the eastern Mediterranean (Trans-Jordan, Levant and Antakia) and Sinai Region. *J. of African Earth Sciences*, vol. n°3, 743-750.

Agard, P; Monie, P; Gerber, W; Omrani, J; Molinaro, M; Meyer, B; Labrousse, L; Vrielynk, B; Jolivet, L; and Yamato, P. (2006)-Transient, synobduction exhumation of Zagros blueschists inferred from P-T, deformation, time, and kinematic constraints: Implications for newtethyan wedge dynamics.*J. Geophysics. Res.* 111, B11401, doi: 10.1029/2005JB004103.

Baker, MA; Menzies, MA; Thirlwall, MF; McPherson, CG. (1997)- Petrogenesis of Quaternary intraplate volcanism, Sana'a, Yemen; implications for plume-lithosphere interaction and polybaric melt hybridization. *Journal of Petrology*, 38, 1359-1390.

Barrier, E;Chamot-Rooke,; and Giordano, G. (2004)- *Carte géodynamique de la Méditerranée.* Commission de la carte géologique du monde.

Bertrand, P. and Mercier J.C.C. (1986)-The mutual solubility of ortho and clinopyroxene : towards an absolute geothermometer for the natural system? *Earth.. planet.. Sci. Lett.* 76, 109-122.

Bilal, A. (1994a)-Remote sensing new tectonic data of the Arabian plate.*IGARSS 94*, California, USA, p58.

Bilal, A. (1994b)- Une nouvelle pensee sur l`origine du complexe ophiolitique, Syria.*15e Renu. Ann. Sci. Terre.* Nancy, p. 105. Livre en depot. Soc. Geol. France.

Bilal, A. (2009a) - Seismicity and volcanism in the rifted zone of western Syria. *C. R. Geosciences*, 341, 299-305.

Bilal, A. (2009b)- Tectono- Seismicity and petrological study of the Syrian rift. *Tishreen University Journal for Research Scientific studies, Basic Sciences*, Vol.31, no (1), pp 127-145.

Bilal, A; and, Sheleh F. (1988) - Pétrologie des enclaves ultrabasiques dans le basalte du sud de la Syrie. *J Univ. Damas*, 109: 49-81.

Bilal, A.and Sheleh, F. (2004)- Un "point chaud" sous le système du rift Syrien : données pétrologiques complémentaires sur les enclaves du volcanisme récent. *Comptes Rendus Géosciences*, 366, 197-204.

Bilal, A. and Mahmoud, M. (1997)-Soil-Structure interaction effect during earthquakes in Syria. *Inter. Post -SMIRT conference on seismic isolation*. Taormina- Italy, August 25-27th.

Bilal, A. and Touret, J.L. (2001)- Les enclaves du volcanisme récent du rift Syrien. *Bull. Soc. Géol. Fr.* Tom 17 2, nO1, 1-14.

Bilal, A. and Ammar, O. (2004)- Field measurements on houses and structures.*Unpublished data*, Syria.

Bilal, A; Jambon, A; Boudouma, O; Badia, D; and Sautter, V. (2011)- Mantel-derived sapphirine in the webstrite xenoliths ofTel Thenoun-Syria. (submitted).

Bohanon, RG; Naeser, CW; Schmidt, LD; and Zimmermann, R.A. (1989)- The timing of uplift, volcanism, and rifting peripheral to the Red Sea; a case for passive rifting ? *Journal of Geophysical Research*, B 94, 1683-1701

Bojoroque, J. and DeRoeck, G. (2007)- Determination of the critical seismic acceleration coefficient in slope stability analysis using finite element methods. (see: *jaime.bojorque @bwk.kuleuven.be*

Bosworth, W; Huchon, P; and McClay, K. (2005)-The Red Sea and Gulf of Aden Basin.*Journal of Africa Tectonophysics*, 209.p115-137.

Brey, GP; and Köhler, T. (1990)- Geothermometry in four phase thermometers and practical assessment of existing thermometers. *J Petrol.* lherzolites II. New 31: 1353-1378.

Butler, L.W; and Spencer, S. (1999)-Landscap evolution and the preservation of tectonic landform along the northern Yammouneh Fault, Lebanon. In: *Smith B.J., Whally W.B. and Warek P.A. (eds), uplift,* Erosion *and Stability: Perspectives on long-term Landscape development.* Geological Society, London, Special Publications, 162, 143-156.

Butler, L.W; Spencer, S; and Griffiths, H.M. (1997)-Transcurrent fault activity on the Dead Sea Transform in Lebanon and its implications for plate tectonics and seismic hazard. *Journal of the Geological Society*, London, 154, 757-760.

Butler, L.W. S; Spenser, S. and Grifffiths, H.M. (1998)- The structural response to evolving plate kinematics during transperssion evolution of the Lebanese restraining bend of the Dead Sea Transform.In: *Continental Transpressional and Transtensionbal Tectonics*. Geological Society, London, Special publications, 135, 81-106.

Camp, V.W. and Roobol, M.J. (1989)- The Arabian continental alkali basalt province: part I. Evolution of Harrat Rahat, Kingdom of Saudi Arabia. *Geological Society of America Bulletin*, 101, 71-95.

Capan, U.Z; Vidal, P. and Cantagrel, J.M. (1987)-K-Ar-Nd-Sr and Pd isotopic study of Quaternary volcanism in :Karasu valley(Hatay).N. *end of Dead Sea Rift zone in SE Turkey*. Yerbilimleri Bulletin of Earth. Sciences, Hacettepe University, Ankara, 14, 165-178.

Cetin, H; Guneyli, H. and Meyer, L. (2003)- Paleoseismology of the Palu-Lake Hazarsegment of the East Anatolia Fault Zone, Turkey.*Tectonophysics*, 374, p.163-197.

Chorowicz, J; DHONT, D; Ammar, O; Rukieh, M.and Bilal, A. (2005)- Tectonics of the Pliocene Homs Basalts (Syria) and implications for the Dead Sea Fault Zone Activity. *Journal of the geological Society*, London, Vol. 162, 259-271.

Coskun, B. (2004)-Arabian-Anatolian plate movements and related trends in southeast turkey`s Oilfields. *Energy Sources*, 26, 978-1003.

Dubertret, L. (1933)-Les grandes nappes basaltiques syriennes :âge et relation avec la tectonique.*Bulletin de la Société Géologique de France*, 3, 178-180.

Dubertret, L. (1962)-Carte géologique du Liban, Syrie et bordure des pays voisins, 1 /1000000.*Muséum d`Histoire Naturelle*, Paris.

Frezzotti, M.L; Andersen, T; Neuman, E.R. and Simonsen, S.L. (2002)- Carbonatite melt CO2 fluid inclusions in mantel xenoliths from Tenerif, Canary Islands: a story of trapping, immiscibility and fluid-rock interaction in the upper mantel.*Lithos*, 64, 77-96.

Giannerini, G; Campredon, R; Ferraud, G. and Abo Zakhem, B. (1998)- Deformation intraplaques et volcanisme associe :exemple de la plaque arabique au Cenozoique. *Bull. Soc.Geol.*, 6, 683-693.

Girdler, RW. (1990)- The Dead Sea transform fault system.*Tectonophysics*, 180, 1-13.

Gregoire M. Moine B.N.O`ReillyS.Y.Cottin j.y. AND Giret A. (2000)-Trace element residence and partitioning in mantle xenoliths metasomatized by alkaline and carbonate-rich melt (Kergulen Islands, India Ocean).*Journal of petrology*, 41, 477-509.

Heimann, A; Rojay B. and Toprak, V. (1998)-Neotectonics characteristics of Karasu fault zone, northern continuation of Dead Sea Transform in Anatolia (Turkey).In: *Third international Turkish Geological Symposium*, Ankara, 99.

Ilani, S. Harlavin, Y. and Trawneh, K et al. (2001)-New K-Ar ages of basalts from the Harrat Ash Shaam volcanic field in Jordan: Implications for the span and duration of the upper-mantle upwelling beneath the western Arabian plate. *Geological Society of America*, 89, p. 1025-1036.

Ismail, M; Delpech, G; Cottin, J-Y; Grégoire, M; Moine, BN. and Bilal, A. (2008)- Petrological and geochemical constraints on the composition of the lithospheric mantle beneath the Syrian rift, northern part of the arabian polate. In Coltorti M and Grégoire M (eds) Metasomatism in Oceanic and Continental Lithospheric Mantle. *Geol Soc London*, Spec. pub. 293: 223-251.

Juteau, T. (1974)- Les ophiolites de la nappe d'Antalya, Turquie, Thèse d'Etat, Univ Nancy 420p.

King, G. (2004)- Les séismes ne se répètent pas. *La recherché*, n 380, 44: 45-60

Khair, K; Karakasis, G.F. and Papadimetriou, E.E. (2000)-Seismic zonation of the Dead Sea Transform Fault area.*Annali di Geophisica*, 43, (1), 61-79.

Kohler, T.P. and Brey, GiP. (1990)-Calcium exchange between olivine and clinopyroxene calibrated as a geothermometer for natural peridotites from 2 to 60 K. bar with applications. *Geochim. Cosmochim. Acta.*, 54, 2375-2388.

Laws, E. and Wilson, M. (1997) Tectonics and magmatism associated with mesozoic passive continental margin development in the Middle-East. *J. Geol. Soc.London*, 154: 459-464.

Lay, T. and Wallace, J.C. (1995)- *Modern Global Seismology*. Academic Press, INC. 521p.

Le Bas, M.J; Le Maitre R.W; Streckeisen, A. and Zanettin, B.A. (1986)-Chemical classification of volcanic rocks based on the total alkali-silica diagram.*Journal of petrology*, 27, 745-750.

Le Béon, M. (2008)- Cinématique d'un segment de faille decrochante à différentes échelles de temps:la faille de Wadi Araba, segment sud de la faille transformante de Levante. *Thèse Doctorat*, Université Paris VI.

Lovelock, PER. (1984) - A review of the tectonics of the northern Middle East region. Geol Mag 121: 577-587.

Maderiaga, R. (2004)- Chaque seisme est unique. *La recherché*, n 275.

McLusky, S; Balassanian, S. and Barka, A. et al. (2000)-Global positining system constraints on plate kinematics and dynamics in the eastern Mediterranean and Caucasus.*Journal of Geophysical Research*, 105, 5695-5719.

Meghraui, M; Gomez, F; Sbeinati, R; Woerd, J.V.D; Mouty, M; AlDarkal, A.N; Radwan, Y; LAYYONS, I; AlNajjjar, H; Darawcheh, R; Hijazi, F; AlGhazzi, R.and Barazangi, M. (2003)-Evidence for 830 years of seismic quiescence from palaeoseismology, archaeology seismology, and historical seismicity along the Dead Sea fault in Syria. *Earth and Planetary Science Letters*, 210, 35-52.

Molinaro, M; Leturmy, P; Guezou, J.C; Frizon de Lamotte, D. and Eshraghi, S.A. (2005)-The structure and Kinematic of the southeastern Zagros fold-thrust belt, Iran: from thin-skinned to thick-skinned tectonics, *Tectonics*, 24, P.TC3007.

Mor, D. (1993)-A time table for the Levant Volcanic Province, according to K-Ar dating in the Golan heights. *J. Afr. Earth Sci.*, 16, p.223-234.

Mouty, M; Delaloye M; Fontignie D; Piskin, O. and Wagner, J.J. (1992)-The volcanic activity in Syria and Lebanon between Jurassic and Actual. *Schweizerische Mineralogische und Petrologische Mitteilungen*, 72, 91-105.

Ozkan, M.Y. (1998)-A review of consideration on seismic safety of embankments and earth and rock-fill dams. *Soil Dynamic and Earthquake Engineering*, 17, 439-458. natol.

Polat, A; Kerrich, R. and Casey, J.F. (1997)-Geochemistry of Quaternary basalts erupted along the East Anatolian and Dead Sea Fault Zone of southern Turkey: Implications for mantle source.*Lithos*, 40, 55-68.

Parrot, J-P. (1977)- Assemblage ophiolitique du Baer-Bassit et termes effusifs du volcano-sédimentaire pétrologique d'un fragment de la croûte océanique téthysienne chariée sur la plateforme syriennne. *O.R.S.T.O.M* Paris. 72.

Ponikarov, V.P. (1967)-The geological map of Syria, Scale 1/50000, Explanatory notes. Syrian Arab Republic, *Ministry of Industry*, Damascus.

Robertson, A.H.F; Clift, P.D; Degman, P. and Jones, J. (1991)- Paleoocenography of the Eastern Mediterranean *Neotehtys. In: Paleogeography and Paleooceanography of Tethys*, Paleogeography, Paleoclimatology, *Paleoecology*, 87, 289-343.

Roedder, E. (1984) - Reviews in mineralogy, vol.12, fluid inclusions. *Mineralogical Society of America*. 645p.

Rojay, B; Heimann, A. and Toprak, V. (2001)-Neotectonic and volcanic charactiristics of the Karasu fault zone (Anatolia, Turkey):The transition between the Dead Sea Transform and the East Anatolian fault zone.*Geodynamica Acta*, 14, 197-212.

Sawaf, T; Al Saad, D; Gebran, A; Barazangi, M; Best, JA. and Chaimove, T. (1993)- Structure and stratigraphy of eastern Syria across the Euphrate depression. *Tectonophysics*, 220: 267-281.

Sbeinati, M. and Darawcheh, R. (1992)- seismological bulletin for earthquakes in and around Syria.*Report international, SAES*, Damascus.

Sharkov, YE.V; Lazko, YE.YE. and Hanna, S. (1993)- Plutonic xenoliths from the Nabi Matta explosive center, Northwest Syria. *Geochemistry international*, 30, (4). 23–24.

Sharkov, YE.V; Shernyshev, I.V; Devyatkine, V. and al. (1994)- Geochronology of Late Cenozoic basalts in Western Syria.*Petrology*, 2(4), 439-448.

Sharkov, YE.V; Chernyshev, I.V. and Devyatken E.V. et al. (1998)-New data on the geochronology of upper Cenozoic plateau basalts from the northeastern periphery of the Red Sea rift area(Northern Syria). *Doklady Earth Sciences*, 358, (1), 19-22.

Shaw, J.E. Baker, J.A. R; Kent, A.J. R; Ibrahim, K.M. and Menzies, M.A. (2007)- The geochemistry of the Arabian lithospheric mantle -a source for intraplate volcanism? *Journal of petrology*, Vol.48, no 8, 1495-1512.

Sheleh, F. (2001)- Étude des enclaves mantéliques associées au rift syrien; Composition et évolution du manteau supérieur en Syrie-Implications géodynamiques et régionales. *Thèse de doctorat* (en Arabe), Université de Damas, 180p.

Sosson, M; Rolland, Y; Corsini, M; Danelian, T; Stephan, J-F; Avagyan, A; Melkonian, R;Jrbashyan, R; Melikian, L. and Galoin, G. (2005)-Tectonic evolution of the Lesser Caucasus (Armenia) revisited in the light of new structural and stratigraphical results.European Geosciences Union. *Geophysical Research Abstracts*, V 7, 06224.

Stein, M. and Hofman, A. (1992)- Fossil plume head beneath the Arabian lithosphere?, *Earth Planet. Sci. Let*, n 114, 193– 209.

Stein, M; Garfunkel, Z.and Jagoutz, E. (1993) Chronometry of peridotite, pyroxenite xenoliths: implications for the thermal evolution of the arabian plate. *Geochim Cosmochim Acta*, 57: 1325-1337.

Stiro, S. (1992)- Epicenters of earthquakes from 1961-1983, after *USGS.Workshop*, Damascus, 32-36

SYRIAN ENGENEERING SYNDICAT, (2004)- *The Syrian seismicity code*. Damascus, Syria.

Taher, M.A. (1979)- Documents historiques des tremblements de terre en Syrie depuis l'Islam jusqu'à XII siècle "hygerique".*Thèse Université*, Paris 1, 300 P.

UNESCO (1983)- Assessement and mitigation of earthquakes risk in the Arab region. *UNESCO/ AFESD/ IDP*.

U.S. Geological Survey (1999)- Special Report:The Hector Mine Earthquake 10/16/99(1999), (see *http//Www.Socal.W.Usgs.Gov/Hector/Report/Html* )

Walley, C.D. (1998)-Some outstanding issues in the geology of Lebanon and their importance in the tectonic evolution of the Levantine region.*Tectonophysics*, 298, 37-62.

Wells, P.R.A. (1977)-Pyroxene thermometry in simple and complex systems. Contr.Mineral.Petrol., 62, 129-139.

Yeats, R.S; Sieh, K. and Allen, C.R. (1997)- *The geology of earthquakes*. Oxford Univ. Press.

Yurtmen, S; Guillou, H; Westaway, R; Rowbotham, G. and Tatar, O. (2002)-Rate of strik-slip motion on the Amanos Fault(Karasu valley, southern Turkey) constrained by Kr-Ar dating and geochemical analysis of Quaternary basalts.*Tectonophysics*, 344, 207-246.

# 9

# Water-Rock Interaction Mechanisms and Ageing Processes in Chalk

Claudia Sorgi[1] and Vincenzo De Gennaro[2]
[1]*INERIS, Verneuil-en-Halatte, Now at Schlumberger, EPRC,*
[2]*Ecole des Ponts ParisTech – CERMES, Now at Schlumberger, EPRC,*
*France*

## 1. Introduction

The microstructures of geomaterials and their evolution under the effects of applied loading and/or environmental conditions can affect the integrity of the solid skeletons and eventually change the mechanical behaviours of the materials at the macroscopic scale. Analyses of geomaterial microstructure and its 'ageing' are therefore critical to the understanding of their mechanical behaviour and performance in engineering environments.

This process of progressive ageing of the microstructure is mainly related to the interaction between the solid skeleton and the fluids that partially or completely saturate the porous network. Water/rock interaction mechanisms and ageing processes in geomaterials are often slow, leading to structural and textural changes that are generally imperceptible to the naked eye but which can significantly affect matrix integrity and cause sudden collapse. This is a particularly well-known phenomenon in shallow abandoned chalk mines, such as those found in France, where the chalk may remain stable over many tens of years but then suddenly break down, leading to collapse of the openings and the ground above them and subsidence at a macroscopic scale (Sorgi, C. & Watelet, J., 2007).

In this chapter we present results from a research program (led by Institut National de l'Environnement Industriel et des Risques (INERIS), France, in collaboration with Ecole des Ponts ParisTech - CERMES, France) that was carried out to evaluate the mechanical behaviour of chalk in the shallow underground mine of Estreux, located in northern France.

The study was conducted at three different scales:

i. site scale (stability analysis)
ii. laboratory scale (standard core testing)
iii. microscopic scale (electron scanning environmental microscope (ESEM) observations and micro-testing)

In situ characteristics of the Estreux mine are first described, including mine geometry, excavation method, overburden lithology, pillar monitoring, and in situ measurements.

Then, petrophysical properties, microstructural characterisation, retention properties, and mechanical behaviour of the chalk investigated at laboratory scale (oedometric and triaxial tests) are presented.

A microstructural analysis of the Estreux chalk is also presented. This part of the study was conducted using an ESEM, a recent technology that allows the observation of

microstructural changes in geomaterials in their natural state, under controlled conditions of temperature and pressure.
Some other aspects of this technology are discussed, along with suggestions for potential development of this tool for further geomechanical applications and analyses of the Estreux chalk.

## 2. Site scale: Stability analysis

The study was carried out on Estreux abandoned underground chalk mine in Northern France, which is located 10 km east of the city of Valenciennes. The Estreux chalk formation is of the Late Cretaceous geological period (89 to 94 Ma years ago).
The Estreux mine, which was principally dug for blocks for dimension stone (for buildings and masonry) at the end of the 18th century, covers an area of 10 hectares. The chalk was excavated at a depth of 20 metres using the pillar-and-stall method, where networks of galleries were dug from access shafts, leaving large pillars in place to support the weight of the roof. This prevents the openings caving in, at least during excavation. The width of the galleries in the Estreux mine vary from 2 to 3 metres, and the pillars measure on average between 1.5 and 4.5 metres square. The excavation ratio (relation between the area exploited and the total area) is around 78%.

### 2.1 Lithology

The lithostratigraphic section of the Estreux mine overburden comprises, from top to bottom:
- a few tens of centimetres of soil
- one metre of clay
- around eleven metres of tuffeau
- three metres of white chalk
- two metres of grey glauconitic clay
- impermeable argillaceous marls

In 1980, Raffoux and Ervel conducted geomechanics laboratory tests on over one hundred core samples from this mine. The mechanical properties obtained are shown in Table 1. From these characteristics and the mining geometry, they estimated a stability factor close to 1 in several zones of the mine. As the mine is not stable in the long term, site monitoring was deemed necessary, and four convergence meters were subsequently installed in a weak zone of the mine.

| State | Elastic Limit, Mpa | Breaking limit, Mpa | Young's Modulus, GPa | Poisson's Ratio |
|---|---|---|---|---|
| Dry | 9 | 10.3 | 2.83 | 0.3 |
| Saturated | 4.6 | 5.3 | 1.61 | 0.3 |

Table 1. Mechanical characteristics of the Estreux chalk (Raffoux & Ervel, 1980)

Since August 2003, new instrumentation has been installed (Fig. 1) to analyse geomechanics phenomena of the rock mass over time. The galleries were monitored and a pillar of 1.4 m square and 1.8 m high was instrumented. The instrumentation comprises:
- an ambient hygrometer
- a water-level sensor

- a convergence meter between the wall and the roof of the mine
- three pore-pressure sensors together with rock temperature sensors, fixed at different points within the pillar (15 cm, 35 cm, and 70 cm from the surface)
- an extension meter measuring lateral displacement from the pillar surface at two points within the pillar

Data were recorded from February 2004 to February 2008, and the following were observed:

- ambient hygrometry was varying between 85% and 100%
- there was (very slight) seasonal variation in the water level in the galleries, with an overall decreasing trend (Fig. 2)
- seasonal variation in pore pressure and rock temperature within the instrumented pillar was occurring (Figs. 3 and 4), indicating cycles of alternating saturation/desaturation
- pore pressure and rock temperature were generally decreasing from the surface to the core of the pillar (Figs. 3 and 4 respectively)
- during each rise in water level there was increasing pore pressure within the pillar (Fig. 3) and lateral extension (swelling) of the pillar (Fig. 5)
- convergence of the wall and roof over time (Fig. 6)

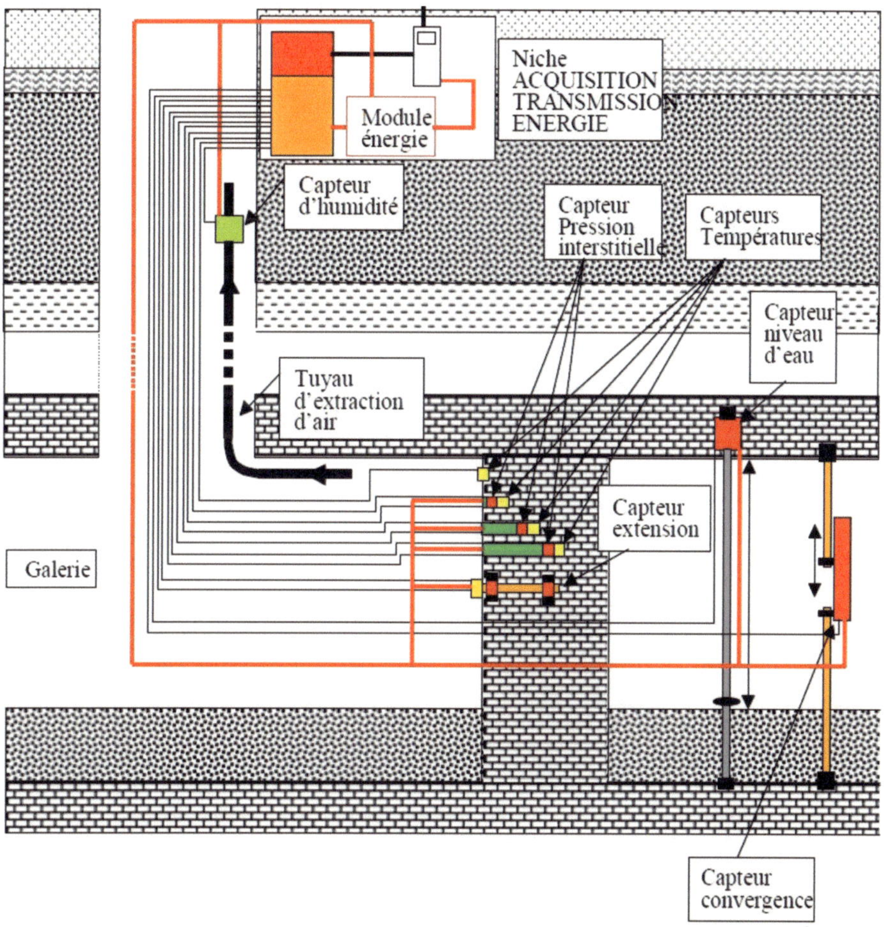


Fig. 1. Diagram of the full instrumentation employed in the Estreux mine

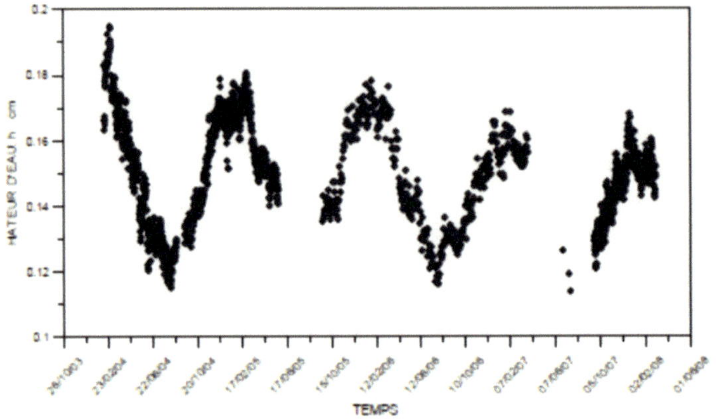


Fig. 2. Evolution of the water level over time

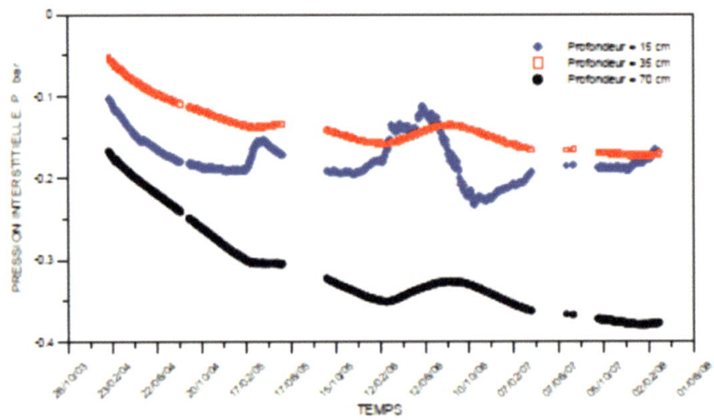


Fig. 3. Evolution of pore pressure at different points within the pillar

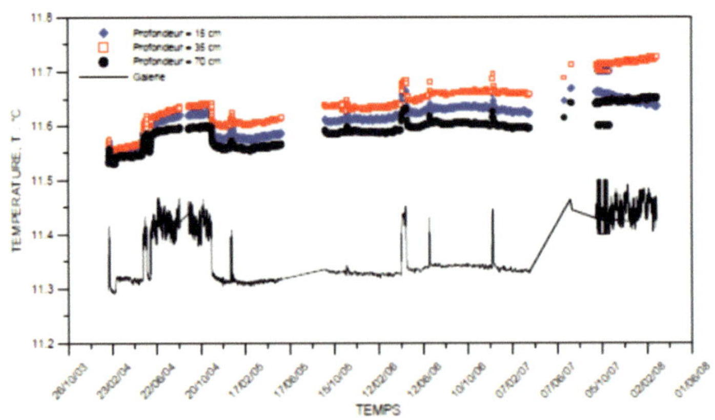


Fig. 4. Changes in rock temperature over time at different points within the pillar and ambient temperature

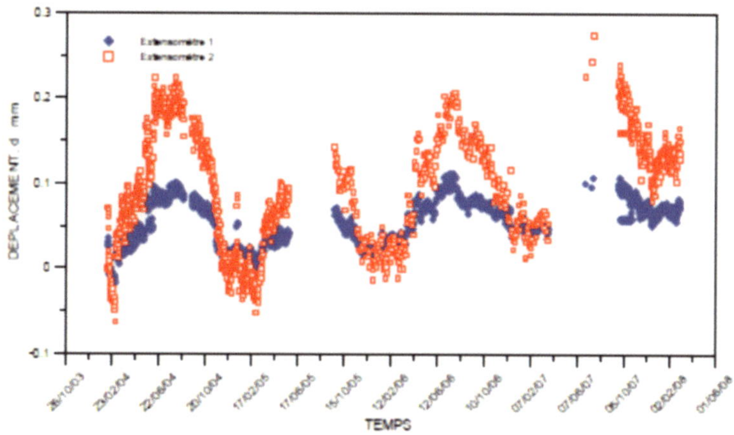


Fig. 5. Changes over time in lateral extension at two points within the pillar

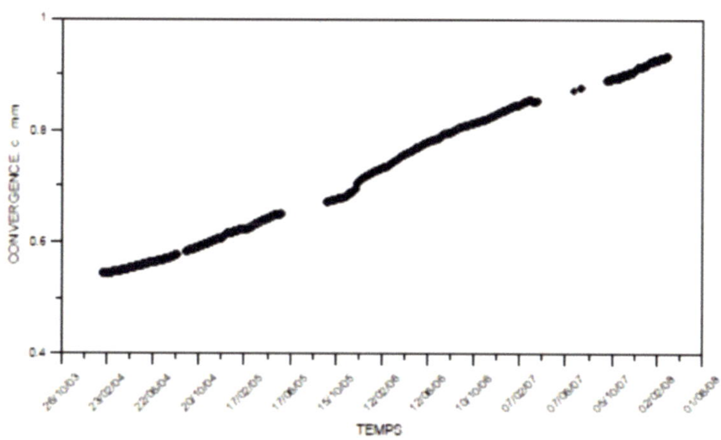


Fig. 6. Convergence over time between the mine wall and roof

In a mine where the water-level shows a generally decreasing trend, the influence of ambient hygrometry can have a major impact on the evolution of the pillars. This has been observed in some gypsum mines (Auvray et al., 2004), where scanning electron microscopy analysis of rock sections from different distances from the surface of the pillar has shown extensive traces of dissolution at the surface of the pillar, reducing towards the centre. This dissolution can induce rock matrix degradation that can lead to progressive spalling (flake-off) of the pillar (Fig. 7). As a direct consequence, the section of the pillar reduces over time, as does the pillar's ability to sustain the stability of the rock mass.

The instrumentation of the underground mine in Estreux shows significant seasonal changes in the water level and ambient hygrometry. These changes induce variations in pore pressure, leading to alternating cycles of saturation/desaturation that can contribute considerably to the degradation of the pillars over time.

To study the consequences of this type of phenomenon on the rock matrix in detail, laboratory tests were carried out reproducing the environmental conditions in the mine.

Fig. 7. Progressive spalling of a pillar in a gypsum mine. From left to right, photos from 1996, 2000, and 2004 (Sorgi & Watelet, 2007)

## 3. Laboratory scale: Core testing

Estreux chalk is a glauconite-rich chalk. Glauconite is an alumino-silicate of iron, potassium, and sodium. Its mineral composition is close to that of illite, although glauconite is not hydrated, with the additional presence of sodium and strong isomorphism by substitution of aluminium atoms with $Fe^{2+}$ and $Fe^{3+}$ iron atoms. Glauconite is often present in chalk deposits in northern France (Masson, 1973).

The porosity of Estreux chalk is about 37%, its specific gravity is $G_s$ = 2.74, and the average water content is equal to 20.7% when the rock is water-saturated. At the microstructural level, the solid matrix is made up of micrometric grains that are principally fragments of coccolithes. Sometimes intact coccolithes also occur. The chalk is then principally made up of calcite (calcium carbonate, $CaCO_3$), which often also constitutes the cementing agent at the intergranular contacts. Microfossils and mineral impurities are also frequently observed.

### 3.1 Retention properties of Estreux chalk

The Estreux chalk samples were completely saturated when extracted; the mine temperature was 11°C and the relative humidity, $h_r$, was ≅ 100% (with 2% accuracy of the hygrometry resistive sensors). Based on Kelvin's law, the change in relative humidity modifies the total air/water suction, $s_t$, the difference between the water vapour pressure (assumed equal to the atmospheric pressure, $p_a$), and the water pressure, $p_w$, according to the following relation:

$$s_t = p_a - p_w = -\frac{\rho_w}{M_v} RT \ln \frac{p_v}{p_{vs}} \tag{1}$$

where $\rho_w$ is the water density, $M_v$ the molar mass of the water vapour, $R$ the universal constant of an ideal gas (8.314 Jmol-1K-1), $T$ the absolute temperature, $p_v$ the vapour pressure and $p_{vs}$ the pressure of the saturating vapour at temperature $T$ ($h_r = p_v/p_{vs}$).

It is well known that any change in total suction induces a change in the degree of water saturation, $S_{rw}$, which can be quantified via the water retention curve (WRC) of the material. The WRC of Estreux chalk is presented in Fig. 8 (De Gennaro et al., 2006). As it can be observed, significant changes in $S_{rw}$ occur when suction varies between 1 and 2 MPa, causing near-total desaturation.

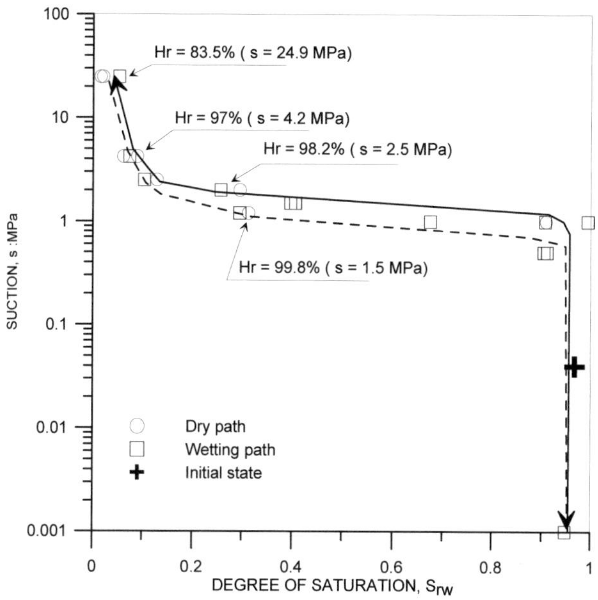


Fig. 8. Water retention curve of Estreux chalk

The slight differences observed between the drying and wetting paths denote a moderate hysteresis effect, also observed in other chalks (Priol, 2005). A possible effect of the glauconite fraction in reducing the hysteresis effect is suspected, although a clear explanation of the slight hysteresis is not straightforward. The drying curve shows that the air entry value of Estreux chalk can be estimated at approximately 1.5 MPa. Following the drying path, the degree of saturation exhibits a dramatic reduction, with a value as low as 30% at 2.5 MPa. At the highest suction ($s$ = 24.9 MPa, $h_r$ = 83.5%) the degree of saturation is as low as 2% to 5%, showing that the chalk is nearly completely desaturated. Based on the water retention curve, the suction of a dry sample can be estimated at 30 MPa. The shape of the water-retention curve of Estreux chalk and the sudden decrease in saturation above 1.5 MPa show that changing values of the ambient relative humidity in the mine (between 80% and 100%) can definitely lead to significantly unsaturated states, at least at the surface of the pillar, directly in contact with the ambient relative humidity. As a consequence, the mechanical properties of the chalk in unsaturated states have to be considered when addressing the long-term stability of the pillars. As a first step, the compressibility properties of the chalk under various controlled suctions are investigated.

### 3.2 Oedometer tests

Oedometer tests involve uniaxial compression of samples that are prevented from expanding laterally. The two independent stress variables commonly used in the investigation of the mechanical behaviour of unsaturated soils are the suction, $s = u_a - u_w$ (where $u_a$ and $u_w$ are the air and water pressure respectively), and mean net stress, $p_{net} = p - u_a$ (where $p$ is the total mean stress).

The loading paths followed during the oedometric tests are presented in Fig. 9. The resulting compressibility curves in the [log $\sigma_v$ : $e$] diagrams are presented in Fig. 10, where $e$ is the void ratio ($e = V_v/V_s$, where $V_v$ is the volume of void and $V_s$ is the volume of solid skeleton).

The testing program comprises four compression tests carried out as follows:

- Test T1 ($e_i$ = 0.575): dry compression ($s \cong$ 30 MPa) up to 39.7 MPa, unload down to 0.44 MPa, water injection under 0.44 MPa, and subsequent loading up to 39.7 MPa.
- Test T2 ($e_i$ = 0.61): dry compression ($s \cong$ 30 MPa) up to 22.41 MPa, unload down to 10.19 MPa, reload to 29.28 MPa, and water injection.
- Test T3 ($e_i$ = 0.602): suction controlled compression ($s$ = 4.2 MPa) up to 39.7 MPa, unload down to 8.82 MPa, and reload to 39.7 MPa.
- Test T4: ($e_i$ = 0.581): saturated compression up to 20.38 MPa, stress release at 0.26 MPa, and reload at 40.76 MPa.

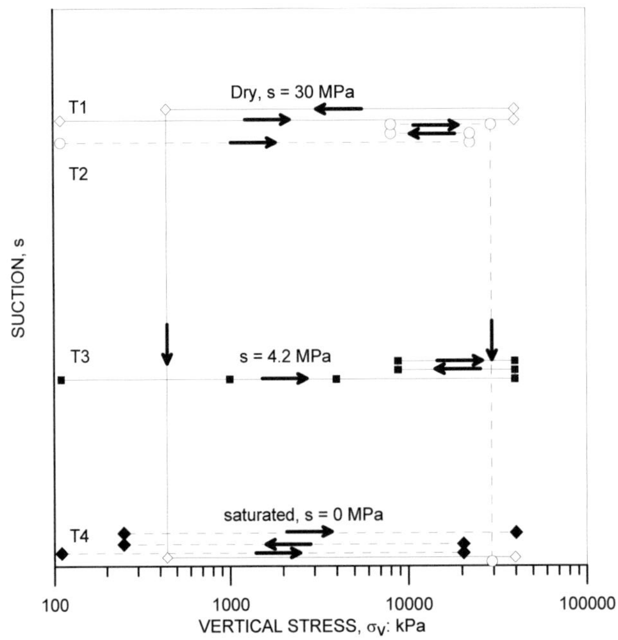


Fig. 9. Loading paths

The compressibility curves in Fig. 10 show some responses typical of unsaturated soils:

- increase in yield stress with increased suction
- increase in compressibility with decreased suction
- slight suction dependency of the pseudo-elastic compressibility module
- slight swelling due to suction release in the elastic zone
- significant collapse during water injection under high constant applied vertical load (pore collapse). Interestingly, the void ratio of the collapsed sample (i.e., after water injection) is close to the saturated compression curves of tests T1 and T4

| State | Compressibility | | Yield Stress, MPa |
|---|---|---|---|
| | Elastic | Plastic | |
| Dry (T1) | 0.0022 | 0.1082 | 16.0 |
| Dry (T2) | 0.0055 | 0.0940 | 13.5 |
| Suction Controlled (T3) | 0.0095 | 0.1137 | 1.4 |
| Saturated (T4) | 0.0039 | 0.1350 | 7.5 |

Table 2. Compressibility and yield stress of Estreux chalk derived from oedometer tests.

The values of the mechanical parameters are given in Table 2. These illustrate the sensitivity of the mechanical response of the Estreux chalk to changes in suction. They are in good agreement with the water-weakening effects described by Matthews and Clayton (1993), and with earlier observations on reservoir chalks (with water and oil as pore fluids) by De Gennaro et al. (2004) and Priol (2005).
Influence of water sensitivity is also denoted by the swelling observed in test T1 (during water injection under 441 kPa of applied vertical load) and by the collapse observed in test T2 under water injection at 29.28 MPa of applied vertical stress. The increase in compressibility and decrease in yield stress with increased degree of saturation (decreased suction) are two other manifestations of the water-weakening effect.

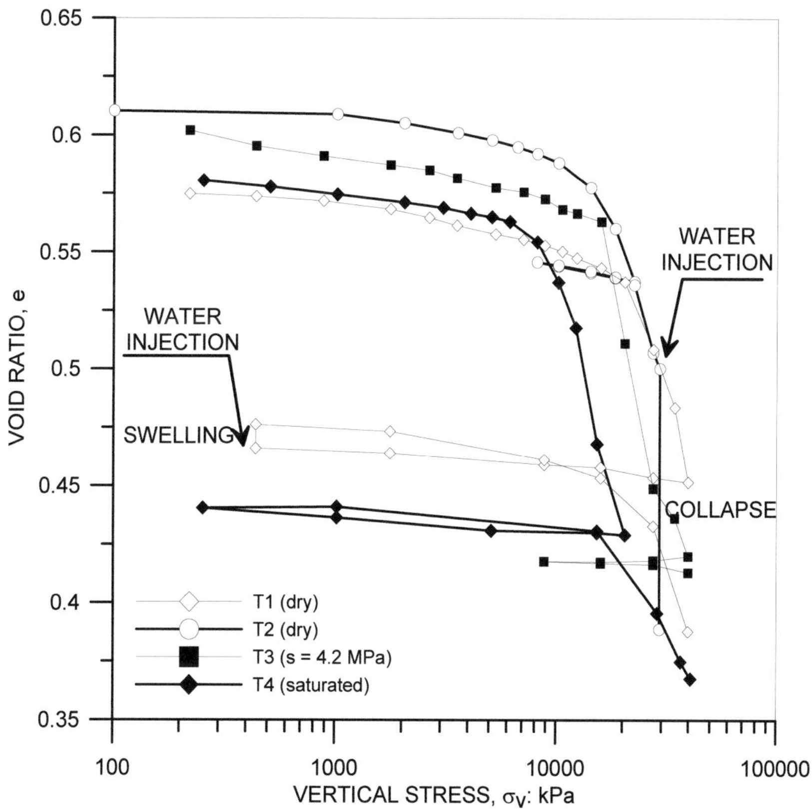


Fig. 10. Compressibility curves obtained with oedometers

## 4. Microscopic scale: Electron Scanning Environmental Microscope (ESEM) observations and micro-testing

The electron scanning environmental microscope (ESEM) allows the observation of microstructural changes of geomaterials in their natural state, under controlled conditions of temperature and pressure. Unlike the traditional scanning electron microscopy (SEM), ESEM technology does not require any preliminary treatment of the observed samples (such as dehydration or conductive coating). This has undeniable advantages in the analysis of the microstructure of geomaterials. In this study, a FEI Quanta 400® ESEM equipped with a Deben® microtesting facility has been used as a tool for the microstructural and micromechanical characterisation of Estreux chalk.
Changes in $S_{rw}$ were reproduced by controlling sample temperature and pressure following the state diagram of water (Fig. 11), being simultaneously correlated to the corresponding microstructural evolutions. A further step of the analysis involved the investigation of the microstructure while the material was subjected to a micromechanical loading, under constant or variable relative humidity, by means of ESEM micromechanical in situ tests.

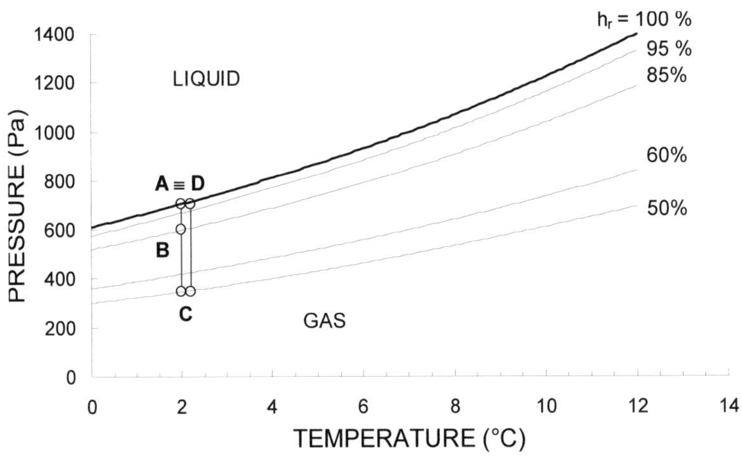


Fig. 11. State diagram of water and imposed changes during observation (ii)

Three types of observations were conducted: (i) changes in microstructure under wetting, (ii) samples submitted to saturation/de-saturation cycles starting from their natural state of saturation (path Fig. 11), and (iii) samples submitted to unconfined axial compression microtests under variable states of water saturation (Sorgi & De Gennaro, 2007).

### 4.1 Sample preparation

Sub-samples were extracted from available blocks of Estreux chalk retrieved from the underground mine, and these were sealed and stored in a thermo-regulated chamber to ensure the preservation of in situ conditions in terms of water content.
Observations (i) and (ii) were conducted on sub-samples having a square section (about 10 mm/side) and a thickness varying from 2 mm to 4 mm. These were fixed on the observation plate inside the ESEM chamber using a carbon conductive glue. Reduced plug thicknesses ensured a more uniform temperature distribution within the samples, and temperature was controlled using a thermo-electric cooler (based on Peltier's effect). The corresponding value

of the pressure in the observation chamber was used to define the level of hygrometry, $h_r$, based on the state diagram of water (Fig. 11).

### 4.2 Microstructural changes under wetting

The changes in microstructure under wetting when passing from $h_r$ = 97% (chalk in its natural state at sampling with $w$ = 20.7%) to $h_r$ = 100% are evident when comparing Figs. 12a and 12b. A reference network has been superposed to the micrograph and the boundary of one characteristic pore has been plotted. Since conditions in the chamber correspond to $h_r$ = 100% ($p$ = 705 Pa, $T$ = 2°C), hydration takes place as time passes. In Fig. 12b, the same pore is visualised after the in-situ hydration. As is seen from the two images, hydration produced a progressive enlargement of the pore boundaries due probably, but not exclusively, to the loss of capillary bridges between the grains. Progressive saturation of smaller pores is also observed on the left side of the photo in Fig. 12b. This observation still remains rather qualitative, though it provides an initial picture of the ongoing phenomena. It should be emphasised that pore enlargement is certainly amplified by the specific condition reproduced in the ESEM environment, namely the absence of any external loading and the observation of the external surface of the sample. It is expected that the extent of this phenomenon could be less for the inner non-visible pores. It is worth mentioning that the occurrence of pore enlargement during saturation at zero external applied load is consistent with the swelling shown in Fig. 10 for test T1 during water injection under low applied vertical stress.

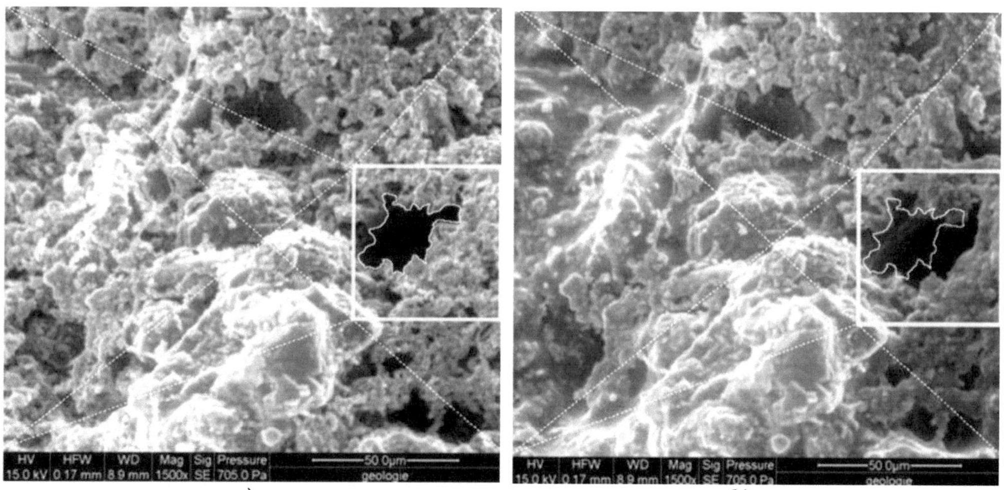


a) b)

Fig. 12. Modifications of the porous network in chalk during wetting: a) initial state, b) intermediate state before complete saturation

### 4.3 Saturation/desaturation cycles with ESEM

A series of tests was carried out on samples submitted to saturation/desaturation cycles following the path indicated in Fig. 11. During these tests a constant temperature condition was chosen ($T$ = 2°C). Relative humidity was then modified, changing the level of vacuum inside the chamber between 705 Pa and 346 Pa, corresponding to an $h_r$ varying between 100% and 50% (path A-B-C-D in Fig. 11). Observations were conducted at 1500 magnifications starting from the saturated state ($h_r$ = 100%). During the pressure changes, images were captured every 2 minutes and later mounted as a video clip. The observed zone was characterised by the presence of a rigid inclusion (crystal) embedded in the chalk porous matrix (Fig. 13a). The analysed cycles included:

- Phase 1: saturation and stabilisation; sample was left 90 minutes at T = 2°C and $p$ = 705 Pa, hence $h_r$ = 100% (Fig. 11). The reference image is captured after 90 minutes of elapsed time.
- Phase 2: desaturation; pressure is decreased instantaneously down to 599 Pa ($h_r$ = 85%, path A-B-C in Fig. 11). Sample is left to stabilise for 60 minutes.
- Phase 3: Second saturation; the pressure inside the chamber is increased up to 705 Pa (Fig. 11, path C-D) and sample is left to stabilise for 60 minutes at $h_r$ = 100 %.

During the first phase of saturation (Phase 1), the initial condition corresponding to full water saturation was reproduced inside the samples (Fig. 13a). The successive drying process (Phase 2) induced a fracture opening at the contact between the crystal and the chalk matrix (indicated by an arrow in Fig. 13b). This fracture was not evident at the beginning of the test (Fig. 13a), and its creation would seem to be associated with the changes in suction induced by wetting and drying cycles, athough it is recognised that capillary effects could also be a cause of this microstructural modification (swelling/shrinkage of the material).

In other words, wetting would have brought on fracture closing whereas drying would cause chalk matrix shrinkage around the crystal, inducing fracture opening. Fracture opening could then be the consequence of increasing capillary bridges (hence air/water interfaces) inside the chalk matrix during drying. In contrast, wetting would decrease the number of air/water menisci between the chalk matrix and the crystal, leading to a progressive fracture sealing (Figs. 13c, 13d). If related to material ageing, the evolution of this phenomenon with time following consecutive wetting and drying cycles could contribute to microstructral features associated with material degradation. The observations made here could also be supplemented and improved by advanced techniques of 2D and 3D image analysis, allowing for a more quantitative characterisation of the morphological modifications induced by changes in water saturation (see Sorgi & De Gennaro, 2007).

### 4.4 Micromechanical in situ testing

The combined use of the ESEM technique with unconfined compression tests was achieved using a micromechanical testing apparatus.

A Deben MICROTEST® loading module allowed the application of a maximal compression load of 5000 N at a constant strain rate of 1x10-5 s-1. A specific setup was developed to carry out micromechanical tests under controlled total suction (controlling the level of relative humidity during the tests). Cylindrical samples of about 8 mm in diameter and 15 mm in height were used. Samples were obtained by means of high-precision coring. End-face parallelism was ensured by means of a high-precision slicer having accuracy of the order of 1 µm. A series of preliminary micromechanical tests was conducted on saturated, partially saturated, and dry samples in order to verify the agreement between the micromechanical test results and the earlier laboratory test results performed on samples with larger (standard) dimensions.

The preliminary results from the unconfined compression microtests are presented in Fig. 14. It can be seen from the test results for the dry samples that there was good reproducibility. The linear slopes of the compression curves after a first tightening phase

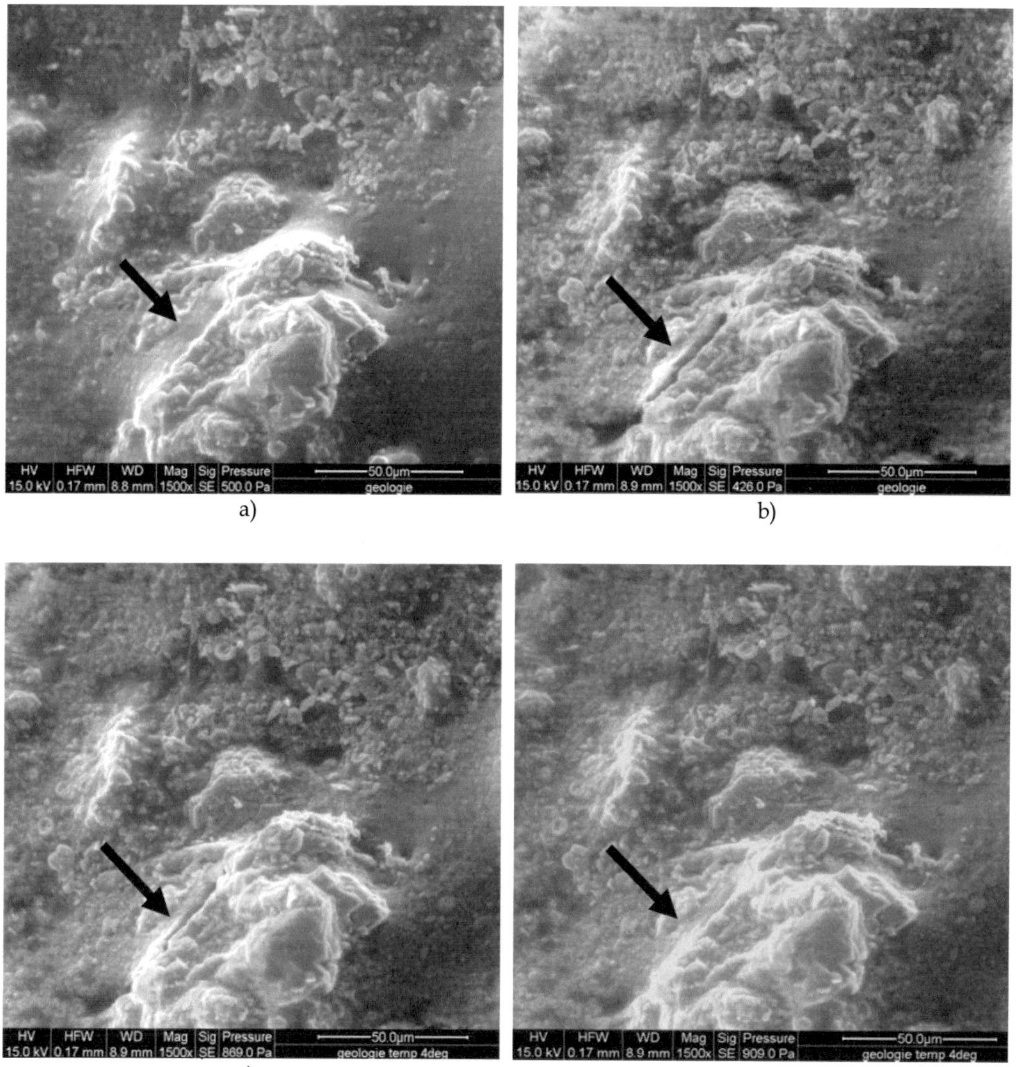


Fig. 13. a) and b) fracture opening in chalk specimen during drying, c) and d) fracture closing following the second saturation

allow the quantification of the Young's modulus. The latter is clearly influenced by the various states of saturation that follow; Young's modulus for dry chalk was $E_{dry}$ = 1.1 GPa, as compared with that of saturated chalk, where $E_{sat}$ = 0.71 GPa. The ratio $E_{dry} / E_{sat}$ = 1.6 is of the same order as that obtained by other researchers by means of standard laboratory unconfined compression tests (e.g. Raffoux & Ervel, 1980). At a suction level, $s_o$, of 4.2 MPa the value of Young's modulus, $E_o$, is between $E_{dry}$ and $E_{sat}$, with a value of 0.78 GPa.

In relation to material strength, the comparison between the unconfined compression strength (UCS) values obtained at saturated and dry states gives a ratio $UCS_{dry}/UCS_{sat} \cong 2$ in agreement with available data on northern French chalk (e.g. Bonvallet, 1979). Results from the sample tested under constant suction equal to 4.2 MPa ($S_r \cong 97\%$, Fig. 10) show that higher suctions strengthen the rock. This is likely to be associated with additional bonding due to capillary effects. The observed behaviour during unconfined compression microtests seems in good agreement with the general behavioural features observed for this chalk in oedometric compression tests under controlled suction conditions (Nguyen et al., 2008).

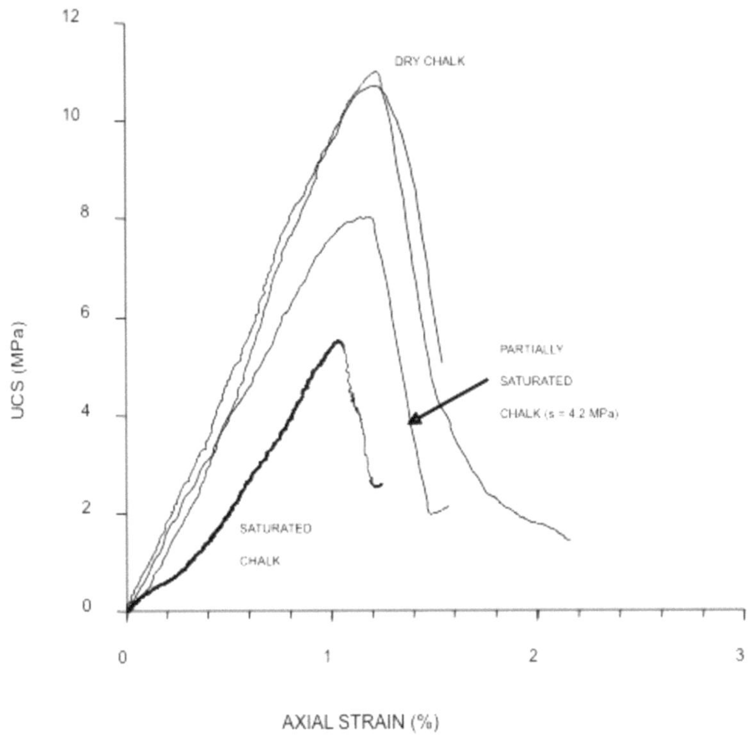


Fig. 14. ESEM in situ unconfined compression tests on dry and water saturated chalk

Note that Nguyen et al. (2008) also found a ratio of 2.1 between the yield stress in oedometric tests in dry and saturated conditions, close to the ratio $UCS_{dry}/UCS_{sat} \cong 2$ found during the ESEM micro-testing performed here. The ratio between the yield stress at a suction level of 4.2 MPa and that at saturated and dry state was 1.5 and 0.7, respectively. Similar ratios obtained by micromechanical testing using ESEM were equal to 1.5 and 0.75, showing a notable agreement with the oedometric test results.

Finally, Fig. 15 shows some preliminary results of ESEM in situ testing with simultaneous visualisation of the deformation pattern and the failure mode. The direction of compression is vertical, as indicated on the ESEM image (A). At peak strength (image B), the sample surface is still apparently unchanged. At about 0.9% axial strain, in the softening regime, a pseudo-vertical fracture is visible (image C), followed by a progressive opening in the post-peak phase (images D and E).

The aim of these preliminary tests was to explore the possibility of obtaining a characterisation of the local strain field during hydro-mechanical loading using ESEM. Some possible developments, like Digital Image Correlation (DIC) (e.g. Vales et al., 2008), might also be considered in order to obtain a quantitative characterisation of the local deformation at microstructural (few hundreds of μm) and mesostructural (some mm) levels.

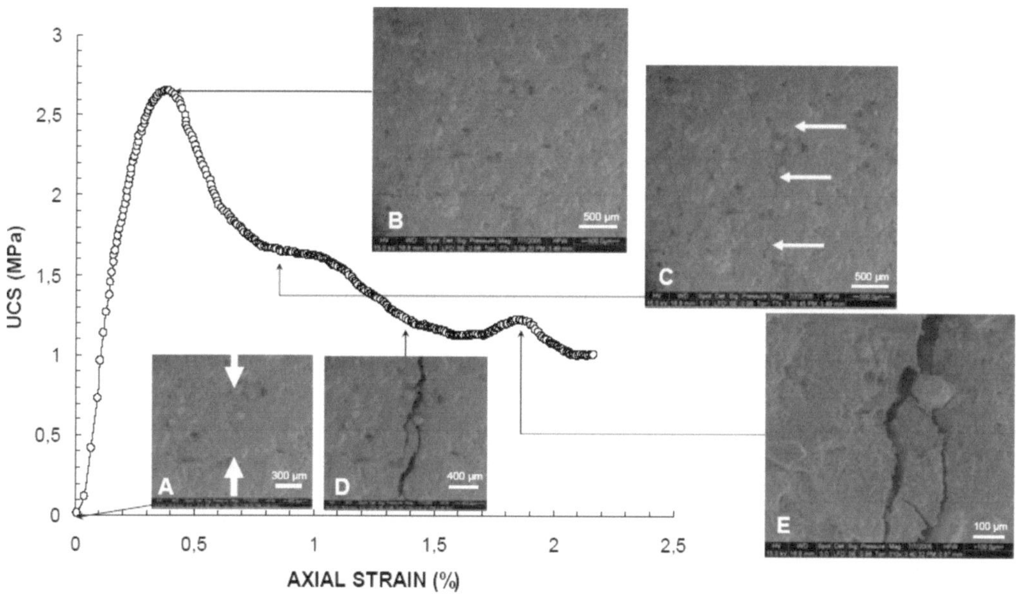


Fig. 15. ESEM failure pattern during ESEM in situ unconfined compression test on water saturated chalk

## 5. Conclusion

A preliminary investigation of the behaviour of chalk samples retrieved from the pillars of the abandoned Estreux mine (northern France), conducted as part of an assessment of the stability of underground chalk mines, has been presented. Due to environmental changes occurring within the mine (such as hygrometry and water table), the pillars are regularly subjected to variations in the degree of water saturation. The potential impact of the evolution of the water saturation on the mechanical behaviour of the chalk has been assessed based on the methods and concepts of the mechanics of unsaturated soils at different scales: site scale, laboratory scale, and microstructural scale.

At site scale, observations conducted during a four-year period (from February 2004 to February 2008) have highlighted that relative humidity in the mine can vary between 85% and 100%. The seasonal changes in the water table have been of reduced extent during the monitoring period, showing a general tendency towards decreasing water table level. However, as the water table rose, both pillar expansion and roof convergence were observed. Pressure and temperature measurements inside the chalk pillars have shown that both parameters vary on a seasonal basis, causing alternating states of saturation/desaturation. As expected, both pressure and temperature are not constant inside the pillars, but rather depends on the position of the point considered (altitude and distance from the pillar surface), on the liquid water inflow at the base of the pillar (which depends on the water table level), and on the evaporation rate at the surface of the pillar. It has been shown that, moving inwards from the free surface to the core of the pillar, both pressure and temperature decrease. The observations at site scale confirmed that water saturation in the chalk, controlled by the evolution of relative humidity and water table, influences the deformation of mine pillars and roof.

At laboratory scale, a detailed experimental program was carried out to better quantify the influence of water saturation on the mechanical behaviour of the Estreux chalk. A significant phenomena identified by the water retention properties is that significant desaturation occurs when the suction increases from 1 to 2.5 MPa, corresponding to a change in relative humidity from 99.3% to 98.2%. In such situations, water saturation decreases from almost full saturation (higher than 90%) down to $S_{rw}$ = 12%. The degree of water saturation of the chalk at equilibrium under a relative humidity of 83.5% (suction of 24.9 MPa) was about 7%. Considering that conditions of relative humidity, $h_r$, of around 80% can occur in the mine, and that for this situation $S_{rw}$ at equilibrium is as low as 5%, the significance of the chalk behaviour in unsaturated conditions is confirmed. Similarly to what has been shown in oil reservoir chalks (De Gennaro et al., 2003, 2004), the methods and concepts of unsaturated soils have been successfully adapted to Estreux chalk. Controlled suction oedometer tests provided more details on the water-weakening effects in chalk, showing increasing compaction during water injection (pore collapse). As observed in unsaturated soils, the yield stress determined in the oedometer increases when suction is increased (i.e. when the degree of saturation is decreased), confirming in more detail the suction-hardening effect as is observed in unsaturated soils.

For the microstructural scale, some applications of the ESEM for the microstructural characterisation of the partially saturated chalk have been presented. ESEM allows the observation of microstructural changes of geomaterials in their natural state under controlled conditions of temperature and pressure. Change in saturation can be easily reproduced in the observation chamber by means of a thermo-electric cooler based on the Peltier effect. This provided a preliminary analysis of the microstructural modifications of the Estreux chalk induced by the saturation/desaturation cycles in the absence of mechanical loading. Suction-controlled micromechanical in situ tests are also feasible. The validation of a specific experimental technique has been presented and results from micromechanical uniaxial unconfined compression tests have been compared with available results from laboratory-scale tests in terms of isotropic Young's moduli and UCS values. Good agreement has been observed between the different tests, confirming the reliability of this technique for further investigation of the micromechanical behaviour of geometerials under controlled saturation states (controlled suction).

Further developments using the ESEM are expected to all quantitatively characterise the effects of the mechanical and physico-chemical processes associated with the water-rock interaction. In the specific case of carbonate rocks these developments could provide a better understanding of some fundamental processes, such as dissolution, precipitation, crystallisation, and solid transport under stress, that lie at the onset and cause the degradation mechanisms of these rocks under the effect of environmental and mechanical agents.

## 6. Acknowledgment

The work presented here has been undertaken within the French National Project BCRD coordinated by INERIS (2005-2008). The financial support of INERIS is gratefully acknowledged. The authors are also indebted to P. Delage, H. D. Nguyen and P. Delalain for their fruitful discussions on a range of issues related to the mechanical behaviour of chalk. The technical support of E. De Laure and J. Thiriat in designing the experimental apparatuses presented is also greatfully acknowledged.

## 7. References

Auvray, C., Homand, F., & Sorgi, C. (2004). The aging of gypsum in underground mines. *Engineering Geology,* Vol. 74, No. 3-4, pp. 183 - 196

Bonvallet, J. (1979). Une classification géotechnique des craies du nord utilisée pour l' étude de stabilité des carrières souterraines. *Revue Française de Géotechnique* Vol. 8, pp. 5-15

De Gennaro, V., Delag,e P., Cui, Y.J., Schroeder, Ch., & Collin, F. 2003. Time-dependent behaviour of oil reservoir chalk: a multiphase approach. *Soils and Foundations,* Vol. 43, No. 4, pp. 131-148

De Gennaro, V., Delage, P., Priol, G., Collin, F., & Cui, Y.J. 2004. On the collapse behaviour of oil reservoir chalk. *Géotechnique,* Vol. 54, No. 6, pp. 415 - 420

De Gennaro, V., Sorgi, C., & Delage P. (2006). Water retention properties of a mine chalk. *Proceedings of the 4th International Conference on Unsaturated Soils (UNSAT 2006),* Phoenix, Arizona, USA, April 2006, pp. 1371-1381

Masson, M. (1973). Pétrophysique de la craie. *La craie, Bulletin des Laboratories des Ponts et Chaussées,* Special Volume, pp. 23-48.

Matthews, M.C. & Clayton, C.R.I (1993). Influence of intact porosity on the engineering properties of a weak rock. *Proceedings of the conference on geotechnical engineering of hard soils - soft rocks,* Vol. 1, pp. 693-702.

Nguyen, H. D., De Gennaro, V., Delage, P., & Sorgi C. (2008). Retention and compressibility properties of a partially saturated mine chalk. *Proceedings of the 1st European Conference, E-UNSAT 2008,* Durham, United Kingdom, July 2008, pp. 283–289

Nguyen, H.D. (2009). Water-rock interaction and long term behaviour of shallow cavities in chalks. Doctoral thesis, Ecole des Ponts ParisTech, Paris

Priol, G. (2005). Comportement mécanique d'une craie pétrolifèrecomportement différé et mouillabilité. Doctoral Thesis, Ecole des Ponts ParisTech, Paris

Raffoux, J. F. & Ervel, C. (1980). Stabilitt ggabilit de la carria c souterraine d'Estreux. Rapport CERCHAR rapport C CTO-CE/JS 80-76-2510/01: 8

Sorgi C. (2004). Contribution méthodologique et expérimentale à l'étude de la diminution de la résistance des massifs rocheux par veillissement. BCRD Rapport Final (2001-01111) INERIS-DRS: 132 pp

Sorgi, C. & Watelet, J. (2007). Fenomeni di degrade e rischio di crollo nelle cave di gesso abbandonate: l'esperienza francese. *Proceedings Dissesti indotti dall'alterazione di rocce evaporitiche,* Patron Editore, Bergamo, Italy, September 2007, pp. 41-59.

Sorgi C. & De Gennaro V. (2007). ESEM analysis of chalk microstructure submitted to hydromechanical loading. *C.R. Géosciences* Vol. 339., No. 7, pp. 468-481.Valès F., Bornert M., Gharbi H., Nguyen, M. D., & Eytard, J.C. (2007). Micromechanical investigations of the hydro-mechanical behaviour of argillite rocks by means of optical full field strain measurement and acoustic emission techniques. *Proceedings of the 11th International Society for Rock Mechanics Congress*, Lisbon, July 2007

# 10

# New Identity of the Kimberlite Melt: Constraints from Unaltered Diamondiferous Udachnaya – East Pipe Kimberlite, Russia

Vadim S. Kamenetsky, Maya B. Kamenetsky and Roland Maas
*University of Tasmania,*
*University of Melbourne,*
*Australia*

## 1. Introduction

Kimberlite magmas are in many aspects unusual compared to other terrestrial magmatic liquids. They are very rare and occur in small volumes, but their intimate relationships with diamonds make them invaluable to the scientific and exploration communities. The association of kimberlite rocks with diamonds and deep-seated mantle xenoliths links the origin of parental kimberlite magmas to the highest known depths (> 150 km) of magma derivation (e.g. Dawson, 1980; Eggler, 1989; Girnis & Ryabchikov, 2005; Mitchell, 1986; Mitchell, 1995; Pasteris, 1984). Kimberlite magmas would have one of the lowest viscosities and highest buoyancies that enable exceptionally rapid transport from the source region (Canil & Fedortchouk, 1999; Eggler, 1989; Haggerty, 1999; Kelley & Wartho, 2000; Sparks et al., 2006) and preservation of diamonds.

Despite significant research efforts, there is still uncertainty about the true chemical identity of kimberlite parental melts and their derivates. Kimberlite magmas are always contaminated by large quantities of lithic fragments and crystals, unrelated to the evolution of the parental melt. In most cases kimberlites are severely modified by syn- and post-magmatic changes that have altered the original alkali and volatile element abundances. These problems are reflected in the definition of the kimberlite rock as "*both a contaminated and altered sample of its parent melt*" (Pasteris, 1984). Numerous other definitions of the kimberlite commonly reflect on ultramafic compositions and enrichment in volatiles ($CO_2$ and $H_2O$; Clement et al., 1984; Kjarsgaard et al., 2009; Kopylova et al., 2007; Mitchell, 1986; Mitchell, 2008; Patterson et al., 2009; Skinner & Clement, 1979) which are supposedly inherited from parental magmas.

The physical properties of a kimberlite magma directly, and occurrence of diamonds indirectly, relate to the enrichment in carbonate components which are represented in common kimberlites by calcite and dolomite. The abundant carbonate component in kimberlite rocks is counter-balanced by a more abundant olivine (ultramafic) component, represented by olivine fragments and crystals that are commonly affected by serpentinisation. The ultramafic silicate compositions of kimberlites are ascribed to abundant olivine macrocrysts and phenocrysts, whereas significant $CO_2$ and $H_2O$ contents are attributed respectively to carbonate minerals (calcite and dolomite) and serpentine (+ other $H_2O$-bearing magnesian silicates). Unfortunately, the masking effects of deuteric and post-magmatic alteration do not permit routine recognition of olivine generations, and so the olivine component originally dissolved in the kimberlite parental melt remains controversial (Brett et al., 2009; Francis & Patterson, 2009; Mitchell & Tappe, 2010; Patterson et al., 2009). Similarly, the original magmatic abundances of volatile and fluid-mobile alkali elements are disturbed by syn- and post-emplacement modifications, thus complicating complicating quantification of the parental melt composition if inferred from bulk kimberlite analyses.

The existing dogma about correspondence between compositions of whole rock kimberlites and their parental melt has been recently challenged by the newcomers to the kimberlite scientific community (e.g., Kamenetsky et al., 2004; Kamenetsky et al., 2007a; Kamenetsky et al., 2007b; Kamenetsky et al., 2008; Kamenetsky et al., 2009a; Kamenetsky et al., 2009b; Kamenetsky et al., 2009c; Maas et al., 2005). A breakthrough into understanding of the kimberlite magma chemical and physical characteristics was made possible by detailed studies of the diamondiferous Udachnaya-East kimberlite pipe in Siberia. Unlike other kimberlites worldwide, severely modified by syn- and post-magmatic changes, the Udachnaya-East kimberlite is the only known fresh rock of this type, and thus it is invaluable source of information on the composition and temperature of primary melt, its mantle source, rheological properties of ascending kimberlite magma. This kimberlite preserved unequivocal evidence for olivine populations, olivine paragenetic assemblages and olivine-hosted melt inclusions, and the role of mantle-derived alkali carbonate and alkali chloride components in the parental melt.

## 2. Udachnaya-East kimberlite: Location and samples

The Udachnaya diamondiferous kimberlite pipe is located in the Daldyn-Alakit region of the Siberian diamondiferous kimberlite province (Fig. 1). Most Siberian pipes are tuff-breccias essentially devoid of unaltered olivine, but some contain large blocks of massive fresh kimberlite. A remarkable characteristic of this region is that it contains more pipes with fresh, unaltered olivine than any other kimberlite region within the Siberian province. About 10% of the intrusions exhibit either two adjacent channelways or repeated intrusion of magma through the same chimney. The Udachnaya pipe, the best known example of these twin diatremes, is located in the northwest part of the Daldyn field (Fig. 1). At the surface it consists of two adjacent bodies (East and West) that are separated at depth >250-270 m. Based on stratigraphic relationships both intrusions formed at the Devonian-Carboniferous boundary (~350 Ma), and the age estimates vary from 389 to 335 Ma (Burgess et al., 1992; Kamenetsky et al., 2009c; Kinny et al., 1997; Maas et al., 2005; Maslovskaja et al., 1983). The eastern and western bodies of the Udachnaya kimberlite pipe are different from each other in terms of mineralogy, petrography, composition, and degree of alteration. As the alteration of the western pipe can be considered typical of this rock type, the rocks of the Udachnaya-East are unique in having lesser alteration, and in some places they are completely unaltered.

At depths greater than 350 m a particularly fresh porphyritic kimberlite has been found. These rocks are described as dark-grey massive kimberlite, characterized by unaltered euhedral-subhedral olivine phenocrysts set in a dominantly carbonate matrix (Marshintsev

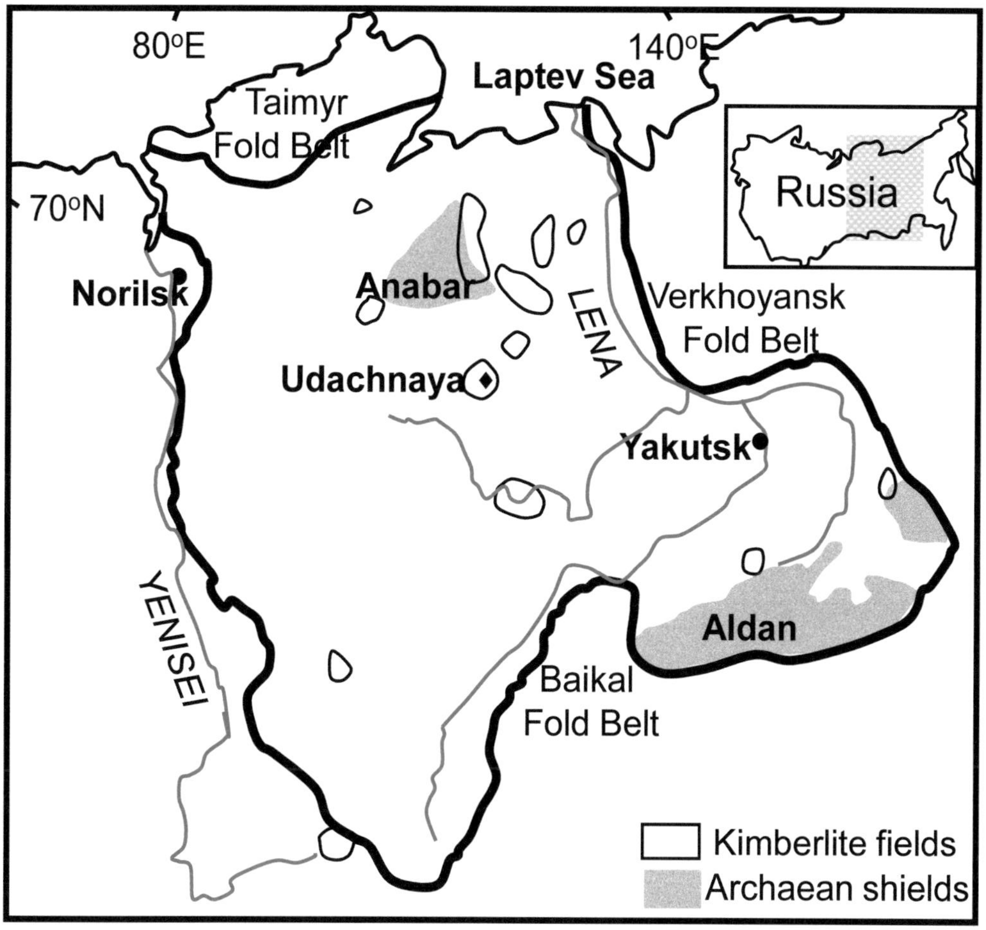


Fig. 1. Map of the Siberian Platform showing the major kimberlite fields after Pearson et al. (1995).

et al., 1976). At deeper levels (> 400 m) of the kimberlite body the amount of serpentine in the groundmass gradually decreases and the amount of carbonate in the groundmass increases. Intensive mining of the Udachnaya pipe revealed widespread chloride minerals (mostly halite) as dispersed masses in the groundmass and massive multi-mineral segregations of halite, serpentine, anhydrite, carbonates and hydrous iron oxides (Pavlov & Ilupin, 1973). The amount of chloride minerals in the groundmass increases with depth, and recently a large number of chloride-carbonate "nodules" were recovered from ~470-500 m depths of the mine.

The studies Udachnaya-East kimberlites are dark massive rocks with porphyroclastic fragmental textures. They are exceptionally olivine-rich (Fig. 2, 12a), a feature shared by the majority of known kimberlites, excluding rare aphanitic kimberlites, such as those from Kimberley, South Africa (Edgar et al., 1988; Edgar & Charbonneau, 1993; le Roex et al., 2003; Shee, 1986) and Jericho, Canada (Price et al., 2000). The large abundance of olivine (45-60 vol%) is reflected in the high MgO content of the bulk rock compositions (28-36 wt%). Olivine is set in a fine-grained matrix of carbonates (Fig. 2, calcite, shortite $Na_2Ca_2(CO_3)_3$ and zemkorite (Na, K)$_2$Ca(CO$_3$)$_2$), chlorides (halite and sylvite), and minor phlogopite and opaque minerals (e.g. spinel group minerals, perovskite, Fe±(Ni,Cu,K) sulphides) (Golovin et al., 2007; Golovin et al., 2003; Kamenetsky et al., 2004; Kamenetsky et al., 2007a; Sharygin et al., 2003; Sobolev et al., 1989). Groundmass olivine is very abundant (up to 40 vol%), completely unaltered, and contains crystal, fluid and melt inclusions (Fig. 3; Golovin et al., 2003; Kamenetsky et al., 2004; Kamenetsky et al., 2007a; Kamenetsky et al., 2008; Kamenetsky et al., 2009a).

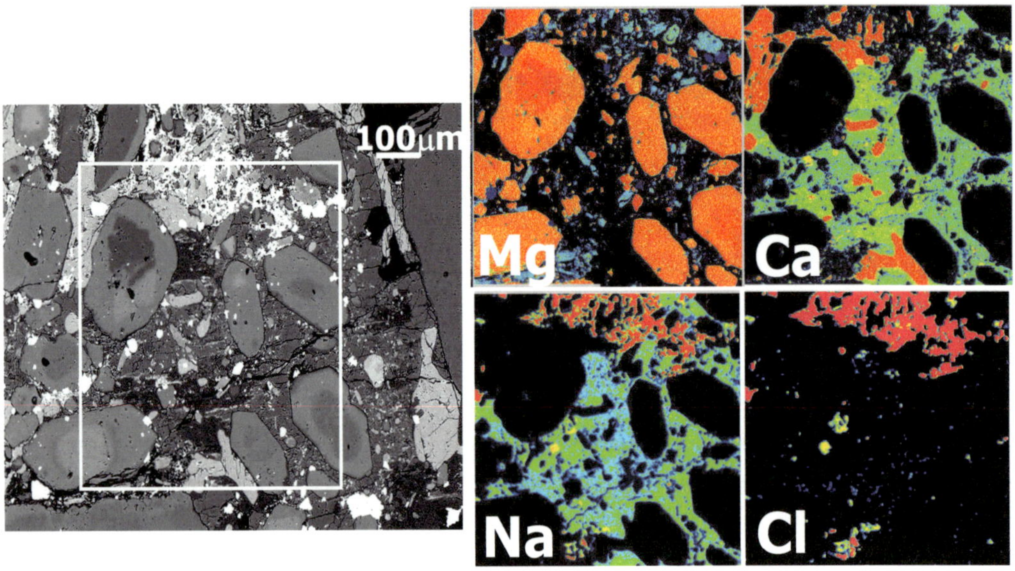


Fig. 2. Backscattered electron image and X-ray element maps showing intimate association of euhedral zoned olivine, Na–K chlorides, alkali carbonates, calcite, and sodalite in the groundmass of Udachnaya-East kimberlite

The bulk rock compositions are also characterised by low $Al_2O_3$ (1.2-2.3 wt%), but high CaO (8.4-18.2 wt%) and $CO_2$ (4-14 wt%) contents. Trace element compositions are similar to those of other kimberlites, having incompatible element enrichment and depletion in heavy rare-earth elements and Y (Fig. 4). The radiogenic isotope data ($^{87}Sr/^{86}Sr_t \approx 0.7047$, □Nd ≈ +4, $^{206}Pb/^{204}Pb_t \approx 18.7$, $^{207}Pb/^{204}Pb_t = 15.53$, $^{208}Pb/^{204}Pb_t = 35.5$–$38.9$, t = 367 Ma; Maas et al., 2005) fall within the field defined by most group-I kimberlites (Fraser et al., 1985; Smith, 1983; Weis & Demaiffe, 1985). The overall petrographic, mineralogical and chemical characteristics of the Udachnaya-East kimberlites suggest that they are common type-I (Mitchell, 1989) or group-I (Clement & Skinner, 1985; Smith, 1983) kimberlite.

However, the studied Udachnaya-East samples are distinctly different from other kimberlites in that they have high abundances of alkali elements (up to 6 wt% $Na_2O$), strong enrichment in chlorine (up to 6 wt%), and extraordinary depletion in $H_2O$ (< 0.5 wt%) correlated with absence of primary or secondary serpentine. Thus unusual for kimberlites low $H_2O$ abundances coupled with extraordinary enrichment in $Na_2O$ and Cl (Fig. 5) may indicate that two of the inferred key characteristics of kimberlitic magmas - low sodium and high water contents (Fig. 5; Kjarsgaard et al., 2009) – unambiguously relate to postmagmatic alteration that affected most kimberlites worldwide.

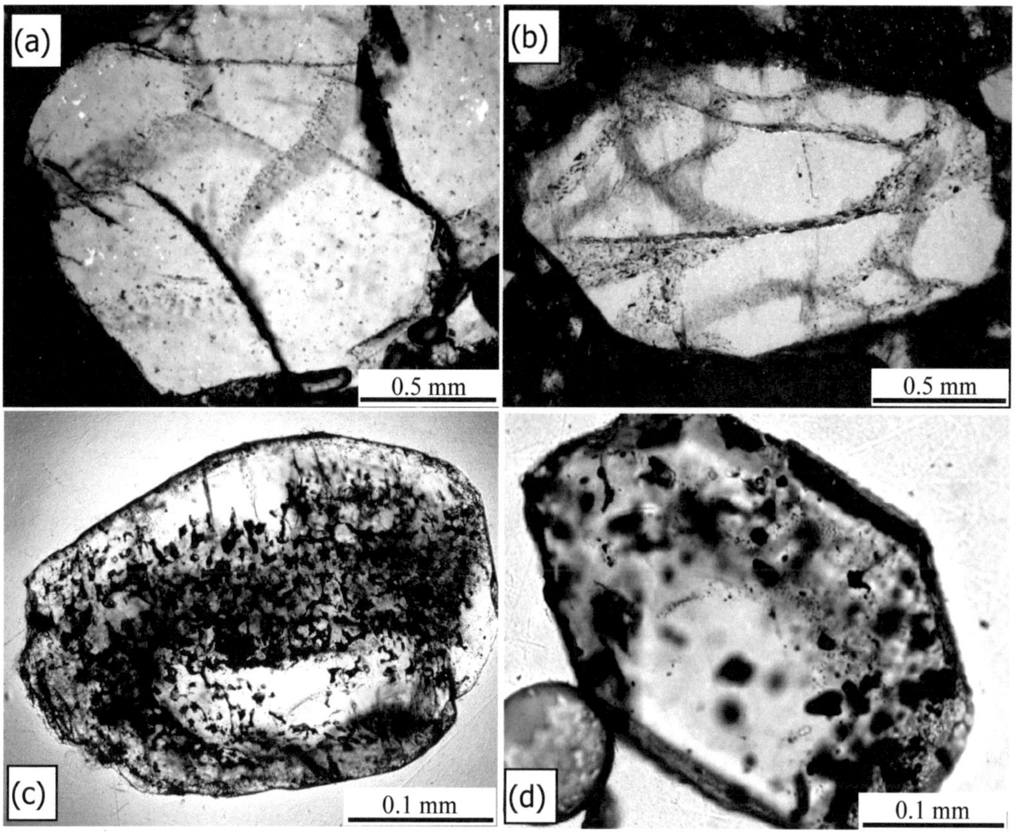


Fig. 3. Photomicrographs in plane-polarised light of individual crystals of olivine-I (a, b) and olivine-II (c, d) showing networks of magmatic inclusions, including crystal, fluid and carbonate-chloride melt inclusions.

## 3. Kimberlite olivine: Morphology and composition

Two populations of olivine in the Udachnaya-East kimberlite can be recognised based on size, colour, morphology, and entrapped inclusions. Consistent with many other studies of kimberlitic olivine (e.g., Boyd & Clement, 1977; Emeleus & Andrews, 1975; Hunter & Taylor, 1984; Mitchell, 1973; Mitchell, 1978; Nielsen & Jensen, 2005; Sobolev et al., 1989) the populations are represented by olivine-I (interpreted by different workers as cognate phenocrysts or xenocrysts) and groundmass olivine-II. However, as it follows from Arndt et al. (2010), Brett et al. (2009) and Kamenetsky et al. (2008) both populations significantly overlap in terms of composition, and possibly origin.

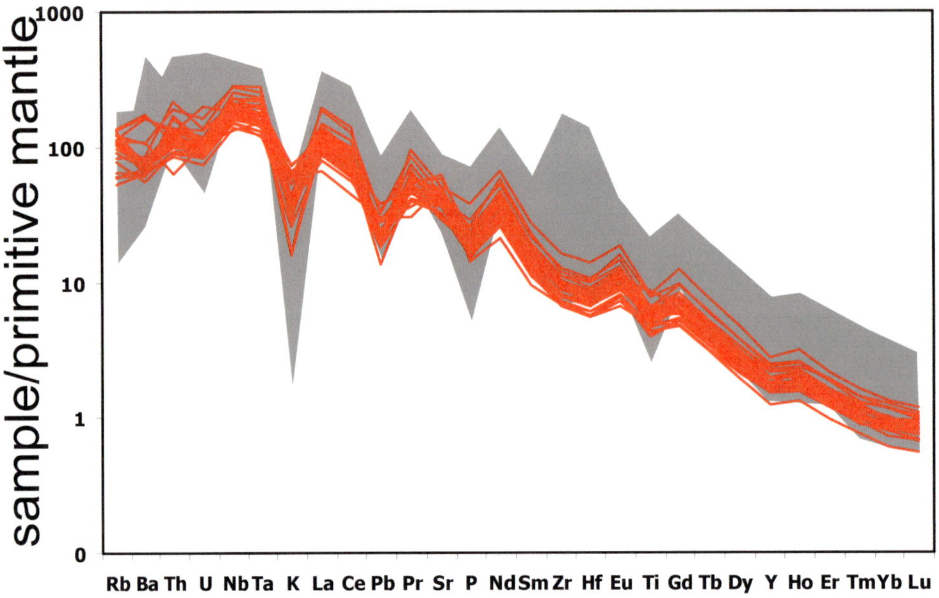


Fig. 4. Trace element abundance patterns of the Udachnaya-East kimberlites (lines) and kimberlites worldwide (field). All compositions are normalised to the "Primitive Mantle" composition of Sun and McDonough (1989).

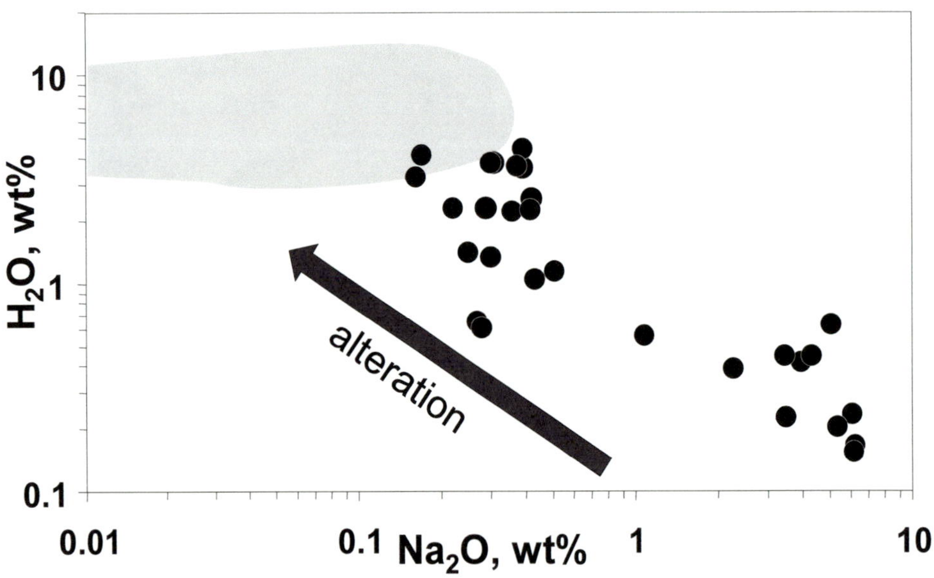


Fig. 5. Compositional co-variations of $Na_2O$ and $H_2O$ (in wt.%) in the Udachnaya-East kimberlites and "archetypal" rocks from South Africa, Greenland and Canada (field). The trend to low sodium and high water contents is considered to be caused by post-magmatic alteration.

### 3.1 Olivine-I

Light-green or light-yellow olivine-I is present as rounded and oval crystals, or more often as angular fragments with smooth edges Fig. 3a, b). Angular olivine-I is characteristically transparent and large (0.5 to 7-8 mm), whereas ovoid grains are smaller (0.7-2 mm) and often 'dusted' with inclusions (Fig. 3a, b). Melt and fluid inclusions occur in angular olivine-I and some round crystals only in "secondary" trails along healed fractures (Fig. 3a, b).
Olivine-I is characterised by variable forsterite content (Fo) from 85 to 94 mol%, although most grains are Fo>91 (Fig. 6a). Most grains appear to be homogeneous, at least in terms of their Fo content, except outermost rims and around healed fractures. Abundances of trace elements Ca, Ni, Cr and Mn in olivine-I vary strongly with Fo (Fig. 6a). The trace element composition trends resembling fractionation can be seen for NiO decreasing (0.43-0.13 wt%) and MnO increasing (0.07-0.17 wt%) as the Fo value decreases (Fig. 6a). However, it should be noted that NiO in the majority of olivine-I is almost constant (0.35-0.39 wt%).

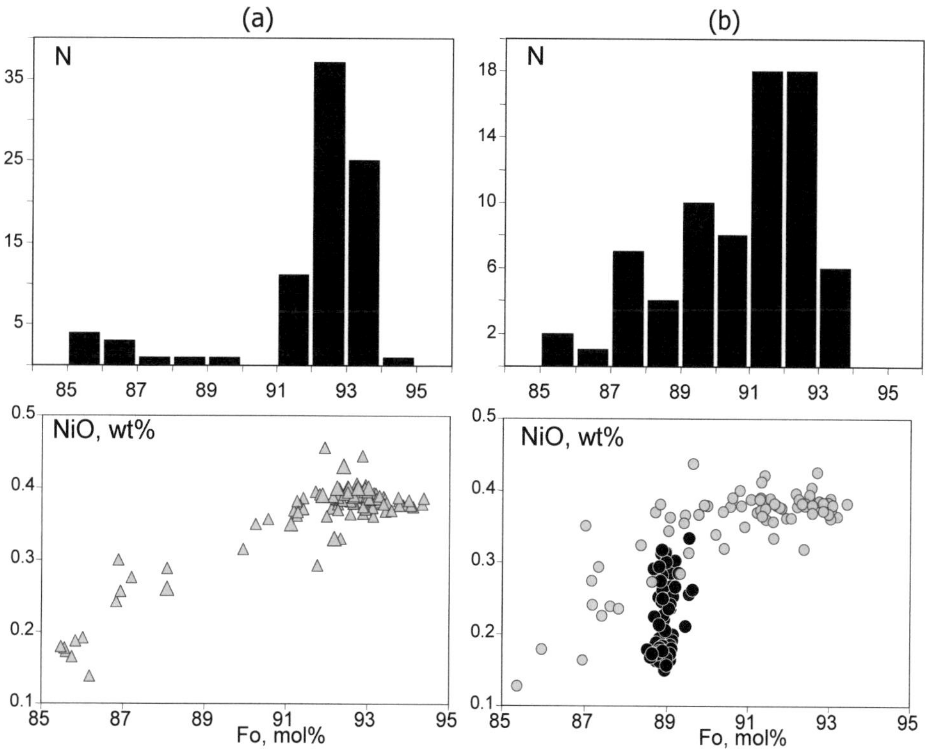


Fig. 6. Forsterite and trace element compositions of olivine-I (a) and olivine-II, 0.3-0.5mm size (b). Grey and black circles represent cores and rims of olivine-II, respectively. N, number of grains. The analytical error (1σ, equals 0.08% for Fo and 2% for NiO) is smaller than the size of the symbols.

### 3.2 Olivine-II: Morphology and zoning

Olivine-II is represented by relatively small (0.05-0.8 mm) euhedral flattened grains (Fig. 3c,d). Crystals display a tabular habit (tablet shape), and crystal growth was preferentially developed in the {100} and {001} directions. Olivine-II is colourless or slightly greenish or brownish, and a large amount of various inclusions is responsible for weak transparency and "cloudy" appearance of their host crystals (Fig. 3c, d).

The BSE images of individual olivine-II grains demonstrate compositional variability in terms of Fe-Mg relationships (higher and lower Fo correspond to *darker* and *lighter* areas, respectively; Fig. 7). Nearly all groundmass olivine crystals, even the smallest, exhibit intra-grain compositional variability (Fig. 7). The commonly used term "zoning" is not appropriate in the case of olivine-II, as evident from the description below. Five main types of olivine "structure" account for most typical Fo variations within single grains (Fig. 7):

1. A single core, euhedral-subhedral in shape, that can be more forsteritic (1a) or less forsteritic (1b) than the rim;
2. A single resorbed core of variable shape, size and composition. The composition of resorbed cores can be more forsteritic (2a) or less forsteritic (2b) than the rim;
3. A single core separated from rims by a thin layer of distinct composition;
4. Two or more cores of different shape and composition;
5. No distinct core - the grains are either compositionally uniform or have a mosaic-like structure (Fig. 7i).

All olivine-II show abrupt change to extremely Mg-rich ($Fo_{96}$) compositions at the very edge of the grains (~5-10 μm thick) in contact with matrix carbonate.

The grains with a single core (types 1-3) are the most abundant (~80%); however, a single core of euhedral or subhedral shape (type 1) is very rare (5%). Some cores have almost perfect olivine crystal shapes, and as a rule the crystallographic outlines of inner cores are parallel to the whole grain outlines (Fig. 7 a, d-g). The majority of olivine grains have corroded core edges (Fig. 7 b-d, f, h), and the degree of irregularity varies even within a single core. In other words, some outlines of the core can be straight and parallel to the crystal's outer rims, whereas other boundaries of the same core appear highly diffuse. These are transitional between types 1 and 2, and are more abundant than type 1.

Type 2 grains have a single core of variable size (relatively to grain size) and degree of resorption. The majority of the type 2 crystals tend to have oval to very irregular outlines of cores (Fig. 7 b, c, f). Some cores exhibit linear features (e.g. cracks) along which the olivine composition changes.

Types 1 and 2 are additionally subdivided into subtypes with normal ($Fo_{core}>Fo_{rim}$) and reverse ($Fo_{core}<Fo_{rim}$) "zoning".

Type 3 grains with a single core are characterised by presence of a compositionally distinct layer, separating cores and rims (Fig. 7 a, e, f). These layers are variable in shape, continuity and width. Even within a single grain the "separating" layer shows significant variability in shape, width and composition. The composition of such layers in the grains with reverse "zoning" is always more Fo-rich than the composition of both cores and rims.

The crystals belonging to type 4 with two or three cores are relatively rare (14%), but can be very important for genetic interpretations. Typically type 4 is an intergrowth of two distinct grains, where the cores with different or similar Fo have the shape and orientation similar to those of the grain's edges (Fig. 7g). In grains with two or more cores of different compositions, the cores are usually separated from each other (Fig. 7h), although a few examples are noted where the cores coalesce. In some crystals, the core has layers of different compositions, manifesting a gradual or abrupt zoning pattern across the olivine crystals. These layers frequently demonstrate well defined crystallographic shapes, parallel to outermost rims of olivine grains.

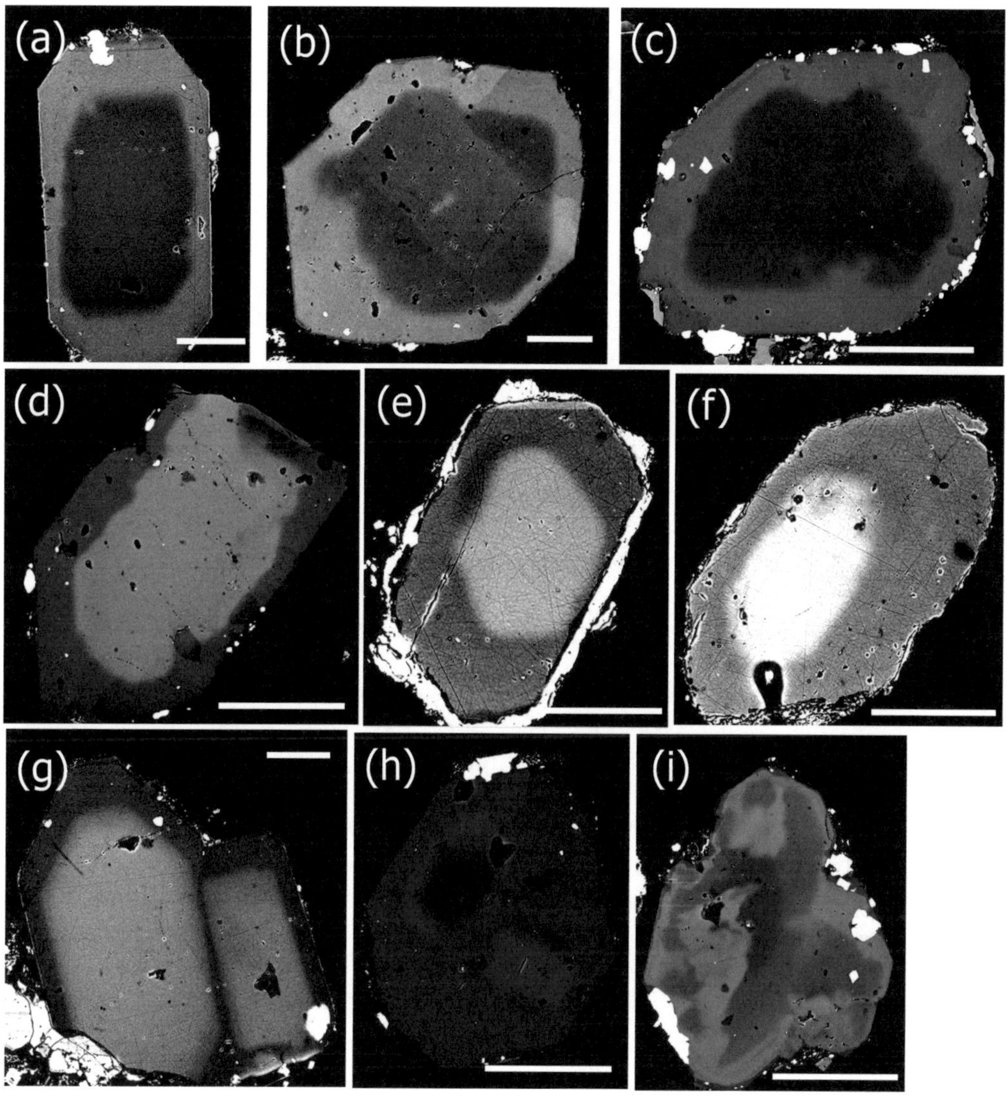


Fig. 7. Back-scattered electron images of olivine-II crystals demonstrating different types of zoning and core and rim relationships. Scale bars represent 100 μm

### 3.3 Olivine-II: Compositional variation

The inner parts ("cores") of olivine-II are strongly variable in Fo content (85.5 - 93.5 mol%), although the compositions 90.5-93 mol% Fo are most common (69%, Fig. 6b). The cores display relatively wide range of NiO (0.13-0.44 wt%, Fig. 6b), CaO (0-0.08 wt%), MnO (0-0.15 wt%), and $Cr_2O_3$ (0-0.09 wt%) contents. NiO contents are the highest and almost constant at Fo >89.5, and then gradually decrease in less magnesian olivine (Fig. 6b).

The outer parts ("rims") of olivine-II, although representing significant volumes of this population, have very constant Fo content 89.0 ± 0.2 mol% (Fig. 6b). In contrast, the trace element abundances in the rims are highly variable (in wt%: NiO 0.15-0.35, CaO 0.03-0.15, MnO 0.11-0.2, $Cr_2O_3$ 0.01-0.11 and $Al_2O_3$ 0-0.04). In general, the rims are richer in MnO, but poorer in NiO than the cores with the same Fo content (Fig. 6b). The outermost forsteritic (Fo96) rims are very enriched in CaO (up to 1 wt%).

## 4. Mineral and melt inclusions in olivine

Inclusions of different composition are present in almost all grains of the Udachnaya-East olivine. They can be very abundant in some grains, but rare in others. Three main types of magmatic inclusions are recognized in the studied samples: crystals, fluid and melt. Inclusion sizes are variable (<1 to ~400 μm) and the distribution of inclusions within a single olivine crystal is very heterogeneous, with some parts totally devoid of inclusions, and some parts so packed with inclusions as to make olivine almost opaque (Fig. 3, 8). The highest density of inclusions is observed along internal fractures and growth planes (Fig. 3). Crystal inclusions in olivine of both populations are always primary. Inclusions of melt and fluid in olivine-I and cores of olivine-II are always restricted to fractures healed with olivine of different composition, and thus are secondary in origin with respect to their host olivine. Similar inclusions in the rims of olivine-II show features reminiscent of both primary and secondary origin (Fig. 8). Melt inclusions in olivine of both populations are predominantly alkali carbonate-chloride in composition (Fig. 8). Silicate melt inclusions have not been found in our studies.

The rims of olivine-II grains contain abundant inclusions of different minerals that are never present in the cores (Kamenetsky et al., 2008; Kamenetsky et al., 2009a). Among them, Cr-spinel, phlogopite, perovskite and rutile are relatively abundant, whereas magnetite and picroilmenite are less common. Inclusions of low-Ca pyroxene (Mg# 88-92) occur in both cores and rims (Fo86-91) in clusters of several (10-30) round and euhedral grains (Kamenetsky et al., 2008; Kamenetsky et al., 2009a). A common association of low-Ca pyroxene in the rims includes numerous melt and fluid inclusions, and $CO_2$-rich bubbles adhered to surfaces of pyroxene crystals. The compositions of low-Ca pyroxene inclusions are characterised by high $SiO_2$ (53.3-58 wt%), $Na_2O$ (0.1-0.9 wt%), elevated $TiO_2$ (0-0.5 wt%), and low $Al_2O_3$ (0.7-1.4 wt%), CaO (0.7-1.7 wt%) and $Cr_2O_3$ (0.1-0.6 wt%), compared to mantle orthopyroxene.

Rare inclusions of high-Ca pyroxene in the Udachnaya-East olivine are restricted to olivine-I and cores of olivine-II (Fig. 9; Kamenetsky et al., 2008; Kamenetsky et al., 2009a). They occur as single crystals or clusters of several crystals. They vary in size (25-400 μm), colour (emerald-green to greyish-green) and shape (round to euhedral-subhedral). Most of them are intimately associated with the carbonate-chloride material, which forms coating on surfaces and inclusions inside clinopyroxene grains (Fig. 9). The clinopyroxene inclusions (Mg# 87.5-94.5 mol%) are in Mg-Fe equilibrium with the host olivine $Fo_{86.3-93}$, and characterised by low $Al_2O_3$ (0.65-2.9 wt%), and high CaO (19.5-23.8 wt%), $Na_2O$ (0.75-2.3 wt%) and $Cr_2O_3$ (0.9-2.6 wt%) contents. Individual crystals show fine-scale compositional zoning, with a general pattern of MgO and CaO increase, and $Na_2O$, $Cr_2O_3$ and in some cases $Al_2O_3$, decrease towards the rims. Major and trace element compositions of high-Ca pyroxene inclusions overlap with compositions of clinopyroxene from lherzolite nodules in the Udachnaya-East kimberlite (Kamenetsky et al., 2009a).

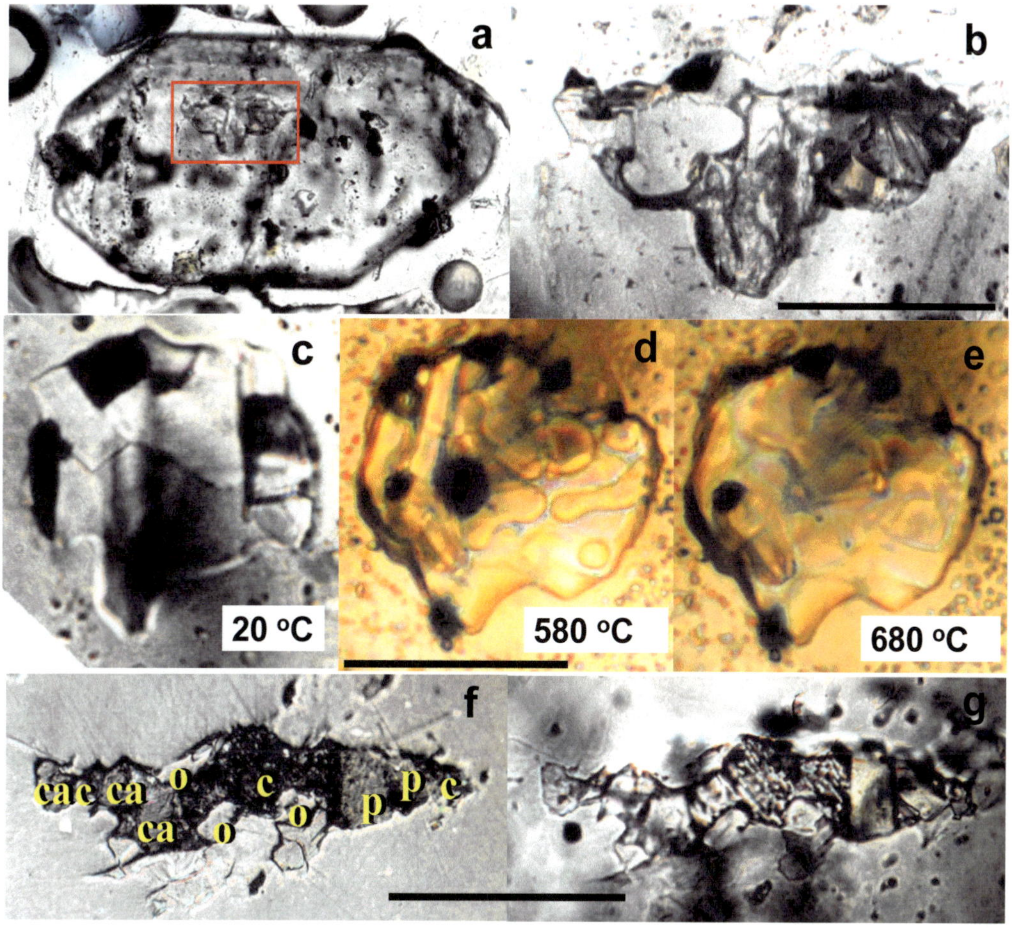


Fig. 8. Photomicrographs of (a) groundmass olivine and (b–g) olivine-hosted melt inclusions. Scale bars represent 50 μm. B: Multiphase melt inclusion hosted in core of olivine (boxed in a). c: Typical melt inclusion at room temperature. d: Same inclusion at 580 °C shows immiscibility between carbonate (matrix) and chloride (globules) melt. e: Same inclusion at 680 °C shows complete miscibility and homogenisation transmitted light). Note sculptured surface of melt inclusion at temperature of homogenisation. f, g: Multiphase melt inclusion in transmitted and reflected light, respectively. Principal daughter phases: c – sodium-potassium chloride; o – olivine; p – phlogopite; ca – sodium-potassium-calcium carbonate

### 4.1 Melt inclusions in groundmass olivine-II

Melt inclusions are trapped either individually within olivine cores and rims, or occur along healed fractures (Fig. 3, 8; Golovin et al., 2003; Kamenetsky et al., 2004; Kamenetsky et al., 2007a). Many inclusions are interconnected by thin channels, and thus modifications of original melt compositions by "necking down" cannot be ruled out. Abundant secondary melt inclusions in fractures connected to the groundmass, and decrepitated inclusions, are assumed to have experienced exchange and loss of material, respectively, after entrapment.

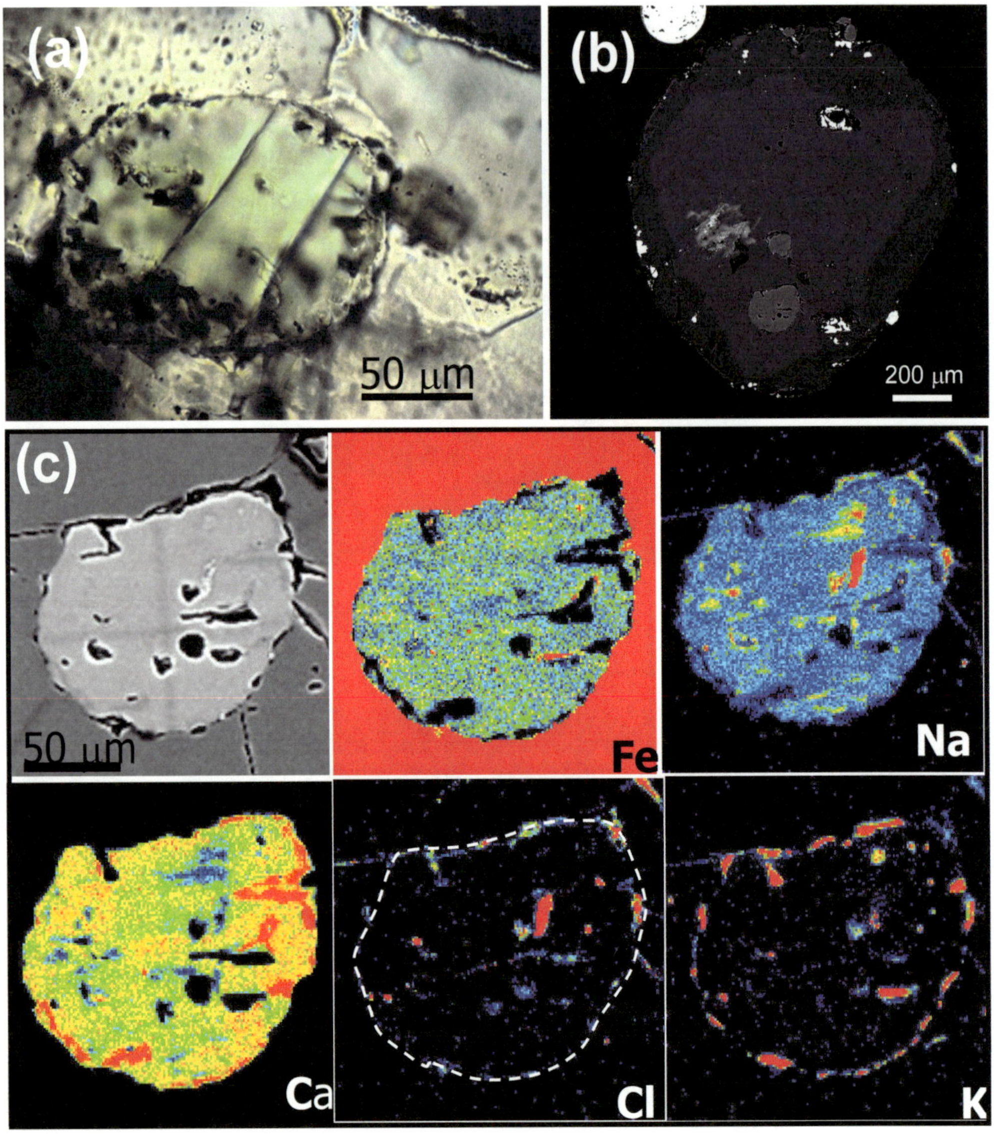


Fig. 9. A plane transmitted light photomicrograph (a) and back-scattered electron image (b) and X-ray element maps (c) of olivine-hosted clinopyroxene inclusions showing carbonate-chloride coatings and melt inclusions associated with clinopyroxene, and fine-scale compositional heterogeneity in terms of Fe, Ca and Na. Clinopyroxene on (c) is a larger inclusion on Fig. 9b. A contour of this inclusion is shown on the Cl Kα map by a dashed line.

"Necking down" can explain variable proportions of fluid and mineral phases in the studied melt inclusions. Fluid components are represented by low-density $CO_2$ bubbles, whereas solid phases are mainly Na-K-Ca carbonates, halite, sylvite, olivine, phlogopite-tetraferriphlopite, calcite, Fe-Ti-Cr oxides, aphthitalite and djerfisherite (Fig. 8f, g; Golovin et al., 2003; Kamenetsky et al., 2004; Sharygin et al., 2003). The inclusions occasionally contain monticellite, humite-clinohumite, northupite, and Ca-Mg-Fe-carbonates.

During heating stage experiments with round, relatively small (40-60 μm) melt inclusions, melting begins at ~160°C, as indicated by jolting movements of either solid phases or vapour bubbles. At 420-580°C bubble movements increase, indicating the appearance of the liquid phase (melt). Daughter phases experience some changes in their relative position, size and shape at 540-600°C. At >600°C we record a number of liquid globules that move freely and change shape continuously (Fig. 8d). The outline of a single globule is always smoothly curved: it can instantaneously change from spherical to cylindrical, embayed or lopsided, similar to an amoeba. With further heating, the number and size of the globules, as well as the number and size of vapour bubbles, gradually decreases. Homogenisation of the inclusions (except some opaque crystals) occurs when the globules and vapour bubbles disappear almost simultaneously (within 20-30°C) at 660-760°C (Fig. 8e).

During slow cooling (5-20°C/min), vapour bubbles nucleate at 690-650°C and then progressively increases in size. Cooling to 610-580°C, the inclusions acquire a 'foggy" appearance for a split second. This process can be best described as the formation of emulsion, i.e, microglobules of liquid in another liquid (melt immiscibility). Microglobules coalesce immediately into elongate or sausage-like pinkish globules. The neighbouring globules ("boudins") are subparallel, and are grouped into regularly aligned formations with a common angle of ~75-80°. A resemblance to a skeletal or spinifex texture is evident for several seconds, after which the original "pinch-and-swell structure" pulls apart giving rise to individual blebs of melt. The latter coalesce and become spherical with time or further cooling. They continue floating, but slow down with decreasing temperature and further coalescence. The exact moment of crystallisation or complete solidification is not detected.

## 5. Chloride-carbonate nodules in kimberlite

The major component of the kimberlite groundmass, carbonate-chloride in composition, sometimes form large segragations ("nodules", Fig. 10; Kamenetsky et al., 2007a; Kamenetsky et al., 2007b). Such samples were collected from fresh kimberlite at the stockpiles of the Udachnaya-East pipe. The assumed depth of their origin in the mine pit is ~500 m. The nodules vary in size from a few cm to 0.5 x 1.5 m, but are commonly 5 to 30 cm across. The shapes are usually round and ellipsoidal, but angular nodules were also encountered. The nodules have very distinct contacts with the host kimberlite, but without any thermometamorphic effects. The contacts are composed of thin (< 1mm) breccia-like aggregate of olivine, calcite, sodalite, phlogopite-tetraferriphlogopite, humite-clinohumite, Fe-Mg carbonates, perovskite, apatite, magnetite, djerfisherite ($K_6(Cu,Fe,Ni)_{25}S_{26}Cl$) and alkali sulphates in a matrix of chlorides. Olivine grains present at the contact with nodules belong to two types: zoned euhedral crystals similar to the Udachnaya-East groundmass olivine-II, and grains with highly irregular shapes and "mosaic" distributions of Fe-Mg.

Based on mineralogy the nodules can be separated into two major groups - chloride (Fig. 10a, b) and chloride-carbonate (Fig. 10 c-e). Chloride minerals are mainly represented by halite with included round grains of sylvite. The grain size, halite colour and transparency are highly variable, ranging from translucent to milky white and from white to all shades of blue. White and blue halite is often randomly interspersed, although in some coarse-grained nodules the interior parts are blue and dark-blue coloured, whereas rims are almost colourless (Fig. 10b). Chloride nodules always contain variable amount of fine-grained silicate-carbonate material (from 1 to 20 vol%) that is either present interstitially among halite crystals or forms irregular compact masses veined by chlorides. Contacts between silicate-carbonate material and chlorides are decorated by euhedral grains of olivine, monticellite, djerfisherite, perovskite, pyrrhotite, shortite and magnetite.

Fig. 10. Occurrence of chloride nodules in kimberlite (a, b) and textural characteristics and mineral relationships in the chloride-carbonate nodules (c-e). c, d – sample UV-5a-03 show texture resembling liquid immiscibility; white zoned sheets are composed of carbonates (Na-Ca±K±S), greyish masses cementing sheets are chlorides (halite-sylvite). e – sample UV-2-03, composed of shortite, northupite and chlorides. Scale: 1 graticule=1 mm

Chloride-carbonate nodules contain roughly similar amounts of chloride and carbonate minerals that are regularly interspersed (Fig. 10 c-e). Carbonates are present as 1-5 mm thick sheets with a bumpy or boudin-like surface. The groups of aligned, subparallel sheets make up rhombohedron formations (2-2.5 cm) that resemble hollow (skeletal) carbonate crystals (~78$^{o}$ angle) in shape. Cross-sections of inflated parts of the sheets show symmetrical zoning that reflects the change from translucent to milky-white carbonate (Fig. 10 c, d). The intra-sheet space and cracks in carbonate sheets are filled with sugary aggregates of chloride minerals. A texturally and mineralogically different variety of the chloride-carbonate nodules is represented by a single sample UV-2-03 (Fig. 10e). In this ~15 cm nodule, carbonates are present as very thin (< 0.2 mm) aligned white calcite-shortite sheets, as well as individual well-formed yellowish crystals of shortite and northupite (up to 1 cm). In the carbonate intergrowths northupite is interstitial and less abundant (25-30%), and can be distinguished from shortite by crystallographic properties and higher transparency.

### 5.1 Mineralogy of chloride-carbonate nodules

The chloride component of the nodules is dominated by halite, whereas individual grains of sylvite are rare. Typically, sylvite is included in halite, making up to 30 vol% of the chloride assemblage, and in places halite is sprinkled with minute sylvite crystals. Sometimes sylvite inclusions in halite show crystallographic outlines, however, round, lens-shaped and ameboid-like blebs of sylvite with different sizes and orientations are a prominent feature of the chloride masses (Fig. 11). Sylvite domains are often extremely irregular in shape, with curved re-entrances and attenuated swellings. Some domains are thin and elongated, and they can be either subparallel or perpendicular to the contacts with the carbonate sheets (Fig. 11 a-c). Chloride minerals also seal fractures in carbonates (Fig. 11).

The carbonate sheets are very heterogeneous in texture and composition (Figs. 11). In some occurrences a patchy distribution of textures and compositions is observed, but commonly a symmetrical zoning across carbonate sheets exists (Fig. 11b). The Na-Ca carbonate (shortite-like) at the rims, near contacts with chlorides forms intergrowths of acicular crystals. The interstitial space between these crystals (at polished surfaces) is either porous or filled with chlorides and Na-K sulphates. The transition from rims to cores is very distinct (Fig. 11b), as the cores do not show crystalline structure and are principally different in composition. On average the carbonate core is characterised by Na-Ca composition with significant $K_2O$ and $SO_3$. Highly variable, but with good correlation, amounts of $SO_3$ (up to 13 wt%) and $K_2O$ (up to 14 wt%) in the individual analyses of core carbonates suggest that Na-Ca carbonates are intermixed with tiny K-(Na) sulphate phases, the presence of which can be identified at high magnification. The Ca/Na in the core carbonate is higher than in the rim carbonate. Another Na-Ca carbonate with the highest Ca/Na is developed along the cleavage planes in the core and at the contacts with the rims.

An alkali sulphate, aphthitalite $(Na_{0.25}K_{0.75})_2SO_4$, is a minor but widespread component of the carbonate-chloride nodules. It is always associated with halite as irregular blebs, fringing the outmost rims of carbonate sheets (Fig. 11d), and filling fractures and interstitial spaces in carbonates (Fig. 11).

Anhydrous and hydrated Na-Ca carbonates with variable Ca/Na ratios are typical in all nodules, but in one sample (UV-2-03, Fig. 10e) an end-member shortite composition $Na_2Ca_2(CO_3)_3$ was found in close association with Cl-bearing Na-Mg carbonate (northupite – $Na_3Mg(CO_3)_2Cl$). Unlike heterogeneous and thus barely transparent carbonates in other nodules, well-formed crystals of shortite and northupite are clear and can be used for the inclusion studies. The mineral assemblage in this nodule is very complex, and includes euhedral crystals of apatite and phlogopite, as well as tetraferriphlogopite, djerfisherite, K-Na and Na-Ca sulphates, Ba-, Ca- and Sr-Ca-Ba- sulphates and carbonates, calcite, perovskite, and bradleyite $Na_3Mg(PO_4)(CO_3)$. The above minerals are present in aggregates within the interstitial chloride cement and as inclusions in shortite.

Maas et al. (2005) concluded that Sr-Nd-Pb isotopic ratios for the silicate, carbonate and halide components in the groundmass of the Udachnaya-East kimberlite support a mantle origin for the carbonate/chloride components. This conclusion relies in part on accurate age corrections to measured 87Sr/86Sr. However, the extreme instability of magmatic halides and alkali carbonates in air, even on the timescale of hours and days (Zaitsev & Keller, 2006), means that Rb-Sr isotope systematics of these kimberlites may have been modified since kimberlite emplacement in the late Devonian. An attempt to use Cl isotopes as a direct tracer of chlorine also proved inconclusive because of the similar $^{37}Cl/^{35}Cl$ ratios in mantle and crustal rocks (Sharp et al., 2007).

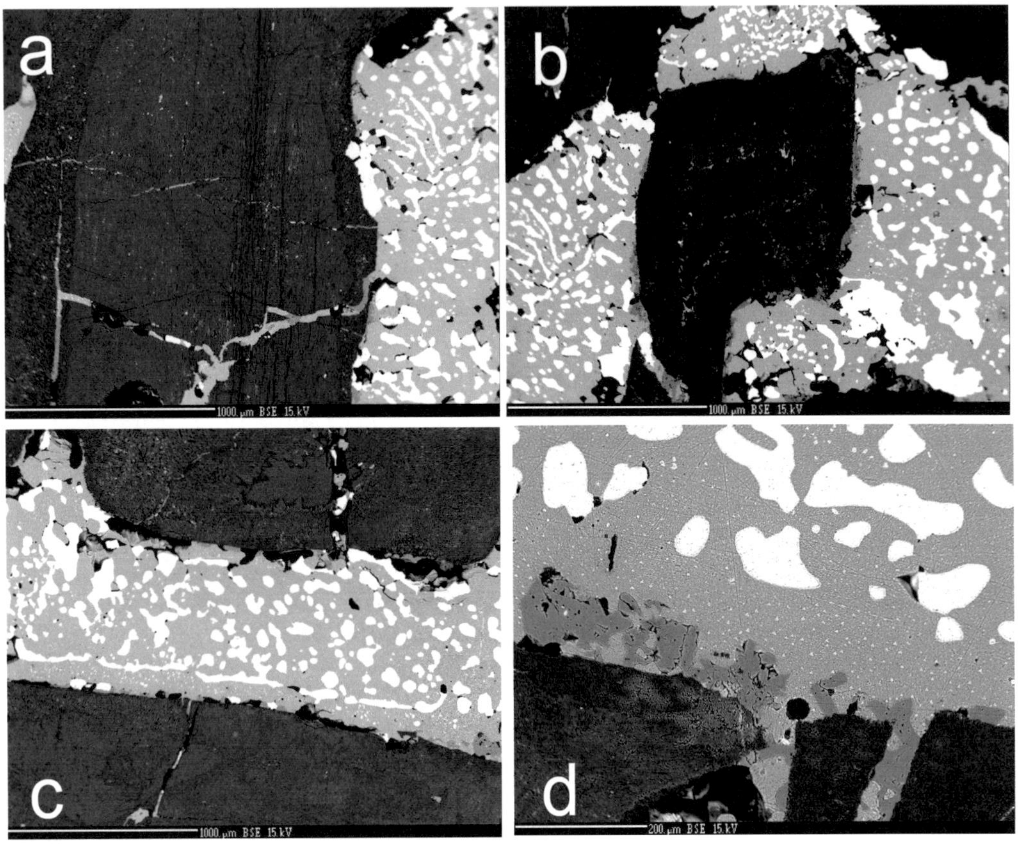


Fig. 11. The nodule texture is determined by a carbonate–chloride grid. Chloride minerals are represented by massive halite (light-grey, hosting amoeboid blebs of sylvite (white). Halite away from large sylvite formations is sprinkled with minute sylvite grains. Often sylvite forms streaks that show distinctive alignment. The overall texture of chloride layers and shape and distribution of sylvite, are reminiscent of liquid immiscibility. Carbonate sheets are symmetrically zoned (a, b). Irregular aphthitalite (grey) is present in halite, always near contacts with carbonate and in veinlets in carbonate (d).

## 6. Radiogenic isotope composition

An alternative approach to tracing relative contribution of mantle and crustal sources to the primary kimberlite melt is based on a study of perovskite, a common late-stage groundmass mineral in kimberlites (Chakhmouradian & Mitchell, 2000). Perovskite ($CaTiO_3$, >1000 ppm Sr, Rb/Sr≈0) should record the $^{87}Sr/^{86}Sr$ of the kimberlite melt at the time of perovskite formation (Heaman, 1989; Paton et al., 2007).

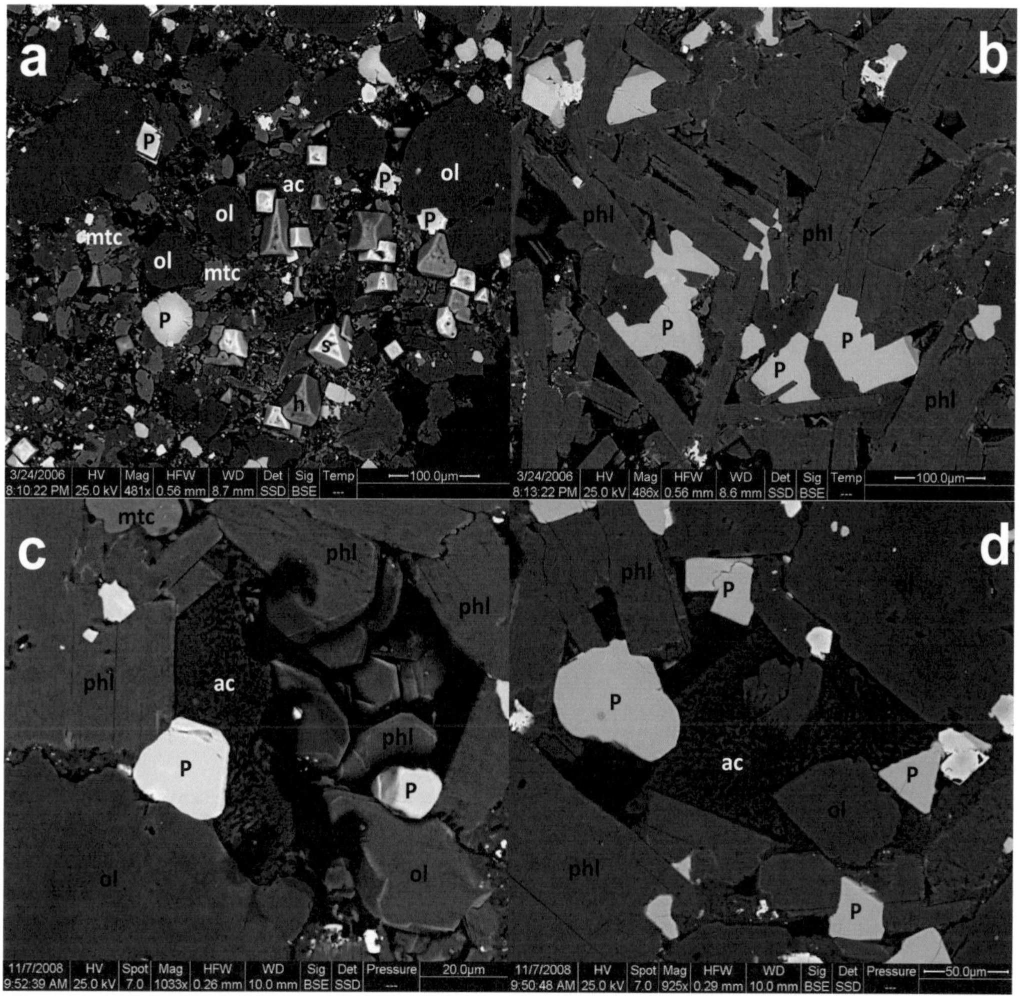


Fig. 12. Backscattered electron images of polished surfaces of the Udachnaya-East kimberlites (a – host kimberlite; b–d – perovskite- and phlogopite-rich clast), showing typical groundmass assemblage of co-crystallised chlorides, halite (h) and sylvite (s), alkali carbonates (ac), perovskite (P), phlogopite (phl), monticellite (mtc) and olivine (ol).

Perovskite is particularly abundant (10%) in sample UV31k-05 (Fig. 12 b-d), an ultramafic (31 wt% MgO), spherical clast ("nucleated autholith" after Mitchell, 1986) found at ~500 m depth in the pipe (Kamenetsky et al., 2009c). Accumulation of perovskite in kimberlite magmas is not unusual and has been reported in other kimberlites (Dawson & Hawthorne, 1973; Mitchell, 1986). The autolith and its host kimberlite are broadly similar in composition, and have the same groundmass assemblage, including interstitial carbonates, chlorides and perovskite (Fig. 2, 12a). Importantly, perovskite is interstitial to phlogopite crystals (Fig. 2b), and thus appears to be later than phlogopite in crystallisation sequence. On the other hand, textural relationships between perovskite and alkali carbonates and chlorides (Fig. 12a, c, d) suggest their co-precipitation from the melt. The melt that crystallised olivine and

phlogopite, the earliest minerals in this assemblage, is recorded in melt inclusions. Phlogopite-hosted melt inclusions in this sample (Kamenetsky et al., 2009c) are identical to olivine-hosted melt inclusions, described in the host kimberlite, in having essentially carbonate-chloride compositions and low homogenisation temperatures (650-700°C).
Three Sr isotope analyses of UV31k-05 perovskite by solution-mode average 0.70305±7 (2σ, age-corrected), similar to laser ablation MC-ICPMS results for 20 individual perovskite grains (average 0.70312±5, 2σ, age-corrected). These $^{87}Sr/^{86}Sr_i$ ratios are lower than those for the host kimberlite (0.7043-0.7049, Kostrovitsky et al., 2007; Maas et al., 2005; Pearson et al., 1995), although results for acid-leached kimberlite (0.7034-0.7037, Maas et al., 2005) provide a closer match. Such offsets between perovskite and host kimberlite were also noticed elsewhere (Paton et al., 2007), and probably reflect minor disturbance of bulk rock Rb-Sr systems. The perovskite-derived Sr isotope ratios are therefore considered a more robust estimate of kimberlite melt $^{87}Sr/^{86}Sr_i$. A ratio of ~0.7031 is the most unradiogenic among bulk rock compositions for group-I archetypal (Nowell et al., 2004; Smith, 1983; Smith et al., 1985) and Siberian kimberlites (Kostrovitsky et al., 2007), and similar to ratios for modern oceanic basalts, including MORB. The Sr isotope data, together with εNd-εHf near +5, are consistent with a parental magma derived from a depleted mantle-like source and suggest an absence of crustal (higher $^{87}Sr/^{86}Sr$, lower εNd) components in the Udachanya-East kimberlite melt, even at the time when perovskite and associated late-stage minerals, including chlorides, crystallised within autolith UV31k-05. This in turn supports a mantle origin of the chlorides in UV31k-05 and similar halides in the host kimberlite.

## 7. Discussion

### 7.1 Major mineral and chemical components of kimberlites

In general, the broad compositional range of kimberlites is defined by two end-members, magnesian silicate (olivine and serpentine) and carbonatitic (calcite). Thus, the kimberlites worldwide form a trend between these two end-members. It is likely that several processes can account for this compositional array. For example, crystallisation of olivine and segregation of carbonatitic melt (Ca increase) is counter-balanced by olivine accumulation and removal of carbonatitic melt (Ca decrease). Whatever the reason for the build-up in Ca, a general consensus exists that the magmatic carbonatitic component is an integral part of all kimberlite rocks, and their parental magmas. What still remains to be understood is why an expected increase in concentrations of alkali elements (Na and K) during the evolution of the kimberlite magmas is not reflected in the compositions of common kimberlites (e.g., $Na_2O$ is invariably <0.3 wt%). Moreover, low abundances of these elements relative to the elements of similar incompatibility are not easily reconciled with expected geochemical characteristics of low-degree mantle melts, even if residual phlogopite is present in the source peridotite (le Roex et al., 2003).
The idea of an alkali element loss and a $H_2O$ gain in kimberlites during post-magmatic processes can be promoted based on the fact that all kimberlites studied to date are inherently altered rocks. The alteration of the carbonate fraction towards essentially alkali-free calcitic compositions has been advocated since the discovery of modern alkali natrocarbonatite lavas from the Oldoinyo Lengai volcano and their altered counterparts (Clarke & Roberts, 1986; Dawson, 1962a; Dawson, 1989; Dawson et al., 1987; Deans & Roberts, 1984; Gittins & McKie, 1980; Hay, 1983). Rapid degradation of alkali carbonates and

dissolution of alkali chlorides in crustal environments (Zaitsev & Keller, 2006) can be responsible for depriving kimberlites (carbonatites) of their original sodium and potassium.

### 7.2 Alkali carbonate-chloride parental melt

The source and origin of alkali carbonates and chlorides in the groundmass of the Udachnaya-East kimberlites is still controversial, given the fact that other group-I kimberlites are devoid of these minerals, but have serpentine. Three possible scenarios of the alkali carbonate-chloride enrichment of the Udachnaya-East rocks can be considered: postmagmatic alteration, contamination of the magma in the crust en route to the surface and derivation from melting of the respectable mantle source. A possibility of post-emplacement ingress of chloride- and carbonate-bearing fluids can be confidently rejected on the basis of petrographic evidence. Any alteration features, typical of kimberlite rocks, are absent in this case; macrocrysts and phenocrysts of olivine bear no serpentine, and the olivine- and phlogopite hosted melt inclusions, trapped at magmatic temperatures (> 660oC) are compositionally similar to the groundmass (Golovin et al., 2007; Golovin et al., 2003; Kamenetsky et al., 2004; Kamenetsky et al., 2007a; Kamenetsky et al., 2009c). Moreover, water-soluble carbonate and chloride minerals in the groundmass were an important factor in preventing ingress of external fluids.

A choice between crustal and mantle origin of the Udachnaya-East unique compositional features is utterly important in deciding whether the Udachnaya-East kimberlite is a "black sheep" in the kimberlite clan or a bearer of the true identity of the primary kimberlite melt, and by inference, the composition of the mantle source and mantle melting process. A potential Na- and Cl-rich contaminant in the form of carbonate–evaporate sedimentary sequence is present in the south and southwest of the Siberian platform, however, it is not confidently recorded in the north, beneath the Daldyn kimberlite field (Brasier & Sukhov, 1998). Moreover, such contaminant is not pronounced in the composition of kimberlites from upper levels of the Udachnaya-East pipe (< 450 m), Udachnaya-West and other pipes from the same field and other kimberlite fields in Siberia. In addition to indirect evidence against likelihood of contamination of the kimberlite magma by evaporites reported in (Kamenetsky et al., 2007a), the deep mantle origin of the carbonate-chloride enrichment of the Udachnaya-East melt is well supported by the isotope composition of Sr in groundmass carbonates and perovskite (Kamenetsky et al., 2009c).

The non-silicate residual kimberlite magma has low temperatures (<650-750oC), as shown by the study of the Udachnaya-East melt inclusions (Kamenetsky et al., 2004), experimental data on the fluorine-bearing Na2CO3-CaCO3 system (Jago & Gittins, 1991) and direct temperature measurements in the halogen-rich (up to 15 wt% F+Cl, Jago & Gittins, 1991) natrocarbonatite lava lakes and flows of the Oldoinyo Lengai volcano (Dawson et al., 1990; Keller & Krafft, 1990; Krafft & Keller, 1989). However, even at these temperatures it is highly fluid. Thus, we envisage that droplets of residual melt separate from a solid aluminosilicate framework of the magma, percolate into weaker, less solidified zones, and finally coalesce, forming melt pockets. The latter are now seen in the kimberlite as chloride-carbonate nodules.

We emphasise that in the Udachnaya-East kimberlite the combination of such features such as extraordinary freshness, high abundances of Na, K and Cl, depletion in H2O, and preservation of water-soluble minerals and chloride-carbonate melt pockets cannot be coincidental. From the analogy with dry carbonatite magmas of Oldoinyo Lengai (Keller &

Krafft, 1990; Keller & Spettel, 1995) and experimental evidence that alkali carbonatite magmas "will persist only if the magma is dry" (Cooper et al., 1975) we conclude that the parental magma of the studied kimberlite was essentially anhydrous and carbonate-rich. This is indirectly supported by the spectroscopic study of micro-inclusions in Udachnaya cubic diamonds that showed that their parental media was a $H_2O$-poor carbonatitic melt (Zedgenizov et al., 2004).

Chlorine and $H_2O$ show opposing solubilities in fluid-saturated silicate melts, as they apparently compete for similar structural positions in the melt. Although Cl does not form complexes with Si in a melt, it may complex with network modifier cations, especially the alkalies, Ca and Mg (Carroll & Webster, 1994). General "dryness" of carbonatites and enrichment of natrocarbonatites in halogens (Gittins, 1989; Jago & Gittins, 1991; Keller & Krafft, 1990) suggest that Cl and $H_2O$ decouple which can be an intrinsic feature of carbonate-rich kimberlite magmas. If this is the case, the conventional role of H2O in governing low temperatures and low viscosities of kimberlite magmas can be readdressed to Cl. Furthermore, the data on carbonate-chloride compositions of melt inclusions in diamonds (Bulanova et al., 1998; Izraeli et al., 2001; Izraeli et al., 2004; Klein-BenDavid et al., 2004), nucleation and growth of diamonds in alkaline carbonate melts (Pal'yanov et al., 2002) and catalytic effect of Cl on the growth of diamonds in the system C–$K_2CO_3$–KCl (Tomlinson et al., 2004) concur with the proposed mantle origin of chloride and alkali carbonate components in the Udachnaya-East kimberlite.

### 7.3 Liquid immiscibility and crystallisation of residual kimberlite magma

Liquid immiscibility is observed in the olivine-hosted melt inclusions at ~600$^{o}$C on cooling (Fig. 8d). The immiscible liquids are recognized as the carbonate and chloride on the basis that these minerals are dominantly present in the unheated melt inclusions (Golovin et al., 2003; Kamenetsky et al., 2004). Remarkable textures, observed in melt inclusions at the exact moment of melt unmixing (Fig. 8d), is governed by the carbonate crystallographic properties. The presence of similar textures in the chloride-carbonate nodules (Fig. 10 c-e) is the first "snapshot" record of the unambiguous chloride-carbonate melt immiscibility in rocks. The previous natural evidence was based on melt and fluid inclusions in the skarn minerals of Mt Vesuvius (Fulignati et al., 2001) and kimberlitic diamonds (Bulanova et al., 1998; Izraeli et al., 2001; Izraeli et al., 2004; Klein-BenDavid et al., 2004). However, the extensive review of experimental studies (Veksler, 2004) points to the lack of data for chloride–carbonate systems.

Given the analogy with the texture of melt inclusions at the onset of immiscibility, the boudin-like shape of the carbonate sheets and their subparallel alignment (Fig. 10c, e), argues for preservation of primary (instantaneous) immiscibility texture. This means that post-immiscibility (< 600$^{o}$C) cooling and crystallisation were fast enough to prevent aggregation of one of the immiscible liquids into ovoid or spherical globules that are more typical of steady-state immiscibility. Occurrence of the chloride-rich veinlets in the carbonate sheets (Fig. 11) testifies to later solidification of the chloride liquid relative to carbonate crystallisation. The round and ameboid-like bleb textures of sylvite in halite (Fig. 11) are also reminiscent of liquid immiscibility. In theory this contradicts the fact of complete miscibility in the system NaCl-KCl above the eutectic point of ~660$^{o}$C. However, the separation of the Na-K chloride melt from the carbonatitic melt, in the case of Udachnaya-East residual melt pockets, occurred at temperatures below the eutectic, and

thus the chloride liquid was supercooled. On the other hand, it was close to the point of solid solution unmixing in the system 75% NaCl - 25% KCl (543°C at 1 atm), and in this case unmixing of liquids rather than solids is more likely.
Crystallisation from a homogeneous chloride-carbonate liquid (i.e., prior to immiscibility) is possible, and very unusual Na-Mg carbonates containing a NaCl molecule (northupite $NaCl*Na_2Mg(CO_3)_2$, Fig. 10e), is an example. Disruption of the melt structure caused by chloride-carbonate immiscibility and followed by reduction in solubility of the phosphate and Fe-Mg aluminosilicate components, prompted rapid crystallisation of zoned and often skeletal micro-crystals of apatite and phlogopite - tetraferriphlogopite. Fibrous aggregates of phlogopite in carbonates and sylvite are common and suggestive of incomplete extraction of the phlogopite component from carbonate and chloride melts by post-immiscibility crystallisation. After chloride-carbonate liquid unmixing the sulphate component of the original melt was largely accommodated within the carbonate melt. It was partially released as an aphthitalite melt at the chloride-carbonate interfaces (Fig. 11d), leaving porous K- and S-free carbonate behind (Fig. 11b), and it was also partially exsolved and re-distributed within the carbonate at subsolidus temperatures.

### 7.4 Rheological properties of kimberlite magmas

Kimberlites, especially those with preserved diamonds (Haggerty, 1999) are undoubtedly fast ascending magmas (>4 m/s; see review in Sparks et al., 2006). Support to this contention also comes from experimentally studied rates of dissolution of garnet in $H_2O$-bearing kimberlite melt (Canil & Fedortchouk, 1999) and Ar diffusive loss profiles of phlogopite in mantle xenoliths (Kelley & Wartho, 2000). Other indirect evidence includes inferred low viscosity of the kimberlite magma and its low density, contributing to high buoyancy (Spence & Turcotte, 1990). The unique physical properties of the kimberlite magma are governed by high abundances of chemical components that reduce melt polymerization (e.g. volatiles). The kimberlite magmas are assigned significant $H_2O$ contents in controlling transport and eruption, and only a few studies cast doubts on magmatic origin of $H_2O$ in kimberlites (e.g., Marshintsev, 1986; Sheppard & Dawson, 1975; Sparks et al., 2006).
Rapid transport and emplacement of the Udachnaya-East kimberlite is supported by the fact that this pipe is one of the most diamond-enriched in the world. However, our study denies the control from H2O on rheological properties of the Udachnaya-East kimberlite magma as the measured H2O abundances are particular low (<0.5 wt%). Instead, we are in position to draw analogy with the Oldoinyo Lengai natrocarbonatite magma, given the observed similarities in temperature (Kamenetsky et al., 2004) and composition. At low eruption temperature (< 600°C) the natrocarbonatite magma has exceptionally low density (2170 $kg/m^3$ ; Dawson et al. (1996), viscosity (0.1-5 Pa s ; Dawson et al., 1996; Keller & Krafft, 1990; Norton & Pinkerton, 1997) and fast flow velocities (1-5 m/s ; Keller & Krafft, 1990). The effect of halogens on reducing apparent viscosity of the carbonatite magma (three orders of magnitude for a three-fold increase in halogen content; Norton & Pinkerton, 1997) makes us confident that enrichment of the Udachnaya-East kimberlite in chlorine (at least 3 wt%) is a key chemical factor responsible for unique rheological properties of kimberlite magmas.

### 7.5 Implications from kimberlites and carbonatites worldwide

The enrichment of the Udachnaya-East kimberlite in alkali carbonates and chlorides, if a primary mantle-derived signature, could have been present in other group-I kimberlites prior to obliteration by common pervasive alteration. Study of melt inclusions trapped in magmatic phenocrysts during crystallisation allows seeing compositions beyond effects of postmagmatic modifications. The study of other least altered kimberlites emplaced into magmatic or metamorphic rocks in the terranes containing little or no sedimentary cover, namely the Gahcho Kué, Jericho, Aaron and Leslie pipes in the Slave Craton (Canada) and the Majuagaa dyke in southern West Greenland, helped to further enhanced the significance of the carbonate-chloride melt composition (Kamenetsky et al., 2009b).

The study of olivine and olivine-hosted melt inclusions in partially altered kimberlites from Canada and Greenland (Kamenetsky et al., 2009b), aimed at comparison with the fresh Udachnaya-East kimberlite and followed by implications of sodium- and chlorine-rich compositions of the parental kimberlite melt, has a precedent in the history of petrological and mineralogical studies of carbonatites. Unlike all ancient intrusive and extrusive carbonatite rocks composed of calcite and/or dolomite, the presently erupting carbonatitic magmas of the Oldoinyo Lengai volcano in Tanzania provides evidence for alkali- and halogen-rich anhydrous melts forming carbonatites. Following the discovery of these natrocarbonatite lavas (Dawson, 1962b) and building on the ideas of von Eckermann (1948), the primary/parental nature of such compositions was defended in a number of empirical (e.g., Clarke & Roberts, 1986; Dawson et al., 1987; Deans & Roberts, 1984; Gittins & McKie, 1980; Hay, 1983; Keller & Zaitsev, 2006; Le Bas, 1987; Schultz et al., 2004; Turner, 1988) and experimental (Safonov et al., 2007; Wallace & Green, 1988) studies. A strong support for the role of alkalies and halogens in magmas parental to mafic silicate intrusions and related carbonatites is further provided by melt/fluid inclusion research (e.g., Andreeva et al., 2006; Aspden, 1980; Aspden, 1981; Kogarko et al., 1991; Le Bas, 1981; Le Bas & Aspden, 1981; Panina, 2005; Panina & Motorina, 2008; Veksler et al., 1998). Syn- and postmagmatic release of alkalies from carbonatite magmas and rocks is recorded respectively in alkaline (mainly soda-dominant) metasomatic "fenitisation halos" around intrusive carbonatite bodies (e.g., (Bailey, 1993; Buhn & Rankin, 1999; Le Bas, 1987; McKie, 1966; Morogan & Lindblom, 1995) and references therein) and rapid decomposition of alkali- and chlorine-bearing minerals in the natrocarbonatites (Dawson, 1962b; Genge et al., 2001; Keller & Zaitsev, 2006; Mitchell, 2006). Same processes can be applicable to kimberlitic magmas in general, during and after their emplacement, as recorded in fenitisation of country rocks (Masun et al., 2004; Smith et al., 2004 and references therein) and gradation from Na-rich "deep" to Na-poor "shallow" kimberlite in the Udachnaya-East pipe.

The groundmass of most kimberlites, including altered kimberlites from the Udachnaya pipe, contain no alkali carbonates and chlorides and have very little $Na_2O$ (<0.2 wt%). We believe that alteration disturbs original melt compositions, with the alkaline elements and chlorine being mostly affected. However, the compositions of melt inclusions and Cl-rich serpentine are indicative of the chemical signature of a melt in which olivine crystallised and accumulated. It appears that enrichment in alkalies and chlorine, as seen in unaltered Udachnaya-East kimberlites, has been significant in other kimberlites prior to their alteration, and thus can be assigned deep mantle origin.

## 7.6 Two populations of olivine in kimberlites: Fellow-travellers or close relatives?

Our work on the uniquely unaltered Udachnaya-East kimberlite concurs with what has been shown in other mineralogical studies of other kimberlites, namely, the presence of morphologically distinct populations of olivine. One population is represented by large rounded grains (olivine-I of disputed origin), whereas another type of olivine is typically smaller but better shaped crystals (olivine-II or groundmass phenocrysts). It has been advocated in the literature that olivine may provide valuable clues to processes of kimberlite formation, transport and emplacement (e.g. Boyd & Clement, 1977; Mitchell, 1973; Mitchell, 1986; Moore, 1988; Skinner, 1989).

Mitchell & Tappe (2010), Mitchell (1973; 1986), and Moore (1988) considered olivine from both populations to be phenocrysts (cognate phenocrysts of olivine-I from high-pressure crystallisation of the kimberlite melt, and groundmass olivine-II), although up to 40 % of olivine was assigned to xenocrystic origin from various mantle and lithospheric sources. A similar conclusion can be endorsed by the extreme diversity of peridotite xenoliths within the Udachnaya-East kimberlite (Shimizu et al., 1997; Sobolev, 1977; Sobolev et al., 2009). The absence of primary melt inclusions and presence of Cr-diopside inclusions in olivine-I also argue against their phenocrystic origin.

Xenocrystic origin of some or all grains of olivine-I does not preclude this olivine being overgrown by the "phenocrystic" olivine. Both types of olivine are transported together, and thus all changes related to chemical and mechanical resorption should be equally imposed on them, making a morphological distinction subjective. Both olivine populations in the studied Udachnaya-East samples demonstrate striking compositional similarity in their Fo values (Fig. 6) and oxygen isotope values (Kamenetsky et al., 2008). Trace elements abundances are also indistinguishable for the olivine-I and core sections of the groundmass olivine (Fig. 6). Moreover, in many cases the olivine-II cores have original crystal faces ground away (Fig. 7), and thus their shapes are similar to those of round olivine-I. It is most likely that crystals that now show as relics in the olivine-II cores were formed at depth and transported upwards in a crystal mush.

Morphological and chemical resemblance between olivine-I and cores of olivine-II can be related to similar chemical and physical conditions exerted during olivine growth (or re-crystallisation) and transport to the surface. If both olivine populations are related, their common origin might be tracked down to the earliest and deepest stages of the kimberlite evolutionary story, i.e. when and where primary (protokimberlite) magma derived and started ascent.

### 7.7 Evolutionary storyline of the kimberlite parental melt

The Udachnaya-East groundmass olivine has a clear compositional structure, where the cores with variable Fo values can be distinguished from the rims with limited range in Fo values (Fig. 6b). It should be emphasised again that the olivine-II rims are essentially uniform with respect to major elements, but minor elements fluctuate strongly, especially Ni abundances which reach maximum near the core-rim boundary, then decrease rapidly towards the outer rims (Fig. 6b). Broadly similar compositional features, namely two groups of olivine with normal and reversed core to rim zonation and similar ranges in Fo and trace element contents, have been previously described in the groundmass olivine in other kimberlites, diamondiferous and barren (Fedortchouk & Canil, 2004; Moore, 1988; Skinner, 1989).

Although the origin of olivine cores (cognate vs exotic) is still debatable, the overall compositional analogy between groundmass olivine from different pipes and different kimberlite provinces argue for that 1) origin of cores and rims of groundmass olivine are intimately linked to kimberlite genesis and evolution; 2) in each case physical and chemical conditions of olivine formation are closely similar; 3) olivine cores and rims originate in different conditions; and 4) variable Fo compositions of cores reflect varying sources or changing conditions, whereas similar Fo values of rims reflect major buffering event.

Composition and zoning of the Udachnaya-East olivine-II are not unique; similar principle compositional characteristics of groundmass olivine phenocrysts (variable and constant Fo of cores and rims, respectively, and variable trace elements at a given Fo of the olivine rims; Fig. 6b) are described in a number of kimberlite suites (e.g., Boyd & Clement, 1977; Emeleus & Andrews, 1975; Fedortchouk & Canil, 2004; Hunter & Taylor, 1984; Kirkley et al., 1989; Mitchell, 1978; Mitchell, 1986; Moore, 1988; Nielsen & Jensen, 2005; Skinner, 1989). Compared to the ambiguous origin of the olivine cores, the rims of olivine-II most certainly crystallised from a melt transporting these crystals to the surface. This is best supported by the cases where several cores of different size, shape and composition are enclosed within a single olivine-II grain (Fig. 7 g, h). As indicated by mineral inclusions, the olivine-II rims formed together with phlogopite, perovskite, minerals of spinel group, rutile and orthopyroxene, i.e. common groundmass minerals (except orthopyroxene) from a melt that is present as melt inclusions in the olivine rims and healed fractures in the olivine-II cores and olivine-I (Fig. 3, 8).

Numerous studies indicate that most common xenoliths in kimberlites are garnet lherzolites, but surprisingly low abundance of orthopyroxene among xenocrysts and macrocrysts has been intriguing (Mitchell, 1973; Mitchell, 2008; Patterson et al., 2009; Skinner, 1989). Low silica activity in the kimberlite magma was offered as an explanation for instability of orthopyroxene, especially at sub-surface pressures (Mitchell, 1973). On the other hand, crystallising groundmass olivine rims and the presence of orthopyroxene inclusions in this olivine (Kamenetsky et al., 2008; Kamenetsky et al., 2009a) seem to be inconsistent with each other. One explanation is that orthopyroxene inclusions (often in groups and always associated with $CO_2$ bubbles) can result from the local reaction of olivine with $CO_2$ fluid ($2SiO_4^{-4} + 2CO_2 \rightarrow Si_2O_6^{-4} + 2CO_3^{-2}$).

A limited range of Fo content in the olivine-II rims, but variable trace element abundances (Fig. 6b) suggest crystallisation over a small temperature range or/and buffering of the magma at a constant Fe/Mg with fractionating Ni, Mn and Ca. In many instances, where the cores are seemingly affected by diffusion (Fig. 7 b, c, f, h) and have a surrounding layer of distinct composition (Fig. 7 a, e, h), the uniform Fo in the rims can reflect attempts by the crystals to equilibrate with a final hybrid magma (Mitchell, 1986). We also propose that the buffering of Fe/Mg can occur if the Mg–Fe distribution coefficient (Kd) between olivine and a carbonate-rich kimberlite melt is significantly higher than for common basaltic systems (i.e. 0.3±0.03). This reflects significantly smaller Mg-Fe fractionation between silicates and carbonate melt, possibly as a result of complexing between carbonate and $Mg^{2+}$ ions (Green & Wallace, 1988; Moore, 1988). The implied higher Kd for carbonatitic liquids, and especially Ca-rich carbonate, has been supported by experimental evidence (Dalton & Wood, 1993; Girnis et al., 2005). Probably an increase in Kd is even more pronounced for alkali-rich carbonatitic liquids.

The melt crystallising the rims of the Udachnaya-East groundmass olivine is represented by the carbonate-chloride matrix of the rocks (Kamenetsky et al., 2004; Kamenetsky et al., 2007a), and by the melt inclusions in olivine (Fig. 3, 8). The composition of this melt is unusually enriched in alkali carbonates and chlorides, but low in aluminosilicate components (Kamenetsky et al., 2004; Kamenetsky et al., 2007a). The crystallisation of olivine from this melt implies saturation in the olivine component, which makes this melt different from the alkali carbonate melt experimentally produced at mantle P-T conditions and low melting extents (Sweeney et al., 1995; Wallace & Green, 1988). How and where is the saturation in olivine acquired?

Study of the olivine populations and complex zoning of the groundmass olivine in the Udachnaya-East and other kimberlites (Kamenetsky et al., 2008; Kamenetsky et al., 2009b) provides evidence that olivine crystals were first entrapped by the melt at depth, then partly abraded, dissolved and recrystallised on ascent, and finally regenerated during emplacement. We suggest that the history of kimberlitic olivine is owed to the extraordinary melt composition, as well as conditions during melt generation and emplacement. In our scenario, a key role is played by the chloride-carbonate (presumably protokimberlite) melt, which forces strong mechanical abrasion and dissolution of the silicate minerals from country rocks in the mantle and lithosphere. Such a melt is capable of accumulating Si and Mg, but only to a certain limit, above which an immiscible Cl-bearing carbonate-silicate liquid appears (Safonov et al., 2007). The amount of forsterite that can be dissolved in the sodium carbonate liquid at 10 kbar and 1300°C is found to be 16 wt% (Hammouda & Laporte, 2000). Dissolution of olivine and other silicate phases at high pressure does not proceed beyond the saturation, and is closely followed by precipitation of olivine (Hammouda & Laporte, 2000). Therefore, ascending kimberlite magma, although being more Si-rich than its parental melt and loaded with xenocrysts and xenoliths, remains buoyant enough to continue rapid ascent. At emplacement, the magma releases the dissolved silicate component in the form of groundmass olivine rims and minor silicate minerals, thus driving the residual melt towards original chloride-carbonate compositions (Kamenetsky et al., 2007a).

## 8. Concluding remarks

Dry, chlorine-bearing alkali minerals in the Udachnaya-East kimberlite are products of crystallisation of the mantle-derived, uncontaminated melt. We suggest that a composition rich in alkalies, $CO_2$ and Cl may be a viable alternative to the currently favoured ultramafic kimberlite magma. A "salty" kimberlite composition can explain trace element signatures consistent with low degrees of partial melting, low temperatures of crystallisation and exceptional rheological properties responsible for fast ascent and the magma's ability to carry abundant high-density mantle nodules and crystals. Evidence for these components, notably Cl and alkalies, is only preserved in an ultrafresh kimberlite such as Udachnaya-East. Nevertheless, Cl-bearing minerals of the type reported here have also been found in the groundmass and melt inclusions in kimberlites from Canada and Greenland (Kamenetsky et al., 2009b). The possible existence of chloride–carbonate liquids within the diamond stability field can be inferred from experiments in the model silicate system with addition of Na-Ca carbonate and K-chloride (Safonov et al., 2010; Safonov et al., 2007; Safonov et al., 2009). These experiments also show that Cl-bearing carbonate-silicate and Si-bearing chloride-carbonate melts evolve towards Cl-rich carbonatitic liquids with decreasing temperature, providing a possible explanation for chlorine- and alkali-enriched microinclusions in some diamonds from Udachnaya-East (Zedgenizov & Ragozin, 2007) and other kimberlites in South Africa and Canada (Izraeli et al., 2001; Klein-BenDavid et al., 2007; Tomlinson et al., 2006). Brine inclusions in diamonds from various kimberlites, and the inferred role of chlorides in diamond nucleation and growth (Palyanov et al., 2007;

Tomlinson et al., 2004) further illustrate the potential significance of mantle-derived "salty" fluids sampled by the kimberlite melt and fortuitously preserved at Udachnaya-East.

## 9. Acknowledgments

We are indebted to Victor Sharygin, Alexander Golovin and Nikolai Pokhilenko who collected and supplied kimberlite samples for these studies and co-authored previous publications. Alexander Sobolev initiated and supervised PhD studies of Maya Kamenetsky and contributed a wealth of provocative ideas for discussion. We thank D. Kuzmin, P. Robinson, S. Gilbert, K. McGoldrick, K. Gőmann, and L. Danyushevsky for providing help with different analyses. The results and ideas were discussed with many researchers at different conferences. In particular, we are grateful to D.H. Green, B. Dawson, G. Brey, C. Ballhaus, G. Yaxley, R. Mitchell, O. Navon, N. Sobolev, O. Safonov, A. Chakhmouradian, S. Kostrovitsky, S. Tappe, S. Matveev, M. Kopylova, L. Heaman, Ya. Fedortchouk and I. Veksler for advice, moral support and friendly criticism. This study was initially (2003-2005) supported by the Alexander von Humboldt Foundation (Germany) in the form of the Wolfgang Paul Award to A. Sobolev and the Friedrich Wilhelm Bessel Award to V. Kamenetsky. Financial support for these studies in 2005-2009 was provided by an Australian Research Council Professorial Fellowship and Discovery Grant to V. Kamenetsky "Unmixing in Magmas: Melt and Fluid Inclusion Constraints on Identity, Timing, and Evolution of Immiscible Fluids, Salt and Sulphide Melts ".

## 10. References

Andreeva, I.A., Kovalenko, V.I., Kononkova, N.N., (2006) Natrocarbonatitic melts of the Bol'shaya Tagna Massif, the eastern Sayan region. *Doklady Earth Sciences*, 408(4), 542-546.

Arndt, N.T., Guitreau, M., Boullier, A.M., Le Roex, A., Tommasi, A., Cordier, P., Sobolev, A., (2010) Olivine, and the origin of kimberlite. *Journal of Petrology*, 51(3), 573-602.

Aspden, J.A., (1980) The mineralogy of primary inclusions in apatite crystals extracted from Alno ijolite. *Lithos*, 13(3), 263-268.

Aspden, J.A., (1981) The composition of solid inclusions and the occurrence of shortite in apatites from the Tororo carbonatite complex of Eastern Uganda. *Mineralogical Magazine*, 44(334), 201-204.

Bailey, D.K., (1993) Carbonate magmas. *Journal of the Geological Society, London*, 150, 637-651.

Boyd, F.R., Clement, C.R., (1977) Compositional zoning of olivines in kimberlites from the De Beers mine, Kimberley, South Africa. *Carnegie Institution of Washington Yearbook*, 76, 485-493.

Brasier, M.D., Sukhov, S.S., (1998) The falling amplitude of carbon isotopic oscillations through the lower to middle Cambrian: northern Siberia data. *Canadian Journal of Earth Sciences*, 35(4), 353-373.

Brett, R.C., Russell, J.K., Moss, S., (2009) Origin of olivine in kimberlite: Phenocryst or impostor? *Lithos*, 112, 201-212.

Buhn, B., Rankin, A.H., (1999) Composition of natural, volatile-rich Na-Ca-REE-Sr carbonatitic fluids trapped in fluid inclusions. *Geochimica et Cosmochimica Acta*, 63(22), 3781-3797.

Bulanova, G.P., Griffin, W.J., Ryan, C.G., (1998) Nucleation environment of diamonds from Yakutian kimberlites. *Mineralogical Magazine*, 62, 409-419.

Burgess, R., Turner, G., Harris, J.W., (1992) $^{40}Ar$-$^{39}Ar$ laser probe studies of clinopyroxene inclusions in eclogitic diamonds. *Geochimica et Cosmochimica Acta*, 56(1), 389-402.

Canil, D., Fedortchouk, Y., (1999) Garnet dissolution and the emplacement of kimberlites. *Earth and Planetary Science Letters*, 167(3-4), 227-237.

Carroll, M.R., Webster, J.D., (1994) Solubilities of sulfur, noble gases, nitrogen, chlorine, and fluorine in magmas. In: M.R. Carroll, J.R. Holloway (Eds.), *Volatiles in magmas. Reviews in mineralogy, 30* (Ed. by M.R. Carroll, J.R. Holloway), pp. 231-279. Mineralogical Society of America, Washington.

Chakhmouradian, A.R., Mitchell, R.H., (2000) Occurrence, alteration patterns and compositional variation of perovskite in kimberlites. *Canadian Mineralogist*, 38, 975-994.

Clarke, M.G.C., Roberts, B., (1986) Carbonated melilitites and calcitized alkalicarbonatites from Homa Mountain, western Kenya: a reinterpretation. *Geological Magazine*, 123, 683-692.

Clement, C.R., Skinner, E.M.W., (1985) A textural-genetic classification of kimberlites. *Transactions of Geological Society of South Africa*, 88, 403-409.

Clement, C.R., Skinner, E.M.W., Scott Smith, B.H., (1984) Kimberlite re-defined. *Journal of Geology*, 32, 223-228.

Cooper, A.F., Gittins, J., Tuttle, O.F., (1975) The system $Na_2CO_3$-$K_2CO_3$-$CaCO_3$ at 1 kilobar and its significance in carbonatite petrogenesis. *American Journal of Science*, 275, 534-560.

Dalton, J.A., Wood, B.J., (1993) The compositions of primary carbonate melts and their evolution through wallrock reaction in the mantle. *Earth and Planetary Science Letters*, 119(4), 511-525.

Dawson, J.B., (1962a) The geology of Oldoinyo Lengai. *Bulletin Volcanologique*, 24, 349-387.

Dawson, J.B., (1962b) Sodium carbonate lavas from Oldoinyo Lengai, Tanganyika. *Nature*, 195(4846), 1075-1076.

Dawson, J.B., (1980) *Kimberlites and their xenoliths*. Springer-Verlag, Berlin.

Dawson, J.B., (1989) Sodium carbonatite extrusions from Oldoinyo Lengai, Tanzania: implications for carbonatite complex genesis. In: K. Bell (Ed.), *Carbonatites. Genesis and evolution* (Ed. by K. Bell), pp. 255-277. Unwin Hyman, London.

Dawson, J.B., Garson, M.S., Roberts, B., (1987) Altered former alkalic carbonatite lava from Oldoinyo Lengai, Tanzania: Inferences for calcite carbonatite lavas. *Geology*, 15(8), 765-768.

Dawson, J.B., Hawthorne, J.B., (1973) Magmatic sedimentation and carbonatitic differentiation in kimberlite sills at Benfontein, South Africa. *Journal of the Geological Society of London*, 129(1), 61-85.

Dawson, J.B., Pinkerton, H., Norton, G.E., Pyle, D.M., (1990) Physicochemical properties of alkali carbonatite lavas: Data from the 1988 eruption of Oldoinyo Lengai, Tanzania. *Geology*, 18(3), 260-263.

Dawson, J.B., Pyle, D.M., Pinkerton, H., (1996) Evolution of natrocarbonatite from a wollastonite nephelinite parent: Evidence from the June, 1993 eruption of Oldoinyo Lengai, Tanzania. *Journal of Geology*, 104(1), 41-54.

Deans, T., Roberts, B., (1984) Carbonatite tuffs and lava clasts of the Tinderet foothills, western Kenya: a study of calcified natrocarbonatites. *Journal of the Geological Society, London*, 141(MAY), 563-580.

Edgar, A.D., Arima, M., Baldwin, D.K., Bell, D.R., Shee, S.R., Skinner, E.M.W., Walker, E.C., (1988) High-pressure-high-temperature melting experiments on a $SiO_2$-poor aphanitic kimberlite from the Wesselton mine, Kimberley, South Africa. *American Mineralogist*, 73(5-6), 524-533.

Edgar, A.D., Charbonneau, H.E., (1993) Melting experiments on a $SiO_2$-poor, CaO-rich aphanitic kimberlite from 5-10 GPa and their bearing on sources of kimberlite magmas. *American Mineralogist*, 78(1-2), 132-142.

Eggler, D.H., (1989) Kimberlites: how do they form? In: J. Ross, et al. (Ed.), *Kimberlites and related rocks: their composition, occurrence, origin and emplacement*, 1 (Ed. by J. Ross, et al.), pp. 489-504. Blackwell Scientific Publications, Sydney.

Emeleus, C.H., Andrews, J.R., (1975) Mineralogy and petrology of kimberlite dyke and sheet intrusions and included peridotite xenoliths from South West Greenland. *Physics and Chemistry of the Earth*, 9, 179-198.

Fedortchouk, Y., Canil, D., (2004) Intensive variables in kimberlite magmas, Lac de Gras, Canada and implications for diamond survival. *Journal of Petrology*, 45(9), 1725-1745.

Francis, D., Patterson, M., (2009) Kimberlites and aillikites as probes of the continental lithospheric mantle. *Lithos*, 109(1-2), 72-80.

Fraser, K.J., Hawkesworth, C.J., Erlank, A.J., Mitchel, R.H., Scott-Smith, B.H., (1985) Sr, Nd, and Pb isotope and minor element geochemistry of lamproites and kimberlites. *Earth and Planetary Science Letters*, 76, 57-70.

Fulignati, P., Kamenetsky, V.S., Marianelli, P., Sbrana, A., Mernagh, T.P., (2001) Melt inclusion record of immiscibility between silicate, hydrosaline and carbonate melts: Applications to skarn genesis at Mount Vesuvius. *Geology*, 29(11), 1043-1046.

Genge, M.J., Balme, M., Jones, A.P., (2001) Salt-bearing fumarole deposits in the summit crater of Oldoinyo Lengai, Northern Tanzania: interactions between natrocarbonatite lava and meteoric water. *Journal of Volcanology and Geothermal Research*, 106(1-2), 111-122.

Girnis, A.V., Bulatov, V.K., Brey, G.P., (2005) Transition from kimberlite to carbonatite melt under mantle parameters: An experimental study. *Petrology*, 13(1), 1-15.

Girnis, A.V., Ryabchikov, I.D., (2005) Conditions and mechanisms of generation of kimberlite magmas. *Geology of Ore Deposits*, 47(6), 476–487.

Gittins, J., (1989) The origin and evolution of carbonatite magmas. In: K. Bell (Ed.), *Carbonatites. Genesis and evolution* (Ed. by K. Bell), pp. 580-599. Unwin Hyman, London.

Gittins, J., McKie, D., (1980) Alkalic carbonatite magmas: Oldoinyo Lengai and its wider applicability. *Lithos*, 13, 213-215.

Golovin, A.V., Sharygin, V.V., Pokhilenko, N.P., (2007) Melt inclusions in olivine phenocrysts in unaltered kimberlites from the Udachnaya-East pipe, Yakutia: Some aspects of kimberlite magma evolution during late crystallization stages. *Petrology*, 15(2), 168-183.

Golovin, A.V., Sharygin, V.V., Pokhilenko, N.P., Mal'kovets, V.G., Kolesov, B.A., Sobolev, N.V., (2003) Secondary melt inclusions in olivine from unaltered kimberlites of the Udachnaya-East pipe, Yakutia. *Doklady Earth Sciences*, 388(1), 93-96.

Green, D.H., Wallace, M.E., (1988) Mantle metasomatism by ephemeral carbonatite melts. *Nature*, 336, 459-462.

Haggerty, S.E., (1999) A diamond trilogy: Superplumes, supercontinents, and supernovae. *Science*, 285, 851-860.

Hammouda, T., Laporte, D., (2000) Ultrafast mantle impregnation by carbonatite melts. *Geology*, 28(3), 283-285.

Hay, R.L., (1983) Natrocarbonatite tephra of Kerimasi volcano, Tanzania. *Geology*, 11(10), 599-602.

Heaman, L.M., (1989) The nature of the subcontinental mantle from Sr-Nd-Pb isotopic studies on kimberlitic perovskite. *Earth and Planetary Science Letters*, 92(3-4), 323-334.

Hunter, R.H., Taylor, L.A., (1984) Magma-mixing in the low velocity zone: kimberlitic megacrysts from Fayette County, Pennsylvania. *American Mineralogist*, 69, 16-29.

Izraeli, E.S., Harris, J.W., Navon, O., (2001) Brine inclusions in diamonds: a new upper mantle fluid. *Earth and Planetary Science Letters*, 187, 323-332.

Izraeli, E.S., Harris, J.W., Navon, O., (2004) Fluid and mineral inclusions in cloudy diamonds from Koffiefontein, South Africa. *Geochimica et Cosmochimica Acta*, 68, 2561-2575.

Jago, B.C., Gittins, J., (1991) The role of fluorine in carbonatite magma evolution. *Nature*, 349(6304), 56-58.

Kamenetsky, M.B., Sobolev, A.V., Kamenetsky, V.S., Maas, R., Danyushevsky, L.V., Thomas, R., Sobolev, N.V., Pokhilenko, N.P., (2004) Kimberlite melts rich in alkali chlorides and carbonates: a potent metasomatic agent in the mantle. *Geology*, 32(10), 845-848.

Kamenetsky, V.S., Kamenetsky, M.B., Sharygin, V.V., Faure, K., Golovin, A.V., (2007a) Chloride and carbonate immiscible liquids at the closure of the kimberlite magma evolution (Udachnaya-East kimberlite, Siberia). *Chemical Geology*, 237(3-4), 384-400.

Kamenetsky, V.S., Kamenetsky, M.B., Sharygin, V.V., Golovin, A.V., (2007b) Carbonate-chloride enrichment in fresh kimberlites of the Udachnaya-East pipe, Siberia: A clue to physical properties of kimberlite magmas? *Geophysical Research Letters*, 34(9), L09316, doi:10.1029/2007GL029389.

Kamenetsky, V.S., Kamenetsky, M.B., Sobolev, A.V., Golovin, A.V., Demouchy, S., Faure, K., Sharygin, V.V., Kuzmin, D.V., (2008) Olivine in the Udachnaya-East kimberlite (Yakutia, Russia): types, compositions and origins. *Journal of Petrology*, 49(4), 823-839.

Kamenetsky, V.S., Kamenetsky, M.B., Sobolev, A.V., Golovin, A.V., Sharygin, V.V., Pokhilenko, N.P., Sobolev, N.V., (2009a) Can pyroxenes be liquidus minerals in the kimberlite magma? *Lithos*, 112, 213-222.

Kamenetsky, V.S., Kamenetsky, M.B., Weiss, Y., Navon, O., Nielsen, T.F.D., Mernagh, T.P., (2009b) How unique is the Udachnaya-East kimberlite? Comparison with kimberlites from the Slave Craton (Canada) and SW Greenland. *Lithos*, 112, 334-346.

Kamenetsky, V.S., Maas, R., Kamenetsky, M.B., Paton, C., Phillips, D., Golovin, A.V., Gornova, M.A., (2009c) Chlorine from the mantle: Magmatic halides in the Udachnaya-East kimberlite, Siberia. *Earth and Planetary Science Letters*, 285(1-2), 96-104.

Keller, J., Krafft, M., (1990) Effusive natrocarbonatite activity of Oldoinyo Lengai, June 1988. *Bulletin of Volcanology*, 52(8), 629-645.

Keller, J., Spettel, B., (1995) The trace element compositions and petrogenesis of natrocarbonatites. In: K. Bell, J. Keller (Eds.), *Carbonatite volcanism: Oldoinyo Lengai and petrogenesis of natrocarbonatites* (Ed. by K. Bell, J. Keller), pp. 70-86. Springer-Verlag.

Keller, J., Zaitsev, A.N., (2006) Calciocarbonatite dykes at Oldoinyo Lengai, Tanzania: The fate of natrocarbonatite. *Canadian Mineralogist*, 44, 857-876.

Kelley, S.P., Wartho, J.A., (2000) Rapid kimberlite ascent and the significance of Ar-Ar ages in xenolith phlogopites. *Science*, 289(5479), 609-611.

Kinny, P.D., Griffin, W.L., Heaman, L.M., Brakhfogel, F.F., Spetsius, Z.V., (1997) SHRIMP U-Pb ages of perovskite from Yakutian kimberlites. *Russian Geology and Geophysics*, 38(1), 91-99.

Kirkley, M.B., Smith, H.S., Gurney, J.J., (1989) Kimberlite carbonates - a carbon and oxygen stable isotope study. In: J. Ross, et al. (Ed.), *Kimberlites and related rocks: their composition, occurrence, origin and emplacement, 1* (Ed. by J. Ross, et al.), pp. 264-281. Blackwell Scientific Publications, Sydney.

Kjarsgaard, B.A., Pearson, D.G., Tappe, S., Nowell, G.M., Dowall, D.P., (2009) Geochemistry of hypabyssal kimberlites from Lac de Gras, Canada: Comparisons to a global database and applications to the parent magma problem. *Lithos*, 112, 236-248.

Klein-BenDavid, O., Izraeli, E.S., Hauri, E., Navon, O., (2007) Fluid inclusions in diamonds from the Diavik mine, Canada and the evolution of diamond-forming fluids. *Geochimica Et Cosmochimica Acta*, 71(3), 723-744.

Klein-BenDavid, O., Izraeli, E.S., Hauri, E.H., Navon, O., (2004) Mantle fluid evolution - a tale of one diamond. *Lithos*, 77(1-4), 243-253.

Kogarko, L.N., Plant, D.A., Henderson, C.M.B., Kjarsgaard, B.A., (1991) Na-rich carbonate inclusions in perovskite and calzirtite from the Guli Intrusive Ca-carbonatite, Polar Siberia. *Contributions to Mineralogy and Petrology*, 109(1), 124-129.

Kopylova, M.G., Matveev, S., Raudsepp, M., (2007) Searching for parental kimberlite melt. *Geochimica et Cosmochimica Acta*, 71, 3616–3629.

Kostrovitsky, S.I., Morikiyo, T., Serov, I.V., Yakovlev, D.A., Amirzhanov, A.A., (2007) Isotope-geochemical systematics of kimberlites and related rocks from the Siberian Platform. *Russian Geology and Geophysics*, 48(3), 272-290.

Krafft, M., Keller, J., (1989) Temperature measurements in carbonatite lava lakes and flows from Oldoinyo Lengai, Tanzania. *Science*, 245(4914), 168-170.

Le Bas, M.J., (1981) Carbonatite Magmas. *Mineralogical Magazine*, 44(334), 133-140.

Le Bas, M.J., (1987) Nephelinites and carbonatites. In: J.G. Fitton, B.G.J. Upton (Eds.), *Alkaline igneous rocks, Special Publication 30* (Ed. by J.G. Fitton, B.G.J. Upton), pp. 53-83. Geological Society of London.

Le Bas, M.J., Aspden, J.A., (1981) The comparability of carbonatitic fluid inclusions in ijolites with natrocarbonatite lava. *Bulletin of Volcanology*, 44, 429-438.

le Roex, A.P., Bell, D.R., Davis, P., (2003) Petrogenesis of group I kimberlites from Kimberley, South Africa: evidence from bulk-rock geochemistry. *Journal of Petrology*, 44(12), 2261-2286.

Maas, R., Kamenetsky, M.B., Sobolev, A.V., Kamenetsky, V.S., Sobolev, N.V., (2005) Sr, Nd, and Pb isotope evidence for a mantle origin of alkali chlorides and carbonates in the Udachnaya kimberlite, Siberia. *Geology*, 33(7), 549–552.

Marshintsev, V.K., (1986) *Vertical heterogeneity of kimberlite bodies in Yakutiya*. Nauka, Novosibirsk.

Marshintsev, V.K., Migalkin, K.N., Nikolaev, N.C., Barashkov, Y.P., (1976) Unaltered kimberlite of the Udachnaya East pipe. *Transactions (Doklady) of the USSR Academy of Sciences*, 231, 961-964.

Maslovskaja, M.N., Yegorov, K.N., Kolosnitsyna, T.I., Brandt, S.B., (1983) Strontium isotope composition, Rb-Sr absolute age, and rare alkalies in micas from Yakutian kimberlites. *Doklady Akademii Nauk SSSR*, 266(2), 451-455.

Masun, K.M., Doyle, B.J., Ball, S., Walker, S., (2004) The geology and mineralogy of the Anuri kimberlite, Nunavut, Canada. *Lithos*, 76(1-4), 75-97.

McKie, D., (1966) Fenitization. In: O.F. Tuttle, J. Gittins (Eds.), *Carbonatites* (Ed. by O.F. Tuttle, J. Gittins), pp. 261-294. John Wiley and Sons, London.

Mitchell, R., Tappe, S., (2010) Discussion of "Kimberlites and aillikites as probes of the continental lithospheric mantle", by D. Francis and M. Patterson (Lithos v. 109, p. 72-80). *Lithos*, 115(1-4), 288-292.

Mitchell, R.H., (1973) Composition of olivine, silica activity and oxygen fugacity in kimberlite. *Lithos*, 6, 65-81.

Mitchell, R.H., (1978) Mineralogy of the Elwin Bay kimberlite, Somerset Island, N.W.T., Canada. *American Mineralogist*, 63, 47-57.

Mitchell, R.H., (1986) *Kimberlites: mineralogy, geochemistry and petrology*. Plenum Press, New York.

Mitchell, R.H., (1989) Aspects of the petrology of kimberlites and lamproites: some definitions and distinctions. In: J. Ross, et al. (Ed.), *Kimberlites and related rocks: their composition, occurrence, origin and emplacement*, 1 (Ed. by J. Ross, et al.), pp. 7-45. Blackwell Scientific Publications, Sydney.

Mitchell, R.H., (1995) *Kimberlites, orangeites and related rocks*. Plenium Press, New York.

Mitchell, R.H., (2006) An ephemeral pentasodium phosphate carbonate from natrocarbonatite lapilli, Oldoinyo Lengai, Tanzania. *Mineralogical Magazine*, 70(2), 211-218.

Mitchell, R.H., (2008) Petrology of hypabyssal kimberlites: Relevance to primary magma compositions,. *Journal of Volcanology and Geothermal Research*, 174, 1-8.

Moore, A.E., (1988) Olivine: a monitor of magma evolutionary paths in kimberlites and olivine melilitites. *Contributions to Mineralogy and Petrology*, 99, 238-248.

Morogan, V., Lindblom, S., (1995) Volatiles associated with the alkaline - carbonatite magmatism at Alno, Sweden: A study of fluid and solid inclusions in minerals from the Langarsholmen ring complex. *Contributions to Mineralogy and Petrology*, 122(3), 262-274.

Nielsen, T.F.D., Jensen, S.M., (2005) The Majuagaa calcite-kimberlite dyke, Maniitsoq, southern West Greenland, pp. 59. Geological Survey of Denmark and Greenland, Report 2005/43.

Norton, G., Pinkerton, H., (1997) Rheological properties of natrocarbonatite lavas from Oldoinyo Lengai, Tanzania. *European Journal of Mineralogy*, 9(2), 351-364.

Nowell, G.M., Pearson, D.G., Bell, D.R., Carlson, R.W., Smith, C.B., Kempton, P.D., Noble, S.R., (2004) Hf isotope systematics of kimberlites and their megacrysts: New constraints on their source regions. *Journal of Petrology*, 45(8), 1583-1612.

Palyanov, Y.N., Shatsky, V.S., Sobolev, N.V., Sokol, A.G., (2007) The role of mantle ultrapotassic fluids in diamond formation. *Proceedings of the National Academy of Sciences of the United States of America*, 104(22), 9122-9127.

Pal'yanov, Y.N., Sokol, A.G., Borzdov, Y.M., Khokhryakov, A.F., (2002) Fluid-bearing alkaline carbonate melts as the medium for the formation of diamonds in the Earth's mantle: an experimental study. *Lithos*, 60, 145-159.

Panina, L.I., (2005) Multiphase carbonate-salt immiscibility in carbonatite melts: data on melt inclusions from the Krestovskiy massif minerals (Polar Siberia). *Contributions to Mineralogy and Petrology*, 150(1), 19-36.

Panina, L.I., Motorina, I.V., (2008) Liquid immiscibility in deeply derived magmas and the origin of carbonatite melts. *Geochemistry International*, 46(5), 448–464.

Pasteris, J.D., (1984) Kimberlites: Complex mantle melts. *Annual Review of Earth and Planetary Sciences*, 12, 133-153.

Paton, C., Hergt, J.M., Phillips, D., Woodhead, J.D., Shee, S.R., (2007) New insights into the genesis of Indian kimberlites from the Dharwar Craton via in situ Sr isotope analysis of groundmass perovskite. *Geology*, 35(11), 1011–1014.

Patterson, M., Francis, D., McCandless, T., (2009) Kimberlites: Magmas or mixtures? *Lithos*, 112, 191-200.

Pavlov, D.I., Ilupin, I.P., (1973) Halite in Yakutian kimberlite, its relation to serpentine and the source of its parent solutions. *Transactions (Doklady) of Russian Academy of Sciences*, 213(6), 178-180.

Pearson, D.G., Shirey, S.B., Carlson, R.W., Boyd, F.R., Pokhilenko, N.P., Shimizu, N., (1995) Re-Os, Sm-Nd, and Rb-Sr Isotope evidence for thick Archaean lithospheric mantle beneath the Siberian craton modified by multistage metasomatism. *Geochimica et Cosmochimica Acta*, 59(5), 959-977.

Price, S.E., Russell, J.K., Kopylova, M.G., (2000) Primitive magma from the Jericho Pipe, NWT, Canada: Constraints on primary kimberlite melt chemistry. *Journal of Petrology*, 41(6), 789-808.

Safonov, O.G., Kamenetsky, V.S., Perchuk, L.L., (2010) Carbonatite-to-kimberlite link in the chloride-carbonate-silicate systems: experimental approach and application to natural assemblages. *Journal of Petrology*, doi:10.1093/petrology/egq034.

Safonov, O.G., Perchuk, L.L., Litvin, Y.A., (2007) Melting relations in the chloride-carbonate–silicate systems at high-pressure and the model for formation of alkalic diamond–forming liquids in the upper mantle. *Earth and Planetary Science Letters*, 253, 112–128.

Safonov, O.G., Perchuk, L.L., Yapaskurt, V.O., Litvin, Y.A., (2009) Immiscibility of carbonate–silicate and chloride–carbonate melts in the kimberlite–$CaCO_3$–$Na_2CO_3$–KCl system at 4.8 GPa. *Doklady Earth Sciences*, 424(1), 142–146.

Schultz, F., Lehmann, B., Tawackoli, S., Rossling, R., Belyatsky, B., Dulski, P., (2004) Carbonatite diversity in the Central Andes: the Ayopaya alkaline province, Bolivia. *Contributions to Mineralogy and Petrology*, 148, 391-408.

Sharp, Z.D., Barnes, J.D., Brearley, A.J., Fischer, T., Chaussidon, M., Kamenetsky, V.S., (2007) Chlorine isotope homogeneity of the mantle, crust and carbonaceous chondrites. *Nature*, 446, 1062-1065.

Sharygin, V.V., Golovin, A.V., Pokhilenko, N.P., Sobolev, N.V., (2003) Djerfisherite in unaltered kimberlites of the Udachnaya-East pipe, Yakutia. *Doklady Earth Sciences*, 390(4), 554-557.

Shee, S.R., (1986) The petrogenesis of the Wesselton mine kimberlites, Kimberley, Cape Province, R.S.A. University of Cape Town.

Sheppard, S.M.F., Dawson, J.B., (1975) Hydrogen, carbon and oxygen isotope studies of megacryst and matrix minerals from Lesothan and South African kimberlites. *Physics and Chemistry of the Earth*, 9, 747-763.

Shimizu, N., Pokhilenko, N.P., Boyd, F.R., Pearson, D.G., (1997) Geochemical characteristics of mantle xenoliths from Udachnaya kimberlite pipe. *Russian Geology and Geophysics*, 38(1), 194-205.

Skinner, E.M.W., (1989) Contrasting Group I and Group II kimberlite petrology: towards a genetic model for kimberlites. In: J. Ross, et al. (Ed.), *Kimberlites and related rocks: their composition, occurrence, origin and emplacement, 1* (Ed. by J. Ross, et al.), pp. 528-544. Blackwell Scientific Publications, Sydney.

Skinner, E.M.W., Clement, C.R., (1979) Mineralogical classification of Southern African kimberlites. In: F.R. Boyd, H.O.A. Meyer (Eds.), *Kimberlites, diatremes and diamonds: their geology, petrology and geochemistry* (Ed. by F.R. Boyd, H.O.A. Meyer), pp. 129-139. American Geophysical Union, Washington, D.C.

Smith, C.B., (1983) Pb, Sr and Nd isotopic evidence for sources of southern African Cretaceous kimberlites. *Nature*, 304, 51-54.

Smith, C.B., Gurney, J.J., Skinner, E.M.W., Clement, C.R., Ebrahim, N., (1985) Geochemical character of the southern African kimberlites: a new approach based on isotopic constraints. *Transactions of Geological Society of South Africa*, 88, 267-280.

Smith, C.B., Sims, K., Chimuka, L., Duffin, A., Beard, A.D., Townend, R., (2004) Kimberlite metasomatism at Murowa and Sese pipes, Zimbabwe. *Lithos*, 76(1-4), 219-232.

Sobolev, A.V., Sobolev, N.V., Smith, C.B., Dubessy, J., (1989) Fluid and melt compositions in lamproites and kimberlites based on the study of inclusions in olivine. In: J. Ross, et al. (Ed.), *Kimberlites and related rocks: their composition, occurrence, origin and emplacement, 1* (Ed. by J. Ross, et al.), pp. 220-241. Blackwell Scientific Publications, Sydney.

Sobolev, N.V., (1977) *Deep-seated inclusions in kimberlites and the problem of the composition of the upper mantle*. American Geophysical Union, Washington, D.C.

Sobolev, N.V., Logvinova, A.M., Zedgenizov, D.A., Pokhilenko, N.P., Malygina, E.V., Kuzmin, D.V., Sobolev, A.V., (2009) Petrogenetic significance of minor elements in olivines from diamonds and peridotite xenoliths from kimberlites of Yakutia. *Lithos*, 112, 701-713.

Sparks, R.S.J., Baker, L., Brown, R.L., Field, M., Schumacher, J., Stripp, G., Walters, A., (2006) Dynamical constraints on kimberlite volcanism. *Journal of Volcanology and Geothermal Research*, 155, 18–48.

Spence, D.A., Turcotte, D.L., (1990) Buoyancy-driven magma fracture: A mechanism for ascent through the lithosphere and the emplacement of diamonds. *Journal of Geophysical Research*, 95(B4), 5133-5139.

Sun, S.-S., McDonough, W.F., (1989) Chemical and isotopic systematics of oceanic basalts: implications for mantle composition and processes. In: A.D. Saunders, M.J. Norry (Eds.), *Magmatism in the Ocean Basins, 42* (Ed. by A.D. Saunders, M.J. Norry), pp. 313-345. Geological Society Special Publication, London.

Sweeney, R.J., Falloon, T.J., Green, D.H., (1995) Experimental constraints on the possible mantle origin of natrocarbonatite. In: K. Bell, J. Keller (Eds.), *Carbonatite volcanism: Oldoinyo Lengai and petrogenesis of natrocarbonatites, pdf* (Ed. by K. Bell, J. Keller), pp. 191-207. Springer-Verlag.

Tomlinson, E., Jones, A., Milledge, J., (2004) High-pressure experimental growth of diamond using C-$K_2CO_3$-KCl as an analogue for Cl-bearing carbonate fluid. *Lithos*, 77(1-4), 287-294.

Tomlinson, E.L., Jones, A.P., Harris, J.W., (2006) Co-existing fluid and silicate inclusions in mantle diamond. *Earth and Planetary Science Letters*, 250(3-4), 581-595.

Turner, D.C., (1988) Volcanic carbonatites of the Kaluwe complex, Zambia. *Journal of the Geological Society*, 145, 95-106.

Veksler, I.V., (2004) Liquid immiscibility and its role at the magmatic-hydrothermal transition: a summary of experimental studies. *Chemical Geology*, 210(1-4), 7-31.

Veksler, I.V., Nielsen, T.F.D., Sokolov, S.V., (1998) Mineralogy of crystallized melt inclusions from Gardiner and Kovdor ultramafic alkaline complexes: Implications for carbonatite genesis. *Journal of Petrology*, 39(11-12), 2015-2031.

von Eckermann, H., (1948) *The alkaline district of Alno Island.*

Wallace, M.E., Green, D.H., (1988) An experimental determination of primary carbonatite magma composition. *Nature*, 335(6188), 343-346.

Weis, D., Demaiffe, D., (1985) A depleted mantle source for kimberlites from Zaire: Nd, Sr and Pb isotopic evidence. *Earth and Planetary Science Letters*, 73, 269-277.

Zaitsev, A.N., Keller, J., (2006) Mineralogical and chemical transformation of Oldoinyo Lengai natrocarbonatites, Tanzania. *Lithos*, 91(1-4), 191-207.

Zedgenizov, D.A., Kagi, H., Shatsky, V.S., Sobolev, N.V., (2004) Carbonatitic melts in cuboid diamonds from Udachnaya kimberlite pipe (Yakutia): evidence from vibrational spectroscopy. *Mineralogical Magazine*, 68(1), 61-73.

Zedgenizov, D.A., Ragozin, A.L., (2007) Chloride-carbonate fluid in diamonds from the eclogite xenolith. *Doklady Earth Sciences*, 415(6), 961-964.

# 11

# Next Generation Geological Modeling for Hydrocarbon Reservoir Characterization

M. Maučec, J.M. Yarus and R.L. Chambers
*Halliburton Energy Services, Landmark Graphics Corporation, USA*

## 1. Introduction

Hydrocarbon reservoir characterization is a process for quantitatively assigning reservoir properties, recognizing geological and geophysical information and quantifying uncertainties in spatial variability (Fowler *et al.*, 1999). It represents an indispensable tool for optimizing costly reservoir management decisions for hydrocarbon field development. In fact reservoir characterization is the first step in the reservoir development program taking into account structural and depositional architecture, pore systems, mineralogy of the reservoir, post deposition diagenesis and the distribution and nature of reservoir fluids. The technologies and tools for reservoir characterization continue to develop and expand, particularly with the aggressive proliferation of three-dimensional (3D) and lately 3D time-lapse (4D) data. However, one of the most critical areas of inquiry remains development of workflows that best capture and represent geological uncertainty and apply that in the integrated environment to optimize reservoir development and production planning (Goins, 2000; Kramers, 1994).

Although the challenges to find, develop and produce ever new hydrocarbon resources are numerous, the ability of petroleum industry to increase the recovery from existing resources has become a global endeavor. It has been generally accepted that the conventional oil production practices produce, on average, approximately one third of the original oil in place (Kramers, 1994) where estimated remaining unrecovered mobile oil varies with different depositional environments. For example, depositional systems with more complicated stratigraphy and facies architecture, such as fluvial systems or deep-sea fans, may demonstrate even larger amounts of unrecovered mobile oil, ranging from 40-80% (Tyler and Finley, 1991; Larue and Yue, 2003). This common industrial knowledge represents a great incentive to increase the overall production, geared by large capital investments in Smart Reservoir Management workflows and Enhanced Oil Recovery (EOR) operations (Alvarado and Manrique, 2010). Still, the use of oversimplified and uncertain geological models based on a sparse data from limited number of widely-spaced wells can render hydrocarbon recovery forecasting as a daunting task. Moreover, inaccurate description of reservoir heterogeneities is probably one of the outstanding reasons for erroneous description of reservoir connectivity leading to the failure in predicting field performance (Damsleth *et al.*, 1992). Overestimating performance could lead to investment disasters, whereas its underestimation could lead to under-designed production facilities that restrict hydrocarbon recovery.

To mitigate and manage such scenarios, the modern workflows optimize field recovery performance by combining major disciplines of reservoir geosciences, *e.g.* geology, geophysics and petrophysics with numerical simulations into integrated flow models for a variety of field development operations. To maximize reservoir profitability, quantification of the effect of stratigraphic and structural uncertainties on its dynamic performance becomes of principal importance (Charles *et al.*, 2001; Seiler *et al.*, 2009). However, it continues to be an issue in geological modeling that the underlying structural frameworks do not correctly portray the true reservoir structure configuration or size, and this uncertainty impacts an accurate evaluation of hydrocarbon gross volumes. It is therefore essential to validate accordingly the role of individual components of high-resolution geological model (HRGM; see Fig. 1) and rank their impact on the uncertainty of HRGM at various levels (Fig. 1).

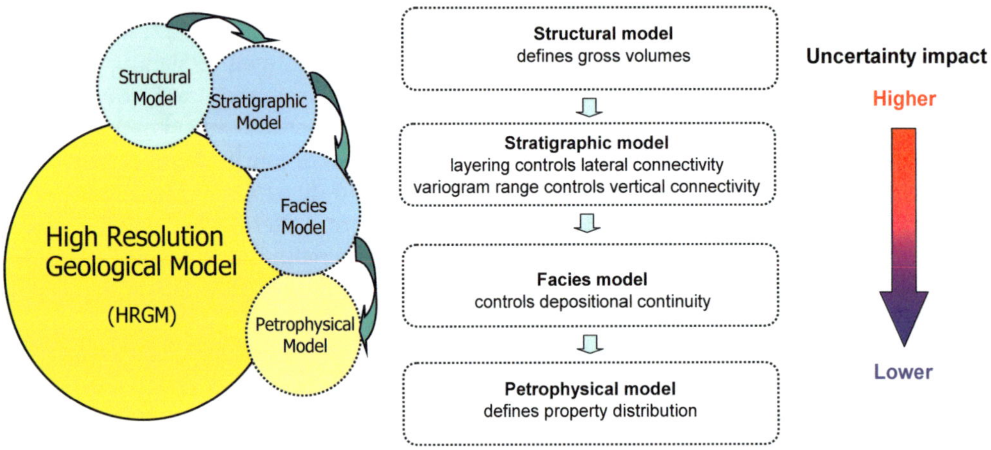


Fig. 1. Generation of High-resolution Geological Model, schematic depiction. The central panel outlines the role of individual phases in overall workflow sequence, with the impact on overall modeling uncertainty ranked to the right.

This paper focuses on recent advances in the technology for building high-resolution, geocellular models and its role in state-of-the-art and future EOR workflows. We first introduce some very basic tools and concepts of geostatistical spatial analysis and modeling, relevant for further understanding of the subject. Furthermore, we highlight what are perceived as some of the outstanding capabilities that differentiate the DecisionSpace® Desktop Earth Modeling, as the next-generation geological modeling tool, from standard industrial approaches and workflows. Current geomodeling practice uses grids to represent 3D reservoir volumes. Estimating gridding parameters is a difficult task and commonly results in artifacts due to topological constraints and misrepresentation of important aspects of the structural framework which may introduce substantial difficulties for dynamic reservoir simulator later in the workflow. We describe a fundamentally novel method that has a potential to resolve most of the common geocellular modeling issues by implementing the concept of interpolation or simulation of reservoir properties using Local Continuity Directions (Yarus *et al.*, 2009; Maučec *et al.*, 2010). Finally, we address some latest developments in the integration of next-generation geological modeling into advanced workflows for quantitative uncertainty assessment and risk management (Maučec and Cullick, 2011) combining two critical steps of reservoir characterization: a) the reconciliation of geomodels with well-production and seismic data, referred to as history-matching (HM) (Oliver and Chen, 2011) and b) dynamic ranking and selection of representative model realizations for reservoir production forecast.

## 2. Methods and techniques

### 2.1 Highlights of geostatistical analysis and modeling

To capture diversity of geologically-complex reservoirs, the next-generation of geological modeling tools are relying increasingly on geostatistical methodologies (Isaaks and Srivastava, 1989; Yarus and Chambers, 1994; Chambers *et al.*, 2000a; Chambers *et al.*, 2000b; Deutsch, 2002; Yarus and Chambers, 2010). The in-depth explanation of geostatistical terminology is available in Olea, 1991. Tailored to identify data limitations and provide better representation of reservoir heterogeneity, geostatistical tools are particularly effective when dealing with data sets with vastly different degrees of spatial density and diverse vertical and horizontal resolution. Classical statistics methods are based on an underlying assumption of data independence in randomly sampled measurements. These assumptions are not true for geosciences data sets, where the data gathered are regionalized (map-able) and demonstrate strong dependence on the distance and orientation. Geostatistical tools provide the unique ability to integrate different types of data, with pronounced variation in scales and direction of continuity. One of the fundamental tools in geostatistics is the *variogram*, a measure of statistical dissimilarity between the pairs of data measurements. It represents a model of spatial correlation and continuity that quantifies the directions and scales of continuity. Variogram analysis can be applied to any regionalized variable $X$ and is used to compute the average square differences between data measurements based on different separation intervals $h$, known as the *lag* interval.

$$\gamma(h) = \frac{\sum_{i=1}^{n}(X_i - X_{i+h})^2}{2n} \quad (1)$$

where $\gamma(h)$ corresponds to the semivariogram (note that denominator $2n$ represents the symmetry relation between data points $X_i$ and $X_{i+h}$) and index $i$ runs over the number of data pairs, $n$. We will in this document refer to semivariogram simply as a variogram. Visualization of a generic variogram $\gamma(h)$ is given in Fig. 2. The left-hand panel corresponds to three distinctive cases of geological continuity: the red curve describes the omni-directional or isotopic variogram that assumes a single, average characteristic direction of continuity, while blue and green curves, correspond to the minimum and maximum direction of continuity, respectively; an anisotropic variogram. The curve of omni-directional variogram will converge to the black line that represents the true variance of the data and is usually referred to as *sill*. One additional parameter of the variogram, which is not depicted in Fig. 2 is the *nugget* effect and corresponds to a discontinuity along the Y-axis resulting in a vertical shift of the variogram curve at the origin. The distance at which the variogram curve levels out at the sill corresponds to the data *correlation range*. The right-hand panel of Fig. 2 is an example of the variogram polar plot, with the major ellipse axis, corresponding to maximum and the (perpendicular) minor ellipse axis, to the minimum direction of continuity.

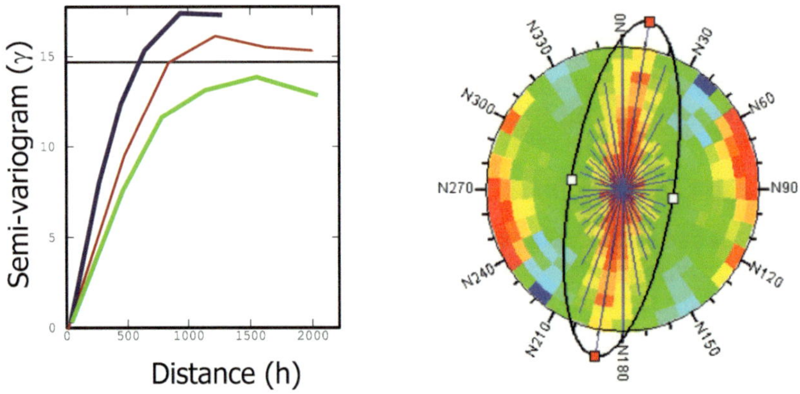


Fig. 2. Visualization of a generic variogram $\gamma(h)$. The left-hand panel corresponds to three distinctive cases of geological continuity: single, average characteristic direction of continuity (red curve), minimum and maximum direction of continuity (blue and green curves, respectively). The black line represents the true variance of the data and is referred to as *sill*. The right-hand panel depicts the variogram polar plot, with the major ellipse axis, corresponding to maximum and the (perpendicular) minor ellipse axis, to the minimum direction of continuity.

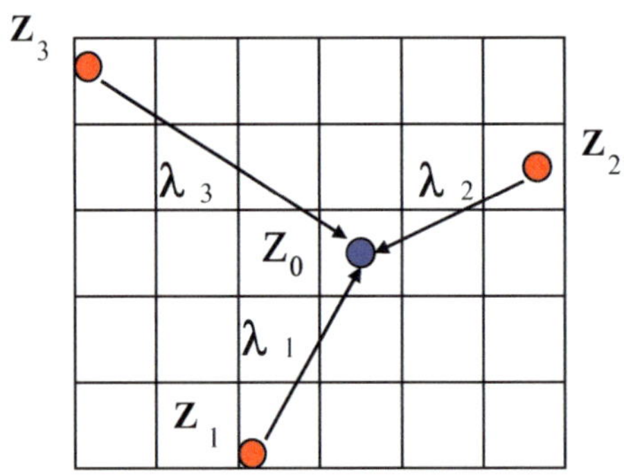


Fig. 3. Principles of kriging, the geostatistical interpolation method. The value of the unsampled location $Z_0$, is estimated based on linear combination of measurements at locations $Z_1$ to $Z_3$, where weights $\lambda_i$ at locations $Z_i$ are calculated from the variogram model.

Among the numerous interpolation methods, the geostatistical kriging algorithm is commonly used in the geosciences. Kriging is an unbiased, linear, least-square regression technique that automatically "de-clusters" data to produce best local or block estimates with minimized error variance. Figure 3 depicts the principles of linear, weighted estimation of the value at location $Z_0$, based on measured values at locations $Z_1$ to $Z_3$:

$$Z_0 = \sum_{i=1}^{n} \lambda_i Z_i \tag{2}$$

where the weights $\lambda_i$ at locations $Z_i$ are calculated from the variogram model. Unlike the more conventional linear weighting estimators, the kriging weights, $\lambda_i$, account for distance and orientation. The constraint for an unbiased estimator is satisfied by maintaining $\sum \lambda_i = 1$.

## 2.2 Definition of lithofacies and mapping of lithotype proportions

High-resolution geomodeling of lithofacies and reservoir properties begins by creating a properly sealed structural framework and then integrating available data from cores, well-logs and seismic surveys. Ideally the lithology model is based on the interpretation of deposition facies from core description. However, for siliciclastic environments, the lithologies are typically electro-facies based on wireline logs calibrated to a few core descriptions. Finally, we often used petro-facies for carbonate reservoir as the primary depositional facies are destroyed by post depositional diagenetic processes. Once the lithology is created, facies are populated with characteristic petrophysical properties.

Micro-logs can identify very thin impermeable shale layers, as thin as 1-2 ft. In order to maintain vertical communication and small scale heterogeneity in the model, particularly in assessing the sweep efficiency of EOR applications (Maučec and Chambers, 2009), *e.g.* $CO_2$ flooding (Culham *et al.*, 2010), it is critical for well blocking to preserve small-scale facies heterogeneities in each interval, assign them correctly to common lithotypes and prevent inadvertently eliminating the essential geological information. The integrity of the blocking and preservation of small scale features is directly proportional to the vertical cells size. If small scale, thin bedding, is critical to sweep efficiencies, then a small vertical cell is required in the model, often resulting in a large multi-million cell geological model, which may be used without upscaling in the dynamic simulator.

Modeling complex geologic environments (*e.g.* fluvial, deltaic) require the ability to control vertical relationships and lateral relationships between the facies. In stratigraphic modeling, which is done on an interval-by-interval basis, the task is to identify the depositional environment and primary depositional facies. Each depositional environment is controlled by physical processes of sedimentation and erosion, which requires the creation of internal bedding geometry, *e.g.* layers representative of the depositional system. These layers act as lines of internal correlation that affect the gathering of statistical information in variogram computation and the distribution of properties in subsequent modeling steps. Once the layering styles are specified for each interval, the well data are re-sampled (coarsened or blocked) at the scale of the layers and a single property value assigned to each layer along the wellbore. For continuous properties, DecisionSpace® Desktop Earth Modeling uses standard averaging methods to assign a value to the gravity center of each grid cell (layer) along the wellbore, biased to the lithology code. For discrete properties, coded by integers (*e.g.* facies) the most commonly occurring facies code is chosen.

Conventional modeling approaches use the average or global facies proportions per interval of interest, which implicitly assumes that the facies proportions are unrealistically the same everywhere throughout the interval and therefore applying a constant lithotype proportion curve (LPC) to the entire interval is inaccurate. Traditional techniques to introduce a geological trend in the data usually require laborious creation of pseudo-wells or application of a generic trend map. The use of a generalized trend map implies that the geological continuity throughout the interval would be fairly similar. In other words, the interpretation of underlying statistics (*e.g.* histograms, variograms) and characteristics such as anisotropy and correlation length, would in mathematical terms, assume the condition of stationarity (Caers, 2005). However, most reservoirs are non-stationary and the introduction of simple trend (vertical or horizontal) could "de-trend" the data, resulting in inadequate solutions and further complicate the computation and interpretation of variograms. The next-generation earth modeling reintroduces a graphical method that generates individual LPCs for each well in the field with the ability to create pseudo-wells to invoke a geological trend that honors the conceptual depositional model.

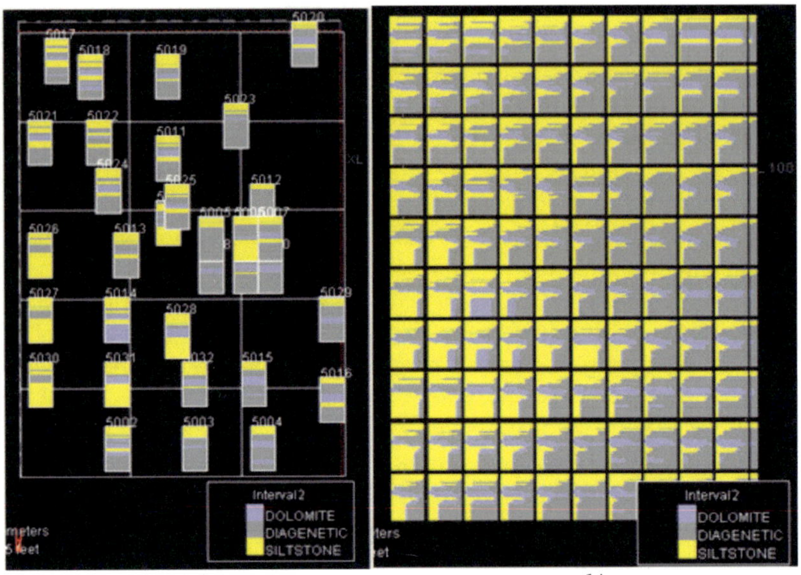


a) b)

Fig. 4. Lithotype proportion mapping: a) example of a map view that uses existing proportion curves, with the ability to create pseudo-wells and impose a trend to honor the conceptual model, b) example of vertical proportion curves organized into vertical proportion matrix (VPM), describing how the facies behave vertically and laterally over the area of the reservoir.

When displayed in the map view (Fig. 4a), the LPCs allow for a quick QC of the vertical variation of the facies distribution and proportions, even before creating the facies model. Editing and copy/move functionality allows the modeler to impose an interpretation to better control trends. The lithotype proportion map (LPM), created from the VPCs, literally consists of hundreds of high resolution trend maps accounting for vertical and lateral non-stationarity (Fig. 4b).

### 2.3 Geologically-driven facies modeling

Facies simulation algorithms, commonly used in earth modeling suffer from a variety of challenges when trying to generate models based on often sparse real data, particularly when attempting to honor depositional facies boundary conditions and proportions, capture depositional overprinting or accounting for geological non-stationarity. For example, the Sequential Indicator Simulation (SIS) (Caers, 2005), lacks the ability to control facies boundary conditions, the Truncated Gaussian Simulation (TGS) (Caers, 2005), provides for only simple facies transitional boundaries and while Object (also termed Boolean) Simulation (OS) can manage most non-overprinted complex facies sets, it is unstable in the presence of high density, closely spaced wells, highly computationally intensive and reliant on the (3D) training image models (Caers, 2005). Multi-Point GeoStatistics (MPS) (Strebelle and Journel, 2001) is rapidly growing in popularity offering the modeler the ability to create geological models with complex geometries, while conditioning to large amounts of well and seismic data. However, as pointed out by Daly and Mariethoz (2011), it is still a relatively new topic, which has had a long academic history and is now just finding its way into commercial software. They also pointed several deficiencies in current implementations related to 1) performance, 2) training image generation, and 3) non-stationarity.

The DecisionSpace® Desktop Earth Modeling approaches facies modeling very differently from most current software offerings. The facies simulation workflow utilizes a powerful combination of describing the geological trends with LPMs created from multiple VPCs, as described previously, integrated with Plurigaussian simulation (PGS), a robust and well-tested algorithm with a long industrial history. The PGS is an expansion of TGS method and uses two Gaussian variograms simultaneously. There are a numerous advantages of PGS over other methods:

- as trends for each facies within each layer and every reservoir interval in the model are calculated, based on the LPMs that account for spatial non-stationarity, the PGS methodology, captures most inter- and intra-facies relationships including post depositional overprinting, such as diagenesis.
- as a pixel-based method, PGS works easily with closely spaced or sparse well control.
- PGS has been available in two commercial software offerings (HERESIM™ and ISATIS™) since the early 1990s, however its use is not necessarily intuitive.
- While it is possible to overcome some of the challenges presented by traditional algorithms through the intervention by experts, the implementation of a LPM with PGS workflow can be presented simply and intuitively.

The next-generation earth model approach to facies modeling introduces a set of facies templates based on the understanding of realistic depositional environments (Walker and James, 1992). The DecisionSpace® Desktop Earth Modeling implements a library of more than forty standard depositional systems, presented with maps and cross-sectional views (see Fig. 5).

The PGS facies modeling requires a set of rules to establish lithotype relationships, where a lithotype is a group of facies sharing common depositional and petrophysical properties. Knowledge of the proportions is not sufficient for accurate modeling of the lithotypes and the depositional system templates help modelers to visualize relationships between the facies and to provide an associated lithotype "rule box" (see: upper left schematics, Fig. 5 - expanded view) that specifies their mathematical relationship. Rules are based on simple or complex lithotype transitions; a) simple transitions corresponds to a strict transition from one lithotype to another and are modeled with one variogram, b) complex transitions require the definition of two lithotype sets, each controlled by its own variogram that can have different anisotropy directions. A lithotype set is a set of lithotypes that share a common spatial model, such as a channel and its associated levy. An example of modeling complex transition of lithotypes, represented by two lithosets and associated variogram models is given in Fig. 6. In this figure the vertical and horizontal sets are schematic representations depicting which variogram controls a lithotype set, when in reality the two variogram models interact to create the final facies model. Because PGS uses lithology proportions, rather than indicators as in SIS, a wider variety variogram model types are available to the modeler. It also provides the ability to model two lithosets, each can exhibit different anisotropic conditions that can be used with calculated LPM to produce realistic models of geological facies distributions.

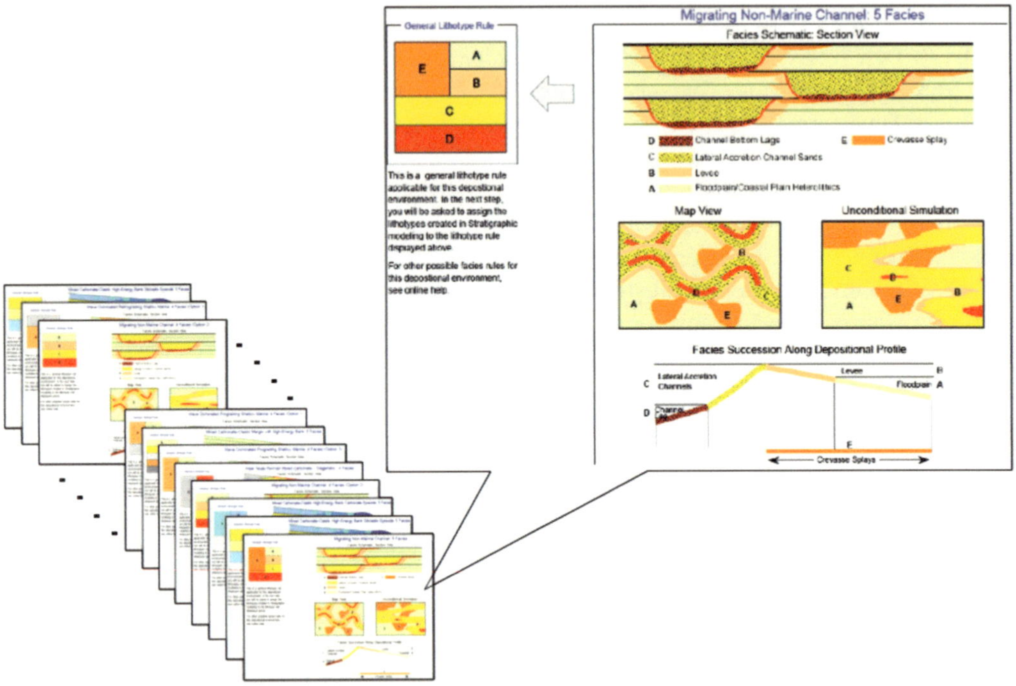


Fig. 5. The DecisionSpace® Desktop Earth Modeling implements a library of more than forty depositional systems, presented with maps and cross-sectional views.

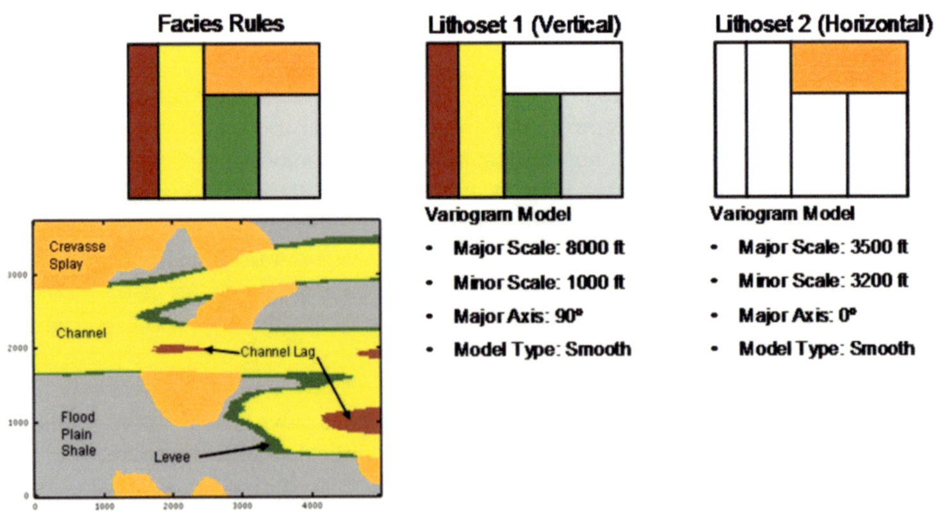


Fig. 6. An example of modeling complex transition of lithotypes, represented by two lithotype sets and associated variogram models

Separate lithology rules are used for each depositional sequence with up to two variograms to controlling different directions and scales. Petrophysical Property Modeling populates facies models with petrophysical properties (porosity, permeability, water saturation, *etc.*) by interval and constrained by facies. The DecisionSpace® Desktop Earth Modeling uses the Turning Bands algorithm as the simulation method with Collocated CoSimulation as one of the options. Kriging and Collocated CoKriging are additional geostatistical algorithms available. Collocated CoKriging is a variation of the classical kriging interpolation, where the variogram is computed from a secondary variable that serves as an additional spatial constraint. Fig. 7, gives an example of facies model realization and facies-constrained model of porosity distribution.

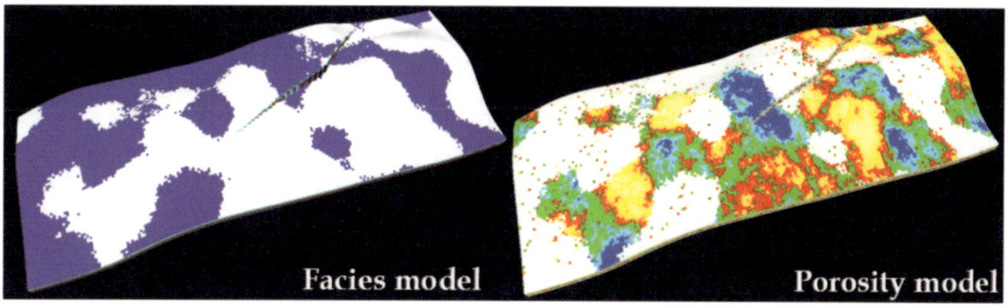


Fig. 7. An example of facies model realization (sand fraction in white, shale fraction in blue) and facies-constrained model of porosity distribution of Brugge synthetic model (Peters e*t al.*, 2009), generated by DecisionSpace® Desktop Earth Modeling.

### 2.4 Novel concepts in local continuity modeling

Traditional reservoir modeling techniques use simplified *two-point statistics* to describe the pattern of spatial variation in geological properties. Such techniques implement a *variogram* model that quantifies the average of expected variability as a function of *distance* and *direction*. In reservoirs, where the geological characteristics are very continuous and easily correlated from well to well, the range (or scale) of correlation will be large while in reservoirs, where the geological characteristics change quickly over short distances, the correlation scales will be shorter. The later phenomenon is very common in sedimentary environments, where the primary mechanism of transport during deposition is water, resulting in highly *channelized* structures (*e.g.* deltaic channels, fluvial deposits, turbidities). These environments usually demonstrate a large degree of local anisotropy and of correlation variation between directions along the channel axis and perpendicular to the channel axis.

Because the principles of conventional (*i.e.* two-point) geostatistical practice still require the nomination of a *single* (average) direction of maximum continuity its use for modeling complex sedimentary environments becomes highly challenging if not impossible. Recently, technology for 3D volumetric modeling of geological properties, using a Maximum Continuity Field (MCF) (Yarus *et al.*, 2009) has been proposed. The new method represents geological properties within a volume of the subsurface by distributing a plurality of data points in the absence of the grid with the notion of geological continuity and directionality represented by MCF, hence entitled as the Point-Vector (PV) method. It introduces several game-changing components to the area of geomodeling:

- Direct control over local continuity directions is controlled using a predefined azimuth map and the local dip (an angle from the horizontal/azimuth plane) of the horizons. The Fault Displacement Field (FDF), annotated in Fig. 8 with symbol FT (fault throw) can be, for example, calculated from the underlying seismic amplitude data (Liang *et al.*, 2010).
- Interactive operation with "geologically intuitive" datasets, such as layering intervals, projection maps and hand drawings via the notion of MCF and the
- Retention of the maximum fidelity of geological model by postponing the creation of grid/mesh until the final stage of (static) model building, immediately before integrating into dynamic model. The reservoir property modeling does not need a standard grid but only the "correct" distance between the points to estimate/simulate the property and data around it.

The key to implementation of these ideas emerges from interpretation of concepts of MCF and their implementation into kriging equations (Eq. 2) for geostatistical estimation. Almost all available geostatistical software restricts the user to certain types of variogram model functions (*e.g.* spherical, exponential, Gaussian etc.) to ensure that a unique set of kriging weights can always be found and to "force" a *single* direction of maximum continuity. However, it is very rare in geology to have a single direction of maximum continuity representative everywhere. Instead, the PV method defines the attributes of Maximum Continuity Vector (MCV) as location, magnitude, direction and length, representing the correlation length (see insert of Fig. 8), along which the magnitude of the geological

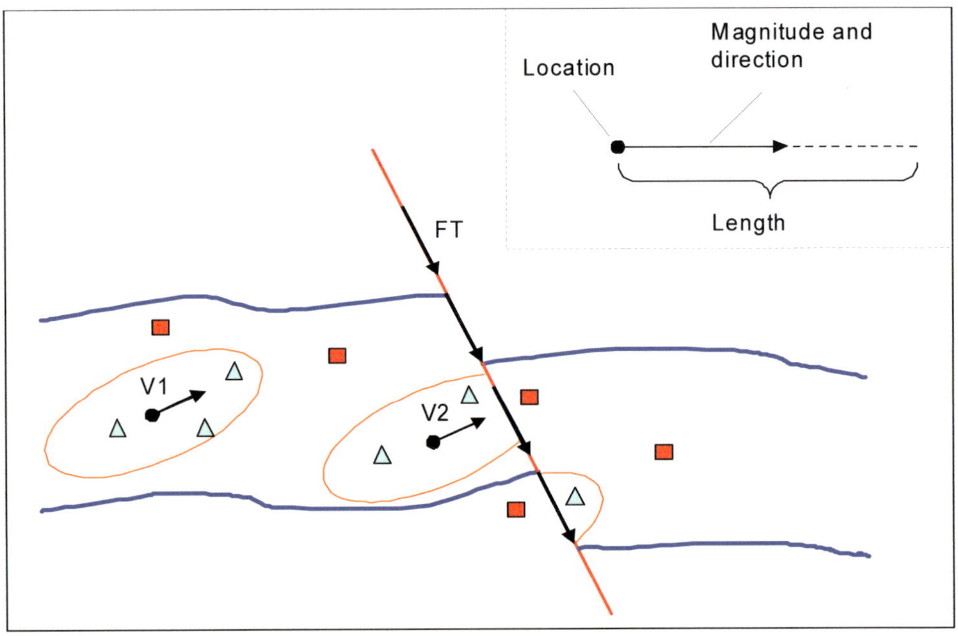


Fig. 8. Interpolation of properties in PV method (2D visualization): the structural framework is represented by top and bottom horizons (in blue) and fault line (in red). The maximum continuity vectors and fault throws are depicted with V and FT, respectively. The data points included in the search neighborhood (ellipse depicted in orange) are represented with triangles in blue, while red squares represent data points excluded from the search.

property remains "substantially the same", *e.g.* within 10%. This allows flexibility in specifying local directions of maximum continuity and interpolation of properties in 3D geological models, aligned with underlying geological structure. The PV property interpolation workflow follows the methodology of Maučec *et al.* 2010:

1. Define structural model and 3D grids where MCF is stored and the property values are interpolated.
2. Define the MCF, using some predefined azimuth map (see Fig. 9b) and the local dip (an angle from the horizontal/azimuth plane) of the horizons.
3. Pre-process of all the fault displacements in the 3D grid (see Fig. 9c).
4. Add known data points (*i.e.* spatially located known reservoir property) to the model, create the covariance neighborhood and the variogram. The covariance calculations take local continuity into account by aligning the axes of the variogram with the local continuity direction.
5. Run ordinary kriging estimator (see Eq. 2), using the created covariance neighborhood. For each point to estimate, the kriging finds the nearest set of known data along shortest geometrical (or Euclidean) distances.

The 2D implementation of property interpolation is schematically depicted in Fig. 8.

The center of the search ellipse (or ellipsoid in 3D) is associated with the location of MCV denoted by V1 and V2. The data points, detected inside the search ellipse ("blue" triangles) are considered in the interpolation along the MCV while the data outside the ellipse ("red" squares) are not included. The relative dimensions of the search ellipsoid, *i.e.* the ratios between major (M), intermediate (I) and minor (m) axis length correspond to "local" anisotropy factor. In a faulted reservoir the property associated with MCV V2 is interpolated across the fault line, following the fault throw vector FT. To validate the PV method, the sealed structural framework, containing top and bottom horizons with a single internal fault, was built from a fluvial reservoir of the Brugge synthetic model (Peters e*t al.*, 2009). Additional details on the validation of PV method are available in Maučec *et al.* 2010.

In Fig. 9a an example of a facies realization for the Brugge fluvial reservoir zone is given. The blue area represents the sand body (pay zone) distributed on shale (non-pay zone in red). Using the facies distribution as a basis or constraint for the generation of a vector field emulates a certain "pre-knowledge" on the geological structure but is in no way a required step. Fig. 9b represents MCF defined on virtually regular mesh of points. No particular continuity information was assumed for the shale zone, only for the discrete sand bodies. The azimuth alone is used from the MCF; the dip angle is calculated as the normal direction relative to the local curvature of the horizon. Fig. 9c visualizes fault displacement vector field with a constant throw of ~50 m.

An example permeability distribution, generated with the MCF-based method of reservoir property interpolation is shown in Fig. 9d, depicting the flattened 2D map view of the major-minor plane and the fault location. The PV method allows the user to define variable sizes of search ellipsoids throughout the model volume of interest (VOI). By this notion, a single-size search sphere was defined for the shale facies, where no particular continuity information was assumed (see Fig. 9b). A variable sized search ellipsoid and different anisotropy factors (*i.e.* ratios between the length (in ft) of the major direction and minor direction of the search ellipsoid in M-m plane) were considered for the sand zone to validate the impact on the interpolation. Qualitatively, the anisotropy ratio of 10:1 (major/minor = 10000/1000) can be interpreted for example as the case with less uncertain (and more trusted) MCF data, used to obtain permeability distribution as depicted in Fig. 9d. The corner insert of Fig. 9d represents the M-I plane cross-section, indicating the fault location. The permeability model effectively represents a VOI with no imposed stratigraphic grid and as such retains the maximum available resolution and information density, limited only by the resolution of underlying data and MCF.

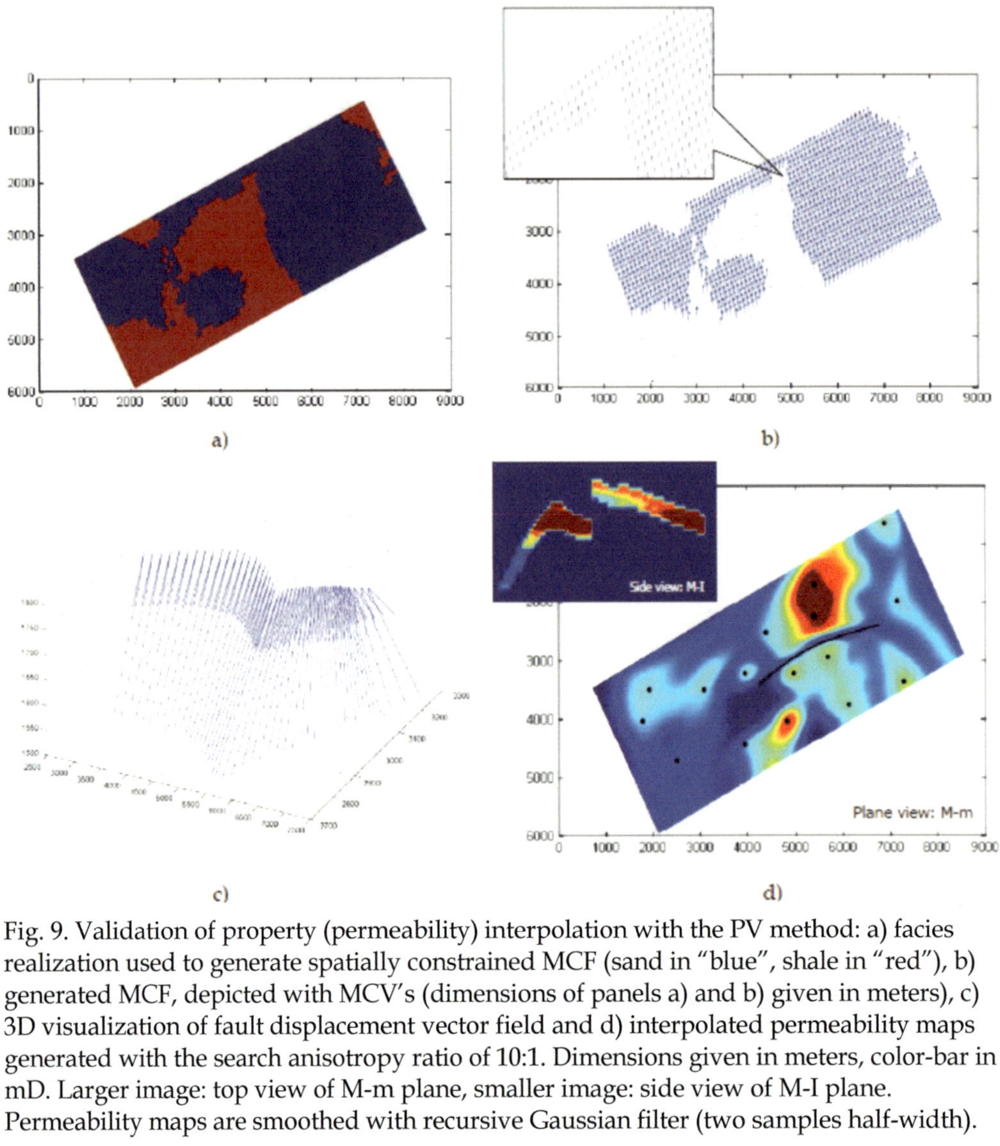


Fig. 9. Validation of property (permeability) interpolation with the PV method: a) facies realization used to generate spatially constrained MCF (sand in "blue", shale in "red"), b) generated MCF, depicted with MCV's (dimensions of panels a) and b) given in meters), c) 3D visualization of fault displacement vector field and d) interpolated permeability maps generated with the search anisotropy ratio of 10:1. Dimensions given in meters, color-bar in mD. Larger image: top view of M-m plane, smaller image: side view of M-I plane. Permeability maps are smoothed with recursive Gaussian filter (two samples half-width).

### 2.5 Conditioning reservoir models to production data

Accurate modeling of the hydrocarbon reservoir production behavior is of fundamental importance in reservoir engineering workflows for optimization of reservoir performance, operational activities, and development plans. A realistic description of geological formations and their fluid-flow-related properties, such as lithofacies distribution and permeability, represents a crucial part of any modeling activity. To quantify and reduce the uncertainty in the description of hydrocarbon reservoirs, the parameters of geological model are usually adjusted and reconciled with pressure and multi-phase production data by history-matching (HM). With an advent of computing capabilities in recent decades, the classical (*i.e.* manual) HM has evolved to so-called computer-Assisted (or Automated) HM (AHM) technology. When we hereafter in this document refer to the history-matching process, the AHM workflow is assumed.

As an inverse problem, HM is highly non-linear and ill-posed by its nature, which means that, depending on the *prior* information, one can obtain a set of non-unique solutions that honor both the prior constraints and conditioned data with associated uncertainty. To assess the uncertainty in estimated reservoir parameters, one must sample from the *posterior* distribution, and the Bayesian methods (Lee, 1997) provide a very efficient framework to perform this operation. Using Bayes' formula, the posterior distribution (*i.e.,* the probability of occurrence of model simulated parameter, **m**, given the measured data values, **d**) is represented as being proportional to the product of prior and likelihood probability distributions of the reservoir model:

$$p_{m|d}(\mathbf{m}\,|\,\mathbf{d}) = \frac{p_{d|m}(\mathbf{d}\,|\,\mathbf{m})\,p_m(\mathbf{m})}{p_d(\mathbf{d})} \tag{3}$$

where, $p_{m|d}(\mathbf{m}\,|\,\mathbf{d})$, $p_{d|m}(\mathbf{d}\,|\,\mathbf{m})$ and $p_m(\mathbf{m})$ represent the posterior, likelihood, and prior distribution, respectively. The normalization factor $p_d(\mathbf{d})$ represents the probability associated with the data and usually treated as a constant.

When the distribution of prior model parameters, $\mathbf{m}$ ($\mathbf{m} \in \Re^M$, where $M$ represents the number of parameters), follows a multi-Gaussian probability density function (pdf), the $p_m(\mathbf{m})$, centered around the prior mean $\mathbf{m}^0$, is given by:

$$p_m(\mathbf{m}) = \frac{1}{(2\pi)^{M/2}|\mathbf{C_M}|^{1/2}} \exp\left[-\frac{1}{2}\left(\mathbf{m}-\mathbf{m}^0\right)^{\mathrm{T}} \mathbf{C}_{\mathbf{M}}^{-1}\left(\mathbf{m}-\mathbf{m}^0\right)\right] \tag{4}$$

with $\mathbf{C_M}$ as the prior covariance matrix ($\mathbf{C_M} \in \Re^{MxM}$). The distribution of likelihood data is defined as the conditional pdf $p_{d|m}(\mathbf{d}\,|\,\mathbf{m})$ of data, **d**, given model parameters, **m**:

$$p_{d|m}(\mathbf{d}\,|\,\mathbf{m}) = \frac{1}{(2\pi)^{N/2}|\mathbf{C}_D|^{1/2}} \exp\left[-\frac{1}{2}\left(\mathbf{d}-\mathbf{g}(\mathbf{m})\right)^{T} \mathbf{C}_D^{-1}\left(\mathbf{d}-\mathbf{g}(\mathbf{m})\right)\right] \tag{5}$$

with $\mathbf{C}_D$ as the data covariance matrix ($\mathbf{C}_D \in \Re^{NxN}$). The relationship between the data and the model parameters is expressed as a non-linear function that maps the model parameters into the data space, $\mathbf{d} = \mathbf{g}(\mathbf{m})$, where **d** is the data vector with $N$ observations representing the output of the model, **m** is a vector of model parameters, and **g** is the forward model operator that maps the model parameters into the data domain. For history-matching problems, **g** represents the reservoir simulator. Using Bayes' theorem (Eq. 3), both prior and likelihood pdfs are combined to define the posterior pdf as:

$$p_{m|d}(\mathbf{m}\,|\,\mathbf{d}) \propto \exp\left[-O(\mathbf{m})\right] \tag{6}$$

where, the objective function $O(\mathbf{m})$ combines prior and likelihood terms:

$$O(\mathbf{m}) = \frac{1}{2}\left(\mathbf{d} - \mathbf{g}(\mathbf{m})\right)^{\mathrm{T}} \mathbf{C}_D^{-1}\left(\mathbf{d} - \mathbf{g}(\mathbf{m})\right) + \frac{1}{2}\left(\mathbf{m} - \mathbf{m}^0\right)^{\mathrm{T}} \mathbf{C}_{\mathbf{M}}^{-1}\left(\mathbf{m} - \mathbf{m}^0\right) \tag{7}$$

The *Maximum A Posteriori* (MAP) $p_{m|d}(\mathbf{m}\,|\,\mathbf{d})$ pdf corresponds to the minimum of $O(\mathbf{m})$, with the set of parameters, $\mathbf{m}$ that minimizes $O(\mathbf{m})$ as the most probable estimate. The HM minimization algorithm renders multiple plausible model realizations and the consequence of non-linearity is that it requires an *iterative* solution. When considering realistic field conditions, the number of parameters of the prior model expands dramatically (*i.e.*, order of $10^6$) and computation of the prior term of the objective function becomes highly demanding and time consuming. A variety of model parameterization and reduction techniques have been implemented in HM workflows, ranging from methods based on linear expansion of weighted eigenvectors of the specific block covariance matrix $\mathbf{C}_{\mathbf{M}}$ (Rodriguez *et al.*, 2007; Jafarpour and McLaughlin, 2009; Le Ravalec Dupin, 2005) to methods, where expensive covariance matrix computations are avoided by generating model updates in wave-number domain (Maučec *et al.*, 2007; Jafarpour and McLaughlin, 2009; Maučec, 2010) that do not require specifying the model covariance matrix, $\mathbf{C}_{\mathbf{M}}$, and performing expensive inversions.

### 2.5.1 Sampling from the posterior distribution

Two methods have been proposed to sample parameters of posterior distribution, for example, sequential Markov chain Monte Carlo (MCMC) algorithms (Neal, 1993) and approximate sampling methods, such as Randomized Maximum Likelihood (RML) (Kitanidis, 1995), both with some inherent deficiencies. Traditional Markov chain Monte Carlo (MCMC) methods attempt to simulate direct draws from some complex statistical distribution of interest. MCMC techniques use the previous sample values to randomly generate the next sample value in a form of a chain, where the transition probabilities between sample values are only a function of the most recent sample value. The MCMC methods arguably provide, statistically, the most rigorous and accurate basis for sampling posterior distribution and uncertainty quantification but they come at high computational costs. On the other hand, the approximate, but faster, RML methods are, in practice, applicable mostly to linear problems.

In an attempt to improve computational efficiency and mixing for the MCMC algorithm, Oliver *et al.*, 1996, proposed a two-step approach in which (1) model and data variables were jointly sampled from the prior distribution and (2) the sampled model variables were calibrated to the sampled data variables, with Metropolis-Hastings sampler (Hastings, 1970) used as the acceptance test. The method works well for linear problems, though it does not hold for non-linear problems, such as HM studied here. To improve on that, Efendiev *et al.*, 2005, proposed a *rigorous* two-step MCMC approach to increase the acceptance rate and reduce the computational effort by using the sensitivities calculated from tracing streamlines (Datta-Gupta and King, 2007). When the sensitivities are known, the solution of the HM inverse problem is greatly simplified. One of the most important advantages of the streamline approach is the ability to analytically compute the sensitivity of the streamline Generalized-Travel-Time (GTT) with respect to reservoir parameters, *e.g.* porosity or permeability. The GTT is defined as an optimal time-shift $\Delta\tilde{t}$ at each well, so as to minimize the production data misfit function $J$:

$$J = \sum_{i=1}^{Ndj} \left[ y_j^{obs}\left(t_i + \Delta t_j\right) - y_j^{cal}\left(t_i\right) \right]^2 = f(\Delta t_j) \quad (8)$$

where $N_{dj}$ stands for the number of observed data (**d**) at well $j$ and $y_j^{obs}$ and $y_j^{cal}$ correspond to observed and calculated production data, respectively, at well $j$.

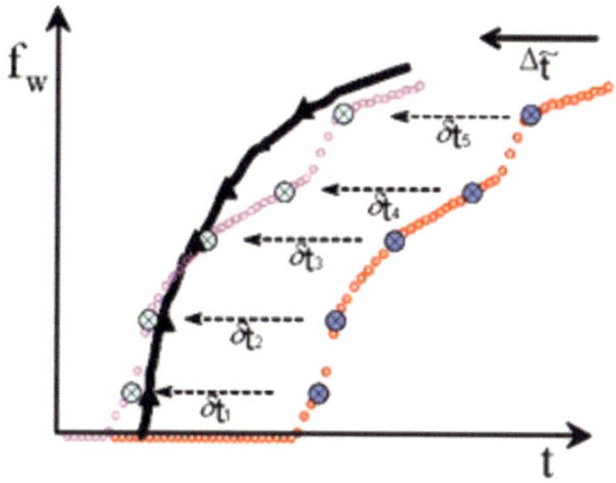


Fig. 10. Illustration of Generalized Travel-Time (GTT) inversion by systematically shifting the calculated fractional flow curve $f_w$ to the observed history (modified from Datta-Gupta and King, 2007). Red and magenta symbols correspond to the initially-calculated and shifted curve, respectively, while the black line represents the observed curve.

This is illustrated in Fig. 10, where the calculated fractional flow response[1] is systematically shifted in small-time increments towards the observed response, every data point in the fractional-flow curve has the same shift time, $\delta t_1 = \delta t_2 = ... = \Delta\tilde{t}$ and the data misfit is computed for each time increment. The misfit function $J$ directly corresponds to the term $\mathbf{d} - \mathbf{g}(\mathbf{m})$, given in Eqs. 5 and 7 that defines the misfit between the observed data and simulated response. The objective of HM inversion workflow is to minimize the misfit in production response by reconciling the geological model with observed (measured) dynamic production data.

The two-step MCMC algorithm (Efendiev *et al.* 2005) uses an approximate likelihood calculation to improve on the (low) acceptance rate of the one-step algorithm (Ma *et al*, 2008). This approach does not compromise the rigor in traditional MCMC sampling, as it adequately samples from the posterior distribution and obeys the *detailed balance* (Maučec *et al.*, 2007), thus, a sufficient condition for a unique stationary distribution. The main steps of the streamline-based, two-step MCMC algorithm are depicted in a flowchart in Fig. 11.

A pre-screening based on approximate likelihood calculations eliminates most of the rejected samples, and the exact MCMC is performed only on the accepted proposals, with higher acceptance rate. The approximate likelihood calculations are fast and typically involve a linearized approximation around an already accepted state rather than an

[1] In water-injection EOR operations, the fractional flow curve frequently corresponds to water-cut curve that represents the water breakthrough at the well as a function of well production time.

expensive computation, such as a flow simulation, using the streamline sensitivities, calculated per geomodel grid block and per production well.

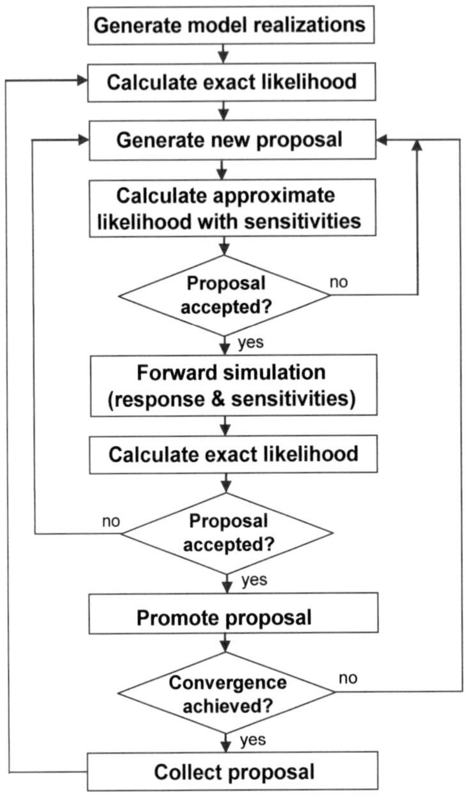


Fig. 11. Flowchart of the streamline-based, two-step MCMC algorithm as implemented in AHM workflow (Maučec *et al.* 2007; Maučec *et al.* 2011a; Maučec *et al.* 2011b).

The streamline-based, two-step MCMC history-matching workflow represents an integral component of the new prototype technology for Quantitative Uncertainty Management (QuantUM) (Maučec and Cullick, 2011) that seamlessly integrates the next-generation Earth Modeling API, VIP® and/or Nexus® reservoir simulators and the code DESTINY (MCERI Research Group, 2008), used as a generator of streamline-based sensitivities. The workflow was validated by reconciling the dynamic well production data with a completely synthetic model of the Brugge field (Maučec *et al.* 2011a). The stratigraphy of Brugge field combines four different depositional environments and one internal fault with a modest throw. The dimensions of the field are roughly 10x3 km. In the original (referred to as the "truth-case") high-resolution model of 20 million grid cells essential reservoir properties, including sedimentary facies, porosity and permeability, Net-To-Gross (NTG), and water saturation were created for the purpose of generating well log curves in the 30 wells (Peters *et al.* 2009). Multiple realizations of high-resolution (211x76x56, *i.e.*, approximately 900k grid cells) of facies-constrained permeability model (referred to as "initial") were generated using the information on the structural model, well properties and depositional environments based on the "truth-case" model.

In general, the real-field HM workflows can be highly computationally demanding, mostly on the account of time-consuming forward reservoir simulations. Furthermore, when the full-fledged uncertainty analysis of the high-resolution geological model is addressed through the generation of multiple (*i.e.*, sometimes in the order of 100s) static model realizations, the multi-iteration AHM workflows may become prohibitively expensive. The QuantUM AHM module addresses the issue of computational efficiency in two ways: a) takes full advantage of parallel execution of VIP® and/or Nexus® reservoir simulator, wherever multi-CPU cores are available and b) uses the option of computational load distribution via standard submission protocol wherever the multi-node computational resources are available.

The proposals generated by the Metropolis-Hastings sampler of the two-step MCMC-based inversion workflow are very likely positively correlated; therefore, the convergence diagnostics ought to be governed by the estimators averaged over the ensemble of realizations. QuantUM AHM workflow implements the *maximum entropy* test (Full *et al.*, 1983), where the (negative) entropy, *S*, of the sampled stationary (posterior) distribution is defined as the expected value of the logarithms of posterior terms of the objective function, $p_{m|d}(\mathbf{m}\,|\,\mathbf{d})$. Further mathematical derivations of the entropy, *S* and its variance, implemented as convergence measures in AHM workflow are given in Maučec *et al.* 2007 and Maučec *et al.* 2011a. Selected results of QuantUM AHM workflow validation are given in Fig. 12, with additional information available in (Maučec *et al.* 2011a; Maučec *et al.* 2011b):

- The behavior of (negative) entropy, *S* (Fig. 12a) and the objective function, defined as the logarithm of transition probability of two-step Metropolis-Hastings sampler (Fig. 12b) demonstrate the convergence rate of the MCMC sequence with the burn-in period of approximately 750 samples (*i.e.*, the total number of processed samples is 1500, a product of 50 model realizations and 30 MCMC iterations).
- Comparison of dynamic well production responses, here defined as the water-cut curves, calculated with prior (Fig. 12c) and posterior (Fig. 12d) model realizations demonstrate the efficiency of water-cut misfit reduction, between the simulated and observed data. To demonstrate the case, the production response of one of the wells with the most pronounced production dynamics over the 10-year period, is depicted. Such non-monotonic behavior is usually most challenging to match.
- The HM workflow demonstrates a significant reduction in the discrepancy of mean dynamic response (Fig. 12e), calculated over ensemble of 50 history-matched models with respect to observed watercut well water-cut curve as well as impressive reduction of the ensemble-averaged variance (Fig. 12f) of history-matched responses with respect to observed production curves.
- Figs. 12g and 12h, depict log-permeability maps for one realization of top layer of Brugge fluvial reservoir, corresponding to prior (*i.e.*, initial, not history-matched) and posterior (*i.e.*, history-matched) models, respectively. The areas where the history-matching algorithm attempts to reconcile the static model with dynamic data, by connecting spatially-separated high-permeability areas to facilitate the fluid flow, as governed by the calculated streamline-sensitivities, are emphasized.

## 2.6 Quantification of reservoir production forecast uncertainty

By its nature, the probabilistic history-matching workflows use multiple equally probable but non-unique realizations of geological models that honor prior spatial constraints and

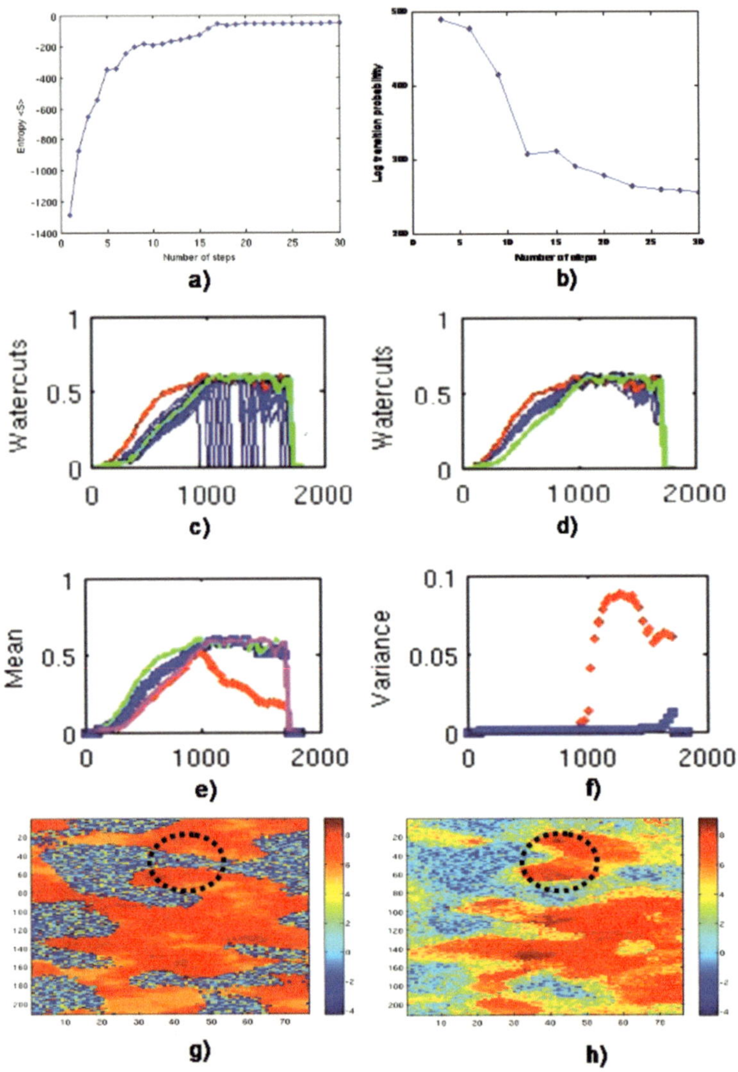


Fig. 12. Validation of QuantUM AHM workflow with dynamic inversion of Brugge synthetic model (Peters *et al.* 2009): a) and b) algorithm convergence diagnostics, (negative) entropy and objective function, respectively, c) and d) well water-cut response curves, obtained with an ensemble of prior and posterior models, respectively, e) and f) ensemble-averaged statistical estimators, mean and variance, respectively and g) and h) prior and posterior log-permeability distribution maps of model top layer, respectively

are conditioned to the data as well as approximate the forecast uncertainty. But the crux of the matter here is two-fold: a) throughout the inversion, some model realizations may have created non-geologically realistic features and b) many of the underlying geological parameters may have an insignificant effect on recovery performance.

In addition to the traditional, single-parameter sensitivity studies to identify the important and geologically relevant parameters, a more sophisticated version uses streamline simulation as a tool for pre-screening geological models (Gilman *et. al.,* 2002) or Experimental Design (ED) techniques that optimally identify values of uncertainty and their variation range (Prada *et. al.,* 2005). Unfortunately, in realistic reservoir forecasting projects, not all of the flow simulations may be necessary but the clear distinction between important and un-important parameters is unknown *a priori*. A workflow aiming to identify and isolate model realizations, relevant to reservoir forecast analysis, out of a full spectrum of statistically probable models has recently been proposed by Scheidt and Caers, 2009 combining the following steps (Fig. 13):

- Computation of a single parameter, namely, pattern-dissimilarity distance, used to distinguish two individual model realizations in terms of dynamic performance. The objective is to identify a set of representative reservoir models through pattern-dissimilarity distance analysis focusing on the dynamic properties of the realizations.
- Computation of pattern-dissimilarity distances are computed via rapid streamline simulations carried out for each ensemble member. Analysis of pattern-distances gives rise to a set of representative models, which are then simulated using a full-physics finite-difference simulator.
- Derivation of the forecast uncertainty from the outcome of these intelligently selected few full-physics simulations.

Further details are available in (Scheidt and Caers, 2009; Maučec *et al.,* 2011b).

### 2.6.1 The concept of dynamic data (dis)similarity

To describe the degree of (dis)similarity between reservoir model realizations in an ensemble it is not required to identify individual reservoir characteristics and corresponding dynamic responses for each ensemble member as knowledge about a representative "difference" (hereafter referred to as "distance") between any two realizations is sufficient. An example of pattern-dissimilarity distance $\theta$, defined as an Euclidean measure, that describes the degree of (dis)similarity between any two of reservoir realizations **m** indexed with $k$ and $l$ within an ensemble of size $I$, in terms of geologic characteristics and pertinent dynamic response, $\tilde{r}_{kl}$, *i.e.* recovery factor, oil production rate, *etc.*

$$\theta_{kl} = \sqrt{\sum_{i=1}^{I} (\tilde{r}_{ki} - \tilde{r}_{li})^2} \tag{9}$$

where, by definition, the pattern-dissimilarity distance honors self-similarity ($\theta_{kk} = 0$) and symmetry ($\theta_{kl} = \theta_{lk}$). Pattern-dissimilarity distances are evaluated using rapid streamline simulation, and assembled into a pattern-dissimilarity matrix, $\mathbf{\Theta} = \{\theta_{kl}\}$, where $\mathbf{\Theta} \in \Re^{IxI}$.

### 2.6.2 Multi dimensional scaling and cluster analysis

Multi-dimensional scaling (MDS) is used to translate the pattern-dissimilarity matrix models into a *p*-dimensional Euclidean space (Borg and Groenen, 2005) where each element of the matrix is represented with a unique point. Hereafter, Euclidean space will be simply referred to as the *E* space, with individual points arranged in such a way that their distances correspond in a least-squares sense to the dissimilarities of individual realizations. Euclidean distances tend to exhibit strong correlation with pattern-dissimilarity distances.

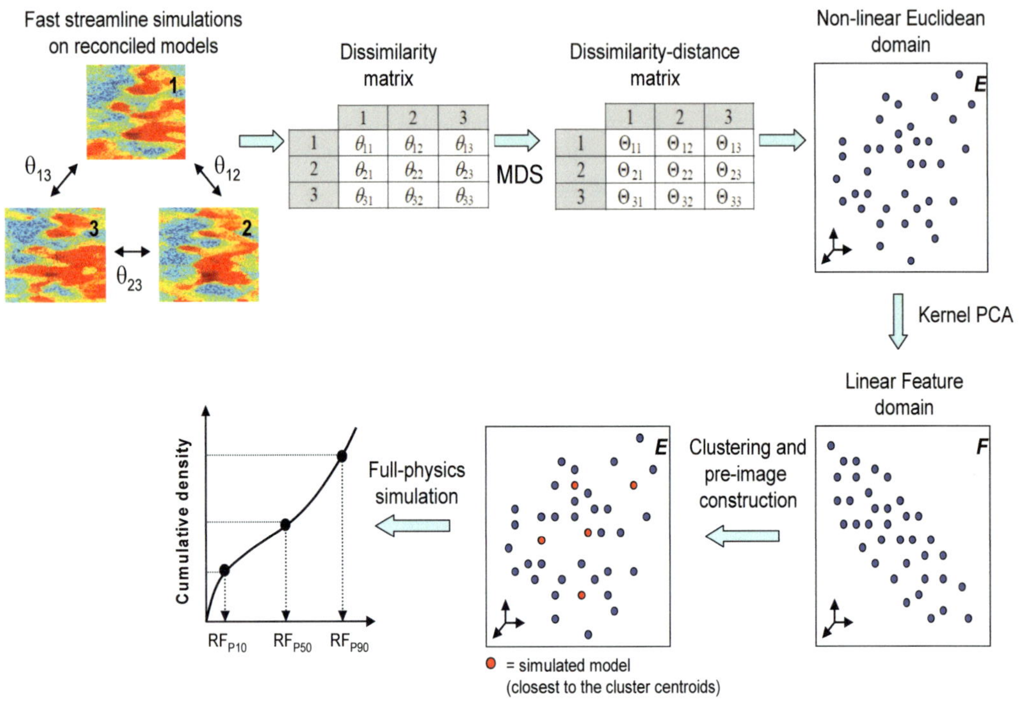


Fig. 13. Workflow diagram using Multi Dimensional Scaling (MDS), k-PCA and k-means clustering (modified from Scheidt and Caers, 2009 for three model realizations).

The workflow of Scheidt and Caers, 2009 incorporates a classical variant of metric MDS where intervals and ratios between points are preserved in a manner of highest fidelity and where it is assumed that the pattern-dissimilarity distances directly correspond to distances in the *E* space. MDS finds the appropriate coordinates consistent with Euclidean distance measure and the resulting map is governed exclusively by the pattern-dissimilarity distances. Subject to the condition, that the distances are strongly correlated with the dynamic response $\tilde{r}$, points within close proximity of each other exhibit similar recovery characteristics, and hence, they are expected to contain similar geologic features. A clustering algorithm is used to compartmentalize the Euclidean space into few distinct clusters and to identify the realizations within the closest proximity of the cluster centroids. The characteristic nonlinear structure of points in the *E* calls for kernel techniques such as kernel Principal Component Analysis (k-PCA) (Schölkopf *et al.*, 1996), where prior to clustering nonlinear domains are mapped into a linear domain. When the MDS data are mapped into a linear, high-dimensional feature domain *F*, using appropriate kernel method *e.g.* k-PCA, cluster analysis, such as *k*-means clustering (Tabachnick and Fidell, 2006) may be applied to further compartmentalize the feature space F in order to identify cluster centroids, where centroid of a cluster corresponds to a point whose parameter values are the mean of the parameter values of all the points in the clusters. The pattern-dissimilarity distances θ, closest to the cluster centroids in the Euclidean space are defined with respect to recovery factor (RF) responses of two individual realizations *k* and *l* at times $t \in T$, following

Eq. 9 as $\theta_{kl} = \sqrt{\sum_{t \in T} (RF_k - RF_l)^2}$ . The geological realizations, corresponding to identified cluster-centroids are selected as the "representative samples" of the entire uncertainty space and simulated with the full-physics simulator as the reference case. Simulation outcome is post-processed to compute the distribution of ultimate recovery factor (URF) after a lengthy period of production, usually selected as the target forecast uncertainty for quantification. The cumulative density functions (CDFs) are finally constructed for the selected model realizations with weights assigned to URFs based on the number of models in each particular cluster and finally quantitatively compared to the reference CDF derived from full-physics simulations.

## 3. Discussion and conclusions

Reservoir characterization encompasses techniques and methods that improve understanding of geological petrophysical controls on a reservoir fluid flow. Presence of a large number of geological uncertainties and limited well data often render recovery forecasting a difficult task in typical appraisal and early development settings. Moreover, in geologically-complex, heavily faulted reservoirs, quantification of the effect of stratigraphic and structural uncertainties on the dynamic performance, fluid mobility and *in situ* hydrocarbons is of principal importance. Although the generation of a sound structural framework is one of the major contributors to uncertainty in hydrocarbons volumes, and therefore risk, in reservoir characterization it often represents a compromise between the actual structure and what the modern modeling technology allows (Hoffman *et al.*, 2007).

In this paper we focus on some of the outstanding features that have the potential to significantly differentiate the DecisionSpace® Desktop Earth Modeling, as the next-generation geological modeling technology, from standard industrial approaches and workflows mainly in the areas of geologically-driven facies modeling, reservoir property modeling in grid-less modality and the state-of-the-art workflows for dynamic quantitative uncertainty and risk management throughout the asset lifecycle. The facies simulation workflow utilizes a powerful combination of describing the spatial geological trends with lithotype proportion curves and matrices, integrated with Plurigaussian simulation (PGS), a robust and widely-tested algorithm with long industrial history. While the implementation of PGS is not unique to DecisionSpace® Desktop Earth Modeling its geologically highly intuitive approach to modeling, based on understanding of realistic depositional environments puts the geologist back into the driver's seat. However, even the most recent advances in geomodeling practice that represents 3D reservoir volumes with high-resolution geocellular grids may only mitigate but not eliminate the fact that estimating gridding parameters commonly results in artifacts due to topological constraints and misrepresentation of important aspects of the structural framework, which may introduce substantial difficulties for dynamic reservoir simulator later in the workflow. Hence we look into the future of building geological models and present and validate the evolving technology, with the truly game-changing industrial potential to utilize Maximum Continuity Fields for 3D reservoir property interpolation, performed in the absence of a geocellular grid. The selection of optimal geocellular parameters with attributes like cell size, number of cells and layering, is postponed throughout the process and rendered at user's discretion only before incorporation into the reservoir simulator. Models, generated in such fashion, retain the maximum available resolution and information density, limited only by the resolution of underlying data and structural continuity.

Regardless of the modeling approach, geoscientists and engineers often select diverse geomodel realizations such that the reservoir simulation outcome will cover a sufficiently large range of uncertainty to approximate the reservoir recovery forecast statistics throughout the asset lifecycle. One of the differentiating attributes of next-generation reservoir characterization is to integrate the reconciliation of geomodels with well-production and seismic data, with dynamic ranking and selection of representative model realizations for reservoir production forecasting. We outline and validate the evolving technology for Quantitative Uncertainty Management that seamlessly interfaces the DecisionSpace® Desktop, VIP® and/or Nexus® reservoir simulators. The highest available adherence to geological detail with respect to structural features that control depositional continuity (*e.g.* facies) is maintained through implementation of advanced model parameterization, based on Discrete Cosine Transform (Rao and Yip, 1990), an industrial standard in image compression. The AHM algorithm utilizes a highly efficient derivative of sequential (Markov chain) Monte Carlo sampling, where the acceptance rate is increased and computational effort reduced, by the utilization of streamline-based sensitivities.

By its nature, the probabilistic HM workflows render multiple equally probable but non-unique realizations of geological models that honor both, prior constraints and production data, with associated uncertainty. Throughout the inversion, however, some model realizations may have created non-geologically realistic features and these models are inadequate for forecasting of recovery performance. We outline the workflow for dynamic quantification production uncertainty that utilizes rapid streamline simulations to calculate data-pattern dissimilarities, Multi Dimensional Scaling to correlate dynamic model responses with pattern dissimilarities and kernel-based clustering methods to intelligently identification and ranking of geo-models, representative in forecasting decisions. When integrated into the DecisionSpace® Desktop suite of reservoir characterization tools, such technology will assist in Smart Reservoir Management and decision making by combining multiple types and scales of data, honoring most first order effects, capturing a full range of outcomes and reducing analysis and decision time.

## 4. Acknowledgement

The authors would like to acknowledge Halliburton/Landmark for the permission to publish this text.

## 5. References

Alvarado, V. & Manrique, E. (2010). Enhanced Oil Recovery: Field Planning and Development Strategies, Elsevier, MO, 208 p.

Borg, I. & Groenen, P.J.F. (2005). *Modern Multi-dimensional Scaling: Theory and Applications*, 2nd Ed., Springer, NY, pp. (207-212).

Caers, J. (2005). *Petroleum Geostatistics*, SPE Primer Series, Richardson, TX, 88 p.

Chambers, R.L Yarus, J.M. & Hird, K.B. (2000a). *Petroleum geostatistics for nongeostatisticians*, Part 1, The Leading Edge, May 2000, p. 474-479.

Chambers, R.L Yarus, J.M. & Hird, K.B. (2000b). *Petroleum geostatistics for nongeostatisticians*, Part 2, The Leading Edge, June 2000, p. 592-599.

Charles, T., Guemene, J.T., Corre, B., Vincent, G. & Dubrule, O. (2001). Experience with the quentification of subsurface uncertainties, SPE-68703, *Proceedings of SPE Asia Pacific Oil and Gas Conference*, 17-19 April 2011, Jakarta, Indonesia.

Culham, W., Sharaf, E., Chambers, R.L. & Yarus, J.M. (2010). New reservoir modeling tools support complex $CO_2$ flood simulation, World Oil, October 2010.

Daly, C. & Mariethoz, G. (2011). Recent Advances and Developments in MPS, Presented at *73rd EAGE Conference & Exhibition incorporating SPE EUROPEC 2011*, Vienna, Austria, 23-26 May 2011.

Damsleth, E., Hage, A. & Volden, R. (1992). Maximum Information at Minimum Cost: A North Sea Field Development Study With an Experimental Design, SPE 23139, *Journal of Petroleum Technology* 44 (12), p. 1350-1356.

Datta-Gupta, A. & King, M.J. (2007). *Streamline Simulation: Theory and Practice*, SPE Textbook Series 11.

Deutsch, C.V. (2002). *Geostatistical Reservoir Modeling*, Oxford University Press, NY, 376 p.

Efendiev, Y., Datta-Gupta, A., Ginting, V., Ma, X. & Mallick, B. (2005). An Efficient Two-Stage Markov Chain Monte Carlo Method for Dynamic Data Integration, *Water Resour.* Res. 41(1), W12423.

Fowler, M. L., Young, M.A., Madden, M.P. & Cole, E.L. (1999). The Role of Reservoir Characterization in the Reservoir Management Process, *Reservoir Characterization – Recent Advances*, Schatzinger, R., Jordan, J. (Eds.), AAPG Memoir 71, p. 3-18.

Full, W.E., Ehrlich, R. & Kennedy, S.K. (1983). Optimal definition of class intervals for frequency tables, *Particulate Science and Technology* 1 (4), pp. (281-293).

Gilman, J.R., Meng, H.-Z., Uland, M.J., Dzurman, P.J., & Cosic, S. (2002). Statistical Ranking of Stochastic Geomodels Using Streamline Simulation: A Field Application, SPE 77374, *Proceedings of the SPE Annual Technical Conference and Exhibition*, 29 September-2 October, San Antonio, TX.

Goins, N.R. (2000). Reservoir Characterization: Challenges and Opportunities, Presented at *2000 Offshore Technology Conference*, 1-4 May 2000, Houston, TX.

Hastings, W.K. (1970). Monte Carlo sampling methods using Markov Chains and their applications, *Biometrika* 57, pp. (97-109).

Hoffman, K.S., Neave, J.W. & Nielsen, E.H. (2007). Building Complex Structural frameworks, SPE-108533, *Proceedings of International Oil Conference and Exhibition*, 27-30 June 2007, Veracruz, Mexico.

Isaaks, E.H. & Srivastava, R.M. (1989). *An Introduction to Applied Geostatistics*, Oxford University press, NY, 561 p.

Jafarpour, B. & McLaughlin, D.B. (2008). History matching with an ensemble Kalman filter and discrete cosine parameterization. *Comput. Geosci.* 12 (2), pp. (227-244).

Kitanidis, P.K. (1995). Quasi-linear Geo-statistical Theory for Inversion, *Water Resour. Res.* 31(10), 2411.

Kramers, J.W. (1994). Integrated Reservoir Characterization: from the well to the numerical model, *Proceedings, 14th World Petroleum Congress*, John Wiley & Sons, 1994.

Larue, D.K. & Yue, Y. (2003). How stratigraphy influences oil recovery: A comparative reservoir database study concentrating on deepwater reservoirs, *The Leading Edge*, April 2003, pp. (332-339).

Lee, P.M. (1997). *Bayesian Statistics*, New York: Wiley & Sons.

Le Ravalec Dupin, M. (2005). *Inverse Stochastic Modeling of Flow in Porous Media: Applications to Reservoir Characterization*, Paris, France: IFP Publications.

Liang, L., Hale, D. & Maučec, M. (2010). Estimating fault displacements in seismic images, *Proceedings of Society of Exploration Geophysicists 2010 Annual Meeting*, paper INT 3.5, 17-22 October, 2010, Denver, CO.

Ma, X., Al-Harbi, M., Datta-Gupta, A. & Efendiev, Y. (2008). An Efficient Two-stage Sampling method for Uncertainty Quantification in History Matching Geological Models, *SPE Journal*, March 2008, pp. (77-87).

Maučec, M., Douma, S., Hohl, D., Leguijt, J., Jimenez, E.A. & Datta-Gupta, A. (2007). Streamline-Based History Matching and Uncertainty: Markov-chain Monte Carlo Study of an Offshore Turbidite Oil Field, SPE-109943, *Proceedings of SPE Annual Technical Conference and Exhibition*, 11-14 November 2007, Anaheim, CA.

Maučec, M. & Chambers, R.L. (2009). Fine-scale heterogeneity models for tertiary recovery: "must-have" to capture sweep recovery mechanisms?, Presented at *SPE Advanced Technology Workshop on History Matching: Improving Reservoir Management*, November 2009, Galveston, TX, USA,

Maučec, M. (2010). Systems and methods for generating updates and geological models, PCT Patent (pending)

Maučec, M., Parks, D., Gehin, M., Shi, G., Yarus & J.M., Chambers, R.L. (2010). Modeling distribution of geological properties using local continuity directions, *Proceedings of Society of Exploration Geophysicists 2010 Annual Meeting*, paper RC 3.6, Denver, CO, 17-20 October, 2010.

Maučec, M. & Cullick, S. (2011). Systems and methods for the quantitative estimate of production-forecast uncertainty, PCT Patent, WO 2011/037580 A1.

Maučec, M., Cullick, S. & Shi, G. (2011a). Quantitative Uncertainty Estimation and Dynamic Model Updating for Improved Oil Recovery, SPE-144092-PP, *Proceedings of SPE Enhanced Oil Recovery Conference (EORC)*, 19-21 July 2011, Kuala Lumpur, Malaysia.

Maučec, M., Cullick, S. & Shi, G. (2011b). Geology-guided Quantification of Production-Forecast Uncertainty in Dynamic Model Inversion, SPE-146748, *Proceedings of SPE Annual Technical Conference and Exhibition*, 30 October -2 November 2011, Denver, CO.

MCERI Research Group (2008). Getting Started with DESTINY, Department of Petroleum Engineering, Texas A&M University, November 7.

Neal, R.M. 1993. Probabilistic Inference using Markov Chain Monte Carlo Methods, Technical Report CRG-TR-93-1, Dept. of Computer Science, University of Toronto.

Olea, R.A. (1991). *Geostatistical Glossary and Multilingual Dictionary*, New York, Oxford University Press, 192 p.

Oliver, D.S., He, N. & Reynolds, A.C. (1996). Conditioning Permeability Fields to Pressure Data, *Proceedings of the 5th European Conference on the Mathematics of Oil Recovery (ECMOR)*, 3-6 September, Leoben, Austria, pp. (1-11).

Oliver, D.S. & Chen, Y. (2011). Recent Progress on Reservoir History Matching: A Review, *Comput. Geosci.* 15, pp. (185–221).

Peters, E., Arts, R.J., Brouwer, G.K., Geel, C., Cullick, S., Lorentzen, R. & Chen, Y. (2009). Results of the Brugge Benchmark Study for Flooding Optimization and History Matching. *SPE Res Eval & Eng*, 13 (3), pp. (391-405).

Prada, J.W.V., Cunha, J.C. & Cunha, L.B. (2005). Uncertainty Assessment Using Experimental Design and Risk Analysis Techniques, Applied to Off-shore Heavy-oil Recovery, SPE-97917-MS, *Proceedings of the SPE/PS-CIM/CHOA International Thermal Operations and Heavy Oil Symposium*, 1-3 November, 2005, Calgary, Alberta, Canada.

Rao, K.R. & Yip, P. (1990). *Discrete Cosine Transform: Algorithms, Advantages, Applications.* Boston, MA: Academic Press.

Rodriguez, A.A., Klie, H., Wheeler, M.F. & Banchs, R. (2007). Assessing Multiple Resolution Scales in History Matching with Meta-models, SPE-105824-MS, *Proceedings of the SPE Reservoir Simulation Symposium*, 26-28 February, 2007, Houston, TX.

Scheidt, C. & Caers, J. (2009). Representing Spatial Uncertainty Using Distances and Kernels, *Math. Geosci.* 41 (4), pp. (397–419).

Schölkopf, B., Smola, A. & Müller, K-R, (1996). Nonlinear Component Analysis as a Kernel Eigenvalue Problem. Max-Planck Institute Technical Report No. 44.

Seiler, A., Rivenaes, J.C. & Evensen, G. (2009). Structural Uncertainty Modeling and Updating by Production Data Integration, SPE-125352, *Proceedings of SPE/EAGE reservoir Characterization and Simulation Conference*, 19-21 October 2009, Abu Dhabi.

Strebelle, S. & Journel, A.G. (2001). Reservoir modeling using multiple-point statistics, SPE-71324, *Proceedings of SPE Annual Technical Conference and Exhibition*, 30 September–3 October, 2011, New Orleans, LA.

Tabachnick, B.G. & Fidell, L.S. (2006). *Using Multivariate Statistics*, 6th Ed., Pearson.

Tyler, N. & Finley, R.J. (1991). Architectural controls on the recovery of hydrocarbons from sandstone reservoirs, In: Miall, A.D., Tyler, N. (Eds.), *SEPM Concepts in Sedimentology and Palaeontology* 3, pp. (1-5).

Walker, R.G. & James, N.P., (Eds.) (1992). *Facies Models: Response To Sea Level Change*, Geological Association of Canada.

Yarus, J.M. & Chambers, R.L. (Eds.) (1994). *Stochastic Modeling and Geostatistics*, AAPG Computer Applications in Geology, No. 3., 379 p.

Yarus, J.M. & Chambers, R.L. (2010). Practical Geostatistics - An Armchair Overview for Petroleum Reservoir Engineers, SPE 103357, *Journal of Petroleum Technology*, November 2006, pp. (78-86).

Yarus, J.M., Srivastava, R.M., Gehin, M. & Chambers, R.L. (2009). Distribution of properties in a 3D volumetric model using a maximum continuity field, PCT Patent, WO 2009/151441 A1.

# 12

# Developing Sediment Yield Prediction Equations for Small Catchments in Tanzania

Preksedis Marco Ndomba
*University of Dar es Salaam, Tanzania*

## 1. Introduction

Sediment yield refers to the amount of sediment exported by a basin over a period of time, which is also the amount which will enter a reservoir or pond located at the downstream limit of the basin (Morris and Fan, 1998). Estimate of long-term sediment yield have been used for many decades to size the sediment storage pool and estimate reservoir life. However, these estimates are often inaccurate especially for small catchments. Besides, it is known from literature that long term period sampling programmes are required to capture the high variability of sediment fluxes in these catchments (Horowitz, 2004; Thodsen *et al.*, 2004). The correlation of sediment yields to erosion is complicated by problem of determining the sediment delivery ratio, which makes it difficult to estimate the sediment load entering a reservoir/pond on the basis of erosion rate within the catchment (Morris and Fan, 1998). Sediment yield from the dam catchment is one of the parameters controlling sedimentation of small dams. This has to be estimated if future sedimentation rates in a dam are to be predicted.

Non Governmental Organizations (NGOs) and Government Agencies (GAs) have constructed thousands of small dams in semi-arid regions of East and Southern Africa including Tanzania to provide water for livestock and small-scale irrigation (Lawrence *et al*, 2004; Faraji, 1995). In Tanzania, in particular, at present it is not known whether the original storage capacities of these dams still exist as a result of many years of operations. Besides, irrigation/water supply schemes ponds/reservoirs are normally draining small catchments. Most of the small catchments are characterized as ungauged. The effective life of many of these dams is reduced by excessive siltation - some small dams silt up after only 2 years. This issue is poorly covered in the many small dam design manuals that are available, which mostly focus on civil engineering design and construction aspects. While a capability to estimate future siltation is needed to ensure that dams are sized correctly, and are not constructed in catchments with very high sediment yields, little guidance is available to small dam planners and designers (Lawrence *et al*, 2004). Therefore, prediction of sediment yields from catchments is very important where water resources sedimentation is a serious problem like Tanzania and construction of dams is needed (Mulengera, 2008).

This chapter discusses also the findings of a few previous representative and related research works in Tanzania and the region at large as this study is a follow up research. These studies were selected in order to cover a wide range of study methods.

Christiansson (1981) made detailed recording of the soil erosion complex within five selected catchments in Dodoma, the semi-arid savannah areas of central Tanzania, 4 of which with reservoirs (Fig. 2.1). The principal methods employed include field surveying and air photo interpretation. The main approach was physical geographical aiming at studying the existing features of soil erosion and sedimentation and analysis of the underlying causes of the processes. Christiansson (1981) estimated sediment yields of 260 – 900 $t/km^2/yr$ or 2.6 – 9 t/ha/yr as averages for the longest periods of available records. Although, Christiansson asserted that the estimated sediment yields from his study were of the same order of magnitude as those recorded in similar environments in other parts of East Africa, the scope of the study was limited in-terms of spatial and climate representation.

Mulengera and Payton (1999) in their review of the soil loss estimation equations noted that most of the countries in the tropics have no appropriate and accurate soil erosion prediction equations, although the Soil Loss Estimation Model for Southern Africa (SLEMSA) and the Universal Soil Loss Equation (USLE) are used in different tropical countries. The SLEMSA( developed in Zimbabwe) still needs some modifications and has, so far, not been widely used or tested outside Zimbabwe and in some instances have shown to give unrealistic soil loss values (Mulengera and Payton,1999). The USLE (developed in the USA) and widely used throughout the world has in most cases been found to be inapplicable in the tropics. This is due to the fact that the equation's soil erodibility nomograph commonly gives unrealistic values for tropical soils. Although derivation of the erodibility equations for the tropical soils have shown that soil erodibility is strongly related to texture-related soil characteristics as has been shown for soils in temperate regions, there are differences in the magnitudes of the characteristics for soils with relatively similar erodibility values in both regions (Mulengera and Payton, 1999). This is due to differences in clay, silt, and sand fractions of the soils and possibly rainfall characteristics found in the two regions. While soils in the temperate region have all the three fractions well distributed, soils in the tropics are mainly composed of clay and (or) sand fractions with a relatively small fraction of silt content (Mulengera and Payton, 1999). So Mulengera and Payton concluded that it is impossible to develop one universal soil erodibility equation. Therefore, the prediction of soil erosion in the tropics using the USLE or its revised version (RUSLE), had been hampered by the common inapplicability of the soil erodibilty nomograph for tropical soils. Furthermore, the table values developed in the U.S.A. for estimating the crop and soil management factor of the equation are not applicable for farming practices and conditions found in the tropics (Mulengera and Payton, 1999). However, some recent studies have shown that it can give good results, especially, when its recent version, the Modified Universal Soil Loss Equation (MUSLE) is used (Ndomba, 2007).

Mulengera and Payton (1999) presented an equation for estimating the USLE -Soil erodibility factor, which resulted from a wider research programme initiated to identify a suitable soil loss prediction equation for use under Tanzanian conditions. The derived equation based on soil texture-related parameters which is technically accurate (i.e. explaining about 84 % of the erodibility variations) for estimating the erodibility factor of the ( R )USLE in the tropics for soils whose physical and chemical characteristics are similar to the soils used in the derivation. The equation is useful for soil conservation planning in these areas currently suffering from severe soil erosion (Mulengera and Payton, 1999). The equation was successfully used by Mtalo and Ndomba (2002) in Pangani basin, in the North-eastern part of Tanzania.

Mtalo and Ndomba (2002) have reported alarming high rate of soil erosion of up to 2,400 t/km$^2$/yr or 24 t/ha-yr in the upstream of Pangani river basin covering parts of Arusha, Kilimanjaro and Tanga regions (Fig. 2.1). Mtalo and Ndomba used USLE equation to map and estimate on site potential soil loss. In the basin high erosion rates can be measured in different perspectives such as increased agricultural and other human activities. For instance, in Arumeru district, which is one of the districts in the Arusha region, soil erosion is one of the major obstacles to increasing or sustaining the agriculture production. The whole district is affected by soil erosion, but the reasons differ from place to place. The amount of livestock in the district is considered to be far above the carrying capacity of the present land area devoted to grazing. Agricultural activities are a contributing factor to increased soil erosion rates in the Pangani basin upstream of Nyumba Ya Mungu reservoir (Mtalo and Ndomba, 2002).

Of recent, there have been attempts to apply complex distributed, physics-based sediment yield models such as Soil and Water Assessment Tool (SWAT) for poor data large catchments, Kagera, Simiyu and upstream part of Pangani River catchment, in Kagera, Mwanza and Arusha/Kilimanjaro regions, respectively, in Tanzania. SWAT model uses the Modified Universal Soil Loss Equation (MUSLE) to estimate sediment yield (Arnold *et al.*, 1995). The model operates at daily time step with output frequency of up to month/annual. However, in order to adopt the model for general applications in watershed management studies, researchers recommended for SWAT model improvements (Ndomba *et al.*, 2005; Ndomba *et al.*, 2008).

One would note that most of the previous sediment yield estimates studies in Tanzania and the region at large were catchment specific. The results could not be transferred easily to other hydrologic similar catchments (Rapp *et al.*, 1972; Mulengera and Payton, 1999; Mulengera, 2008; Ndomba, 2007, 2010). In order to estimate catchment yield researchers were forced to use uncertain factor such as sediment delivery ratio (Ndomba *et al.*, 2009). In some studies attempts were made to develop only a simple procedure which would distinguish between dams that will silt up rapidly from dams that will have a sedimentation lifetime well in excess of twenty years (Lawrence *et al*, 2004). The estimation tools used were either complex for operational and wider application or data intensive (Ndomba *et al.* 2008). In some cases due to limitation in data the developed Sediment yield predictive tools could not be validated (Rapp *et al.*, 1972; Lawrence *et al*, 2004). As acknowledged by Faraji (1995) and others, at present, there is very scanty knowledge about reservoir sedimentation in Tanzania. Previous studies were done on few reservoirs/dams. The studies gave some guidelines on the rate of sedimentation of the respective areas (Rapp *et al.*, 1972). However, this knowledge should be backed with further extensive surveys and resurveys to get improved relationships. Critical tools in this context include sediment yield and/or reservoir life estimation. The country has limited resources in terms of funding and human capital for developing the planning tools (Mulengera, 2008). The latter problem might be common to most of the developing countries.

Based on the discussions above and literature, generally, sediment yield models may vary greatly in complexity from simple regression relationships linking annual sediment yields to climatic physiographic variable such as regional regression relationships to complex distributed simulation model (Garde and Ranga Raju, 2000). Modelling as one of the approaches for estimating catchment sediment yields, if properly applied, can provide information on both the type of erosion and its spatial distribution across the catchment. Sediment mobilized by sheet and rill erosion may be deposited by a variety of mechanisms prior to reaching stream channels. Six major factors which influence the long-term sediment yields/delivery from a catchment based on Renfro (1975) as reported in Morris and Fan (1998) and critically reviewed by Ndomba (2007) are: i) Erosion process - the sediment delivered to the catchment outlet will generally be higher for sediment derived from channel-type erosion which immediately places sediment into the main channels of the transport system, as compared to sheet erosion; ii) Proximity to catchment outlet - sediment delivery will be influenced by the geographic distribution of sediment sources within the catchment and their relationship to depositional areas. Sediment is more likely to be exported from a source area near the catchment outlet as compared to a distant sediment source, since sediment from the distant sources will typically encounter more opportunities for re-deposition before reaching the catchment outlet; iii) Drainage efficiency- hydraulically efficient channels networks with a high drainage density will be more efficient in exporting sediment as compared to catchments having low channel density; iv) Soil and land cover characteristics - finer particles tend to be transported with greater facility than coarse particles. Because of the formation of particle aggregates by clays, silts tend to be more erosive and produce higher delivery ratios than clay soils; v) Depositional features - the presence of depositional areas, including vegetation, ponds, wetlands, reservoirs and floodplains, will decrease the sediment yields at the catchment outlet. Most eroded sediments from large catchments may be re-deposited at the base of slopes, as outwash fans below gullies, in channels or on floodplains; vi) catchment size and slope - a large, gently sloping catchment will characteristically have a lower delivery ratio than a smaller and steeper catchment.

This chapter is therefore reporting the developed sediment yield-fill equations as categorized as regional regression relationships for various climatic regions in Tanzania as simple and efficient planning tools of water supply schemes small reservoirs/pond with limited data. In this study additional new data on dams is used. It should be noted that requirements for data and computational modelling skills rule out the use of more sophisticated methods to predict sediment yields, and as a result a simple regional sediment yield predictor was chosen for this application. The equations are developed from readily available data on catchment area and reservoir sediment fill from Ministries and Government Agencies. The size of the area is very important factor in respect to the total yield of sediment from a catchment. However, it should be noted that its relative importance to the influence of the sediment delivery ratio and sediment production rate is subject to questioning. It is suggested in the literature that sediment production rates declined with increasing catchment area (Morris and Fan, 1998). This theory is supported by the fact that the probability of entrapment and lodgment of a particle being transported downstream increases as the drainage area increases. Besides, for the same length of a river network, the smaller the catchment area the higher the drainage density as well the sediment yields. The proximity to the catchment outlet may also be indirectly related to catchment size. The preceding discussions suggest that catchment area size may sometimes be directly or indirectly linked to various factors controlling sediment delivery to the outlet of the catchment. However, catchment area size could not be directly related to erosion process, soil type and land cover. Such limitations would render the general relationships between catchment area size as independent variable and the sediment yield-fill be used only for preliminary planning purposes or as a rough check. It is anticipated that small dam designers/planners would be able to use these tools /methods; they typically need to carry out assessments rapidly using limited local data, and may not have software skills or access to computers.

## 2. Materials and methods

### 2.1 Description of the study area

The study area, Tanzania, is situated in East Africa just south of the equator (Fig. 2.1). Tanzania lies between the area of the great Lakes—Victoria, Tanganyika, and Nyasa—and the Indian Ocean. It contains a total area of 945,087 km$^2$, including 59,050 km$^2$ of inland water (Fig. 2.1). It is bounded on the North by Uganda and Kenya, on the East by the Indian Ocean, on the South by Mozambique and Malawi, on the South West by Zambia, and on the West by Democratic Republic of Congo, Burundi, and Rwanda, with a total boundary length of 4,826 km, of which 1,424 km is coastline.

Tanzania has a tropical climate with 3 major climatic zones (Fig. 2.2) viz. dry, moderate and wet. Moderate Climatic zone occupies a large area of Tanzania (URT, 1999). It receives rainfall for 1 to 3 months in a year. The administrative regions which fall under this climatic zone are Tabora, Mwanza, Mara, Iringa, Morogoro, Arusha, Tanga, Moshi, Lindi and Bukoba. In the highlands, temperatures range between 10 and 20 °C (50 and 68 °F) during cold and hot seasons, respectively. The rest of the country has temperatures rarely falling lower than 20 °C (68 °F). The hottest period extends between November and February (25 - 31 °C) while the coldest period occurs between May and August (15 - 20 °C ). Tanzania has two major rainfall regions. One is unimodal (December - April) and the other is bimodal (October - December and March - May). The former is experienced in southern, south-west, central and western parts of the country, and the latter is found to the north and northern coast. In the bimodal regime the March - May rains are referred to as the long rains or "Masika", whereas the October - December rains are generally known as short rains or "Vuli".

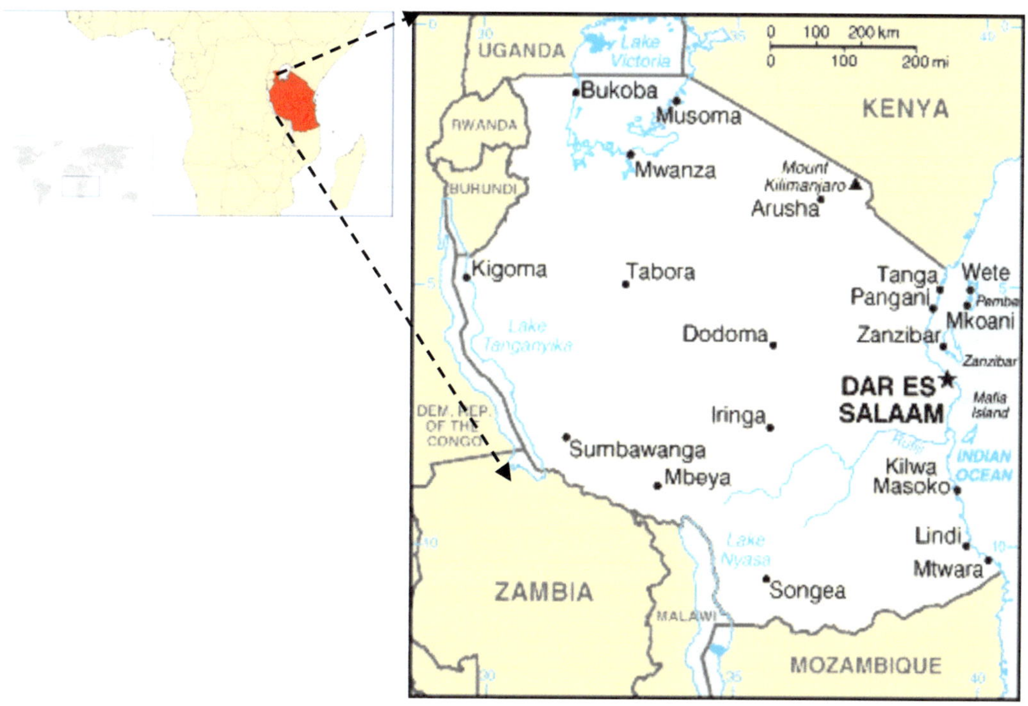


Fig. 2.1. Location map of Tanzania

Fig. 2.2. Map showing Climatic zones of Tanzania (as adopted from URT, 1999)

According to PLDPT (1984) Tanzanian soils are very varied, a simplified classification follows: a) Volcanic soils: are of high agricultural potential and livestock production tends to be restricted to zero-grazing systems. They predominate in Arusha, Kilimanjaro and South west Highlands, Kitulo plateau. At high and medium altitudes they are notable for the production of forage for dairy production; b) Light sandy soils: predominate in the coastal areas. Grazing is available during the rains but the soils dry out rapidly thereafter and the forage has little worth; c) Soils of granite/gneiss origin: are poor and occur mainly in mid-west especially in Mwanza and Tabora; d) Red soils: occupy most of central plateau. They produce good grazing in the limited rainy seasons and the quality of herbage persists into the dry seasons; e) Ironstone soils: found in the far west, mainly in Kagera, Kigoma and Sumbawanga. They are poor and acidic but can be productive with inputs, *i.e.*, mulching and manuring; and f) The mbuga black vertisols are widespread and an important source of dry season grazing.

The following are the descriptions of the respective regions from which the dam data for this study were collected. These are Dodoma, Shinyanga, Singida, Tabora, and Arusha (Tables 2.2.1 & 2.2.2).

| Serial No. | Names Of dams | Catchment Area ($km^2$) | Full Supply Level, FSL,(m) | Capacity at FSL (Million $m^3$) | Year of Construction | Sed Fill ($m^3$) | Year(Data Collected) | Sed Fill rate ($m^3$ Per Year) |
|---|---|---|---|---|---|---|---|---|
| **Dodoma region** | | | | | | | | |
| 1 | Mambali | 2.7 | *NA* | | 1958 | 21500 | 1978 | 1075 |
| 2 | Mabisilo | 25 | *NA* | | 1956 | 31500 | 1978 | 1431.8 |
| 3 | Kakola | 3.7 | *NA* | 0.202 | 1954 | 19700 | 1978 | 820.8 |
| 4 | Manolea | 3 | *NA* | 0.033 | 1956 | 15400 | 1978 | 700 |
| 5 | Kasisi | 3.2 | *NA* | 0.02 | 1968 | 7300 | 1978 | 730 |
| 6 | Mbola | 6.4 | *NA* | 0.016 | 1972 | 7500 | 1978 | 1250 |
| 7 | Igingwa | 10 | *NA* | 0.20 | 1958 | 2800 | 1978 | 1400 |
| 8 | Matumbuhi | 16.8 | 59.74 | 0.31 | 1960 | | 1978 | 1322.4 |
| 9 | Buigiri | 10.3 | 35.26 | 0.48 | 1969 | | 1978 | 1410 |
| 10 | Imagi | 2.2 | *NA* | 0.1695 | 1934 | | 1971 | 950 |
| 11 | Matumbulu | 15 | *NA* | 0.333 | 1949 | | 1962-74 | 1374.5 |
| 12 | Msalatu | 8.5 | *NA* | 0.42 | 1944 | | 1974 | 1225.8 |
| 13 | Kisongo | 9.3 | *NA* | 0.1265 | | | 1969-71 | 1228 |
| **Shinyanga region** | | | | | | | | |
| 14 | Bubiki | 11 | 2.75 | 0.35 | | | | 1584 |
| 15 | Ibadakuli | 10 | 4.9 | 0.37 | | | | 1472 |
| 16 | Malya | 15 | 5.8 | 1.49 | | | | 2011 |
| 17 | Nguliati | 12 | 3.35 | 0.49 | | | | 1593 |
| 18 | Sakwe | 9 | 3.65 | 0.28 | | | | 1357.4 |

Note: "NA" and "blank" imply No Any data.

Table 2.2.1. Sedimentation data of small dams in Dry climatic zone for regions of Dodoma and Shinyanga

Dodoma region is characterized by long dry seasons (April to December) and short rainy seasons (December to March). Mean annual rainfall in the area range from 500 to 600 mm/annum and potential evaporation ranges between 2000 to 2500 mm per annum (Christiansson, 1981). The topography of the central semi arid Tanzania is characterized by plains and scattered inselberg or ridge. The soils appear in catena sequence where the upper slopes of inselberg have thin stony soils. The valley bottoms and flood plains have black and grey deposits. Natural vegetation of dense thicket or "miombo" woodland has generally been replaced by semi natural vegetation of grasses and herbs. The inhabitant of the study area are cultivating pastoralist. They mainly practice shifting cultivation where no manure is applied. They maintain large heards of cattle, sheep, and goats. The staple crops grown are sorghum and bulrush millet, maize also grown on significant areas.

| Tabora region | | | | | | | | |
|---|---|---|---|---|---|---|---|---|
| **Serial No.** | **Names of Dams** | **Cat Area ($km^2$)** | **Full Supply Level (m)** | **Capacity At FSL (Million $m^3$)** | **Year of Construction** | **Year (Data Collected)** | **Sed Fill ($m^3$)** | **Sed Fill rate ($m^3$ Per Year)** |
| 1 | Malolo | 15 | 35.7 | 0.936 | 1962 | 1975 | 19300 | 1485 |
| 2 | Itambo | 25 | 29.3 | 0.234 | 1947 | 1978 | 96200 | 3103 |
| 3 | Magulya | 15.8 | 35 | *NA* | 1957 | 1978 | 35200 | 1676 |
| 4 | Ulaya | 8.3 | 30.8 | 0.298 | 1947 | 1978 | 31600 | 1019 |
| 5 | Igurubi | 1.2 | 26.7 | 0.113 | 1959 | 1978 | 20900 | 1100 |
| 6 | Charo | 6.7 | 28.5 | 0.015 | 1969 | 1978 | 11700 | 1300 |
| 7 | Kakolo | 3.7 | 33.8 | 0.202 | 1954 | 1978 | 19700 | 820 |
| 8 | Manolea | 3 | 30.6 | 0.033 | 1956 | 1978 | 15400 | 700 |
| 9 | Kasisi | 3.2 | 13.2 | 0.02 | 1968 | 1978 | 7300 | 730 |
| 10 | Mbola | 6.4 | 31.8 | 0.016 | 1972 | 1978 | 7500 | 1250 |
| 11 | Usoke Mission | 1.9 | 28.2 | 0.02 | 1971 | 1978 | 3150 | 450 |
| 12 | Kalangali | 2.8 | 1200.7 | 0.84 | 1958 | 1978 | 88400 | 4420 |
| 13 | Uchama | 11 | 36.9 | 1.322 | 1955 | 1978 | 27700 | 1204.3 |
| 14 | Mambali | 2.7 | 24.9 | *NA* | 1956 | 1978 | 21500 | 977.3 |
| 15 | Mabisilo | 25 | 35.0 | *NA* | 1956 | 1978 | 31500 | 1431.8 |
| 16 | Nkinazawa | 180 | 29.6 | 0.75 | 1956 | 1978 | 110000 | 5000 |
| 17 | Iduduma | 267 | 32.9 | 0.86 | 1959 | 1978 | 140000 | 7368 |
| 18 | Mwamashimba | 16.5 | 25.0 | *NA* | 1973 | 1978 | 21500 | 4300 |
| 19 | Bulenya Hills | 194 | 36 | 1.62 | 1961 | 1978 | 96400 | 5670.5 |
| 20 | Sorefu | 7.5 | 28.1 | 0.045 | 1970 | 1978 | 9400 | 1175 |
| 21 | Urambo | 38 | 18.2 | *NA* | 1956 | 1978 | 75717 | 3441.7 |
| 22 | Tura | 105 | 30.8 | 0.27 | 1948 | 1978 | 108003 | 3600 |
| 23 | Utatya | 4 | *NA* | 35.4 | 1959 | 1978 | 11050 | 581.6 |
| 24 | Igingwa | 10 | 26.2 | 0.2 | 1958 | 1978 | 28000 | 1400 |
| **Arusha region** | | | | | | | | |
| 25 | Moita Bwawani | 97 | | 1.556 | | | | 31120 |
| 26 | Moita Kiloriti | 115 | | 1.257 | | | | 37710 |

| Serial No. | Names of Dams | Cat Area (km²) | Full Supply Level (m) | Capacity At FSL (Million m³) | Year of Construction | Year (Data Collected) | Sed Fill (m³) | Sed Fill rate (m³ Per Year) |
|---|---|---|---|---|---|---|---|---|
| 27 | Leken | 183 | | 6.972 | | | | 223104 |
| 28 | Kimokouwa | 110 | | 0.950 | | | | 40850 |
| 29 | Lossimingori | 94 | | 1.640 | | | | 57400 |
| 30 | Moriatata | 120 | | 1.342 | | | | 34892 |
| 31 | Losirwa | 65 | | 0.670 | | | | 20783.95 |
| 32 | Ngamuriak | 76 | | 1.104 | | | | 18768 |
| 33 | Bashay | 78 | | 1.420 | | | | 8520 |
| 34 | Meserani | 56 | | 0.568 | | | | 17608 |
| 35 | Lepurko | 87 | | 0.856 | | | | 24837.05 |

Table 2.2.2. Sedimentation data of small dams in Moderate climatic zone for regions of Tabora and Arusha

Shinyanga region is characterized by a tropical type of climate with clearly distinguished rainy and long dry seasons. According to meteorological statistics the average temperature for the region is about 28°C. The region experiences rainfall of 600 mm as minimum and 900 mm as maximum per year. The rainy season usually starts between mid-October and December and ends in the second week of May. Normally it has two peak seasons. The first peak occurs between mid- October and December, while the second one, the longer season, falls between February and mid-May. As such, the whole rainy season covers a total of almost 6 months, with a dry spell which usually occurs in January. The dry season begins in mid-May and ends in mid-October. This is a period of about 5 months. The dry season is the worst period for the Shinyanga region. The topography of Shinyanga region is characterized by flat, gently undulating plains covered with low sparse vegetation. The North-Western and North- Eastern parts of the region are covered by natural forests which are mainly "miombo" woodland. The Eastern part of the region is dominated by heavy black clay soils with areas of red loam and sandy soil. It is observed that most of the Shinyanga region is dry flat lowland. Thus its agro-economic zones are not well pronounced as it is with some regions in the country. The soils are hard to cultivate, pastures become very poor, and availability of water for domestic use and livestock become acute. The amount and distribution pattern of rainfall in the region is generally unequal and unpredictable. This implies that rainfall as a source of water for domestic and production purposes in the region is less reliable for sustainable water supply. Despite of the recent mushrooming of mine industry, agriculture has continued to dominate the livelihood and economic performance of Shinyanga region. The sector contributes about 75 percent to the regional economy and employs about 90 percent of the working population in the region. Agriculture is dominated by peasantry farming. Main cash crops are cotton and tobacco while the main food crops are maize, sorghum, paddy, sweet potatoes, millet and cassava. The region has the largest planted area of maize and second largest for paddy and sorghum than other regions in

Tanzania. Besides farming, livestock keeping is also a major activity in the region. Cattle, goats and sheep are the major domesticated animals. Modern diary farming and poultry keeping are confined to urban centers

Tabora region is among the areas of Tanzania which are in moderate climatic zones. Tabora region is located in the mid-western part of Mainland Tanzania. Tabora is characterized as tropical type of climate with clearly distinguished rainy and dry seasons. According to meteorological statistics the average temperature for the region is about $27^0$C. Tabora receives Mean Annual Rainfall of 892 mm/annum or 74 mm/month. In Tabora, about 76% of the population are farmers, and thus agriculture is the largest single sector in the economy directory producing about 80 percent of Tabora region's wealth of goods and services. Main cash crops grown are tobacco, cotton and paddy. Tobacco and cotton are mainly grown for export markets. Principal food crops are maize, sorghum, cassava, sweet potatoes and legumes.

Arusha region lies in moderate climatic zone. With the exception of a few spots the region is in the high altitudes ranging from 800 to 4,500 meters above sea level. Because of the high altitude the region experiences moderate temperatures with rainfall varying with the altitude. The average annual temperature is $21^0$C in the highlands and $24^0$C in the low lands. Arusha region has two types of rainfall patterns: unimodal and bimodal. The southern district of Karatu normally enjoys unimodal rainfall which usually starts in November and ends in April. The rainfall in this district is usually reliable, ranging from 800 to 1,000 mm/annum. The major crop produced is cereals. Soils have been classified by colour into grey, brown and red brown. The extensive soils which originate from recent volcanic ash are found to the north-western parts of the region, west of the rift valley and in the Ngorongoro massif. Brown soils cover large areas in the central part and western side of the region. The southern- eastern areas are characterised by grey brown and red-brown soils. Soil erosion is particularly severe in the heavily settled central part of the region. Generally soil erosion is widespread throughout the region and is deemed to be an environmental disaster in the making.

### 2.2 Data type and sources

This study collected readily available secondary data from reports of Ministry of Water and Irrigation (Husebye and Torblaa, 1995) as adopted in Tables 2.2.1 & 2.2.2. However, data from Arusha in Table 2.2.2 were sourced from a recent study by Malisa (2007). They include name of the dam, full supply level of the dam, capacity of the dam at full supply level, year of construction, accumulated sediment volume in the dam (Sed Fill ), dam survey Year (Year Data Collected), volumetric rate of sediment accumulation in the dam (Sed Fill per Year), catchment area, and Sediment yield. It should be noted also that important data such as geographical locations of these dams are missing. The dams were built for various purposes, including and not limited to irrigation, domestic water supply, livestock watering, flood control and fishing.

It should be noted that most of these sedimentation data presented in Tables 2.2.1 and 2.2.2 were collected using mainly two approaches, namely, direct measurement of transported materials (*i.e.*, suspended sediment concentration) and measurement of the rate of siltation of reservoirs/dams. The author would like to note that the first procedure has some setbacks, especially, when practiced in tropics. For instance, a majority of sediment would be transported in one or two days. Typically for Tanzania, most of sediment samples had been taken at medium or low stage. The high stages were hardly sampled. This might have made it difficult to establish a linear relationship between sediment load and river discharge. Surveys of reservoirs/dams whose relevant technical maps were available were done. Profiles available from old/design maps were compared with the new sounding.

### 2.3 Data analysis

Although the data collected are scarce (*i.e.*, only 53 dams) but one could see that based on climatic zonation about seventy percent (70%) of Tanzania land is represented. Besides, the data coverage represents a wide range of dam and catchment physical characteristics viz. catchment area 1 - 267 $km^2$; dam capacity at FSL, 0.02 - 35 Million $m^3$ (Table 2.3). The mean values are 41.7 $km^2$ and 1.5 Million $m^3$ for catchment area and dam capacity, respectively. Notwithstanding the high spatial variability of data as captured by high Coefficient of variation (Cv), the data represent the population as demonstrated by low Standard Error of the Mean (SEM).

| Serial No. | Statistics | Catchment area ($km^2$) | Dam capacity at Full Supply Level (Million $m^3$) |
|---|---|---|---|
| 1. | Lowest | 1.20 | 0.02 |
| 2. | Maximum | 267.00 | 35.40 |
| 3. | Mean | 41.66 | 1.52 |
| 4. | Standard Deviation (STD) | 59.62 | 6.41 |
| 5. | Coefficient of Variation, Cv (%) | 143.12 | 422.76 |
| 6. | Standard Error of the Mean, SEM | 8.19 | 1.17 |

Table 2.3. Summary statistics of catchment size and dam capacity of data used in this study

### 2.4 Development of sediment yield-fill equations

Sediment yield-fill equations were developed by regression analysis approach. It should be noted that if data on sediment yield-fill and catchment characteristics are available from many sites, it may be possible to develop a regression relationship which describes the sediment yield within the region as a function of independent variables such as catchment area, slope, land use, and rainfall erosivity (Morris and Fan, 1998). The only independent variable used for this study is the catchment area as it was readily available. This study assumed the following: i) the sample is representative of the population for the inference prediction; ii) the error is a random variable with a mean of zero conditional on the explanatory variables; iii) the independent variables based on low standard error of the mean (SEM) as presented in Table 2.3 were measured with no error; iv) the predictors are linearly independent, *i.e.* it is not possible to express any predictor as a linear combination of the others; v) the errors are uncorrelated, that is, the variance-covariance matrix of the errors is diagonal and each non-zero element is the variance of the error; and vi) the variance of the error is constant across observations (homoscedasticity). These are sufficient conditions for the least-squares estimator to possess desirable properties, in particular, these assumptions imply that the parameter estimates will be unbiased, consistent, and efficient in the class of linear unbiased estimators. Besides, sediment yield is assumed as equal to sediment fill due to the uncertainty involved in estimating trap efficiency of small dams in the study area. It should be noted that previous researchers such as Mulengera (2008) adopted a similar approach. However, the author is aware that the actual data rarely satisfies the assumptions. That is, the method is used even though at some points the assumptions are not necessarily true.

Firstly, the analysis was conducted to choose type of regression equation forms. Two candidate's forms of equations were investigated, which are straight line and power function (Equations 2.4.1 & 2.4.2). This was achieved by comparing the strength of correlation between sediment fill and catchment area, and corresponding log-transformed values (2.4.3). A power relationship is confirmed when the correlation of log-transformed is high, otherwise, a linear model is chosen.

$$y_i = \beta_o + \beta_1 x_i + \varepsilon_i, \mathrm{i} = 1, \ldots, \mathrm{n}. \tag{2.4.1}$$

Where $n$ is a number of observations, $x_i$, is independent variable (catchment area), $y_i$, dependent variable (Sediment yield-fill), and two parameters, $\beta_0$ and $\beta_1$, and $\varepsilon_i$ is an error term and the subscript $i$ indexes a particular observation.

$$\mathrm{y_i} = \alpha\, \mathrm{x_i}^{\beta} \tag{2.4.2}$$

where $a$ and $\beta$ are coefficient and exponent of the equation, respectively, $y_i$ and $x_i$ are as defined above.

$$\mathrm{Log}\, \mathrm{y_i} = \mathrm{Log}\, \alpha + \beta\, \mathrm{Log}\, \mathrm{X_i} \tag{2.4.3}$$

Secondly, the parameter values were estimated under Excel 2007's Regression Analysis Tool using 70% of the data set, where applicable. The splitting of data was possible for cases where the sample size was adequate for 2 independent variables (*i.e.*, α, β) as presented above. As recommended by Statsoft (2011) at least 10 to 20 times as many observations (cases, respondents) as variables, should be used for stable estimates of the regression line and replicability of the results. The tool outputs, among others; the *t* statistic (a measure of how extreme a statistical estimate is); a *p*-value (a measure of how much evidence we have against the null hypothesis, *Ho*, no change or no effect; confidence interval (an interval in which a measurement or trial falls corresponding to a given probability, the best confidence interval used is 95%); degrees of freedom (the minimal number of values which should be specified to determine all the data points), *df*; the standardized residual value (observed minus predicted divided by the square root of the residual mean square), Coefficient of determination, $R^2$ (this is the square of the product-moment correlation between two variables -It expresses the amount of common variation between the two variables); Multiple R (is the positive square root of R-square - this statistic is useful in multivariate regression when you want to describe the relationship between the variables); The standard error ( is the standard deviation of a mean). The developed equations were validated using independent data set (30%), where appropriate.

## 3. Results and discussions

### 3.1 Selected regression model

As a result of conducting correlation analysis as described under section 2.4 above and qualitative analysis of scatter plots (Figs. 3.1a,b) below, the power function was chosen as the best regression model for this study. It can be seen from the plots that the strength of correlation increases substantially with log-transformation of selected data set, *i.e.*, from $R^2$ equal to 0.037 to 0.665.

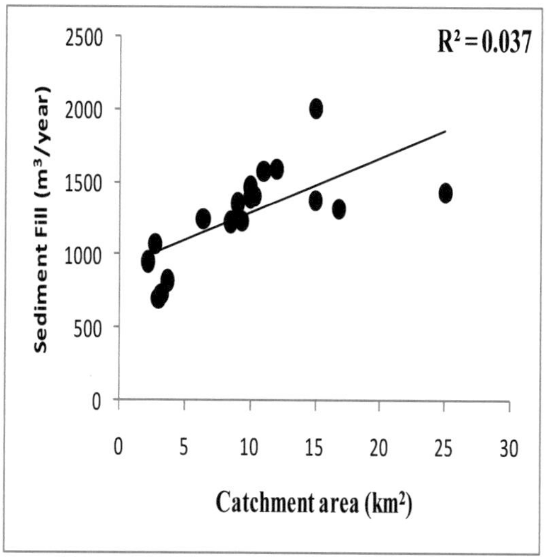


Fig. 3.1. (a) Scatter diagram of Sediment Yield-Fill in $m^3$/year against catchment area in $km^2$ for dry climatic zone.

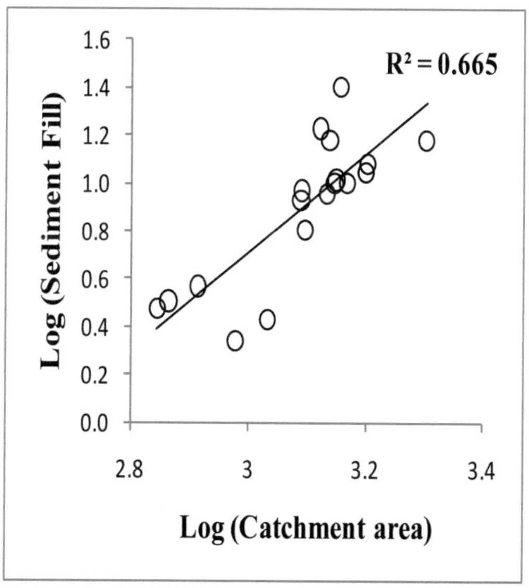


Fig. 3.1. (b) Scatter diagram of log-tranformed values of sediment yield-fill and catchment area for dry climatic zone

### 3.2 Developed sediment yield prediction equations

Sample regression analysis result for dry climatic zone is presented in Tables 3.2.1 & 3.2.2 below:

| *Regression Statistics* | |
|---|---|
| Multiple R | 0.8081451 |
| R Square | 0.6530985 |
| Adjusted R Square | 0.6442036 |
| Standard Error | 0.1655968 |
| Observations | 41 |

ANOVA

| | *df* | *SS* | *MS* | *F* | *Significance F* |
|---|---|---|---|---|---|
| Regression | 1 | 2.013449766 | 2.01345 | 73.423838 | 1.67761E-10 |
| Residual | 39 | 1.0694693 | 0.027422 | | |
| Total | 40 | 3.082919067 | | | |

| | *Coefficients* | *Standard Error* | *t Stat* | *P-value* | *Lower 95%* | *Upper 95%* |
|---|---|---|---|---|---|---|
| Intercept | 2.7452 | 0.05577 | 49.21851 | 1.01E-36 | 2.632384636 | 2.858019 |
| X variable | 0.4313 | 0.04885 | 8.568771 | 1.678E-10 | 0.319783073 | 0.5174043 |

Table 3.2.1. Regression statistics in log scale for both dry and moderate zones data points

From Table 3.2.1 results we have, Log $\alpha$ = 2.7452 ± Standard Error, and thus $\alpha$ would range from 439.20 to 632.27. And if you take the average you will get $\alpha$ = 556.16. Also the value of $\beta$ obtained as $\beta$ = 0.4313± Standard Error. $\beta$ would therefore range from 0.3824 to 0.4802 with the average value 0.4313. The resulting equation will look like (Equation 3.2)

$$SF = 556.2\ A^{0.4313} \tag{3.2}$$

Where; SF= Sediment fill ($m^3/yr$); A = Catchment area ($km^2$); $\alpha$ = constant; and $\beta$ = scaling exponent.

The corresponding graph of sediment fill-yield versus catchment area in dry and moderate climatic zone (all regions analysed) is presented in Fig. 3.2.

One would note from Fig. 3.2 below that the developed equations satisfactorily predict the sediment fill in small catchment at 95 % confidence interval. It is worth noting that a few of the observed sediment fill-yield data points, for instance, for a catchment sizes of 2.8 and 16.5 $km^2$ from Tabora region (Table 2.2.2) plotted outside the prediction range (*i.e.*, outlier). The author is attributing it to uncertainty in field measurements. As reported in literature, all techniques for estimating reservoir volume incorporate errors (Morris and Fan, 1998). An estimated error of about ±30% in determining reservoir capacity volumes have been reported in Morris and Fan (1998) by various workers. These discussions may suggest that the error observed above could be explained by uncertainty in determining actual sediment fill (reservoir sedimentation rate), (Ndomba, 2007). Such errors are also acknowledged by Mulengera (2008). A set of sediment fill-yield prediction equations for various climatic zones

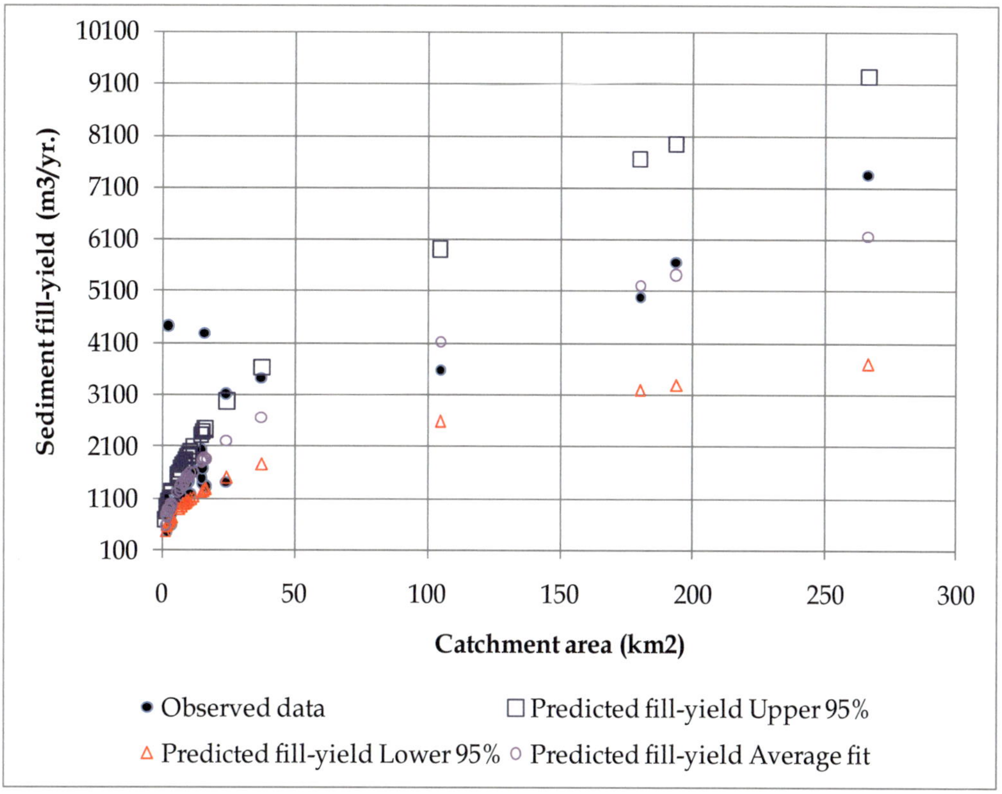


Fig. 3.2. Relationship between sediment fill-yield and catchment area for both dry and moderate climatic zones of Tanzania

were developed in the form of a power function as proposed in section 3.1 above (Table 3.2.2) and as illustrated in Equation 3.2. Besides, the estimated parameter uncertainty bounds are presented in the table. Specific equations for administrative regions are also included.

One would note from Table 3.2.2 below that the performances, as measured by coefficient of determination, $R^2$, of the developed prediction equations are higher in moderate climate zone than in dry climatic zones. The strength of correlations between sediment yields and catchment area sizes could be categorized as high, moderate and low for Arusha and Tabora, Dodoma, and Dodoma and Shinyanga and Singida regions, respectively. These results suggest that large variance in sediment yields remains unexplained by the developed regressions in dry climatic zone. This may be attributed to high uncertainty in representations of long term sediment yields in catchment with high temporal variability of rainfall intensity, runoff and sediment (Mulengera, 2008). However, independent analysis indicates that sediment fill-yield data used for the dry climatic zones are good in space representation with coefficient of variation, CV, between 23 and 26 in percent. The corresponding values for the moderate and all climatic zones (*i.e.*, dry and moderate) range from 64 to 128 in percent. The high sediment yields for Arusha region were expected as the soils in the region are mostly recent volcanic ash and/or highly erodible. The overall

| Climatic zone | Region (s) | No. of data points, n | Range of α=Coefficient ± Standard Error | Average values of α | Range of β =Coefficient ± Standard Error | Average values of β | Sediment Fill/Yield prediction Equation (SF or SY=$\alpha A^{\beta}$) | $R^2$ |
|---|---|---|---|---|---|---|---|---|
| Dry climatic zones | Dodoma, Shinyanga, & Singida | 18 | 508.42 – 663.8 | 580 | 0.3712 – 0.4914 | 0.4313 | $SF = 580A^{0.4313}$ | 0.4586 |
| | Dodoma | 13 | 471 - 815 | 619.6 | 0.0636 – 0.3288 | 0.1962 | $SF = 619.6A^{0.1962}$ | 0.5708 |
| Moderate climatic zone | Tabora | 17* | 505 - 802.8 | 637.2 | 0.2989- 0.4745 | 0.3867 | $SF = 637.2A^{0.3867}$ | 0.7591 |
| | Arusha | 11 | 333.43 – 31974.22 | 3264.37 | 0.4609 – 0.5613 | 0.511 | $SF = 3264.4A^{0.511}$ | 0.7689 |
| Dry and Moderate zone | All regions analyzed** | 41 | 439.2 – 632.27 | 556.16 | 0.3824 – 0.4802 | 0.4313 | $SF = 556.2A^{0.4313}$ | 0.7318 |

Note:

* Only 70% of data points for Tabora region were used to fit the regression relationship. Thirty percent (30%), i.e. 7 data points were used for validation purposes.

** The data for Arusha region was not included in fitting the regression for all regions (i.e. dry and moderate climatic zone) as it presents itself with unique soil/erodibility characteristics as discussed under section 2.1 of this chapter..

Table 3.2.2. A summary of developed sediment fill-yield prediction equations for small catchments in Tanzania relationship developed from data collected from both dry and moderate climatic regions combined has an improved performance according to $R^2$ of 0.732 as compared to 0.451 for dry climatic zone alone. This could be partly due the fact that the number of data points used to fit the regression is adequate for robust relationship as recommended by Statsoft (2011). However, the author would like to recommend the use of a specific equation for particular purpose/climate, especially Arusha region, as they have been developed. Validation result for the developed prediction equation was much better than during model training phase with $R^2$ of 0.873. This was attempted only to regions/climatic zone/region where splitting of data into 70% and 30% for calibration and validation was possible, that is Tabora. The number of observations used for this purpose was 7 (serial numbers 18 through 24 in Table 2.2.2).

Although the performances of the developed Regional Regression Relationships are satisfactory, the author caution the reader, as supported by Morris and Fan (1998) that these equations express only the general relationships between independent variables and the sediment yield-fill and should therefore be used only for preliminary planning purposes or as a rough check. Because these equations reflect regional average conditions, the actual yields will tend to be higher (or much higher) than predicted in erosive areas and lower than predicted in areas of undisturbed catchments. Local site-specific conditions can influence sediment yield much more than drainage area or runoff, for instance.

## 4. Conclusions and recommendations

### 4.1 Conclusions

This study uses readily available data on catchment area and reservoir sediment fill and/or sediment yield to calibrate the prediction equations' parameters by regression analysis approach. The influence of rainfall and/or runoff as important input variables were indirectly captured by developing and grouping the equations with respect to their climatic zones. The equations were validated and parameter uncertainty bounds estimated. The data set was split into 70% and 30% proportions for calibration and validation purpose, respectively. The measured and predicted reservoir sediment fill-yield rates have satisfactory to good correlations with Coefficient of determination , $R^2$, between 0.46 and 0.77 with a degree of freedom of, $n$, 11 to 41 at probability level of significance, $p$, of 5%. $R^2$ of 0.87 at $n$ equals 7 was achieved in one of the validation experiments in moderate climatic zone. Although the performances of the developed Regional Regression Relationship are satisfactory, the author would like to caution the reader that these equations express only the general relationships between independent variables and the sediment yield-fill and should therefore be used only for preliminary planning purposes or as a rough check.

### 4.2 Recommendations

It should be noted that in this work the study area was ill-defined as the wet climatic zone of Tanzania was not adequately represented. Notwithstanding a satisfactory performance achieved, this chapter recommends both extending the data set to cover the wetter regions and incorporating other parameters affecting sediment yield for processes studies and better prediction in the follow up research. This can be achieved if more data such as sediment Particle Size Distribution (PSD), bulky density, dam trap efficiency, catchment environmental variables (*i.e.*, land cover/use, slope, slope length, runoff, rainfall, etc), and operational data and geographic locations of the small dams could be supplemented.

## 5. Acknowledgment

I acknowledge the cordial cooperation rendered to me by staff of the then Ministry of Water and Irrigation for availing previous reports on the subject matter. I'm also indebted to Mr. Kipagile, D., an undergraduate student of University of Dar es Salaam, for collecting and providing some data. FRIEND/Nile project funded by UNESCO Cairo Office is also acknowledged for co-sponsoring the write-up of this chapter.

## 6. References

Arnold, J.G., Williams, J.R., and Maidment, D.R., (1995). Continuous-time water and sediment-routing model for large basins. *Journal of Hydraulic Engineering Vol.121(2): pp171-183.*

Christiansson, C., (1981). *Soil erosion and sedimentation in semi-arid Tanzania*: studies of environmental change and ecological imbalance. Scandinavian Institute of African Studies, Uppsala, Sweden. 208p.

Faraji, S.A., (1995). Implementation of Reservoir Sedimentation Monitoring in Tanzania. *In the proceedings of Results of dam safety and reservoir sedimentation workshop* in Tanzania, edited by Husebye, S. and Torblaa, E. (1995).

Garde, R.J., and Ranga Raju, K.G., (2000). *Mechanics of Sediment Transportation and Alluvial Stream Problems*. Third Edition, 5p. New Age International (P) Limited, Publishers. New Delhi, India.

Horowitz, A.J., (2004). Monitoring suspended sediment and associated trace element and nutrient fluxes in large river basins in the USA. In: Sediment Transfer through the Fluvial System (Proceedings of Moscow Symposium, August 2004), 419-428, IAHS Publication no. 288.

Husebye, S. and Torblaa, E., (1995). Results *of dam safety and reservoir sedimentation workshop in Tanzania*. A proceedings report.

Lawrence, P., Cascio, A., Goldsmith, O., & Abott, C.L. (2004). *Sedimentation in small dams – Development of a catchment characterization and sediment yield prediction procedure.* Department For International Development (DFID) Project R7391 HR Project MDS0533 by HR Wallingford.

Malisa, J., (2007). *Dam Safety Analysis Using Physical And Numerical Models For Small Dams In Tanzania.* A Case Study Of Arusha Region. A PhD Thesis (Water Resources Engineering) of University of Dar es Salaam.

Morris, G. & Fan, J., (1998) *Reservoir sedimentation Handbook*: Design and Management of Dams, Reservoirs and Watershed for Sustainable use. McGrawHill, New York. Chapter 7, 7.1-7.44.

Mtalo, F.W. & Ndomba, P.M., (2002). Estimation of Soil erosion in the Pangani basin upstream of Nyumba ya Mungu reservoir. *Water Resources Management: The case of Pangani Basin. Issues and Approaches; Workshop Proceedings Edited by J. O. Ngana.* Chapter 18, pp.196-210. Dar es Salaam University Press.

Mulengera, M.K., and Payton, R.W., (1999). Estimating the USLE-Soil erodibility factor in developing tropical countries. *Trop. Agric. (Trinidad) Vol. 76 No. 1, pp17-22.*

Mulengera, M.K. (2008). Sediment Yield Prediction in Tanzania: Case Study of Dodoma District Catchments. *Tanzania Journal of Engineering and Technology (TJET) Vol. 2 (No.1)* March 2008, pp63-71.

Ndomba, P.M., Mtalo, F.W., and Killingtveit, A., (2005). The Suitability of SWAT Model in Sediment Yield Modelling for Ungauged Catchments. A Case of Simiyu Subcatchment, Tanzania. Proceedings of the 3rd International SWAT conference, pp61-69. EAWAG-Zurich, Switzerland, 11th -15th July 2005, Sourced at http://www.brc.tamus.edu/swat.

Ndomba, P.M., Mtalo, F.W., and Killingtveit, A., (2008). A Guided SWAT Model Application on Sediment Yield Modeling in Pangani River Basin: Lessons Learnt. *Journal of Urban and Environmental Engineering, V.2,n.2,p.53-62.* ISSN 1982-3932, doi:10.4090/juee.2008.v2n2.053062.

Ndomba, P.M., Mtalo, F.W., and Killingtveit, A., (2009). Estimating Gully Erosion Contribution to Large Catchment Sediment Yield Rate in Tanzania. *Journal of Physics and Chemistry of the Earth 34 (2009) 741 – 748.* DOI: 10.1016/j.pce.2009.06.00. Journal homepage: www.elsevier.com/locate/pce.

Ndomba, P.M., (2007). *Modelling of Erosion Processes and Reservoir Sedimentation Upstream of Nyumba Ya Mungu Reservoir in the Pangani River Basin.* A PhD Thesis (Water Resources Engineering), University of Dar es Salaam, November 2007.

Ndomba, P.M. (2010). Modelling of Sedimentation Upstream of Nyumba Ya Mungu Reservoir in Pangani River Basin. *Nile Water Science & Engineering Journal, Vo.3, Issue 2, 2010; p.25-38. ISSN: 2090-0953. Journal home page: http://www.nbcbn.com/index.php/knowledge-services/nwse-journal.html.*

PLDP, *Proposal for livestock development programme for Tanzania* 1984. (Electronic Version): StatSoft, Inc. (2011). *Electronic Statistics Textbook.* Tulsa, OK: StatSoft. WEB: http://www.statsoft.com/textbook/, Sourced on July 1, 2011.

Rapp, A., Axelsson, V.,Berry, L., and Murray Rust, D. H., (1972). Soil erosion and sediment transport in the Morogoro river catchment, Tanzania. *Georgr. Ann., Vol. 54 A (3-4) : 125-155.*

Renfro, G.W., (1975). Use of Erosion Equations and Sediment-Delivery Ratios for predicting Sediment Yield. In: *Present and Prospective Technology for Predicting Sediment Yields and Sources.* Pp33-45. ARS-S-40. USDA Sedimentation Lab., Oxford, Miss.Thodsen, H., Hasholt, B., and Pejrup, M., (2004). Transport of phosphorous, wash load and suspended sediment in the River Varde Å in Southwest Jutland, Denmark. In: *Sediment Transfer through the Fluvial System (Proceedings of Moscow Symposium, August 2004)*, 466-473, IAHS Publication no.288.

URT, United Republic of Tanzania, (1999). Highway Design Manual Ministry of Works Tanzania.

# 13

# Suitability of SWAT Model for Sediment Yields Modelling in the Eastern Africa

Preksedis Marco Ndomba[1] and Ann van Griensven[2]
[1]University of Dar es Salaam, Dar es Salaam,
[2]UNESCO-IHE, Institute for Water Education, Delft,
[1]Tanzania
[2]Netherlands

## 1. Introduction

Sediment yield refers to the amount of sediment exported by a basin over a period of time, which is also the amount that will enter a reservoir located at the downstream limit of the basin (Morris and Fan, 1998). The subject of sediment yield modelling has attracted the attention of many scientists but lack of data, resources and widely accepted methods to predict/estimate sediment yields are some of the barriers against this direction of research (Summer *et al.*, 1992; Wasson 2002; Lawrence *et al.*, 2004; Ndomba, 2007; Ndomba *et al.*, 2005, 2008b, 2009; Shimelis *et al.*, 2010).

The sediment yield model evaluated in this paper is the Soil and Water Assessment Tool (SWAT). It is hypothesized in the presented study cases that distributed and process based mathematical models such as SWAT could be a potential tool in predicting and estimating sediment yield especially at a catchment scale. Application of the distributed and process-based models could minimize the uncertainty resulting from assuming lumped, stationary and linear systems. Besides, the SWAT model has particular advantages for the study of basin change impacts and applications to basins with limited records (Bathurst, 2002; Ndomba, 2007). In principle, their parameters have a physical meaning and can be measured in the field, and therefore model validation can be concluded on the basis of a short field survey and a short time series of meteorological and hydrological data (Bathurst, 2002).

SWAT was originally developed by the United States Department of Agriculture (USDA) to predict the impact of land management practices on water, sediment and agricultural chemical yields in large ungauged basins (Arnold *et al.*, 1995). The SWAT model has a long modelling history since it incorporates features of several Agriculture Research Service (ARS) models (Neitsch *et al.*, 2005). The SWAT model is a catchment-scale continuous time model that operates on a daily time step with up to monthly/annual output frequency. The major components of the model include weather, hydrology, erosion, soil temperature, plant growth, nutrients, pesticides, land management, channel and reservoir routing. It divides a catchment into subcatchments. Each subcatchment is connected through a stream channel and further divided into a Hydrologic Response Unit (HRU). The HRU is a unique combination of a soil and vegetation types within the subcatchment. Sediment yield is estimated for each HRU with the Modified Universal Soil Loss Equation (MUSLE) (Williams, 1975) (Equation 1).

$$Sed = 11.8\left(Q_{surf}q_{peak}Area_{hru}\right)^{0.56} K_{USLE}C_{USLE}P_{USLE}LS_{USLE}CFRG \qquad (1)$$

Where *Sed* is defined as Sediment yield rate (tones/day), $Q_{surf}$ is the surface runoff volume (mm/day), $q_{peak}$ is the peak runoff rate ($m^3/s$), $Area_{hru}$ is the area of the HRU (ha), $K_{USLE}$ is the USLE soil erodibility factor (0.013 metric ton $m^2$ hr/($m^3$-metric ton cm)), $C_{USLE}$ is the USLE crop management factor or cover management factor, $P_{USLE}$ is the USLE support practice factor, $LS_{USLE}$ is the USLE topographic factor, and CFRG is the coarse fragment factor.
The runoff component of the SWAT model supplies estimates of runoff volume and peak runoff rate using the curve number method (SCS, 1972) and modified rational method, respectively, which, along with the subbasin area, are used to calculate the runoff erosive energy variable. The crop management factor or cover management factor is recalculated every day that runoff occurs. It is a function of above-ground biomass, residue on the soil surface and the minimum cover factor for the plant. The $K_{USLE}$ factor is estimated using an equation proposed by Mulengera and Payton (1999) for tropics. Other factors of the erosion equation are estimated as described by Neitsch *et al.* (2005). The current version of the SWAT model uses the simplified stream power equation of Bagnold's (1977) to route sediment in the channel. The maximum amount of sediment that can be transported from a reach segment is a function of the peak channel velocity. Sediment transport in the channel network is a function of two processes, degradation and aggradation (*i.e.* deposition), operating simultaneously in the reach (Neitsch *et al.*, 2005).
The SWAT model includes an automated calibration procedure. The calibration procedure is based on the Shuffled Complex Evolution-University of Arizona algorithm (SCE-UA) as proposed by Duan *et al.* (1992). The autocalibration option in SWAT provides a powerful, labour saving tool that can be used to substantially reduce the frustration and uncertainty that often characterizes manual calibration (Van Liew *et al*, 2005). In one of the study cases other calibration tools such as the 'Sequential Uncertainty Fitting Algorithm' (SUFI-2) program (Abbaspour *et al.*, 2004, 2007) were used.
Although general SWAT applications have shown that the model performs satisfactorily (Ndomba and Birhanu, 2008), its suitability for specific applications such as sediment yield modelling has yet to be ascertained. In this paper, various sediment yield modelling issues involved with using SWAT such as data requirements and analysis, calibration, sensitivity and uncertainty are critically evaluated in three well-studied cases, the Nyumba Ya Mungu (NYM) Reservoir subcatchment located in the upstream part of the Pangani River catchment (PRC), (a trans-boundary catchment shared between Kenya and Tanzania); the Simiyu River catchment (SRC), (a Lake Victoria Basin subcatchment in Tanzania); and the Koka Reservoir catchment (KRC) in Ethiopia. Growing population, growing demand of cultivated land, mostly inaccurate traditional land usage and dangerously increasing deforestation have increased soil erosion. Erosion has a major impact on nature and diminished the agriculture potential of the selected study cases. Excessive exploitation increases the susceptibility of the soil to fluvial and upland erosion, which is responsible for the increased sediment transport and deposition into the reservoirs.

## 2. Materials and methods

### 2.1 Description of the study cases

Case study 1, the Koka Reservoir catchment (KRC) lies within the western part of the Awash Basin and has an area of approximately 11,000 $km^2$ (Figure 1(a-d) & Table 1). The Awash

River begins in the Central Ethiopian Highlands at an altitude of 3000 m to the west of Addis Ababa. After flowing through Koka Reservoir, it flows northeastwards along the rift valley until eventually discharging into Lake Abbe. The Koka Reservoir is situated about 90 km southeast of Addis Ababa at a longitude of 39° 10' E and latitude of 8° 25' N. The erosion rates in the KRC and in the Awash Basin as a whole are high with values generally exceeding 6,000 t/km$^2$/yr and occasionally as high as 20,000 t/km$^2$/yr.

The climate of the Awash Basin is characterized by the Inter-Tropical Convergence Zone (ITCZ). The Mean annual rainfall (MAR) and temperature vary spatially across the catchment between 170 and 1978 mm/annum and 12.8 and 31.5°C, respectively. The area is dominated by a bimodal rainfall pattern. According to the National Meteorological Services Agency, the study area is characterized by a quasi-double maxima rainfall pattern with a small peak in April and maximum peak in August. The rainfall in the highlands shows a strong correlation with altitude (Lemma, 1996). The mean annual wind speed in the KRC is 1.9 m/s.

Two major relief features are found in the Awash Basin: the highlands of the Ethiopian Plateau and the lowlands of the Rift Valley. The bedrock and soil in the area determine the amount and composition of transported sediments in the river. The geology of the basin is dominated by sedimentary rocks such as limestone and sandstone. The alluvium deposits consist of clay, sand and tuff. The long rains occur between June and September.

Case study 2, the Nyumba Ya Mungu (NYM) Reservoir subcatchment, has runoff that is highly regulated by the man-made Nyumba Ya Mungu (NYM) Reservoir of some 140 km$^2$. The NYM Reservoir subcatchment is located in the upstream part of Pangani River catchment (PRC) (Figure 1 & Table 1). The PRC is located between coordinates 36°20′ E, 02°55′ S and 39°02′ E, 05°40′ S in the northeastern part of Tanzania and covers an area of about 42,200 km$^2$, with approximately 5% in Kenya (Figure 1). The Pangani River has two main tributaries, the Kikuletwa (1DD1) and the Ruvu (1DC1) (Figure 1), which join at the Nyumba Ya Mungu (NYM) reservoir.

The catchment of NYM occupies a total land and water area of about 12,000 km$^2$ (Ndomba, 2007). It is located between coordinates 36°20'00'' E, 3°00'00'' S and 38°00'00'' E, 4°3'50'' S. This area has a Mean Annual Rainfall (MAR) of about 1000 mm/annum. The rainfall pattern is bimodal with two distinct rainy seasons, long rains from March to June and short rains from November to December. Rohr and Killingtveit (2003) indicated that the maximum precipitation on the southern hillside of Mt. Kilimanjaro takes place at about 2,200 m.a.s.l., which is 400 - 500 m higher than previously assumed. The mean annual wind speed is 1.87 m/s. The altitude in the study area ranges from 700 and 5,825 m.a.s.l. with the peak of Mt. Kilimanjaro as the highest ground. Based on the Soil Atlas of Tanzania, the main soil type in the study area is clay with good drainage (Hathout, 1983). The landcover of the catchment is dominated by actively-induced vegetation, forest, bushland and thickets with some alpine desert. The majority of the population in the basin directly or indirectly depends on irrigated agriculture. Agriculture is concentrated in the highlands with area coverage less than 20% (Ndomba, 2007). Lowlands are better suited for pastoralism. The main runoff-sediment generating subcatchments in the study area upstream of NYM reservoir are Weruweru, Kikafu, Sanya, Upper Kikuletwa and Mt. Meru. The basin is also important for hydropower generation, which is connected to the national grid. Hydropower plants, which are downstream of NYM Reservoir are NYM (8 MW), Hale (21 MW), and New Pangani falls (66 MW).

| Case study | Catchment area ($km^2$) | Topography | Land use/cover | Geology and Soil type | Climate |
|---|---|---|---|---|---|
| Nyumba Ya Mungu (NYM) Reservoir catchment in the upstream part of the Pangani River catchment | 12,000 | Plains - mountainous | Actively induced vegetation, forest, bushland and thickets with some alpine desert and agricultural land for farming | Neogene Volcanic and pre-Cambrian metamorphic rocks extensively covered by superficial Neogene deposits including calcareous tuffaceosus; Clay with moderate to good drainage | Semi arid - humid |
| Koka Reservoir catchment (KRC) | 11,000 | Rugged | Agricultural land for farming, Acacia and eucalyptus trees are prevailing ones | Sedimentary rocks such as limestone and sandstone; clay, sand and tuff | Tropical |
| Simiyu River catchment (SRC) | 10,659 | Relatively flat | Agricultural land for farming, grassland for grazing, and bushland. | Dominated by Precambrian rocks and some quaternary sediments; There are also some extensive areas overlain by recent alluvial deposits; sandy loam covers a large part of the catchment | Warm tropical savannah climate/ Diverse |

Table 1. Major characteristics of the three study cases in Eastern Africa

Case study 3, the Simiyu River catchment (SRC), is located in the northern part of Tanzania southeast of Lake Victoria (Figure 1 & Table 1). It covers an area of 10,659 $km^2$ and is located between the coordinates 33°15′00″ E, 02°30′00″ S and 35°00′00″ E, 03°30′00″ S. The SRC is occupied by about one million inhabitants. The catchment is mainly covered by agricultural land for farming, grassland for grazing, and bushland. The Simiyu River flows from the Serengeti National Park Plains to Lake Victoria in the downstream region. The two major tributaries of the Simiyu River are Simiyu-Duma and Simiyu-Ndagalu and they merge shortly before the Simiyu River enters Lake Victoria. The river is characterized as ephemeral

(Ndomba *et al.*, 2005) and normally stays dry in the months of August, September and October. During the long rainy season, discharge from the river reaches as high as 331$m^3/s$.

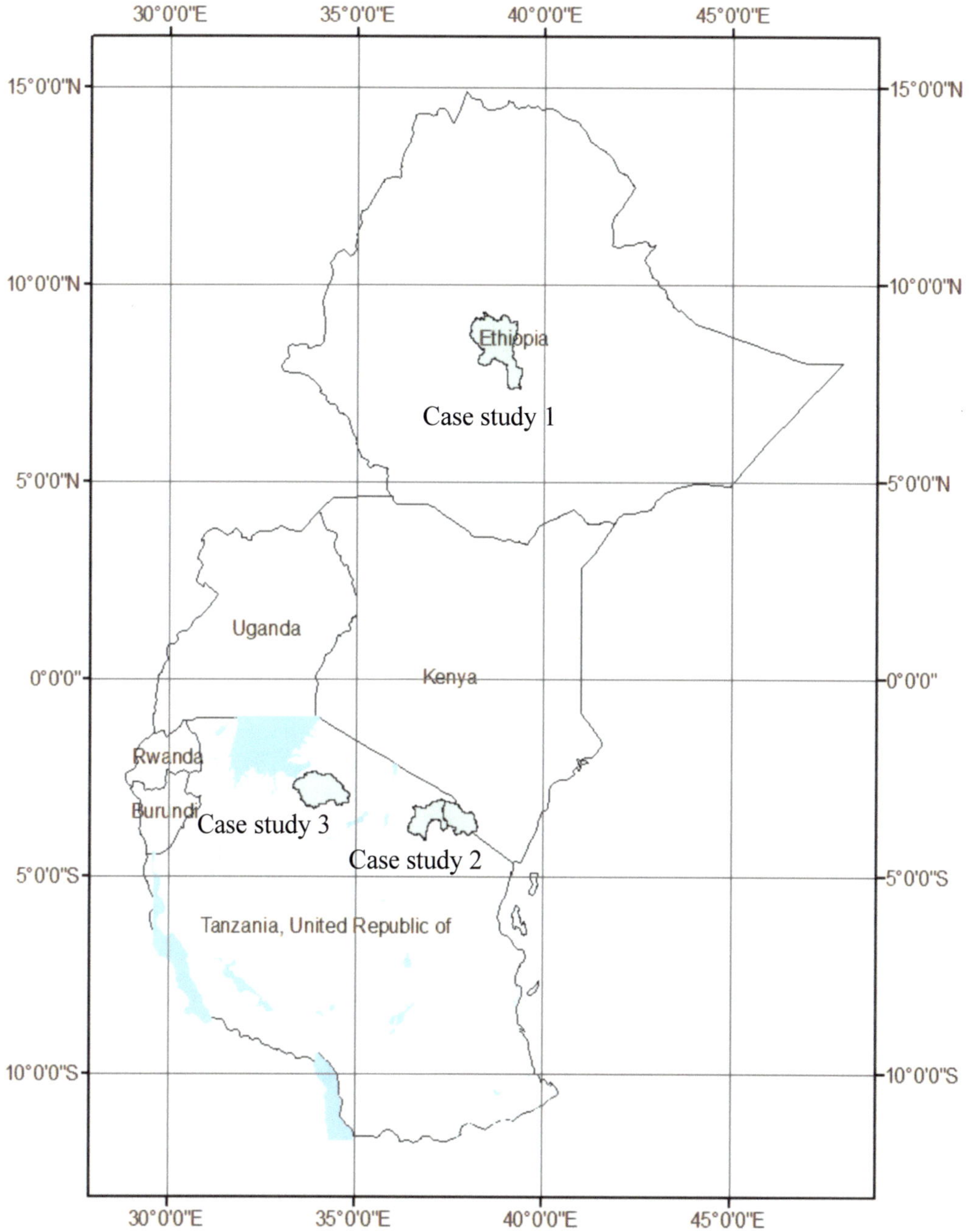


Fig. 1a. A map showing the three case studies in Eastern Africa

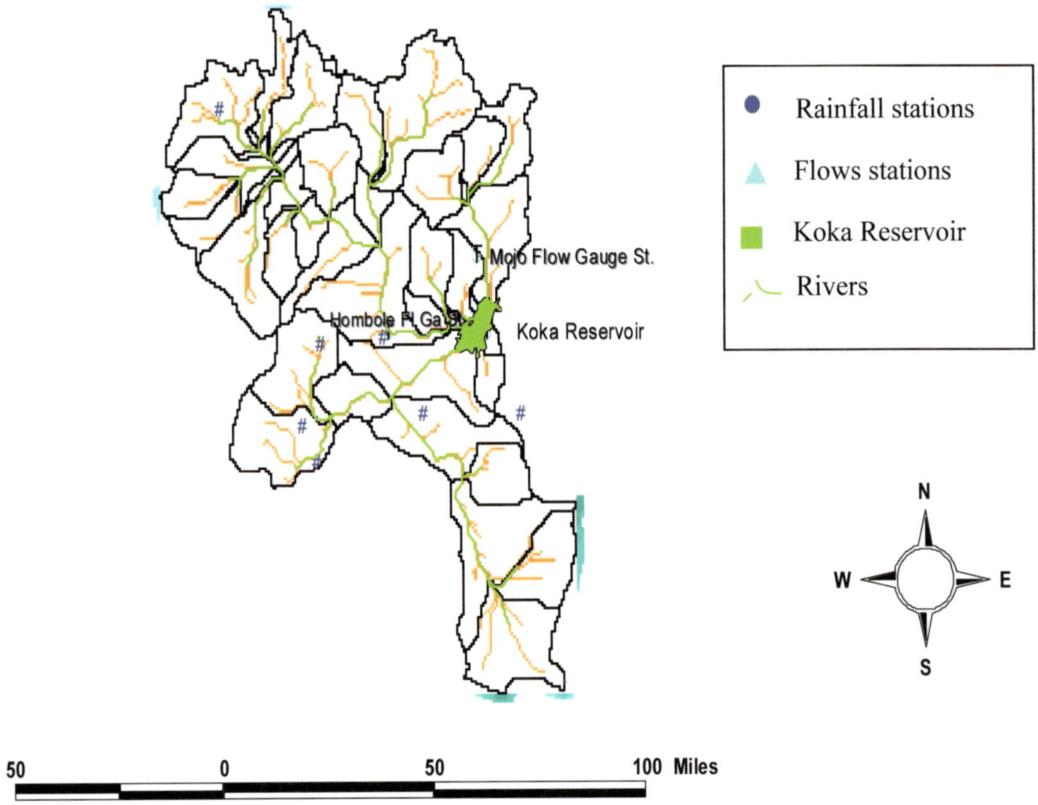


Fig. 1b. A location map of Case study 1, Koka Reservoir catchment (KRC), as adopted from Endale(2008)

The catchment has a warm tropical savannah climate with an average temperature of about 23oC. The total average annual precipitation varies between 700 and 800 mm/annum. The mean annual wind speed is 1.60 m/s. The SRC is considered to be one of the main contributors to the deterioration of water quality (*i.e.*, sediments and nutrients) in the Lake Victoria. This is because of its relatively large size, its large inflow contribution and its many agricultural activities using agrochemicals which generate high yields of sediments.

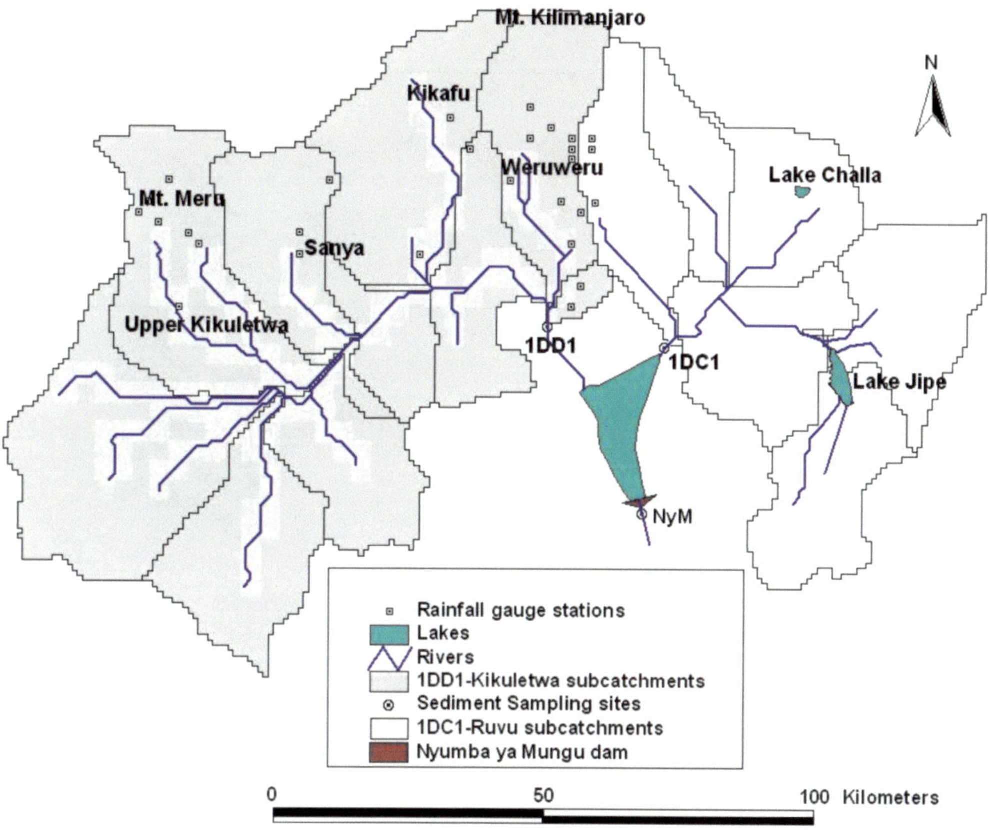


Fig. 1c. A location map of Case study 2, Nyumba Ya Mungu Reservoir, in the upstream part of the Pangani River catchment as adopted from (Ndomba, 2007)

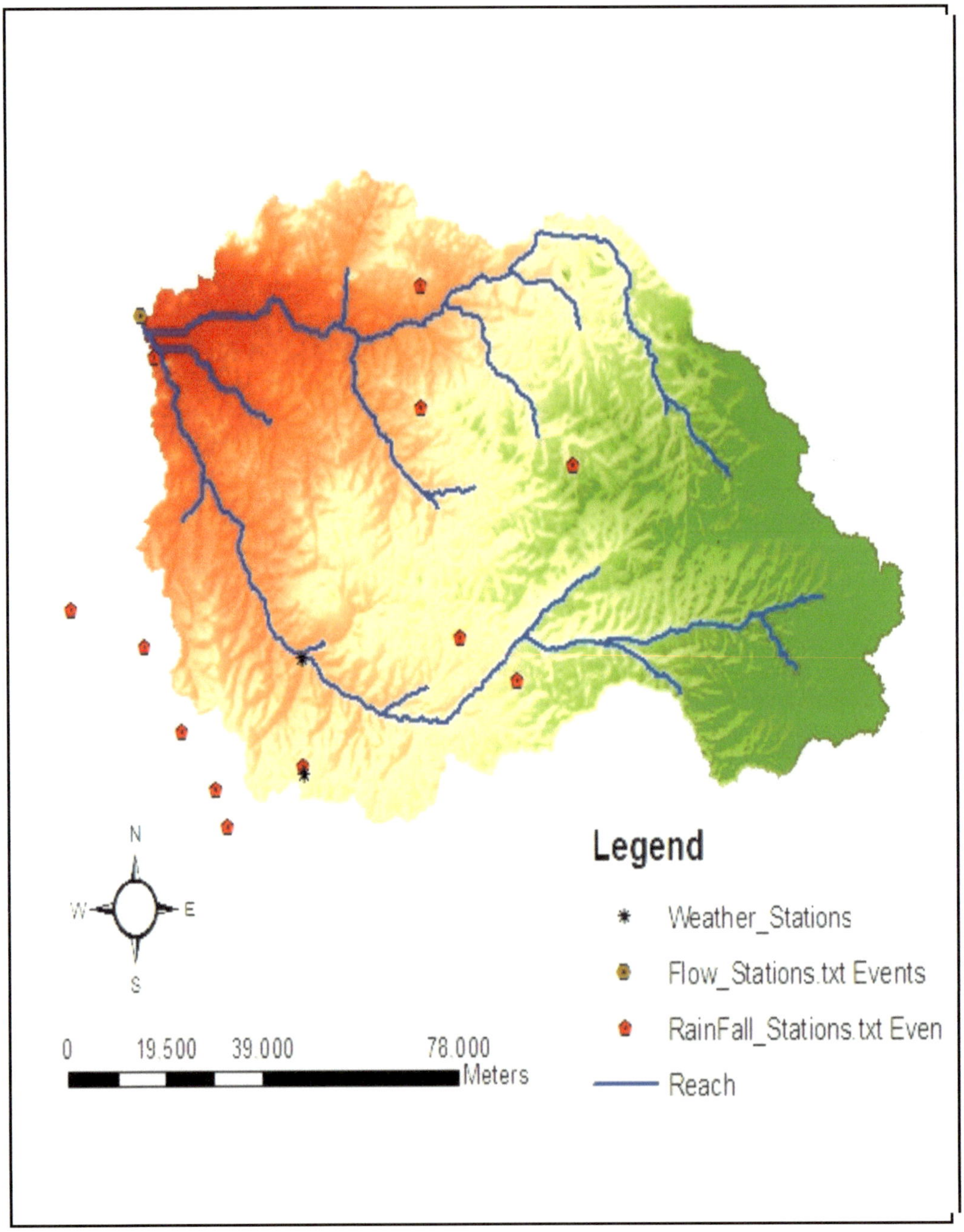


Fig. 1d. A location map of Case study 3, Simiyu River catchment, Tanzania as adopted from Abdelhamid (2010)

### 2.2 SWAT concept

The SWAT model uses the Modified USLE (MUSLE) equation developed by Williams (1975) (Equation 1) to simulate the sediment yield from the upland catchments (Neitsch *et al.*,

2005). The surface runoff ($Q_{surf}$) as input to the MUSLE equation is simulated by the runoff component of SWAT. The SWAT model uses water balance Equation 2 as a driving force behind everything that happens in the watershed (Neitsch *et al.*, 2005).

$$SW_t = SW_{t-1} + \sum_{i=1}^{t}\left(R_{day\_i} - Q_{surf\_i} - E_{a\_i} - w_{seep\_i} - Q_{gw\_i}\right) \quad (2)$$

$SW_t$ is the final soil water content (mm), $SW_{t-1}$ is the initial soil water content on day i (mm), t = 1, 2, 3,…,n where "n" is the total number of days during the simulation (days), $R_{day\_i}$ is the amount of precipitation on day i (mm), $Q_{surf\_i}$ is the amount of surface runoff on day i (mm), $E_{a\_i}$ is the amount of evapotranspiration on day i (mm), $W_{seep\_i}$ is the amount of water entering the vadose zone from the soil profile on day i (mm), and $Q_{gw\_i}$ is the amount of return flow on day i (mm).

SWAT uses Manning's equation to define the rate and velocity of flow. Water is routed through the channel network using the variable storage routing method or the Muskingum River routing method. Both the variable storage and Muskingum routing methods are variations of the kinematic wave model (Neitsch *et al.*, 2005).

### 2.3 Data and data analysis

The sediment flow data are readily available (Table 2a-c). The quality and adequacy of data varies from one catchment to the other. In the two study cases, the Simiyu River and Koka Reservoir catchments, secondary data on streams flows, climate, sediment flow and spatial data were used to setup, calibrate and validate the model. These are typical data types used in most of the SWAT applications elsewhere in the region (Andualem and Yonas, 2008; Shimelis *et al.*, 2010). Most of the sediment flow data are intermittent instantaneous sediment flow data. In one of the cases, the NYM Reservoir subcatchment, primary data on sediment flow was collected to complement the analysis. These are continuous subdaily sediment concentrations data plus multi-temporal reservoir survey information. As Table 2 stipulates, various sources of data were explored. The data preparation and analysis task involved analyzing statistics such as season mean, percent missing data, identifying outliers, length of the records, temporal and spatial variability of rainfall, and wet years' period. The wet years' period is defined as the period when the annual total rainfall is above the long term annual average. The analysis was meant to guide and provide data for the SWAT modelling. For instance, spatial variability justified the need for distributed modelling. The derived statistics were also used as inputs to weather generator module of SWAT. The module generates climatic data or fills in gaps in measured records. As presented in Tables 2a-1, 2b-1 and 2c-1, the input spatial data included base maps such as readily available topographic maps in the Ministries and global spatial thematic maps (*i.e.* Digital Elevation Models, DEM; Soil, and Landuse-cover) of various resolutions. One of the case studies reviewed in this paper (Mulungu and Munishi, 2007), used high-resolution data on land use from the 30 m LandSat TM Satellite, the 90 m Digital Elevation Model and the Soil and Terrain Database for Southern Africa (SOTERSAF). In some cases such as NYM, the soil types were extracted from Pauw (1984) digital map and complemented by the Soil Atlas of Tanzania (Hathout, 1983). Similarly, climatic data included rainfall data from the regular ground monitoring network.

| SN | Data type | No. of stations | Data availability | % missing | Source | Resolution | |
|---|---|---|---|---|---|---|---|
| | | | | | | Spatial | Temporal |
| 1. | Rainfall | 31 | 1922-2005 | 1-60 | MoWI/ TMA/PBWO | - | Daily/ hourly |
| | Climate | 9 | 1958-1999 | 0.1-62 | MoWI/ TMA/PBWO | - | Daily |
| 2. | Flows | 3 | 1952-2005 | 1-54 | MoWI/ /PBWO | - | Daily |
| 3. | DEM | - | 2006 | - | USGS/ HDRO 1K | 1 km | - |
| 4. | Landuse | | 1990's | - | IRA/Landsat | 30 m | - |
| 5. | Soil | - | 1983-1984 | - | Pauw(1984) & Hathout, (1983) | - | - |
| 6. | Sediment load | 3 | 2005-2006 | 0 | UDSM | - | Subdaily continuou s |
| 7. | Reservoir bed contour | 1 | 1968 & 2005 | - | UDSM | 1.53/1.0m | 37 years |

Table 2a-1. Data types and sources for the Nyumba Ya Mungu Reservoir catchment located in the upstream part of the Pangani River catchment
Note: MoWI: Minstry of Water and Irrigation; TMA: Tanzania Meteorological Authority; PBWO: Pangani Basin Water Office; IRA: Institute of Resources Assessment based at University of Dar es Salaam (UDSM)

| Statistic | Subdaily suspended sediment concentration [mg/l] | Gauge Height [m] | Streamflow discharge [m$^3$/s] |
|---|---|---|---|
| No. Of data points | 291 | 291 | 291 |
| Maximum | 9110.0 | 4.44 | 256.53 |
| Minimum | 16.0 | 0.89 | 12.19 |
| Mean | 282.5 | 1.32 | 34.79 |
| Standard Deviation, STD | 801.7 | 0.49 | 30.02 |
| Coefficient of Variation, Cv (%) | 283.8 | 36.69 | 86.27 |
| Standard Error of the Mean, SEM | 47.0 | 0.03 | 1.76 |

Table 2a-2. A summary of sediment flow data for the Nyumba Ya Mungu Reservoir catchment in the upstream part of the Pangani River catchment as sampled between March 18 and November 10, 2005 by an ISCO 6712 machine at 1DD1 site

| Statistic | Suspended sediment concentrations [mg/l] | Base Gauge Height [m] | Fall, [m] | Stream flow discharge [$m^3/s$] | Sediment load, [t/day] |
|---|---|---|---|---|---|
| No. Of data points | 288 | 288 | 288 | 288 | 288 |
| Average | 54.9 | 0.724 | 0.098 | 4.025 | 19.1 |
| Minimum | 2.8 | 0.420 | 0.025 | 2.885 | 1 |
| Maximum | 830.0 | 2.270 | 0.190 | 12.227 | 359 |
| Standard deviation | 75.5 | 0.28 | 0.014 | 1.12 | 34.91 |
| Standard Error of the Mean (SEM) | 4.45 | 0.02 | 0.001 | 0.07 | 2.06 |

Table 2a-3. Daily suspended sediment flow data for the Nyumba Ya Mungu Reservoir catchment in the upstream part of the Pangani River catchment at the 1DC1-Ruvu site sampled using DH-48 sampler sampled at 9.00 hrs between April 19, 2005 and January 31, 2006

Note: (1) Fall signifies measured water gauge height difference between Base and Auxiliary gauging stations; and (2) Stream flow discharge data were derived from complex rating curve (Ndomba, 2007).

| SN | Data type | No. of stations | Data availability | % missing | Source | Resolution | |
|---|---|---|---|---|---|---|---|
| | | | | | | Spatial | Temporal |
| 1. | Rainfall | 13 | 1928-2003 | 8.8-69.8 | MoWI/ TMA | - | Daily/ hourly |
| 2. | Climate | 2 | 1970-1984 | 3.4-21 | MoWI/ TMA | - | Daily |
| 3. | Flows | 1 | 1969-2000 | 46 | MoWI | - | Daily |
| 4. | DEM | - | 2010 | - | ASTER | 30 m | - |
| 5. | Landuse | | 2010 | - | GLCC | 1 km | - |
| 6. | Soil | - | 2010 | - | FAO/SOTER SAF | 10 km | - |
| 7. | Sediment load | 1 | 1999-2003 | 0 | UDSM | - | Daily intermi-ttent |

Table 2b-1. Data types and sources for the Simiyu River catchment

Note: ASTER: Advanced Space borne Thermal Emission and Reflection Radiometer; GLCC: Global Land Cover Charactersization; SOTERSAF: Soil Terrain Database for Southern Africa; FAO: Food and Agricultural Organization.

| Statistic | Suspended sediment concentrations [mg/l] | Daily Stream flow discharge [m³/s] | Gauge Height [m] | Sediment load, [t/day] |
|---|---|---|---|---|
| No. | 102 | 70 | 13 | 70 |
| Max | 5067 | 213 | 3.250 | 20,707 |
| Min | 5.0 | 2.0 | 0.250 | 13.0 |
| Mean | 981.0 | 19.925 | 1.813 | 2,615 |
| Standard Deviation, STD | 1108.6 | 29.240 | 1.051 | 4,325 |
| Coefficient of Variation, Cv (%) | 113 | 146.748 | 57.949 | 165 |
| Standard Error of the Mean, SEM | 109.8 | 3.495 | 0.291 | 517 |

Table 2b-2. A summary of intermittent daily sediment flow data for the Simiyu River catchment as sampled between June 30, 1999 and May 29, 2004 at the Main Bridge site, the outlet

| SN | Data type | No. of stations | Data availability | % missing | Source | Resolution | |
|---|---|---|---|---|---|---|---|
| | | | | | | Spatial | Temporal |
| 1. | Rainfall | 3 | 1990-2004 | 0.12-0.24 | NMSA | - | Daily/hourly |
| 2. | Climate | 4 | 1990-2004 | 1.32-1.42 | NMSA | - | Daily |
| 3. | Flows | 3 | 1990-2004 | 0.15-0.27 | DH-MoWR | - | Daily |
| 4. | DEM | | 2008 | - | USGS/ HDRO 1K | 1 km | - |
| 5. | Landuse | | 2008 | - | GLCC | 1 km | - |
| 6. | Soil | | 2008 | - | FAO/SOTERSAF | 10 km | - |
| 7. | Sediment load | 3 | 1990-2004 | 2.2-2.8 | DH-MoWR | - | Daily intermittent |
| 8. | Reservoir bed contour | | 1959, 1981, 1988, 1999 | - | DH-MoWR | - | 22, 7, 11 years, respectively |

Table 2c-1. Data types and sources for the Koka Reservoir catchment
Note: DH-MoWR: Department of Hydrology, Ministry of Water Resources, Ethiopia; NMSA: National Meteorological Service Agency, Ethiopia; and "-" Not applicable.

| Statistic | Mean monthly Stream flow discharges [$m^3/s$] | Sediment load, [t/month] |
|---|---|---|
| No. | 180 | 180.00 |
| Max | 421.930 | 179,841 |
| Min | 1.000 | 200 |
| Mean | 47.710 | 12,596 |
| Standard Deviation, STD | 78.010 | 18,178 |
| Coefficient of Variation, Cv (%) | 163.510 | 144 |
| Standard Error of the Mean, SEM | 5.810 | 1355 |

Table 2c-2. A summary of continuous monthly sediment flow data for the Koka Reservoir catchment for the period from January 1990 to December 2004 at the Koka Reservoir as adopted from Endale (2008)

### 2.4 SWAT model applications procedures and assumptions

It should be noted that SWAT, if not properly applied, may result in parameter uncertainty problems. Therefore, elaboration of the rationale of each application step is necessary.

In these study cases the model was set up to represent the spatial variability of the main runoff-sediment yield controlling features such as soils, land use/cover, terrain (*i.e,* slope and slope length), river channels and reservoirs. The distributed nature of the sediment yield and erosion representation (lumped, semi and fully distributed) depended on the availability of data and computation resources.

The Latin Hypercube One-factor-At-a-Time (LH-OAT) design as proposed by Morris (1991) implemented in SWAT was used as a sensitivity analysis tool. Sensitivity analysis of hydrology and sediment transport components parameters were conducted without and/or with observed data before and after calibration. Various lengths of simulations (*i.e.* 2, 4, 6, 8 yrs and greater) were tested in order to capture model input (*i.e.,* parameter and data) uncertainty. These analyses were used to identify the sensitive parameters.

Manual calibration, expert knowledge and automatic calibration techniques were tested for the calibration procedures. The autocalibration routine based on the Shuffled Complex Evolution-University of Arizona (SCE-UA) that is incorporated in the SWAT model has been used very often (Duan *et al.*, 1992). In one of the study cases, SRC, the SUFI-2 program which combines calibration and uncertainty analysis (Abbaspour *et al.*, 2004, 2007) was used. This tool is widely used in the region (Shimelis *et al.*, 2010). The sensitive model parameters were adjusted within their feasible ranges during calibration to minimize model prediction errors for daily and monthly flow and sediment loads. In one of the study cases, *i.e.* the NYM Reservoir catchment, soil erodibility ($K_{USLE}$) (a MUSLE factor) was estimated according to the equation proposed by Mulengera and Payton (1999) for tropics. Bias-corrected rating curves were developed and used to interpolate or extrapolate sediment loads (Ndomba *et al.*, 2008b). It should be noted that to date there is no consensus on how to develop an excellent rating curve, especially from a short period of records. Ndomba *et al.* (2008b) developed the rating curve from continuous subdaily suspended sediment data (*i.e.* 2 to 12 samples a day) collected by an automatic pumping sampler (ISCO 6712). The ISCO 6712 sampler data were calibrated by daily-midway and intermittent-cross section sediment samples collected by a depth-integrating sampler (D-74). The sediment loads from rating curve were bias corrected. Sediment load correction factors were derived from both statistical bias estimators (Ferguson, 1986) and actual sediment load approaches (Ndomba *et al.*, 2008a). It was important to do this as it is known that uncorrected rating curves, developed by Ordinary Least Square (OLS) tend to underestimate sediment loads (Ferguson, 1986; Ndomba *et al.*, 2008b). The SWAT model simulation was validated with long term reservoir sediment accumulation and/or sediment loads.

Parameter uncertainty in this study was reduced by placing emphasis on the most sensitive parameters and reformulating the model. This was achieved in one case by estimating some important parameters outside the model using proposed equations/estimators for the Eastern Africa and tropics. This approach was also suggested by Melching (1995). In some cases, the degree of parameter estimation uncertainty of the catchment sediment yield model was reduced by calibrating the parameters during the wet years' period for which most of the hydro-climatic and sediment flow data required by the model are available, as suggested by Yapo, *et al.*, (1996) who observed that the hydrographs of wet years produce more identifiable parameters.

The performance of the model using filled and raw rainfall was evaluated in order to assess input data uncertainty (Ndomba *et al.*, 2008b). Model performances were mainly evaluated based on Nash-Sutcliffe Coefficient of Efficiency (CE), Relative Error (RE), and Total Mass Balance Controller (TMC). CE provides a normalized estimate of the relationship between the observed and predicted model values. The simulation results were considered good for CE values greater than 0.75, while for values of efficiency between 0.75 and 0.36 the simulation results were considered to be satisfactory. CE values between 0.36 and 0 were considered to be fair. A value of zero would indicate that the fit was as good as using the average value of all the measured data. RE was estimated as the ratio of the absolute error to the true value and expressed in percentage. RE of less than twenty percent (20%) is considered acceptable for most scientific applications.

SWAT applications in this study assume a number of things. Although the principal external dynamic agents of sedimentation are water, wind, gravity and ice (Vanoni, 1975) only the hydrospheric forces of rainfall, runoff and streamflow were considered. The computed sediment yield in SWAT is solely a result of sheet erosion processes in the catchment (Shimelis *et al.*, 2010).

## 3. Results and discussions

### 3.1 Sensitive parameters controlling sediment generation and routing

Seven (7) out of nine (9) SWAT parameters that directly govern the sediment yield and transport in the study cases NYM, SRC and KRC were found to be sensitive (Table 3). It should be noted that rank 10 signifies that a parameter is not sensitive/influential at all. These parameters can be categorized into two groups: upland and channel factors. The former group includes parameters such as $P_{USLE}$, $C_{USLE}$, $K_{USLE}$, Biological mixing efficiency (BIOMIX), and Initial residual cover (RSDIN); whereas Linear re-entrainment parameter for channel sediment routing (Csp), Channel cover factor (CCH), Channel erodibility factor (KCH) and Exponential re-entrainment parameter for channel sediment routing (SPEXP) parameters belong to the latter group.

However, it should be noted that only channel routing parameters with serial numbers 1, 2, 4, and 5 in the Table 3 were calibrated in all cases. As described in Neitsch *et al.* (2005), SWAT upland factors according to the MUSLE equation are formed based on regression analysis of runoff plots-based data. Mulengera and Payton (1999) attempted to define $K_{USLE}$ values for tropical regions. One may note from Table 3 that the same set of important parameters for all the cases is retained. There are few cases of swapping of the ranking of the parameter importance, *e.g.*, for BIOMIX in SRC. The influence of these parameters in sediment yield is well documented in Neitsch *et al.* (2005) and Ndomba (2007). Besides, the independently performed simulation results for catchment sediment management scenarios in the study cases indicate that all sorts of farming practices as captured by the $P_{USLE}$ and $C_{USLE}$ parameters (Table 3) are the main determinants in reducing soil loss and sediment yield in the upland catchments and subsequent sedimentation problems in the downstream reservoirs (Endale, 2008). The results also suggest that micro river channels also act as important sources of sediment as represented by the high rank of the linear re-entrainment parameter for channel sediment routing factor, Csp. It should be noted that many other factors affect the sediment yield estimations. For instance, the hydrological parameters seem to have a high effect as well on the sediment computations (Ndomba, 2007). This may be explained by the fact that one of the parameters/factors in the sediment yield equation, the MUSLE, used in the SWAT model is the surface runoff, $Q_{surf}$. Experience also shows that the resolution of DEM and the monthly average rainfall intensities, which are provided in the weather generator database are fundamental and crucial. The results in Table 3 below compare well with that of Shimelis *et al.* (2010) who worked in Angeni Gauged watershed, Ethiopia. In their case the ranking for Csp and CCH are first and second, respectively, with $C_{USLE}$ ranked in the sixth position.

| SN | Parameter | Description of parameter | NYM Rank | SRC Rank | KRC Rank |
|---|---|---|---|---|---|
| 1. | Csp | Linear re-entrainment parameter for channel sediment routing | 1 | 2 | 1 |
| 2. | CCH | Channel cover factor | 2 | 5 | 2 |
| 3. | $P_{USLE}$ | USLE support practice factor | 3 | 3 | 3 |
| 4. | KCH | Channel erodibility factor [cm/h/Pa] | 4 | 6 | 4 |
| 5. | SPEXP | Exponential re-entrainment parameter for channel sediment routing | 5 | 4 | 5 |
| 6. | $C_{USLE}$ | Minimum USLE cover factor | 6 | 7 | 6 |
| 7. | BIOMIX | Biological mixing efficiency. | 7 | 1 | 7 |
| 8. | $K_{USLE}$ | USLE soil erodibility factor [t.ha.h./(ha.MJ.mm] | 10 | 10 | 10 |
| 9. | RSDIN | Initial residue cover [kg/ha] | 10 | 10 | 10 |

Table 3. Sensitivity analysis results of sediment component of SWAT for three study cases, i.e., NYM, SRC and KRC

### 3.2 Model performance

The discussion in this section focuses on the study cases where there was relatively adequate data and where more modelling efforts were applied (Table 4). For instance, in KRC the results of the model performance according to CE for flow calibration and validation are 68 and 63%, respectively. For sediment calibration and validation, CE is 66 and 68%, respectively. The calibration and validation results have shown that measured and simulated values were closely related with RE of 7.5 %.

The sediment yield calibration results for SRC for the period from 1970 to 1975 in daily and monthly aggregated outputs are presented in Table 4 and Figure 2. The result from the SWAT model at a daily time step is fair with the model performance of CE= 24%. The sediment loads in the peak flood events such as those in 1970, 1972 and 1974 are over-predicted. However, the performance of the model when looking at the monthly sediment loads is good, with CE= 83%. The sediment modelling was validated for the period between 1976 to 1978 (Figure 3). Daily sediment load is fairly simulated with a model performance of CE = 16%. Sediment loads during the peak flood events are over-predicted such as in the months of April, 1976 and November, 1978. On the other hand, it under-predicted the loads in January and April of 1977. However, the performance of the model in simulating monthly sediment loads is good with CE = 80%. The improvement of the SWAT model performance when aggregating the outputs over a longer time period has also been observed by other researchers in the region and elsewhere (Schmidt and Volk, 2005; Shimelis *et al.*, 2010).

Since the routing models for the SWAT model as well as the coupled SWAT-SOBEK models have shown to give acceptable and comparable results, the uncertainty of the sediment routing component has been evaluated to be minimal (Abdelhamid, 2010). The performance was measured against total sediment loads transported to the main outlet of the catchment. A similar result was obtained by Andualem and Yonas (2008) in Ethiopia where they applied SWAT and CCHE1D sediment transport models together in tandem. In Figure 4 shows that simulated and observed annual sediment loads are comparable for Case study 1, the Koka Reservoir catchment.

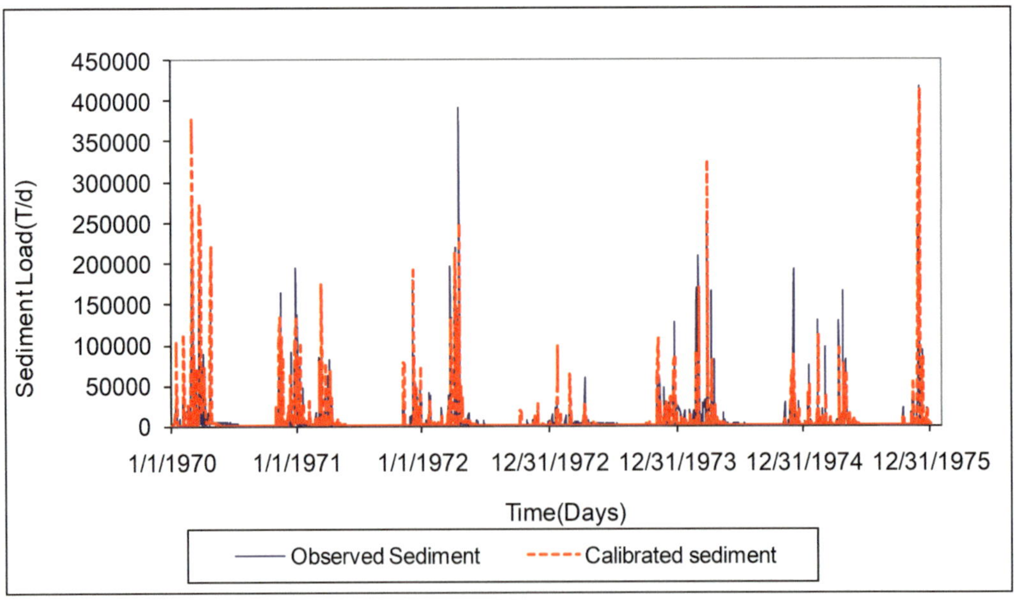


Fig. 2. Comparison between "observed" and simulated Simiyu daily sediment for calibration period from 1970and 1975 at Main Bridge outlet (Abdelhamid, 2010).

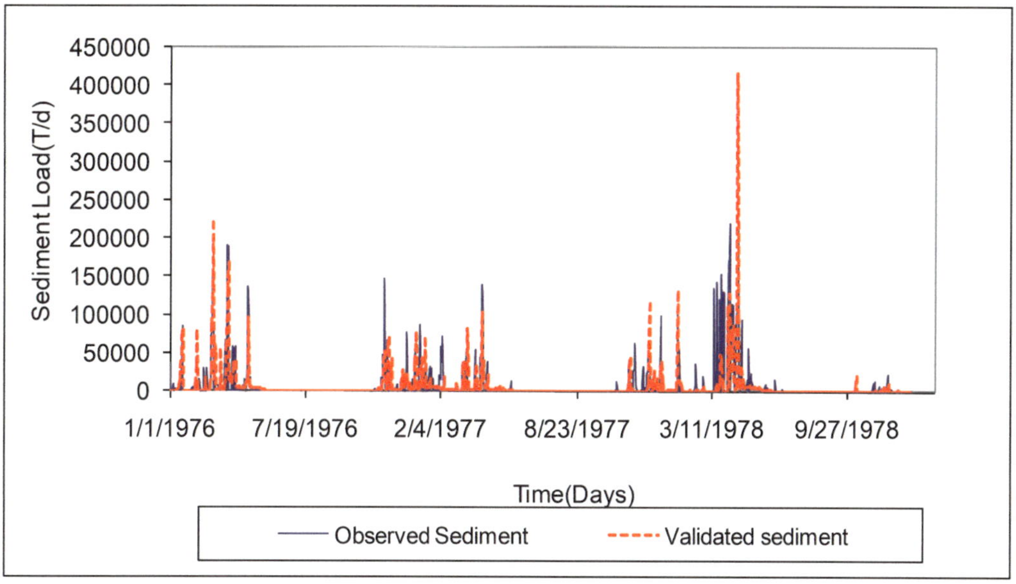


Fig. 3. Comparison between "observed" and simulated Simiyu daily sediment for the validation period from 1976 and 1978 at Main Bridge outlet (Abdelhamid, 2010)

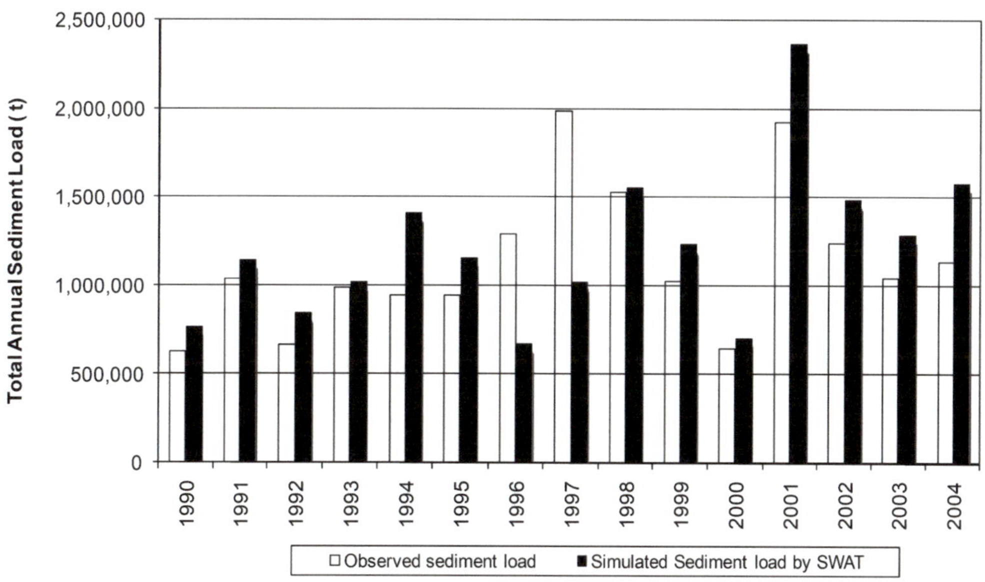


Fig. 4. SWAT simulations vs observed annual sediment loads at Case study 1, the Koka Reservoir, between 1990 and 2004 (As adopted from Endale, 2008)

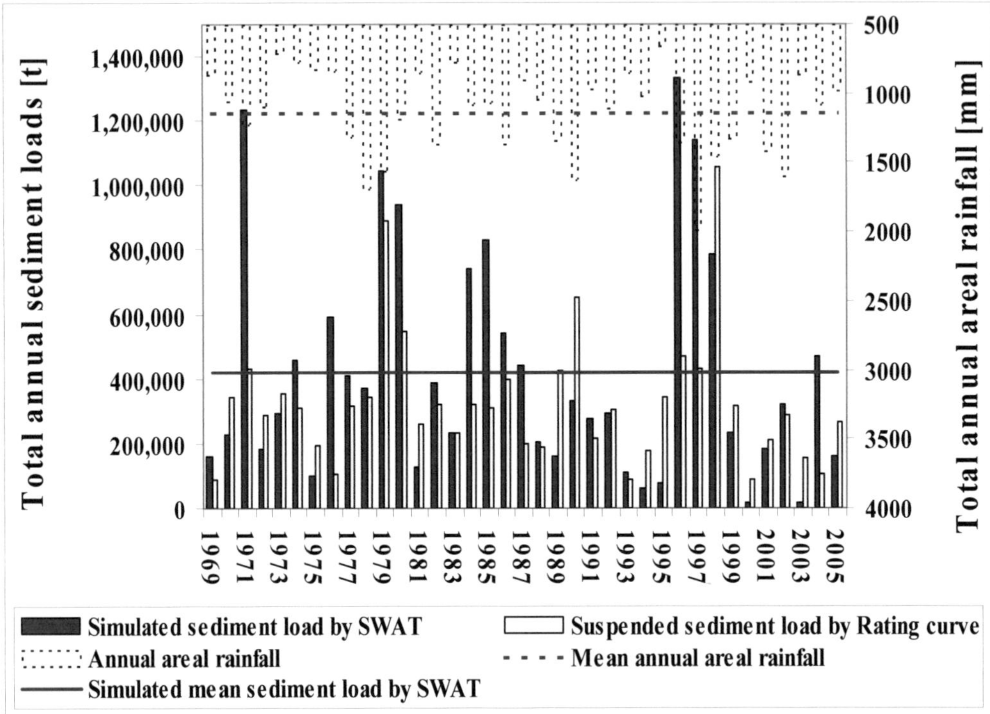


Fig. 5. SWAT simulations vs "observed", rating curve based sediment loads at 1DD1 (annually), between January, 1969 - December, 2005 (Ndomba *et al.*, 2008b)

For the case of the NYM Reservoir catchment the SWAT model captured 56 percent of the variance (CE) and underestimated the observed daily sediment loads by 0.9 percent according to TMC performance indices during a normal wet hydrological year, *i.e.*, with a calibration period set between November 1, 1977 and October 31, 1978 (Table 4). The SWAT model predicted satisfactorily the long-term sediment catchment yield with a relative error of 2.6 percent (Table 4 & Figure 5). It should be noted that the "observed" sediment loads were estimated from the bias-corrected suspended sediment rating curve. Authors are aware that the SWAT model simulates bed-material load (*i.e.* bed and suspended load) while the rating curve computes only suspended sediment load. However, it should be noted that the sediment loads delivered to streams are characterized as fine (Ndomba, 2007). They would mostly be transported as suspended sediments loads. The accuracy achieved by using the SWAT model was expected because it was hypothesized that correct estimation of surface runoff would lead to a better prediction of the sediment yield. Other researchers such as Garde and Ranga Raju (2000) are of similar opinion. Also, the model has identified erosion sources spatially and has replicated some erosion processes as determined in other studies and field observations in the NYM (Ndomba *et al.*, 2008b). This result suggests that for catchments where sheet erosion is dominant, the SWAT model may substitute other sediment yield estimation methods such as the sediment-rating curve. However, SWAT model simulations results for sediment storage in a reach such as the 1DD1 gauging station in the Myumba Ya Mungu Reservoir catchment differ from the findings based on other approaches, , *i.e.*, field observations and analysis of field-based sediment flow data. It was observed that the SWAT model could not capture the dynamics of sediment load delivery in some seasons to the catchment outlet (Ndomba *et al.*, 2008b). The reach is known to transport most of the sediment loads delivered from the catchment (*i.e.* equilibrium river reach). The particular study linked the latter problem to model deficiency. The authors would like to note that it is difficult to compare the model performance objectively as the quality and quantity of data used are different. Notwithstanding, there is a general agreement on the performance based on the cumulative sediment yield amount as measured by a relative error below 20 percent for both (Table 4). It could also be observed that the catchment size and climate are relatively similar. However, there are differences in catchment characteristics, *i.e.* geology, soils type, and topography.

| Variables | Performance indicators | Time step | Study cases | | |
|---|---|---|---|---|---|
| | | | KRC | NYM | SRC |
| Runoff | Calibration, CE (%) | Daily | 68 | 54.6 | 38 |
| | | Monthly | - | 65 | 82 |
| | Validation, CE (%) | Daily | 63 | 68 | 30 |
| | | Monthly | - | 77.4 | 81 |
| | IVF (%) | | - | 100 | 104 |
| Sediment yield rate | Calibration, CE (%\) | Daily | 66 | 56 | 24 |
| | | Monthly | | - | 83 |
| | Validation, CE (%) | Daily | 68 | - | 16 |
| | | Monthly | - | - | 80 |
| | Relative Error, (RE) (%) | | 7.5 | 2.6 | 0.76 |

Table 4. SWAT model performance for the 3 study cases, i.e., KRC, NYM and SRC
*Note: "-" not evaluated as a result of missing data.*

From the engineering perspective, sediment yield information is critical to estimating the design life of a reservoir as a result of sedimentation. A lumped spatial and temporal scales model could serve this purpose. However, if further insights into erosion processes and sediment sources are sought then finer temporal and spatial scales are important, *e.g.*, to evaluate best management practices, effects of land use and effects of land cover change. Besides, as the overall objective of this paper is to critically assess the suitability of the SWAT model for sediment yield modeling, various components and/or functionalities were evaluated. As presented earlier in this chapter, the SWAT model predicted satisfactorily the cumulative long-term sediment catchment yield, and the performance was measured using Relative Error (RE) in percent.

The performances of the SWAT model in the study cases and others conducted in catchments of the Eastern Africa as reported in literatures based on CE and IVF and Relative

Errors (RE) criteria suggest that the model can fairly/satisfactorily estimate sediment yield for even poorly gauged catchments (Ndomba *et al.*, 2005; Ndomba, 2007; Mulungu and Munishi, (2007); Ndomba and Birhanu, 2008; Ndomba *et al.*, 2008b; Andualem and Yonas, 2008; Shimelis *et al.*, 2010). The latter mainly proves that the models are able to capture the dynamics in time, but it provides little validation on the identification of the processes and/or the parameters in space which would require observations at several internal points in the catchment.

## 4. Conclusions and recommendations

### 4.1 Conclusions

This chapter presents three study cases in Eastern Africa where the SWAT model was applied extensively using the available data. In few cases primary data on sediment loads were explored. These cases represent various climatic conditions within the equatorial and/or tropical region. Based on the results of this study, the SWAT model seems to be robust and can be relied upon as a tool for catchment sediment management in the tropics. However, the model could not capture dynamics of sediment load delivery (i.*e*. equilibrium river regime) in some seasons in one catchment. The particular study linked the latter problem to model deficiency. Based on the simulation results the study has found that all sorts of farming practices captured by $P_{USLE}$ and $C_{USLE}$ parameters are the main determining management techniques in reducing soil loss/sediment yield and subsequently sedimentation problems in the reservoir. Besides, the performances of the SWAT model in these study cases as well as others conducted in Eastern Africa suggest that the model can satisfactorily estimate sediment yield for even poorly gauged catchments. The temporal variability is quite well captured. It should be noted that the calibration of the distributed parameters was typically done in a "lumped" way using sediment observations at the outlet instead of using observations at interior locations in the river basin. Therefore, the physical meaning of these parameters as well as the spatial representativeness could be questioned. The performance of the model suggests that the model can be used as a research tool in reservoir sedimentation and sediment yield modelling studies in the region.

The results of these study cases are not conclusive enough because some challenges have not been addressed. Although the input data varies in type and quality, from coarse-resolution to high-resolution measured spatial and climate data, there is a general lack of high-resolution spatial input data. More high-resolution spatial input data may not necessarily improve the performance of the model, but it may contribute to a better representation of the spatial variability. For erosion modelling, a high DEM resolution is especially important because the DEM is used to compute the slopes. Slopes have dual role: they affect the runoff processes in the hydrology that directly influence the erosion computations, and they are also directly used in the MUSLE equation.

It should be noted that the authors are aware that the performance of the SWAT model applications in the study cases can not be compared objectively because the performance is affected by modelling efforts and techniques, input data quality and catchment representation of important hydrological features.

### 4.2 Recommendations

A general recommendation is that more attention needs to be given to the spatial representativeness of the processes, the process parameters and the input data. The latter involves:

i. Move from a rather "lumped calibration" on data at the outlet to a more distributed calibration by using internal gauging data for both flow, sediments and rainfalls. This can also be achieved through using higher resolution data, especially for DEM but also the land use and soil maps;
ii. Improving the representation of important hydrological features, especially the water ponds, wetlands/marsh or swamps;
iii. Improve the routing component of the model. In some cases the SWAT model simulations indicated sediment storage in a river reach to be unlike the findings based on other approaches.

It is important to continue efforts in applying SWAT in Eastern Africa. However, the authors appeal to those who want to apply SWAT in their studies is not to apply it blindly. They need to consult experience from previous studies in Eastern Africa.

## 5. Acknowledgements

This work was co-funded by The Norwegian Programme for Development, Research and Education (NUFU) - Water Management in Pangani River Basin Tanzania Project at University of Dar es Salaam, Nile Basin Capacity Building Network-River Engineering Initiative, FRIEND/Nile based at UNESCO - Cairo Office in Cairo, Egypt and UNESCO-IHE Partnership Research Fund (UPaRF) through Adaptation to Climate Change Impacts on the Nile River Basin (ACCION) project. In addition, the authors wish to express their gratitude to Mr. F. Mashingia, a PhD fellow at University of Dar es Salaam, for preparing study area maps, as well as to anonymous reviewers, who helped to improve this chapter through their thorough review

## 6. References

Abbaspour, K.C., Johnson, C.A., van Genuchten, M.T., (2004). Estimating uncertain flow and transport parameters using a sequential uncertainty fitting procedure. *Vadose Zone J.* 3, 1340–1352.

Abbaspour, K.C., Yang, J., Maximov, I., Siber, R., Bogner, K., Mieleitner, J., Zobrist, J., Srinivasan, R., (2007). Modelling of hydrology and water quality in the pre-alpine/alpine Thur watershed using SWAT. *J. Hydrol.* 333, 413–430.

Abdelhamid, M.R., (2010). Sediment Transport Modelling in Simiyu Catchment of Lake Victoria Basin, Tanzania. MSc Thesis WSE-HI.10-06. UNESCO-IHE, Institute for Water Education, Netherlands.

Arnold, J.G., Williams, J.R., & Maidment, D.R., (1995). Continuous-time water and sediment-routing model for large basins. *Journal of Hydraulic Engineering,* Vol. 121(2): pp. 171-183.

Bagnold, R.A., (1977). Bedload transport in natural rivers. Water Resour. Res. 13:303-312.

Bathurst, J.C., (2002). *Physically-based erosion and sediment yield modelling: the SHETRAN concept.* In: Wolfgang Summer and Desmond E.Walling (ed.), Modelling erosion, sediment transport and sediment yield. IHP-VI Technical Documents in Hydrology, No.60, pp47-68.

De Pauw, E. (1984). *Soils, Physiography and Agro ecological zones of Tanzania Publication*: Crop monitoring and early warning systems Project GCPS/URT/047/NET. Ministry of

Agriculture, Dar es Salaam Food and Agriculture Organization of the United Nations.

Duan, Q.D., Gupta, V.K. and Sorooshian, S., (1992). Effective and efficient global optimization for conceptual rainfall-runoff models. *Water Resources Research* 28(4), pp 1015-1031.

Endale, B.A., (2008). Application of SWAT model in studying sedimentation problems in the reservoirs and proposing possible mitigation measures. A case of Koka reservoir, Ethiopia. MSc Dissertation. University of Dar es Salaam.

Ferguson, R.I., (1986) River loads underestimated by rating curves. Water Resour. Res. 22(1), 74–76.

Hathout, S.A., (1983). *Soil Atlas of Tanzania*. Tanzania Publishing House, Dar es Salaam

Lawrence, P., Cascio, A., Goldsmith, O., & Abott, C.L., (2004). Sedimentation in small dams - Development of a catchment characterization and sediment yield prediction procedure. Department For International Development (DFID) Project R7391 HR Project MDS0533 by HR Wallingford.

Lemma G., (1996), Climate Classifications of Ethiopia. National Meteorological Services Agency, Addis Ababa, Ethiopia.

Melching, C.S., (1995). *Reliability estimation*. In Computer Models of Watershed Hydrology by V.P. Singh, (eds), Water Resources Publications, pp69-118.

Morris, G. & Fan, J., (1998). *Reservoir sedimentation Handbook*: Design and Management of Dams, Reservoirs and Catchment for Sustainable use. McGrawHill, New York. Chapter 7, pp7.1-7.44.

Morris, M.D., (1991). Factorial Sampling Plans for Preliminary Computational Experiemnts. *Technometrics*, 33(2), 161-174.

Mulengera, M.K., and Payton, R.W., (1999). Estimating the USLE-Soil erodibility factor in developing tropical countries. *Trop. Agric. (Trinidad)* Vol. 76 No. 1, pp17-22.

Ndomba, P.M., Mtalo, F. & Killingtveit, A., (2005). The Suitability of SWAT Model in Sediment Yield Modelling for Ungauged Catchments. A Case of Simiyu Subcatchment, Tanzania. *Proceedings of the 3rd International SWAT conference, pp61-69. EAWAG-Zurich, Switzerland,* 11th –15th July 2005, Sourced at http://www.brc.tamus.edu/swat.

Ndomba, P.M., (2007). *Modelling of Erosion Processes and Reservoir Sedimentation Upstream of Nyumba ya Mungu Reservoir in the Pangani River Basin*. A PhD Thesis (Water Resources Engineering) of University of Dar es Salaam.

Andualem G. & Yonas M. (2008). Prediction of Sediment Inflow to Legedadi Reservoir Using SWAT Watershed and CCHE1D Sediment Transport Models. *Nile Basin Water Engineering Scientific Magazine*, Vol.1, (2008), pp 65-74.

Ndomba, P.M., Mtalo, F.W., and Killingtveit, A., (2008a). Developing an Excellent Sediment Rating Curve From One Hydrological Year Sampling Programme Data: Approach. *Journal of Urban and Environmental Engineering, V.2,n.1,p.21-27.* ISSN 1982-3932, doi:10.4090/juee.2008.v2n1.021027

Ndomba, P.M., Mtalo, F.W., and Killingtveit, A., (2008b). A Guided SWAT Model Application on Sediment Yield Modelling in Pangani River Basin: Lessons Learnt. *Journal of Urban and Environmental Engineering, V.2,n.2,p.53-62.* ISSN 1982-3932, doi:10.4090/juee.2008.v2n2.053062.

Ndomba, P.M. & Birhanu, B.Z., (2008). Problems and Prospects of SWAT Model Applications in Nilotic Catchments: A Review. *Nile Basin Water Engineering Scientific Magazine*, Vol.1, (2008), pp 41-52.

Ndomba, P.M., Mtalo, F., and Killingtveit, A., (2009). Estimating Gully Erosion Contribution to Large Catchment Sediment Yield Rate in Tanzania. *Journal of Physics and Chemistry of the Earth 34 (2009) 741 – 748. DOI: 10.1016/j.pce.2009.06.00. Journal homepage:* www.elsevier.com/locate/pce.

Neitsch, S.L., Arnold, J.G., Kiniry, J.R., & Williams, J.R., (2005). *Soil and Water Assessment Tool Theoretical Documentation Version 2005*. Grassland, Soil and Water Research Laboratory; Agricultural Research Service 808 East Blackland Road; Temple, Texas 76502; Blackland Research Research Center; Texas Agricultural Experiment Station 720 East Blackland Road; Temple, Texas 76502, USA.

Rohr, P.C., & Killingtveit, A., (2003). Rainfall distribution on the slopes of Mt. Kilimanjaro. *Journal of Hydrological Sciences*, 48(1):65-77

SCS (USDA Soil Conservation Service), (1972). *National Engineering Handbook* Section 4: Hydrology, Chapters 4-10.

Shimelis G. Setegn, Bijan Dargahi, Ragahavan Srinivasan, and Assefa M. Melesse (2010). Modelling Of Sediment Yield From Anjeni-Gauged Watershed, Ethiopia Using SWAT Model. *Journal of The American Water Resources Association*, Vol. 46, No. 3, pp 514-526.

Schmidt, G. & Volk, M., (2005) Effects of input data resolution on SWAT simulations - A case study at the Ems river basin (Northwestern Germany). Proc. 3rd Int. SWAT conf., 241–250. EAWAG-Zurich, Switzerland, 11–15 July 2005, Sourced at http://www.brc.tamus.edu/swat.

Summer, W., Klaghofer, E., Abi-zeid, I., & Villeneuve, J.P., (1992). Critical reflections on long term sediment monitoring programmes demonstrated on the Austrian Danube. *Erosion and Sediment Transport Monitoring Programmes in River Basins. Proceedings of the Oslo Symposium, August 1992*. IAHS Publ. No. 210, pp.255-262.

Van Liew, M.W., Arnold, J.G., & Bosch, D.D., (2005). Problems and Potential of Autocalibrating a Hydrologic Model. *Soil & Water Division of ASAE*, Vol.48 (3):pp1025-1040.

Van Griensven, A., and Srinivasan, R., (2005). AVSWATX SWAT-2005 Advanced Workshop workbook. SWAT2005 3rd international conference, July 11-15, 2005, Zurich, Switzerland.

Vanoni, V.A. (eds), (1975). Sedimentation Engineering. Prepared by the ASCE Task Committee for the preparation of the manual on sedimentation of the sedimentation committee of the Hydraulic Division. Copyright 1975 by the American Society of Civil Engineers. Library of Congress Catalog Card Number: 75-7751. ISBN_ 0-87262-001-8.

Wasson, R.J., (2002). What approach to the modelling of catchment scale erosion and sediment transport should be adopted? In: *Wolfgang Summer and Desmond E.Walling (ed.), Modelling erosion, sediment transport and sediment yield.* IHP-VI Technical Documents in Hydrology, No.60, pp1-12

Williams, J.R., (1975). Sediment-yield prediction with universal equation using runoff energy factor. In: *Present and prospective technology for predicting sediment yield and sources: Proceedings of the sediment-yield workshop,* USDA Sedimentation Lab., Oxford, MS, November 28-30, 1972. ARS-S-40.

Yapo, P.O., Gupta, H.V., & Sorooshian, S., (1996). Automatic calibration of conceptual rainfall-runoff models: sensitivity to calibration data. *Journal of Hydrology,* vol.181, pp23-48.

# 14

# Fuzzy Hydrologic Model in Tropical Watershed

Aurélio Azevedo Barreto-Neto
*Federal Institute of Espírito Santo,*
*Brazil*

## 1. Introduction

Natural elements such as soil, geology and vegetation are usually represented as map classes, whose boundaries are sharply defined. However not all entries in geo-objects datasets are sharp and this is true both for their attribute values as well as for their spatial distribution (Burrough, 1996).

This traditional representation between geo-objects is considered as an oversimplification of a more complex pattern. In some conditions, these boundaries are recognized more easily because are associated to significant and abrupt land changes, such as situations in which the boundaries are located in river banks, in geologic phenomenon (intrusions, flaws, fractures) or associated with sudden relief variations (Burrough, 1986). Apart from these situations, the boundaries are associated to uncertainties caused by limited observations (Hadzilacos, 1996). In all these cases, fuzzy methods are more suitable than boolean logic.

Zadeh (1965) developed fuzzy set theory allowing the mathematical modeling in zones of imprecisions and uncertainties. Fuzzy set theory is a generalization of the boolean logic to situations where data are modelled by entities whose attributes have zones of gradual transition, rather than sharp boundaries. Studies of natural phenomena and natural objects demonstrated that the use of boolean logic is an inadequate method and brings much inferior results (Burrough, 1986).

The objective of this study was to develop a fuzzy rule-based modelling to predict runoff in a watershed using the Soil Conservation Service Curve Number (SCSCN) model (SCS, 1972).

Although the SCSCN model was developed primarily based on small watersheds, it can be applied in medium and large watersheds, with a diversified variety of soils and vegetation, if integrated to a geographical information system (GIS) (Johnson & Miller, 1997; Thompson, 1999).

## 2. Study area

The study area is the Quilombo River watershed, located in Ribeira Valley, South of the State of São Paulo, Brazil (Fig. 1). The land-cover of the area is composed of Atlantic forest (dominant) and pasture. The choice to study this watershed was driven by the availability of soil map, rain record gage, and stream discharge record gage.

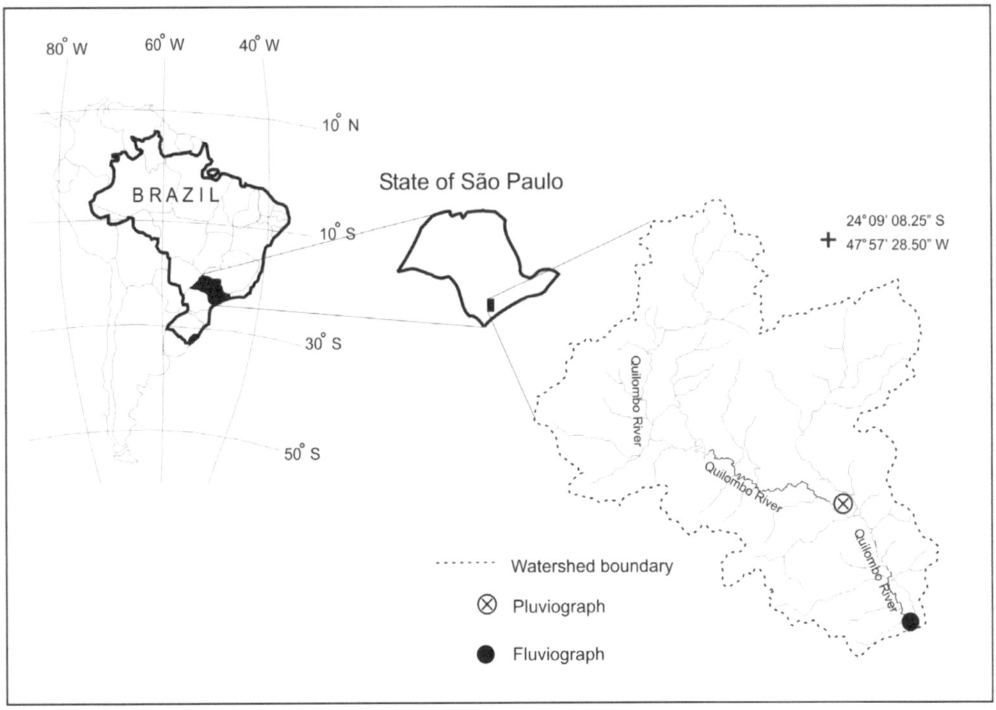


Fig. 1. Locality map

## 3. Fuzzy theory

A fuzzy set is defined mathematically as follows: if X = {x} is a finite set (for space) of points, then the fuzzy set A in X is the set of ordered pairs:

$$A = \{x, \mu_A(x)\} \quad x \in X \tag{1}$$

where $\mu_A(x)$ is known as the grade of membership of x in A and $x \in X$ mean that x is contained in X. For all A, $\mu_A(x)$ represents the grade of membership of x in A and is a real number in the interval [0, 1], with 1 representing full membership of the set and 0 non-membership (Zadeh, 1965). In practice, $X = \{ x_1, x_2 \ldots, x_n)$ and the Eq. (1) can be written as:

$$A = \{x_1, \mu_A(x_1); x_2, \mu_A(x_2); \ldots ; x_n, \mu_A(x_n)\} \tag{2}$$

## 4. The Soil Conservation Service Curve Number (SCSCN) hydrologic model

The SCSCN model is a well known archetype for estimating the storm runoff depth from storm rainfall depth for watershed and thus, stream flow, infiltration, soil moisture content and transport of sediments. Therefore, the model can assist hydraulic projects, soil conservation projects and flood control (SCS, 1972; Engel et al., 1993; Mack, 1995; Johnson & Miller, 1997; Thompson, 1999; Pullar & Springer, 2000;

In the SCSCN model, the physical characteristics of the watershed, such as hydrologic soil group (HGS), land cover and antecedent moisture conditions, are important because these characteristics determine the curve number (CN) parameter that estimate the runoff from a rain event. The CN ranges from 0 to 100, where larger CN represents greater proportion of surface runoff.

Basically, four steps are necessary to evaluate runoff from a rainfall by the SCSCN model: (i) to determine the hydrologic soil group (Table 1); (ii) to determine the five-day antecedent moisture condition of the soil from the precipitation record; (iii) to determine the runoff CN (on the basis of land cover, soil treatment, plus hydrologic condition and hydrologic soil group of the soil); and (iv) to calculate the runoff volume for one rain event. Concepts related to these four main steps are given below.

| HSG | Characteristics |
|---|---|
| A | Soils with high infiltration rates |
| B | Soils with moderate infiltration rates |
| C | Soils with low infiltration rates |
| D | Soils with very low infiltration rates |

Table 1. Hydrologic soil group (HSG) according to SCSCN model

In this model, the soils are classified to one of four HSG (A, B, C or D) defined by the SCS. This classification was accomplished by the analysis of the infiltration capacity of the soil. The description of each group, according to SCS (1972) and Rawls et al. (1992), is listed in Table 1.

In the SCSCN model, the watershed surface setting is assessed as a function of land cover, type of soil treatment and soil hydrologic condition. Land cover varies with landuse and can include key categories such as forests, swamps, pasture, bare soil, impermeable areas, etc. The soil treatment is related to automated farming practices (plantation along topographic contour lines and terraces) and management practices (pasture control, crop rotation and reduction). The association between landuse and the type of soil treatment is named class. Some examples of classes are: cereal plantations on topographic contour lines; dense forests; dense pasture, flat bare soil, paved highways, etc.

The association between specific HSG, land cover and type of soil treatment is referred to as soil-cover hydrologic complex, for which the CN attribute can be derived from the specialized literature (SCS, 1972; Rawls et al., 1992; Pilgrim & Cordery, 1993).

Antecedent Moisture Conditions (AMC) are related to the soil moisture due to accumulated rain, but considering the five last days that precede a particular rain event. There are three types of AMC: AMC I = soil is dry; AMC II = soil moisture is medium; and AMC III = soil is saturated in water.

The CNs were firstly obtained by measures made in a great number of watersheds for AMC II. The CN derived for AMC II can be converted to AMC I or AMC III through a transfer table provided by SCS (SCS, 1972).

The runoff begins when the portion of lost rain by infiltration, evapotranspiration, interception and depression storage, denominated initial abstractions, is less than the total precipitation. The runoff equation defined by SCS and detailed on the National Engineering Handbook (SCS, 1972) is the following:

$$Q = \frac{(P-0.2S)^2}{P+0.8S} \quad (3)$$

where Q is the direct runoff or excess precipitation, P is the precipitation, S is potential maximum storage in the watershed after beginning of the runoff .
The CN parameter relates to S (mm) as:

$$S = \frac{25,400}{CN} - 254 \quad (4)$$

## 5. Model implementation

### 5.1 Landuse map

The land use map was obtained by image processing of Advanced Spaceborne Thermal Emission and Reflection Radiometer (ASTER) data (Abrams, 2000). Firstly, the ASTER image was compensated for atmospheric effects and converted into surface reflectance, through the Atmospheric Correction Now (ACORN) software, which involves a MODTRAN4-based method for radiative transfer calculation (Imspec, 2001). The Leaf Pigment Index (LPI) (Almeida & Souza Filho, 2004) was then calculated using ASTER reflectance data to represent the continuous surface associated to the vegetation coverage of the study area (Fig. 2). The LPI was calculated by:

$$LPI = (ASTER\ 1) / (ASTER\ 2) \quad (5)$$

where ASTER 1 is the band 1 (0.52-0.60 m - visible green) and ASTER 2 is the band 2 (0.63-0.69 m - visible red). The LPI indicates the amount of chlorophyll in plant foliage – higher index values highlight areas in the image where photosyntetically active vegetation is denser. Other vegetation indices such as the Normalized Difference Vegetation Index (NDVI) (Rouse et al., 1974) and the Moisture Stress Index (MSI) (Rock et al., 1986) were also tested, but the LPI showed to best represent the vegetation cover of the study area when the results were confronted with field observations. The map generated with LPI was converted to ASCII format, compatible with PCRaster EML.

### 5.2 Soil map

Soil data of the Quilombo River watershed were extracted from the soil map of Ribeira do Iguape Region at 1/100,000 scale (Sakai et al., 1983). Basically, the watershed is composed of four soil types: latosol, podzolic, inceptisol and organic soils. The soil map, originally in paper format, was converted to digital vector data. These vector data were transformed to raster data at 15 m resolution. The raster map was further converted to ASCII format.
The runoff estimate was obtained through the SCSCN model based on the hydrologic soil groups defined by the USA Soil Conservation Service, where the soil is classified into one of four different categories, ranging from A to D.
An important characteristic of the tropical soils in the São Paulo State is the fact that the clay-rich soils provide high infiltration rates (Lombardi-Neto et al., 1991). Another particular aspect of the studied watershed is that the organic soils are found in the bottom of the valleys and have high moisture content (Barreto-Neto, 2004). Based on these soil characteristics, the soil map was reclassified in agreement with the hydrologic soil groups (Table 2 and Fig. 3).

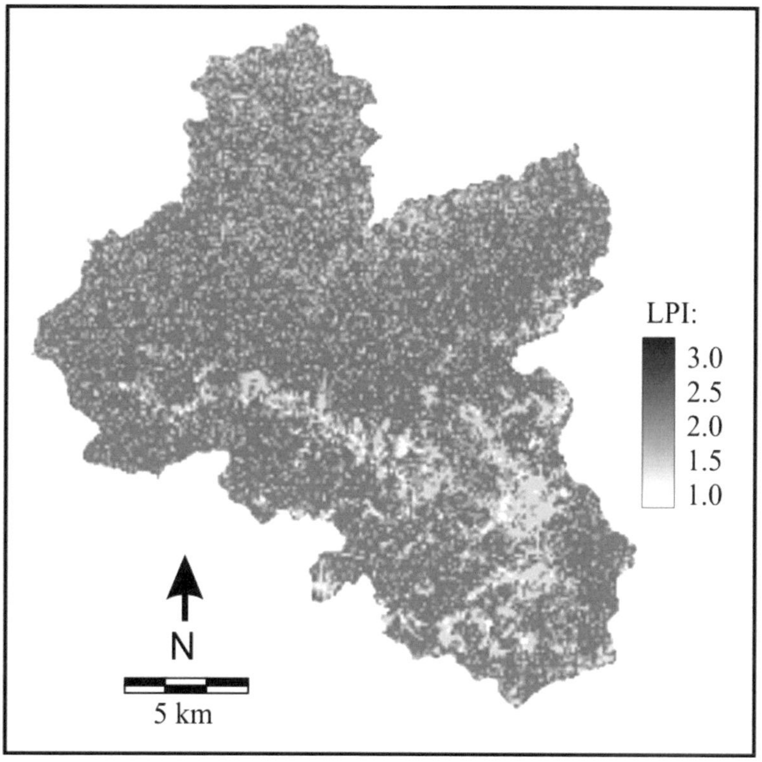


Fig. 2. Leaf Pigment Index (LPI) map

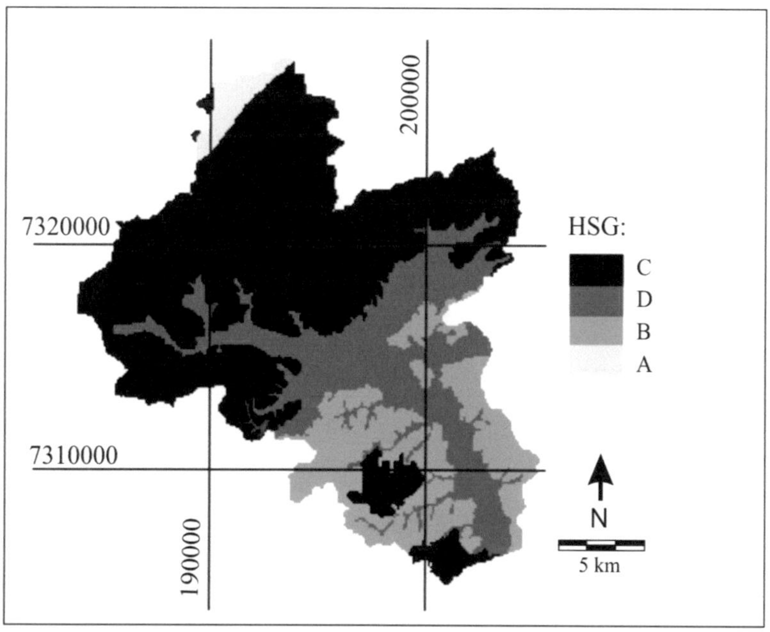


Fig. 3. Hydrologic Soil Group (HGS) map

| Soil types | % of area | HSG |
|---|---|---|
| Latosol | 2.3 | A |
| Podzolic | 17.7 | B |
| Inceptisol | 56.3 | C |
| Organic | 23.7 | D |

Table 2. Soil types, % area in the watershed and their respective Hydrologic Soil Group (HGS) according to Lombardi- Neto et al. (1991) and Embrapa (1999)

### 5.3 Fuzzy SCSCN model

For the developed model, each input variable was coded, fuzzified and, subsequently, input into the fuzzy inference system for decision making, using the PCRaster EML (Wesseling et al., 1996). The implementation of the computer model followed three steps: (i) the soil and cover maps were transformed in a fuzzy set using the membership functions (linear and bell-shaped); (ii) using the fuzzy inference system, the CN map was generated based in the fuzzy soil map and the fuzzy cover map (both developed in the previous steps); (iii) runoff calculation.

#### 5.3.1 Fuzzy soil map

Using the methods of fuzzy logic on polygon boundaries makes it simple to incorporate information about the nature of the boundaries. In this paper, the map-unit approach described for Burrough and McDonnell (1998) was employed. This approach assumes that the width of the transition zone is the same in all map boundaries. Information about the type of boundary was converted to parameters for two fuzzy membership functions (linear and bell-shaped) (Fig. 4), which were applied to the distance from the drawn boundary.

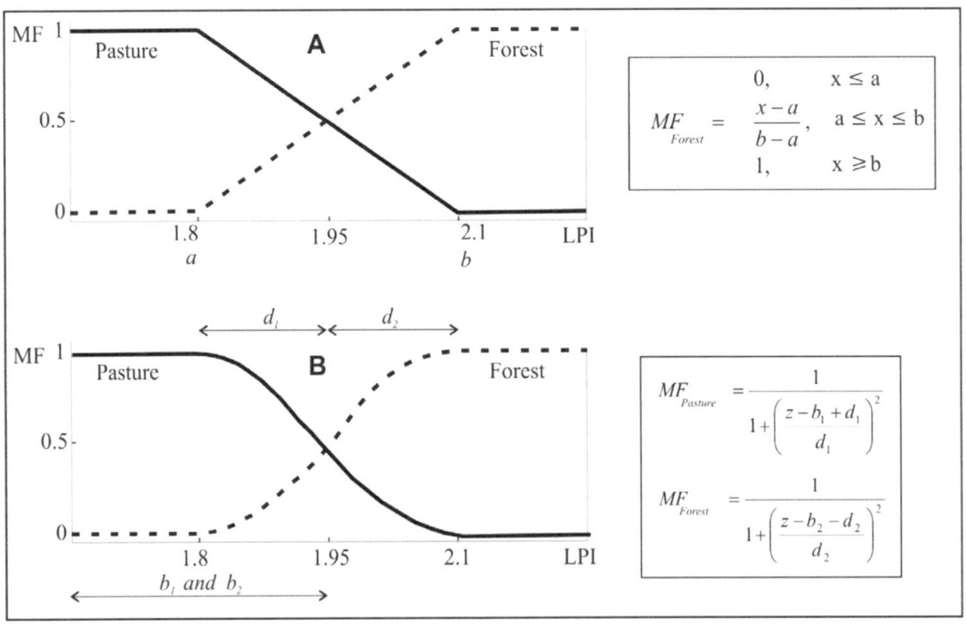


Fig. 4. Membership functions: (A) linear; (B) bell-shaped

The width of the transition zones was chosen based in the scale of the map, according to indications provided by Lagacherie et al. (1996) and Burrough and McDonnell (1998). Burrough and McDonnell (1998) exemplified that a sharp boundary drawn as a 0.2 mm-thick line on a 1:25,000 scale map covers 50 m (25 m to the right and 25 m to the left from drawn boundary) and a diffuse boundary at the same scale might extend over 500 m. In this study, sharp boundaries were drawn as a 1 mm-thick line on a 1:100,000 scale map, so that the width of the spatial transition zone centered over the drawn boundary location was 200 m. Only the uncertainty related to the drawn boundaries of the map was used, although the transition zones verified in the field show larger extensions.

In order to model fuzzy transition zones, the computer model involved the following steps: (i) separation of each soil unit (polygon boundary) in different map layers; (ii) isotropic spread of the boundary of each polygon (inside and outside the polygon); (iii) application of a membership function. Each soil unit was considered as a fuzzy set $A = \{x, \mu_A(x)\}$. In this case, x denotes a point in geographic space that belongs to A, and $\mu_A(x)$ is a number that ranges from 0 to 1 and reflects the grade of membership of x in A. The fuzziness of the boundary between soil units A and B were indicated by both distributions of grades of membership, $\mu_A(x)$ and $\mu_B(x)$ (Fig. 5). Points located far enough from the boundary have either $\mu_A(x) = 1$ and $\mu_B(x) = 0$ (if x is contained in A) or $\mu_B(x) = 1$ and $\mu_A(x) = 0$ (if x is contained in B). Points located close to the boundary $\mu_A(x)$ and $\mu_B(x)$ have values between 0 and 1; (iv) the procedure was repeated for each soil unit, yielding a soil boundary fuzziness map for each soil unit. The width of the transition zone can be defined by the user before the computer program is run. Fig. 6 and Fig. 7 show the boundary of organic soil using the fuzzy and boolean model, respectively.

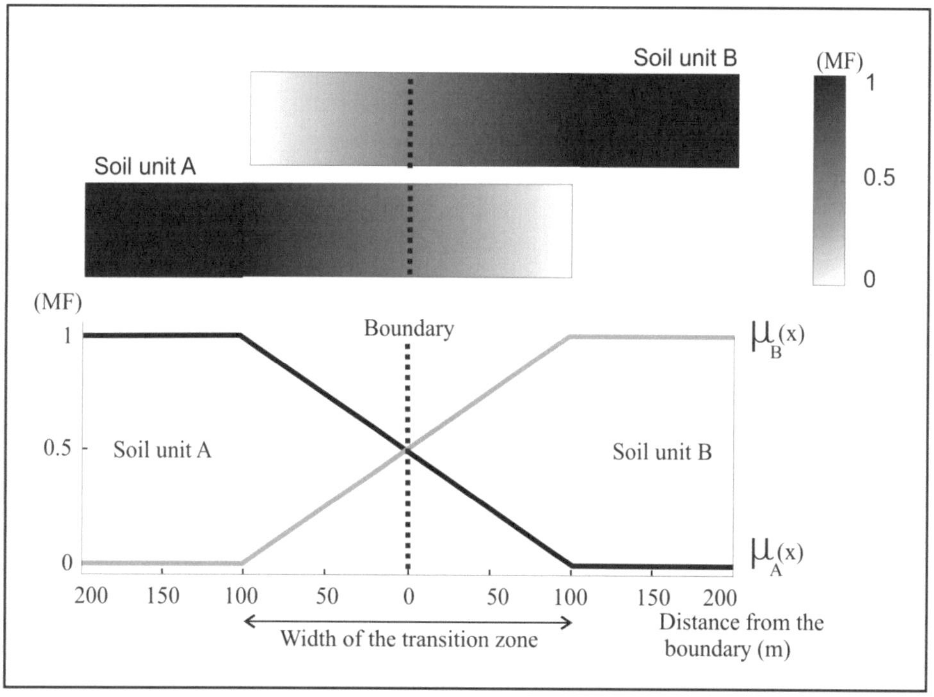


Fig. 5. Linear membership function and illustration of the methodology employed in the conversion of crispy soil data to fuzzy soil data

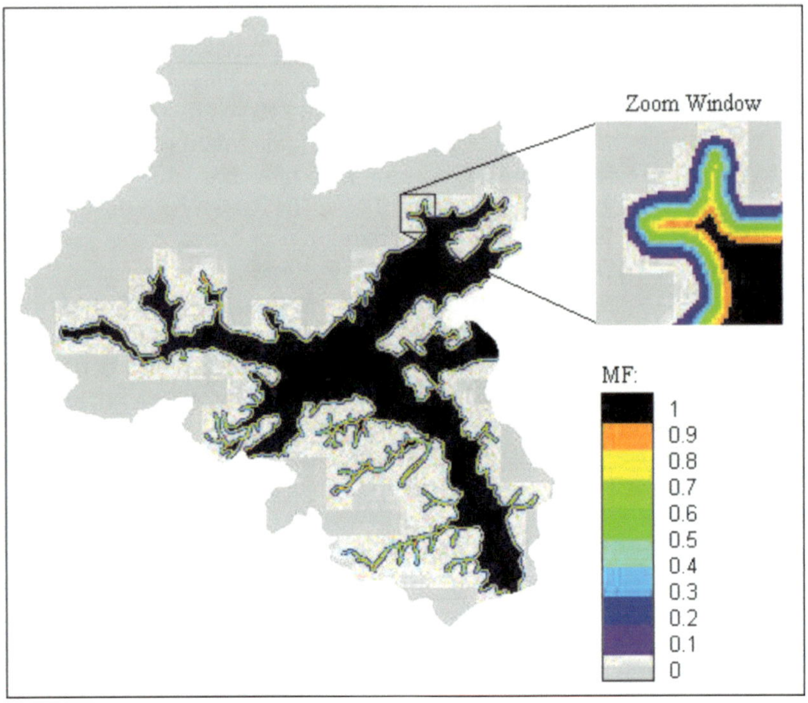


Fig. 6. Organic soil map with fuzzy boundaries

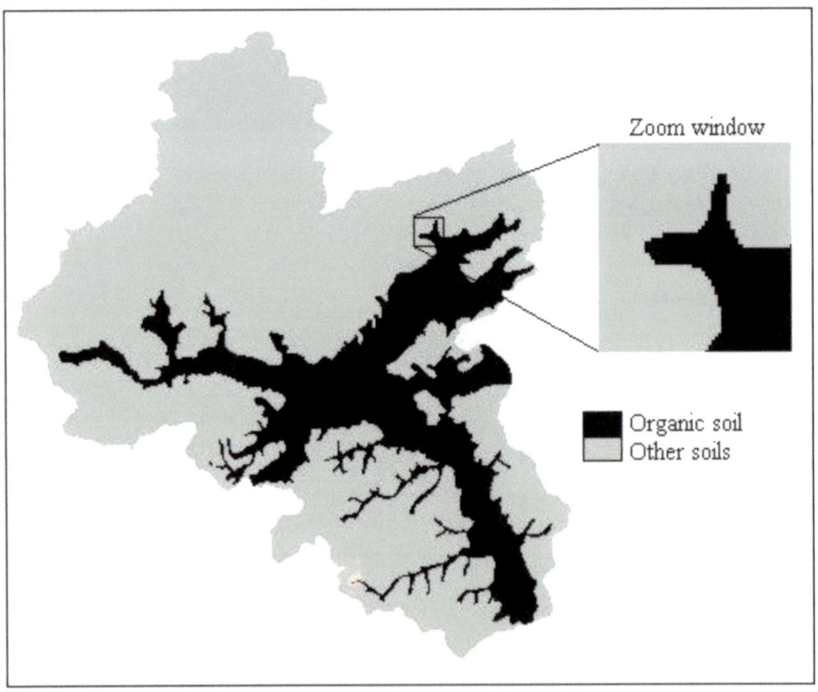


Fig. 7. Organic soil map with Boolean boundaries

### 5.3.2 Fuzzy land cover map

The fuzzy feature of the LPI map (land cover map) was calculated by the membership functions illustrated in Fig. 4. Field observations allowed the identification of transition zones on the vegetation cover (forest and pasture). The transition zones are covered by brushwood, as well as by degraded forest with grass fields. The diffuse boundaries observed in field were identified on the LPI map, so allowing proper membership function parameters to be used. This procedure generated four fuzzy maps: forest and pasture fuzzy maps using linear and bell-shaped membership functions.

### 5.3.3 Fuzzy rule-based modeling

In the fuzzy rule-based modeling, the relationships between variables are represented by means of fuzzy **if-then** rules that assume the form:

$$\textbf{If } x \text{ is } A \textbf{ then } y \text{ is } B \tag{6}$$

where x and y are linguistic variables, A and B are linguistic constants. The if-part of the rule "x is A" is named the antecedent, while the then-part of the rule "y is B" is named the consequent.

In this study, the Sugeno's method of fuzzy inference (Sugeno, 1985) was used to calculate the CN of all cells in the watershed map. In this method the antecedent is a fuzzy proposition and the consequent is a crisp function. Two typical fuzzy rules used in a Sugeno fuzzy model will be demonstrated as an example:

$$\textbf{If } x_1 \text{ is } A_{11} \textbf{ and } x_2 \text{ is } A_{12} \textbf{ then } y \text{ is } B_1 \tag{7}$$

$$\textbf{If } x_1 \text{ is } A_{21} \textbf{ and } x_2 \text{ is } A_{22} \textbf{ then } y \text{ is } B_2 \tag{8}$$

where $x_i$ ($i = 1, 2$) is an input variable (e.g. soil, vegetation), y is an output variable (e.g. CN parameter), $A_{ij}$ ($i = 1, 2$ and $j = 1, 2$) is a fuzzy set (e.g. high infiltration capacity, forest), and $B_i$ is a number that represents the consequent of the rule.

If $x_1^0$ and $x_2^0$ are values assumed by $x_1$ and $x_2$ and $A_{ij}(x_i^0)$ the grade of pertinence then the consequent value (crisp function) is $W_1$ and $W_2$ :

$$W_1 = \min(A_{11}(x_1^0), A_{12}(x_2^0)) \tag{9}$$

$$W_2 = \min(A_{21}(x_1^0), A_{22}(x_2^0)) \tag{10}$$

where "min" denotes "minimum value of". The global output $y^0$, that can be the CN parameter, is calculated by equation (11) (Kruse et al, 1994; Burrough, 1998):

$$y^0 = (W_1B_1 + W_2B_2)/(W_1 + W_2) \tag{11}$$

The fuzzy inference system of the Fuzzy SCSCN model was accomplished through the following steps: (i) transformation of the input data in a fuzzy set; (ii) application of the fuzzy rules (Table 3); (iii) computation of the information associated to transition zones on different soil and vegetation map units, using the Sugeno's method; (iv) generation of CN raster maps with the CN values of all pixels of the studied watershed (Fig. 8); and (v) runoff calculation.

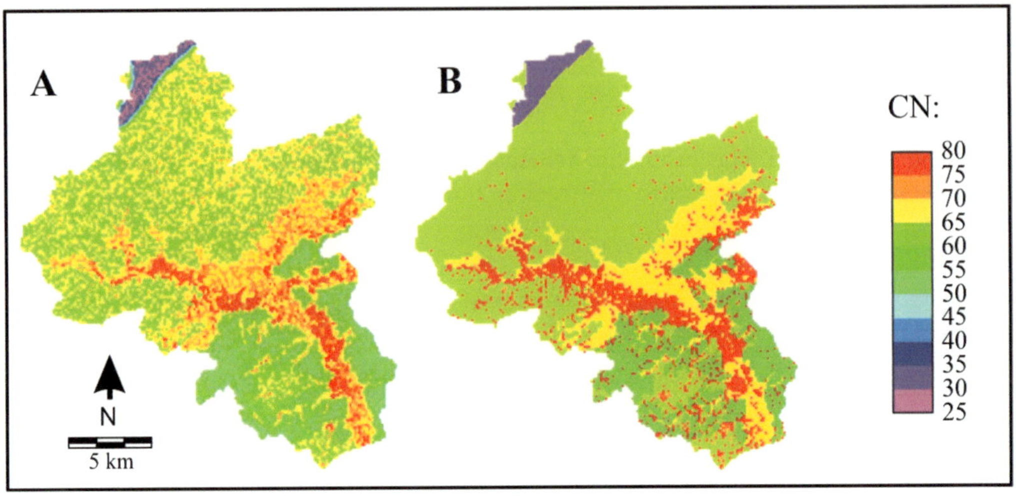


Fig. 8. CN maps obtained by the Fuzzy SCSCN model (A) and by the standard SCSCN model (B)

| Rule no. ($R_i$) | ***If*** | HSG | ***and*** | LPI | ***Then*** | CN |
|---|---|---|---|---|---|---|
| $R_1$ | ***If*** | D (*v. low inf.*) | ***and*** | pasture | ***then*** | 80 (*v. high)* |
| $R_2$ | ***If*** | D (*v. low inf.*) | ***and*** | forest | ***then*** | 69 *(medium-high)* |
| $R_3$ | ***If*** | C *(low inf.)* | ***and*** | pasture | ***then*** | 74 *(High)* |
| $R_4$ | ***If*** | C *(low inf.*) | ***and*** | forest | ***then*** | 62 *(Medium)* |
| $R_5$ | ***If*** | A (*high inf.*) | ***and*** | pasture | ***then*** | 39 *(medium-low)* |
| $R_6$ | ***If*** | A (*high inf.*) | ***and*** | forest | ***then*** | 26 *(Low)* |
| $R_7$ | ***If*** | B (*moderate inf.*) | ***and*** | pasture | ***then*** | 61 *(Medium)* |
| $R_8$ | ***If*** | B (*moderate inf.*) | ***and*** | forest | ***then*** | 52 *(medium-low)* |

Table 3. Fuzzy rule-based model for providing the *CN* parameters for the study area.

## 6. Result discussions

The CNs used here were selected on the basis of calibrations between modeled and observed runoffs. Key characteristics of the watershed, chiefly the hydrologic soil group, land cover and antecedent moisture conditions, plus CN tables available in the literature (e.g., SCS 1972; Thompson 1999), guided the CN selection. Once the CNs were selected, the runoff modelling was tested through a comparison between the modeled runoff depth and the recorded runoff depth observed in field. The validation of the CNs for the Quilombo River watershed was carried out for 16 rain events (Table 4). The results indicate that the modeled and the observed runoffs are akin and, therefore, the employed CNs proved suitable.

| Event | rain (mm) | Recorded runoff (mm) | Runoff for Boolean SCSCN model (mm) | Runoff for Fuzzy SCSCN model (mm) | |
|---|---|---|---|---|---|
| | | | | Linear membership function | Bell-shaped membership function |
| rain 1 | 5.5 | 2.1 | 0 | 0 | 0 |
| rain 2 | 14 | 2.4 | 0 | 0 | 0 |
| rain 3 | 21.2 | 3.4 | 0 | 0 | 0 |
| rain 4 | 27.4 | 4.5 | 0.2 | 0 | 0 |
| rain 5 | 32.8 | 5 | 0 | 1 | 1 |
| rain 6 | 35.5 | 5 | 1.0 | 1 | 1 |
| rain 7 | 45 | 7 | 2 | 3 | 3 |
| rain 8 | 46 | 7 | 2 | 3 | 3 |
| rain 9 | 56.5 | 12.6 | 5 | 6 | 6 |
| rain 10 | 71.6 | 14 | 10 | 12.3 | 12.3 |
| rain 11 | 87.0 | 19 | 16.6 | 18.4 | 18.6 |
| rain 12 | 122.6 | 40 | 31 | 37 | 37 |
| rain 13 | 134 | 49 | 43.2 | 50 | 50 |
| rain 14 | 140.2 | 56 | 53 | 57 | 57 |
| rain 15 | 150 | 59 | 54 | 58 | 58 |
| rain 16 | 162 | 62 | 59 | 61.5 | 61.7 |

Table 4. Simulated runoff with the Boolean SCSCN and with the Fuzzy SCSCN using sixteen rain events

Runoff simulations with the Fuzzy SCSCN model were accomplished using recorded precipitation data in the watershed (Table 4). Runoff modeling was also carried out using soil and vegetation cover data in Boolean format. The notion here was to compare the results derived from conventional SCSCN model and the Fuzzy SCSCN model, as presented in Table 4.

Figure 8 portrays two maps with the spatial distribution of the CNs calculated for the Fuzzy SCSCN model and for the boolean SCSCN model. These CN maps represent the capacity of the land to produce surface runoff from a rain event. It is clear that there is a greater variety of values of the parameter CN when it is calculated by the Fuzzy SCS model. Using the boolean SCSCN model it was possible to achieve only 8 CNs, whereas a larger range of CNs was yielded with the fuzzy SCSCN.

The simulated runoff values derived from the Fuzzy SCSCN model were closer to measured runoff values in the watershed than the simulated runoff values yielded from the Boolean SCSCN model (Table 4). The better performance reached by the fuzzy model signifies that it can conveniently express natural phenomena, including zones of imprecision and/or uncertainties like transition zones among soil types and vegetation cover. Table 4 shows that the runoff data calculated by the models (from rain 1 to rain 9) is not in agreement with the runoff recorded in the field. This can be explained by the fact that the SCSCN model is inappropriate for estimating the storm runoff depth from small storm rainfall depth (SCS, 1972).

The choice of membership function employed in the Fuzzy SCSCN model, either linear or bell-shaped, showed no significant variation in the simulated runoff. The observed equivalence in the model using these membership functions can be explicated by four factors: (i) both functions are very similar in shape (Fig. 4); (ii) the map scale (1/100,000) is small; (iii) the watershed is medium-sized (270 $km^2$) and this imparted a low runoff variability; and (iv) the transition area is of limited width (200 m).

## 7. Conclusions

A methodology for runoff modeling using fuzzy sets, fuzzy membership functions and fuzzy rules was presented in this paper. The computer model was created within a GIS environment and its use can be extended to other watersheds in Brazil by simple changes on the database.

Fuzzy logic has a great potential in hydrologic sciences. The incorporation of the fuzzy theory to the SCSCN model allowed a better representation of natural phenomena because fuzzy theory considers the transition zones among geo-objects, which differs from the boolean logic that considers such boundaries as crisp. The calculated runoff by fuzzy model was closer from the measured runoff than the calculated runoff by the boolean model, confirming the adequacy of the fuzzy theory in modeling natural phenomena.

The Fuzzy SCSCN model can be used as a tool for predicting runoff and, consequently, soil erosion and quality of water in watersheds. The model is relatively inexpensive because the PCRaster program, where the script is run, is freeware. The program developed here can produce fuzzy boundaries with different widths and can be used with numerous membership functions by simple changes in program script.

## 8. References

Abrams, M. (2000). The Advanced Spaceborne Thermal Emission and Reflection Radiometer (ASTER): data products for the high spatial resolution imager on NASA´s Terra platform. *International Journal of Remote Sensing*, Vol.21, pp. 847-859, ISSN 0143-1161

Almeida, T.I.R. & de Souza Filho, C. R. (2004). Principal Component Analysis Applied to Feature-Oriented Band Ratios of Hyperspectral Data: A Tool for Vegetation Studies. *International Journal of Remote Sensing*, Vol.25, No.22, pp. 5005-5024, ISSN 0143-1161

Barreto-Neto, A. A. (2004). *Modelagem dinâmica de processos ambientais*. Ph.D. Thesis, Universidade Estadual de Campinas, Campinas, Brazil.

Burrough, P. A. (1986). *Principles of Geographical Information Systems for Land Resources Assessment* (1st), Oxford University Press: Oxford, ISBN 0-19-854592-4

Burrough, P. A. (1996). Natural Objects with Indeterminate Boundaries, In: *Geographic Objects with Indeterminate Boundaries*, P. A. Burrough & A. U. Frank, (Eds), Taylor & Francis, ISBN 0748403876, London

Burrough, P. A. & McDonnell, R. A. (1998). *Principles of Geographical Information Systems* (1st), Oxford University Press, ISBN 0198233655, USA

Engel, B. A., Srinivasan, R. & Rewerts, C. (1993). A spatial decision support system for modeling and managing agricultural non-point-source pollution, In: *Environmental Modeling with GIS*, M. F. Goodchild, B. O. Parks & L. T. Steyaert, (Eds.), Oxford University Press, ISBN 0195080076, USA, New York

Hadzilacos, T. (1996). On Layer-based Systems for Undetermined Boundaries, In: *Geographic Objects with Indeterminate Boundaries,* P. A. Burrough & A. U. Frank, (Eds), Taylor & Francis, ISBN 0748403868, London

Imspec (2001). ACORN User's Guide. Analytical Imaging and Geophysics. Boulder, CO, USA

Johnson, D. L. & Miller, A. C. (1997). A spatially distributed hidrologic model utilizing raster data structure. *Computers & Geosciences*, Vol.23, No.3, (April 1997), pp. 267-272 ISSN 0098-3004

Kruse, R., Gebhardt, J. & Klawonn, F. (1994). *Foundations of Fuzzy Systems* (1st), John Wiley & Sons, ISBN 047194243X, England

Lagacherie, P., Andrieux, P. & Bouzigues, R. (1996). Fuzziness and Uncertainty of Soil Boundaries: From Reality to Coding in GIS, In: *Geographic Objects with Indeterminate Boundaries,* P. A. Burrough & A. U. Frank, (Eds), Taylor & Francis, ISBN 0748403868, London

Lombardi-Neto, F., Junior, R. B., Lepsh, I. G., Oliveira, J. B., Bertolini, D., Galeti, P. A. & Drugowich, M. I. (1991). *Terraceamento Agrícola.* Boletim técnico 206, Secretaria de Agricultura e Abastecimento/CATI/IAC, Campinas, Brazil

Mack, M. J. (1995). HER-Hidrologic evaluation of runoff: the soil conservation service curve curve number technique as an interactive computer model. *Computer & Geosciences,* Vol.21, No.8 (October 1995), pp. 929-935, ISSN 0098-3004

Pilgrim, D. H. & Cordery, I. (1993). Flood Runoff, In: *Handbook of Hydrology*, D. R. Maidment, (Ed.), McGraw-Hill, ISBN 9780070397323, New York

Pullar, D. & Springer, D. (2000). Tawards integrating GIS and catchment models. *Environmental Modelling & Software*, Vol.15, No.5, (July 2000), pp. 451-459, ISSN 1364-8152

Rock, B. N., Vogelmann, J. E., Williams, D. L., Voglemann, A. F. & Hoshizaki, T. (1986). Remote Detection of Forest Damage. *BioScience*, Vol.36, No.7, (July-August 1986), pp. 439-445 ISSN 0006-3568

Rouse, J. W., Haas, R. H., Schell, J. A. & Deering, D. W. (1974). Monitoring Vegetation Systems in the Great Plains with ERTS, *Proceeding of Third Earth Resources Technology Satellite-1 Symposium*, pp. 310-317, Washington, December 10-14, 1973

Sakai, E., Lepsch, I. F. & Amaral, A. Z. (1983). *Levantamento Pedológico de Reconhecimento semidetalhado da Região de Ribeira do Iguape no Estado de São Paulo*, SAA/IAC, São Paulo.

SCS, Soil Conservation Service (1972). National Engineering Handbook. USDA.

Sugeno, M. (1985). An introductory survey of fuzzy control. *Information Sciences*, Vol.36, No.1-2, (July-August 1985) pp. 59-83, ISSN 0020-0255

Thompson, S. A. (1999). *Hydrology for water management* (1st), A.A. Balkema, ISBN 90-5410-436-8 Rotterdam, The Netherlands

Wesseling, C. G., Karssenberg, D., Van Deursen, W. P. A. & Burrough, P. A. (1996). Integrating dynamic environmental models in GIS: The development of a Dynamic

Modelling language. *Transactions in GIS*, Vol.1, No.1, (January 1996), pp. 40-48, ISSN 1467-9671

Zadeh, L. A. (1965). Fuzzy sets. *Information and Control*, Vol.8, No.3, (June 1965), pp. 338-353

15

# Analysis of Rocky Desertification Monitoring Using MODIS Data in Western Guangxi, China

Yuanzhi Zhang[1,*], Jinrong Hu[1], Hongyan Xi[1], Yuli Zhu[1] and DongMei Chen[2]

[1]*The Yuen Yuen Research Center for Satellite Remote Sensing, Institute of Space and Earth Information Science, The Chinese University of Hong Kong, Shatin,*
[2]*Department of Geography, Queen's University, Kingston, Ontario,*
[1]*Hong Kong*
[2]*Canada*

## 1. Introduction

Rocky desertification (RD) is the process of land degradation characterized by soil erosion and bedrock exposure. It is one of the most serious land degradation problems in Karst areas especially in western Guangxi of southwest China, which is usually regarded as an obstacle to the local sustainable development. Recent investigations suggest that the RD is mainly caused by direct human activities, but some researchers also take account the climate change into a key factor of RD (e.g., Yao et al., 2001; Jing et al., 2003; Liao et al., 2004; Hu et al., 2004; Wang et al., 2004; Xiong et al., 2008).

Most areas in the western part of Guangxi Province belong to the Karst region. Similar to the other Karst areas in the southwest China, the geological environment is fragile and sensitive, with a high density of population but a low degree economic development. These aspects make the environment degraded quickly. RD is one of the most serious weaknesses of sustainable development in western Guangxi of southwest China. The investigation of the RD and its change monitoring are very significant and also necessarily meaningful.

In Guangxi, the area of Karst regions is about 89,500 $km^2$, which takes up 37.8% of the total area of Guangxi Province. Among them, the exposed surface of the Karst area is about 78,800 $km^2$, at 88% of the total Karst area in Guangxi. In the past few decades, due to deforestation, over cutting and grazing, the forest in the mountains was severely damaged, which led to serious soil erosion. According to the recent survey, the RD land is more than 233×104 $hm^2$, about 10% of the total area of Guangxi; the potential RD land area is more than $186\times10^4$ $hm^2$, about 8% of the total area. The RD land is mainly distributed in the middle of Guangxi: Red River Basin, Liujiang Basin; and western Guangxi: Left, Right River Basin; northeast of Guangxi: the two sides of middle and lower reaches of Li River. There are 32 counties (cities, districts), and Karst areas take up more than 60% of the administrative areas. The typical characteristics of RD areas are lack of soil, water, food and fuel with a lower economic level. 28 counties are designated to be poor ones, and 23 of them located in the Karst Rock Hill areas take up more than 30% of the administrative areas. RD in Guangxi has become a main cause of disaster and poverty, which constrains the regional economic and social development.

In China, studies on RD have been paid a lot of attention by many researchers since 1980s. Remote sensing and GIS technique always plays an important role in this research field. Landsat TM image, topographic map, geological map and GPS in-situ data were applied to produce a RD classification distribution map in Du'an Yao Autonomous County of Guangxi (Jiang et al., 2004) and to monitor the RD area in Wenshan County of Yunnan Province (Wu, 2009). ASTER image was used to study the situation of RD and its change trend from 2000 to 2005 in the Karst area of Guizhou Province (Chen et al., 2007). In addition, NOAA/AVHRR and MODIS data were used to monitor land desertification (Liu et al., 2007), in which humidity index was used to define the desertification area and two suitable classification methods were established to monitor the desertification dynamics from 1995 to 2001.

MODIS data was first applied in the western Guangxi of southwest China to monitor the rocky desertification with the change of land cover types from 2000 to 2010. The study area covers 30 counties in the western Guangxi. The study tends to give some suggestions to the local governments on the reconstruction of the rocky desertification and defense on new desertification in order to sustain the balance of the whole eco-geo-environment in western Guangxi of southwest China in the near future.

## 2. Study area

The study area is located in the western Guangxi province of southwest China (see Fig. 1), which is adjacent to Vietnam. The study area contains 30 counties with the total area of about $7.4\times10^4 km^2$, 31% of the whole province's area of Guangxi ($23.76\times10^4$ $km^2$). Its geographic location is north latitude 21°36'N to 25°40'N, and east longitude 104°20'E to 108°31'E. The area is mountainous region and belongs to subtropical zone with enough rainfall and rich natural resources.

The conflict between human and land in Guangxi is very sharp, but in the western Guangxi, it is even more severe. The land resources in this area have the following characteristics:

a. the area of land is large but the farmland is very limited due to mountainous and rocky landform;
b. a high density of population in Guangxi results in the small farmland per capita, which is less than 0.1 $hm^2$. In the study area it is much smaller;
c. the land cannot be utilized adequately as the whole agricultural productivity with a very low land quality. Most of the farmland belongs the second or third class. Moreover, the land is difficult to be utilized in the Karst area;
d. soil fertility was lost due to a serious soil erosion.

The study area is mainly composed of carbonate rocks, granite, purple sandstone and shale with weak anti-erosion properties. Climate in the study area is complex and changeable, and sunlight and rain is abundant all over a year, which may accelerate soil erosion. Additionally, with the increasing population and development of economy, human activities impact the probability of soil erosion. All of the characteristics of the study area made the Karst rocky desertification more seriously.

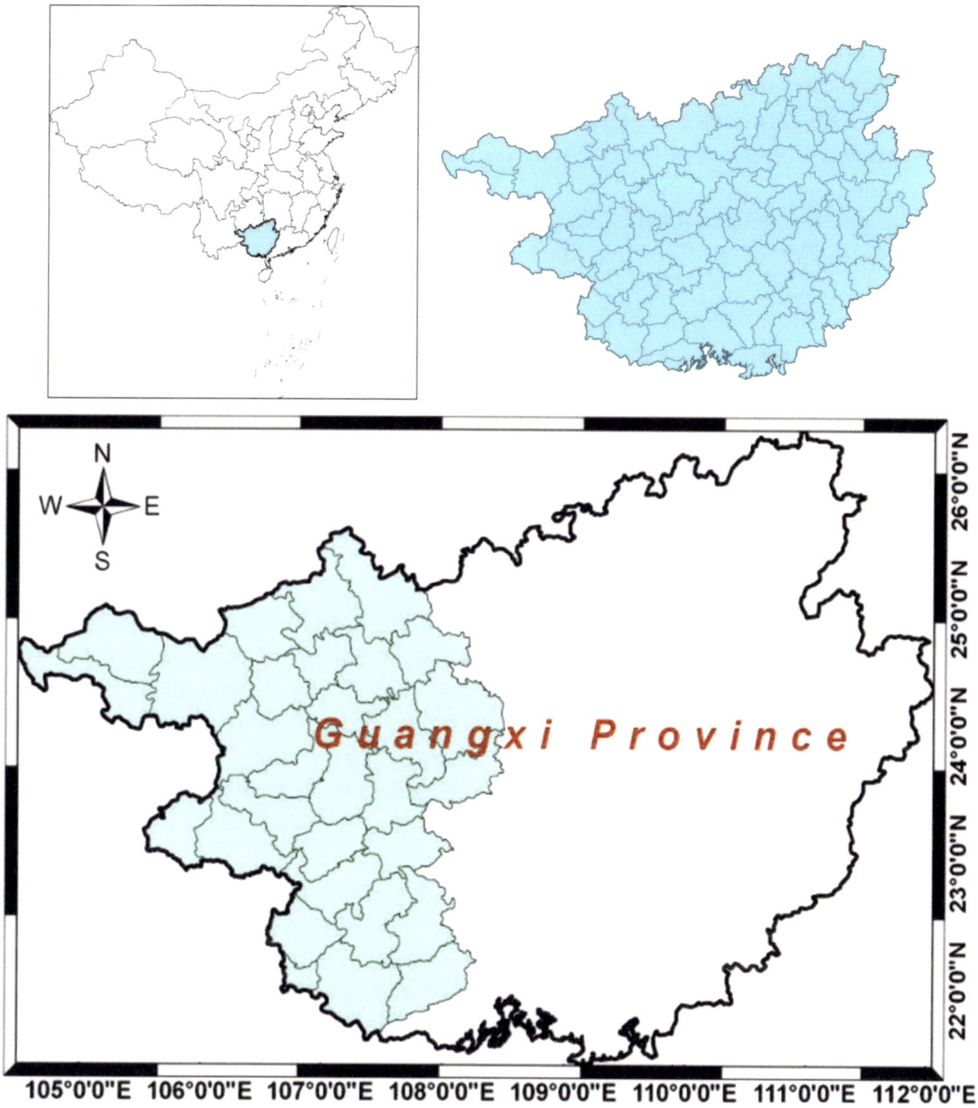


Fig. 1. Location map of the study area in western Guangxi, China.

## 3. Data and methods

### 3.1 Data

In this study, MODIS L1B data were used because of its large covered area with a coarse resolution. Other data sources included vector data, administrative maps, some information from previous research and the yearbook of Guangxi.

#### 3.1.1 MODIS data

The MODIS instrument is operating on both the Terra and Aqua spacecraft. It has a viewing swath width of 2,330 km and views the entire surface of the Earth every one to two days, its detectors measure 36 spectral bands, 0.405μm~14.385μm, covering the range of the electromagnetic spectrum. Among these bands, the 1-19 and 26 bands are for the visible and near-infrared channels, and the remaining 16 bands are thermal infrared channels. In addition, MODIS data have three spatial resolutions: 250 m (2 bands), 500 m (5 bands) and 1 km (29 bands). Compared with NOAA/AVHRR and MODIS data are of high spatial, temporal and spectral resolution. Therefore, MODIS data have been widely used in a lot of studies on land use land cover (LULC) mapping and LULC change detection at both global and local scales (Perera and Tsuchiya, 2009; Friedl et al., 2002).

MODIS Level 1B data with 250 m resolution were used in this study. Although the number of bands is limited, the two bands are in the red and near-infrared wavelengths, which are among the most important spectral regions for remote sensing of vegetation. MODIS L1B 250 m radiance data have been utilized for detection of vegetative cover conversion caused by recent significant natural events (burning and flooding) and human activities (deforestation) (Zhan et al., 2002). MODIS L1B data with 250 m resolution in November of 2000, 2003, 2006, 2008 and 2010 were downloaded to detect changes, because the weather in this month does not change too much and it is easier to get clear and cloudless images.

#### 3.1.2 Other supporting data

The boundary vector data of the study area is from the National Fundamental Geographic Information System with a scale of 1:4,000,000. It contains the information of borders (national, province, city, county), rivers (the first, second, third class), main roads, main railways, and residences (e.g., city and county). In this study, the border and residence data are mainly used. In addition, local administrative maps, information from previous research and the yearbooks of Guangxi are used as supporting data in the study.

### 3.2 Methods

The processing steps of the study are shown in Fig. 2. Firstly, the MODIS L1B data were pre-processed and the study area of western Guangxi was retrieved using the border vector data. Secondly, the images and other data were projected to the same coordinate system and spatial resolution after geo-reference calibration. MODIS data with 250 m spatial resolution in 2000, 2003, 2006, 2008, and 2010 were obtained with two bands of red and infrared bands, respectively. Thereafter, two methods were used to monitor the RD: a) NDVI calculation to identify the extent of RD; and b) analysis on land cover change after classification on the two images. Finally, the changed information was extracted and compared. Through a statistical analysis, the RD results were quantitatively analyzed.

#### 3.2.1 RD identified by NDVI calculation

NDVI (Normalized Difference Vegetation Index) is a simple numerical indicator that can be used to analyze remote sensing measurements. NDVI provides a crude estimate of vegetation health and a means of monitoring changes in vegetation over time. Vegetation index is extracted from the multi-spectral remote sensing data, it can quantized reflect the plants situation and helps strengthen our interpretation of remote sensing images. As a means of remote sensing, it is widely used in monitoring land-use cover, vegetation cover, density assessment, crop identification and crop forecasting. It has enhanced the ability of the classification in the topic mapping (Du, 2008).

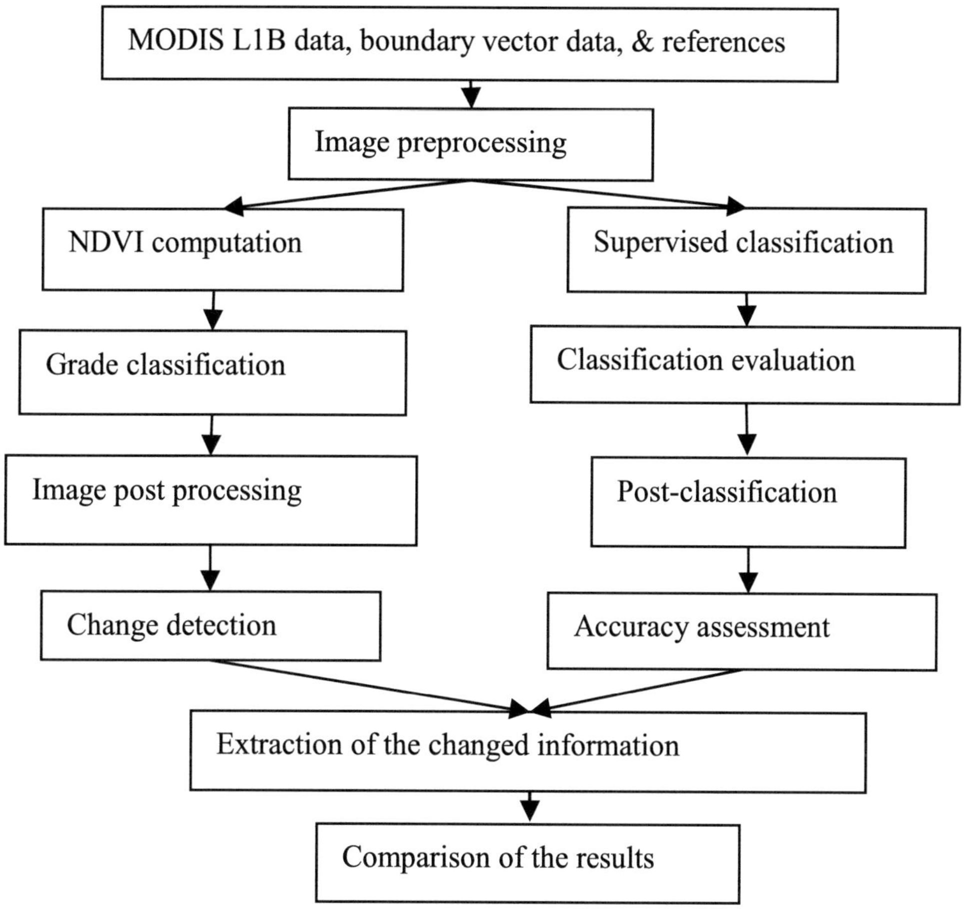


Fig. 2. The flowchart of data processing

The vegetation index is linear correlation to vegetation distribution density, the bigger of NDVI, the better of vegetation cover. The formula for NDVI calculation can be expressed as follows:

$$NDVI = (R_{nir} - R_{red}) / (R_{nir} + R_{red}) \quad (1)$$

Where, $R_{nir}$ in the formula is the reflectance of near infrared band and $R_{red}$ the reflectance of the red band, corresponding to the second band and the first band of MODIS L1B data with 250 m spatial resolution, respectively.

| NDVI value | <0.2 | 0.2-0.4 | 0.4-0.6 | >0.6 |
|---|---|---|---|---|
| The extent of RD | Intensity | Moderate | Mild | Good protected |

Table 1. The relationship of NDVI and the extent of RD (adopted from Hu et al., 2004)

There is a relationship between NDVI and the extent of RD (Hu et al., 2004) as shown in Table 1. From the table, one can see that if NDVI value is below 0.2, it means there is little vegetation cover on this area, and much rocky land exposed to the air, so the RD here is intense; if the NDVI value is between 0.2 and 0.4, the extent of RD is moderate; if NDVI is between 0.4 and 0.6, the extent of RD is mild; when the NDVI is above 0.6, it means these areas are good protected.

### 3.2.2 Land cover classification

After the detection of NDVI change, it is still needed to know the specific changes of land cover types in the study area. Generally, there are two methods to distinguish and interpret the remote sensing image: supervised classification and unsupervised classification. Supervised - image analyst "supervises" the selection of spectral classes that represent patterns or land cover features that the analyst can recognize. Unsupervised - statistical "clustering" algorithms used to select spectral classes inherent to the data, more computer-automated.

Supervised classification was used in this study. It is much more accurate for mapping classes, but depends heavily on the cognition and skills of the image specialist. The strategy is simple: the specialist must recognize conventional classes (real and familiar) or meaningful (but somewhat artificial) classes in a scene from prior knowledge, such as personal experience with what is present in the scene, or more generally, the region it is located in, by experience with thematic maps, or by on-site visits. This familiarity allows the individual(s) making the classification to choose and set up discrete classes (thus supervising the selection) and then, assign them category names.

Training ground and training sample selection is very important in supervised classification, the classification result will have a big different in supervised classification if the training sample is different. So it should be careful to select the training ground and choose the represented training sample correctly. These are the key points to produce a good classification result. In this study, supporting data and local land cover maps were used to help distinguish the land cover types. Due to the coarse resolution of MODIS data, it is not credible to classify many detailed land cover types. Thus based on information from supporting data and local land cover maps, six types of land cover were to be classified: water, wood, grassland, residence, farmland and unused land. After the classification, post-processing of image classification should be performed to get more reliable land cover maps, whilst accuracy assessment would be done.

## 4. Results

### 4.1 NDVI distribution and change of RD from 2000 to 2010

NDVI values were calculated by equation (1) and their distribution maps of 2000 and 2010 were obtained using ENVI software. Difference can be seen in the NDVI maps of the two years. In north part of the study area, NDVI values decreased in November 2008 compared with that in November 2000. However, NDVI maps can not show these changes distinctly. To distinguish the extent of the RD from 2000 to 2010, a decision tree upon Table 1 was produced as shown in Fig. 3. The Decision Tree classifier performs multistage classifications by using a series of binary decisions to place pixels into classes. Each decision divides the pixels in a set of images into two classes based on an expression. Based on these rules, the extent distribution of RD from 2000 to 2010 can be mapped clearly (see Fig. 4).

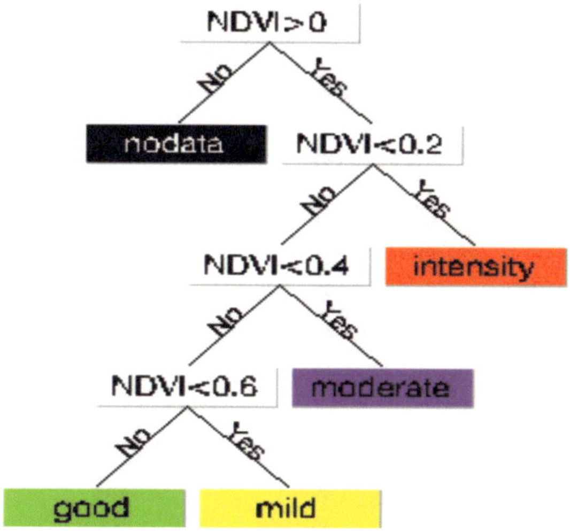


Fig. 3. Level slicing for the extent of RD (This is a segmentation method called level slicing).

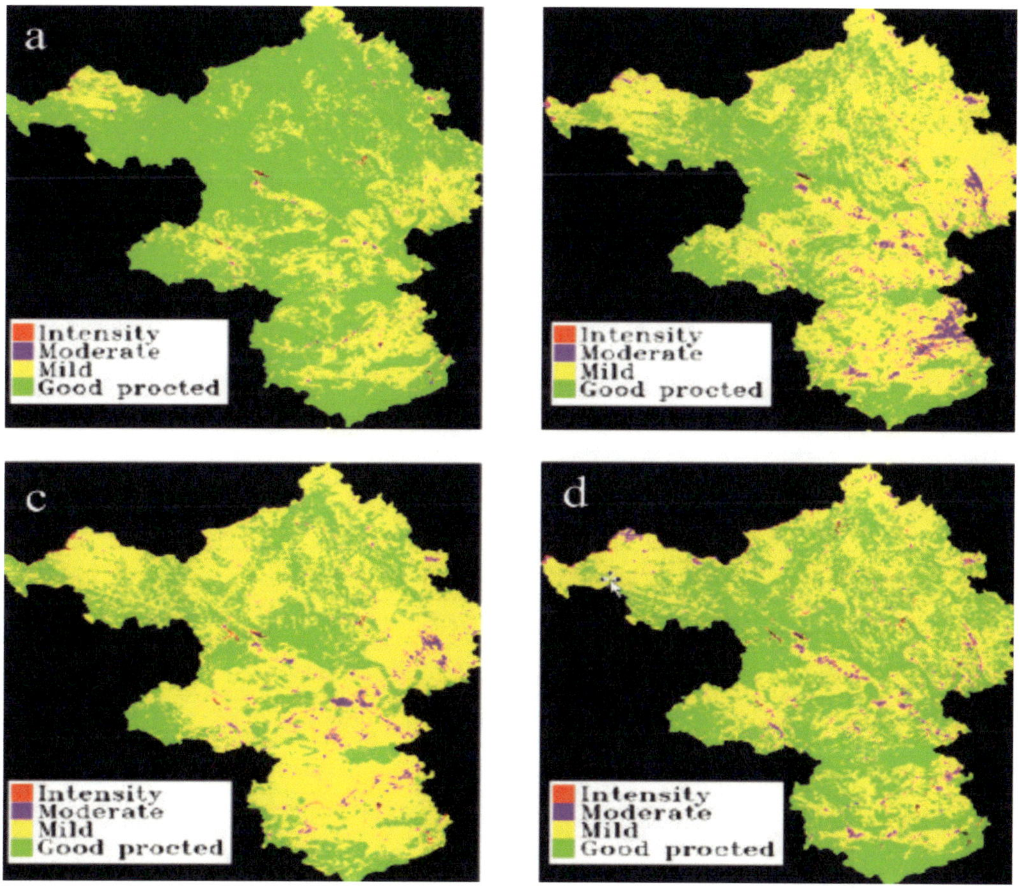


Fig. 4. (continued)

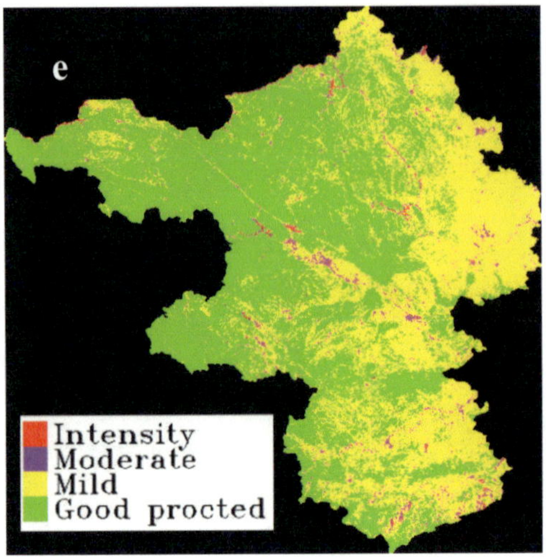


Fig. 4. The RD extent mapping in 2000 (a) 2003 (b), 2006 (c), 2008 (d) and 2010(e)

Compared the results from 2000, 2003, 2006, 2008, and 2010, it is clearly found that the good protected area has decreased from 2000 to 2006, and increased again from 2006 to 2010 in west part of the study area; the mild and moderate RD area is increasing dramatically in middle and north part of the study area. Nevertheless, intense RD area is seldom noted in all years, which may indicate that the environment of the study area does not deteriorate very badly. A change table of class statistics was made as shown in Table 2, which gives the percentages of each class in the whole area. It is clear that good protected area decreased from 2000 to 2006 and increased again from 2006 to 2010, while intense and moderate areas are relatively stable, and mild area decreased remarkably from 2006 to 2010. This indicates that many mild areas have been converted to good protected lands since 2006 due to governmental land protection policies.

| | | intensity | moderate | mild | good protected |
|---|---|---|---|---|---|
| 2000 | percent | 0.12% | 1.15% | 37.98% | 60.76% |
| | area(km²) | 85.71 | 847.40 | 28104.77 | 44962.13 |
| 2003 | percent | 0.21% | 6.15% | 66.79% | 26.86% |
| | area(km²) | 151.98 | 4548.10 | 49422.80 | 19877.12 |
| 2006 | percent | 0.20% | 3.91% | 72.45% | 23.44% |
| | area(km²) | 145.59 | 2892.45 | 53613.76 | 17348.20 |
| 2008 | percent | 0.35% | 3.73% | 57.49% | 38.44% |
| | area(km²) | 257.39 | 2756.87 | 42542.88 | 28442.86 |
| 2010 | percent | 0.33% | 1.68% | 46.47% | 51.52% |
| | area(km²) | 247.24 | 1242.65 | 34387.04 | 38123.07 |

Table 2. The RD extent in 2000, 2003, 2006, 2008, and 2010

### 4.2 Supervised classification and change analysis

MODIS L1B images in 2000 and 2010 were classified by supervised classification, in which the maximum likelihood classifying algorithm was employed as the most typical and wide method. After the ground training and selection of training samples, the image classification of the MODIS data was performed under ENVI environment. Afterwards, the post-processing classification was also made to reduce or eliminate the effect of noise caused by mixed scattered point features. Therefore, a filter kernel 3×3 matrix was used to make cluster analysis, which can smooth the classification maps and combine the similar areas to the neighbor region. The final results of the classification maps in 2000, 2003, 2006, 2008 and 2010 were shown in Fig. 5.

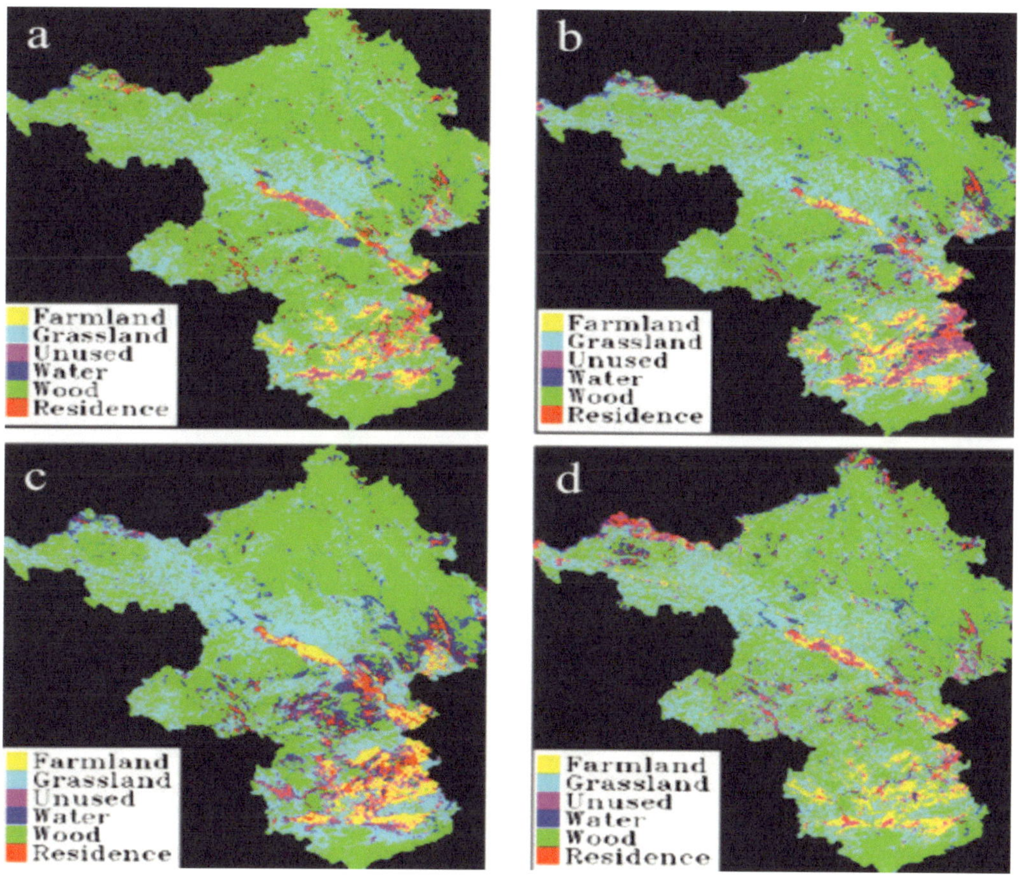


Fig. 5. (continued)

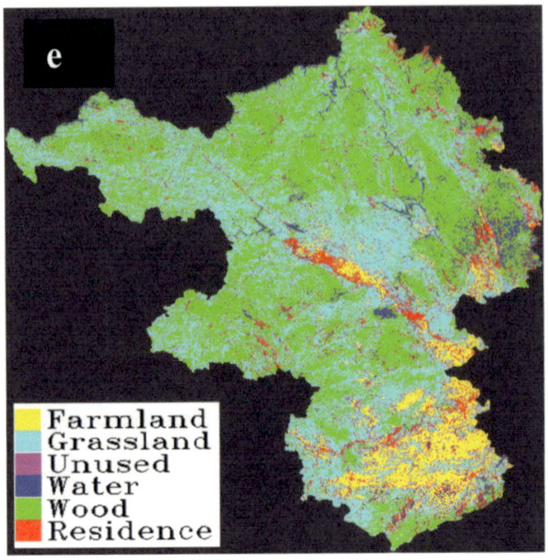


Fig. 5. Land cover classification from MODIS in 2000 (a), 2003 (b), 2006 (c), 2008 (d) and 2010(e)

A lot of change of land cover types can be found during the period from 2000 to 2010. In the classification map of 2000, woodland is the main class, which takes up to 47.68% of the study area; grassland is about 31.82% and farmland takes up to 8.32%, while other classes are relatively small. However, in 2008, the woodland only takes 38.69%, and grassland and farmland are up to 37.03% and 10.97%, respectively, which may suggest that these woodland areas were degenerating into grassland and farmland areas in a large region with the intensifying degree of RD issues. But this trend stops from 2008 to 2010. From 2008 to 2010 the woodland has increased to 44.95% and grassland and farmland decreased to 35.50% and 8.23%, respectively.

Table 3 and Fig. 6 show the total percentage and area of each land cover type from 2000 to 2010. It is clear that residential areas, farmland and grassland increased remarkably, whereas woodland decreased dramatically. In comparison, unused land areas decreased quite smaller, while water areas retains relative stable.

| | water | wood | grassland | farmland | residence | unused land |
|---|---|---|---|---|---|---|
| 2000 | 3.24% | 47.68% | 31.82% | 8.32% | 6.16% | 2.78% |
| 2003 | 4.05% | 45.03% | 33.52% | 7.55% | 3.20% | 6.66% |
| 2006 | 5.72% | 44.09% | 33.04% | 6.16% | 4.74% | 6.24% |
| 2008 | 3.30% | 38.69% | 37.03% | 10.97% | 3.74% | 6.26% |
| 2010 | 3.40% | 44.95% | 35.50% | 8.23% | 3.80% | 4.13% |

Table 3. The percentages of each land cover type in 2000, 2003, 2006, 2008 and 2010

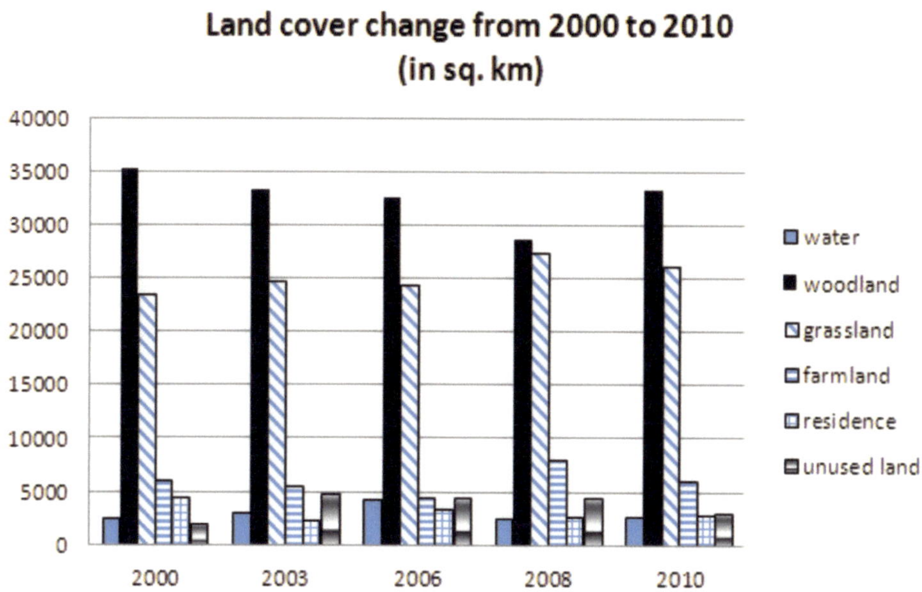


Fig. 6. The areas of each land cover type and its change from 2000 to 2010

## 5. Discussion and conclusion

In this study, MODIS L1B data with 250 m resolution were used to monitor the RD change in western Guangxi from 2000 to 2010. Two methods of NDVI calculation and supervised classification were performed to detect the RD extent.

From the above results, the distribution of RD areas extended from 2000 to 2008 and reduced from 2008 to 2010. The first method is based on the relationships between NDVI and RD. In general, if NDVI values of a region are high, it means the vegetation cover is well protected with a rare extent of RD. Otherwise the lower of the NDVI, the more serious of RD. Based on this assumption, the RD extent was determined in the study area from 2000 to 2010.

However, the RD areas were not only identified by NDVI. Some other factors may also affect the RD extent which can be extracted from MODIS data. To compare the RD extent with the change of land cover types, supervised classification was performed to determine six types of land cover in the study area (see Fig. 5). With the reference of previous studies of local land cover types (Li et al., 2006; Nong, 2007) and the yearbooks of Guangxi, the training sites and samples of six classes were selected and determined. Although there are misclassification errors involved, the results of image classification are reasonable to agree well with the previous results of land cover types and the statistic data in the yearbooks of Guangxi. Comparatively, the NDVI calculation is better and easier to be utilized to detect the RD extent than image classification in the study area.

It is reported that 37.6% RD is resulted from natural factors, while 62.4% of the RD area is caused by direct human activities (Nong, 2007). In this study, the natural factors may include:

- Climate effects: Guangxi is located in the subtropical climate with a long sunshine, much heat and rainfall. The average rainfall is usually 1400 ~ 1800mm per year, sometimes even more than 3000 mm. All these factors lead to serious soil loss, especially in the heavy rain season, in which the erodible soil is strongly rinsed off and only bare bedrocks remain;
- The impact of geological conditions: the southeast of Guangxi is granite collapse Kong area, and the northwest region is limestone area. Both of these two types of geological rocks are more prone to form RD or potential RD areas;
- Vegetation influence: Rare vegetation cover is an important factor to result in soil erosion and form RD (see Fig. 9 and Fig. 10);
- Topography effects: Soil erosion may also be accelerated by hilly and flat ground, steep slope, broken terrain and cutting deep ravines in western Guangxi (Wei, 2002).

On the other hand, direct human activities possibly cause RD or potential RD areas in the following ways:

- Excessive deforestation makes the RD area enlarge seriously;
- Human activities of production and life have an inextricably relationship with RD. The impact of anthropogenic factors is always through various forms. Along with the rapid growth of population, land and energy demand is increasing. This makes the original forest resource dissipated very quickly. Moreover, the inappropriate farming methods, such as the cultivation in high slope land, overgrazing, even excessive exploitation, as well as the quick exploration of local small mines, road building projects, and other industrial projects, make the ecological Karst areas brittle and weak, and showing a rapid trend of RD (Li et al., 2006).

The prevention and control measures of RD from the natural factors can be carried out by the following ways (Luo, 2007):

- Water storage construction. Water storage project is an effective way to control RD extending. This can reduce the seepage of rainwater, then reduce the soil erosion, and can also satisfy the industrial and agricultural water demand as much as possible;
- Forest planting. Making full use of solar thermal and water resources in the small gap of the Karst land and planting more at these areas are both effective ways to reduce the rock surface temperature and water consumption. They improve the micro-climate conditions and slow down the rock desert process;
- Development of three-dimensional ecological agriculture. In such extreme degradation ecosystems of RD areas, the natural recovery of vegetation is very difficult. So it needs biological, engineering and management measures by adjusting the irrigation system and transformation of soil quality to improve soil fertility and improve the ecological environment, in order to make the RD area back to normal.

There are also various means on prevention and control measures of RD from the human activity factors, such as development of the biogas construction, population growth controlling, industrial pollution prevention (Xu, 2006). In addition, it is effective to propagate scientific knowledge on environment protection in the Karst area (Tang et al., 2003). The obvious increase of good protected area and woodland from 2008 to 2010 indicates these propagation and prevention policies have produces positive results of reducing RD in this area. However, there is still a long rough way to go for the public and the government to bring the RD under control in western Guangxi of southwest China in the future.

## 6. Acknowledgements

The MODIS Level 1B data downloaded from the website of NASA MODIS products and the vector data from National Fundamental Geographic Information System of China are highly appreciated. The authors would like to thank Mr. Xianzhi Hu for his help of image pre-processing. The research was partially supported by the Yuen Yuen Remote Sensing Scholarship at ISEIS of CUHK and CUHK Direct Grants.

## 7. References

Chen, Q., Xiong, K., and Lan, A., 2007. Analysis the Karst Rocky Desertification situation and change trend in Guizhou base on "3S". China Karst, 26(1), 37-42.

Du, L., 2008. Preprocessing and NDVI calculation of MODIS 1B data. Desert and Oasis Meteorology, 2(1), 26-28.

Friedl, M.A., McIver, D.K., Hodges, J.C.F., Zhang, X.Y., Muchoney, D., Strahler, A.H., Woodcock, C.E., Gopal, S., Schneider, A., Cooper, A., Baccini, A., Gao, F., and Schaaf, C., 2002. Global land cover mapping from MODIS: algorithms and early results. Remote Sensing of Environment, 83(1), 287–302.

Hu, B., Liao, C., Yan, Z., Jiang, S., Huang, Q., and Li, S., 2004. Diving Mechanism Diagnosis of Karst Rocky Desertification in Du'an Yao Autonomous County of Guangxi based on RS and GIS. Journal of Mountain Science, 22(5). 583-590.

Li, S., Shu, N, Wang, G., and Liao, S., 2006. The Origination Analysis and Progress of the Rocky Desert of Land in Guangxi. Journal of Guangxi Academy of Sciences, 22(3), 193-196.

Liu, A., Wang, C., Wang, J., and Shao, 2007. Method for remote sensing monitoring of desertification based on MODIS and NOAA/AVHRR data. Journal of Agricultural Engineering, 2007, 23(10), 145-150.

Liu, L., Jing, X., Wang, J., and Zhao, C., 2009. Analysis of the changes of vegetation coverage of western Beijing mountainous areas using remote sensing and GIS. Environ. Monit. Assess, 153, 339–349.

Luo, G., 2007. Rocky desertification and climatic factors in Guangxi. Journal of Meteorological Research and Application. 28(S1), 74-75.

Nong, S., 2007. Rock desertification status analysis in karst areas of Guangxi and control measures. Guangxi Forestry Science, 36(3), 170–172.

Perera, K., and Tsuchiya, K., 2009. Experiment for mapping land cover and it's change in southeastern Sri Lanka utilizing 250m resolution MODIS imageries. Advances in Space Research, 43, 1349–1355.

Tang, X., He, X., Peng, H, and Chen, C., 2003. The causes and control measures of the Rocky Desertification in Guangxi Karst areas. Resource Development & Market, 19(3), 154-156.

Wang, G., 2008. Features and preparation of MODIS data. China's scientific and technical papers online, http://www.paper.edu.cn, published on 27th August, 2008.

Wei, M., 2002. Conditions and countermeasures of rocky desertification in Guangxi. Journal of Guangxi University (Philosophy and Social Science), 24(2), 42–47.

Wu, N., 2009. TM Image Based Monitoring on Rocky Desertification in Karst Area of Wenshan Prefecture, Yunnan Province. Journal of Southwest Forestry University, 29(1), 62-66.

Xiong, Y., Qiu, G., Mo, D., Lin, H., Sun, H., Wang, Q., Zhao, S., and Yin, J., 2009. Rocky desertification and its causes in karst areas: a case study in Yongshun County, Hunan Province, China. Environ. Geol., 57, 1481-1488.

Zhan, X., Sohiberg., R.A., Townshend, J.R.G., DiMiceli, C., Carroll, M.L., Eastman, J.C., Hansen, M.C., and Defries, R.S., 2002. Detection of land cover changes using MODIS 250 m data. Remote Sensing of Environment, 83(1), 336-350.

# 16

# Lower Eocene Crustacean Burrows (Israel) Reflect a Change from K- to r-Type Mode of Breeding Across the K-T Boundary Clarifying the Process of the End-Cretaceous Biological Crisis

Zeev Lewy, Michael Dvorachek,
Lydia Perelis-Grossowicz and Shimon Ilani
*Geological Survey of Israel,
Israel*

## 1. Introduction

Crustacean burrows filled with chalk were found at the lowermost part of the Lower Eocene sequence in southwestern Israel. They are exposed on both sides of a road-cut about 40 m long (Fig. 1) on the way leading from the city of Be'er Sheva to the Israeli-Egyptian border (Fig. 2A; site A; N 30$^0$ 57′ 36″, E 34$^0$ 39′ 20″). The road-cut exposes grey-greenish clay of the upper part of the Paleocene Taqiya Formation, overlain by white chalk with chert nodules of the Lower Eocene Mor Formation (Fig. 2B). The burrow system consists of horizontal galleries leading to heart-shaped flattened casts of chambers embedded in 12-cm-thick argillaceous chalk (Fig. 2B). This single type of chamber cast preserves on its surface ovoid blister-like elevations with transversal fine scratches. These peripheral structures are identical to those on the phosphatic casts of Campanian crustacean burrow-system chambers described from another exposure along the same road, some 18 km to the east where two types of chamber fillings were found (Lewy & Goldring, 2006). One is circular (D=45 mm) and replicates the arched ceiling with finely scratched elevated ovoid structures, whereas the cast of the floor comprises rings of about eight tubercles replicating pits 4 mm in diameter and 3.0-3.5 mm in depth. The pits were interpreted to host and protect large eggs in a brood chamber. The second kind of chamber changes shape and dimensions from circular (D=45 mm) to arrowhead-shaped up to 100 mm in length with one end rounded and tapering towards the opposite end. This gradually enlarged chamber was suggested to host the young (nursery chamber) and perhaps store food or provide gardening sites. The Lower Eocene burrow system lacks the brood chamber, whereas the heart-like chamber looks as a shortened modification of the Campanian arrowhead chamber, preserving the wall structure of the Campanian chambers. The diameter of the galleries of the Campanian burrows is about 14-17 mm compared to 10-15 mm of the Lower Eocene ones, and the greatest width of the Campanian chambers (about 7 cm) is similar to that of the Lower Eocene ones. Accordingly, the Campanian and Lower Eocene crustacean burrows into pelagic chalk have a similar structure of horizontal galleries connecting between chambers of similar dimensions and wall sculpture. These common features attest to a similar body structure. The main difference between these two burrow systems is the lack of brood chambers during the Lower Eocene, which reflects a change in breeding strategy sometime between the Campanian and the Early Eocene. Both burrow systems are in pelagic chalk and thus external ecological factors turned the specially constructed brood chamber useless. These local ecological changes were probably associated with the global biological turnover at the K-T boundary. Their evaluation in comparison with other biological changes clarifies the natural processes which resulted in the end-Cretaceous biological crisis.

Fig. 1. Road-cut exposing Taqiya Fm. greenish clay overlain by Mor Fm. chalk (M. Kitin pointing at the fossiliferous bed).

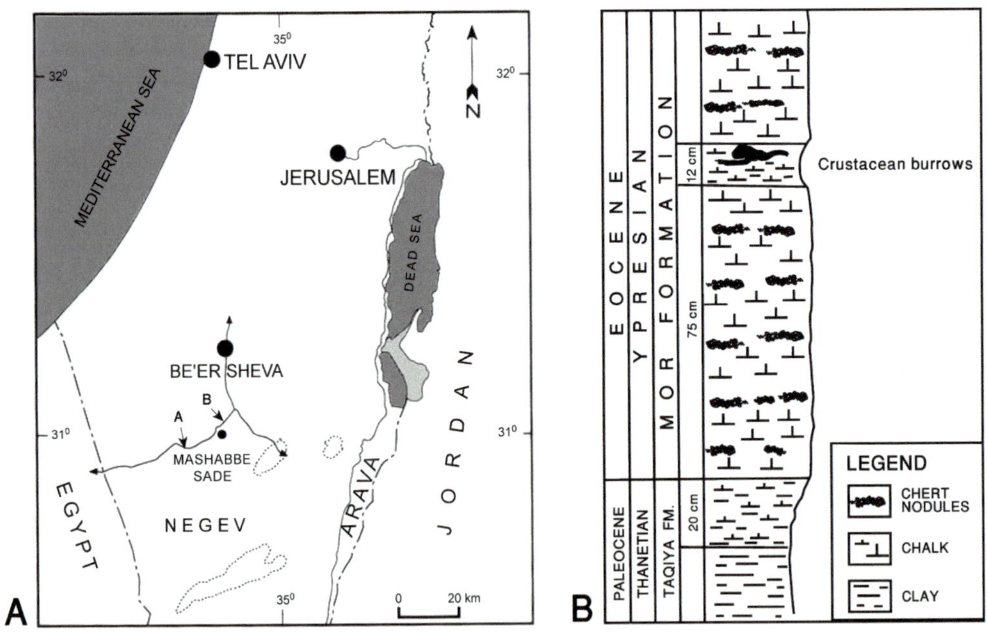


Fig. 2. A. Location map. B. Columnar section of the Upper Paleocene-Lower Eocene fossiliferous interval.

## 2. Crustacean burrows in pelagic chalk

### 2.1 Lower eocene burrows

The road cut exposes the upper part of the greenish clay of the Taqiya Formation (Fig. 2) containing the latest Paleocene (Thanetian) planktonic foraminifer *Morozovella velascoensis* (Cushman). It ranges into the overlying 20 cm of argillaceous chalk which forms a lithological transition to chalk of the Lower Eocene (Ypresian) Mor Formation. The formation begins with 75 cm of lithified chalk with chert nodules containing the Lower Eocene *Morozovella formosa formosa* Bolli and *Morozovella aragonensis* (Nuttall). The layer above is comprised of 12 cm of argillaceous chalk with secondary gypsum at the base and the crustacean burrow system in the upper part. The overlying sequence consists of 0.5-1.0 m thick units of hard chalk with chert nodules alternating with 10-15 cm thick beds of argillaceous chalk. The chalk-filled crustacean burrows consist of horizontal galleries connecting between chambers. Vertical shafts are not preserved. The present elliptical shape of the horizontal galleries attests to sediment compaction to 60-65% of the original vertical dimensions of the galleries as well as of the chambers. The periphery of these chalky casts is friable and part of the external features of the chamber cast is erased. None of the gallery fillings shows any scratches and their walls seem to have been smooth, forming tubes of 1.0-1.5 cm in diameter (Fig. 5G). About 27 heart-shaped casts were collected, probably representing the only kind of chamber fill. Their orientation in the layer is with the heart-shape on the horizontal plane. A gallery enters into the middle of the floor and another one is connected close to the constricted end of the chamber ceiling, pointing a little upward and continues horizontally (Figs. 3, 4B). The chamber casts vary a little in dimensions and proportions whereby the longitudinal length (along the connecting galleries) may be shorter than the transversal width in some specimens. Length ranges between 55-75 mm and width between 55-73 mm. Despite sediment compaction, the thickness (chamber height) of all these casts decreases toward the end with the gallery opening (Figs. 4C, F, H, 5C, E) strengthening the pear-shape of the chamber in side view. Well-preserved casts show ovoid blister-like elevations 5-6 mm broad and 5-9 mm long covering the whole cast, being compressed on the flanks by later compaction. These ovoid structures on the chamber ceiling (upper surface without a gallery entrance) of some specimens tend to orient into transversal lines forming slightly arched ribs about 7 mm in width (Fig. 5B). The horizontal galleries and the flattened chamber casts are concentrated in the upper half of the 12-cm-thick layer.

Fig. 3. Lower Eocene burrowing crustacean chamber and connecting gallery within the chalky sediment (80% of natural size).

The absence of any relict of vertical shaft indicates the truncation of the unconsolidated sediment reaching close to the horizontal burrows. This sediment removal from the deep sea bottom by deep marine currents could have occurred before the fill of the burrows by the following deposition of foraminiferal ooze, or after sediment filled the burrows and formed a stabilized level at which sediment removal stopped.

Fig. 4. Lower Eocene, burrowing crustacean, chamber casts (85% of natural size). A: Upper side with relic of broken gallery at the lower (narrow) end. B: Lower side with the gallery exiting from the middle; C: Side view with relics of both galleries. D: Upper side; E: Lower side; F: Side view; G-H: Smallest chamber cast; G: Upper side; H: side view; (A-C sample GSI 8990:1; C-F sample GSI 8990:2; G-H: sample GSI 8990:5).

Fig. 5. Lower Eocene, burrowing crustacean, chamber casts (85% of natural size). A: Lower side; B: Upper side showing ovoid blister-like elevations arranged 80% of transversal direction; C: Side view. D: Lower side of heart-shaped specimen; E: Upper side; F: side view; G: Gallery fragment without any distinct sculpture (A-C sample GSI 8990:3; D-F sample GSI 8990:4; G Sample 8990:6).

### 2.2 Campanian burrows

The Campanian crustacean burrows into chalk filled with granular phosphorite were described by Lewy & Goldring (2006). They comprise straight and slightly bent casts of galleries covered by longitudinal scratches. Most of them were compressed by sediment compaction but some preserve the circular cross-section of 14-17 mm in diameter. One of the two kinds of chamber casts occurs in a constant size and a rounded shape 45 mm in diameter. The biscuit-like chamber cast has a gently arched ceiling, with circular to ovoid blister-like elevations covered by transversal fine ribs replicating scratches. The lower surface of the cast comprises rings of about 8 tubercles 4 mm in diameter and 3.0-3.5 mm high in a honeycomb pattern extending over the whole surface (Fig. 6 C, D). The tubercles are smooth, but in an unfinished state of chamber the few tubercles bear fine scratches. On opposite sides of the cast are relics of the pair of connecting galleries. This structure construction replicates a pitted chamber-floor in which the honeycomb configuration of the pits enables crustaceans to cross the chamber by stepping on the floor in the center of the pit rings. This carefully constructed chamber floor comprises 60-70 pits which were interpreted as individual sites for a large egg in a brood chamber within the network of the burrow system. The other chamber type is represented by different shapes and dimensions suggesting its continuous enlargement. The smallest chamber cast is flattened and biscuit-like of a diameter of 4-5 cm, with both sides covered by the finely scratched blister-like elevations. Further enlargement changes the round periphery into an arrowhead shape with one end broad and rounded, tapering toward the opposite end (Fig. 6B). Above a length of 9-10 cm the enlargement of the chamber is vertical (Fig. 6A). The scratched elevations cover the whole cast up to the longitudinally scratched gallery casts on both ends of the elongated cast (Fig. 6B). Only movable objects could be stored in these gradually enlarged chambers. Therefore they were interpreted as nursing chamber for the young, for storing food, and perhaps also as gardening sites.

### 2.3 Comparison between the campanian and the lower eocene burrow systems

Both crustacean burrow systems were dug into pelagic chalk suggesting rather deep marine bottom conditions of several hundred meters in depth. Both occur in the same region, which at the time of deposition were on the seaward flank of the anticlinal structures of the Syrian Arc fold system (Krenkel, 1924), which has been operating and intensifying the folded structures from the Late Coniacian to the Middle Eocene. The chambers in both systems posses the same wall structure carved by the crustacean appendages, and the similar diameter of the galleries in both systems suggest morphological similarity of the producers. The crustaceans living in the Campnian burrow system had to cross the chambers. Therefore the brood chamber was carefully constructed to avoid damage to the eggs. The size of the pits (D=4 mm) attests to the rather large size of the eggs hosted in the brood chamber, probably being cared for until hatching. It seems that only the large eggs laid by the females were kept to assure total recovery (K-type breeding) whereby the number of the young of each hatching phase was the same, keeping more or less a constant size of the community. The construction of the brood chamber, exclusively for egg development, required to transfer the hatchling (larvae) to a nursery chamber where the young developed.

The Lower Eocene burrow system lacks the brood chamber and the single type of chamber has a nearly constant flattened heart-like shape as if it was a concise shortened modification of the Campanian nursery chamber. The entrance and exit at the opposite ends of the Campanian nursery chamber, which required the crustaceans to cross the whole length of

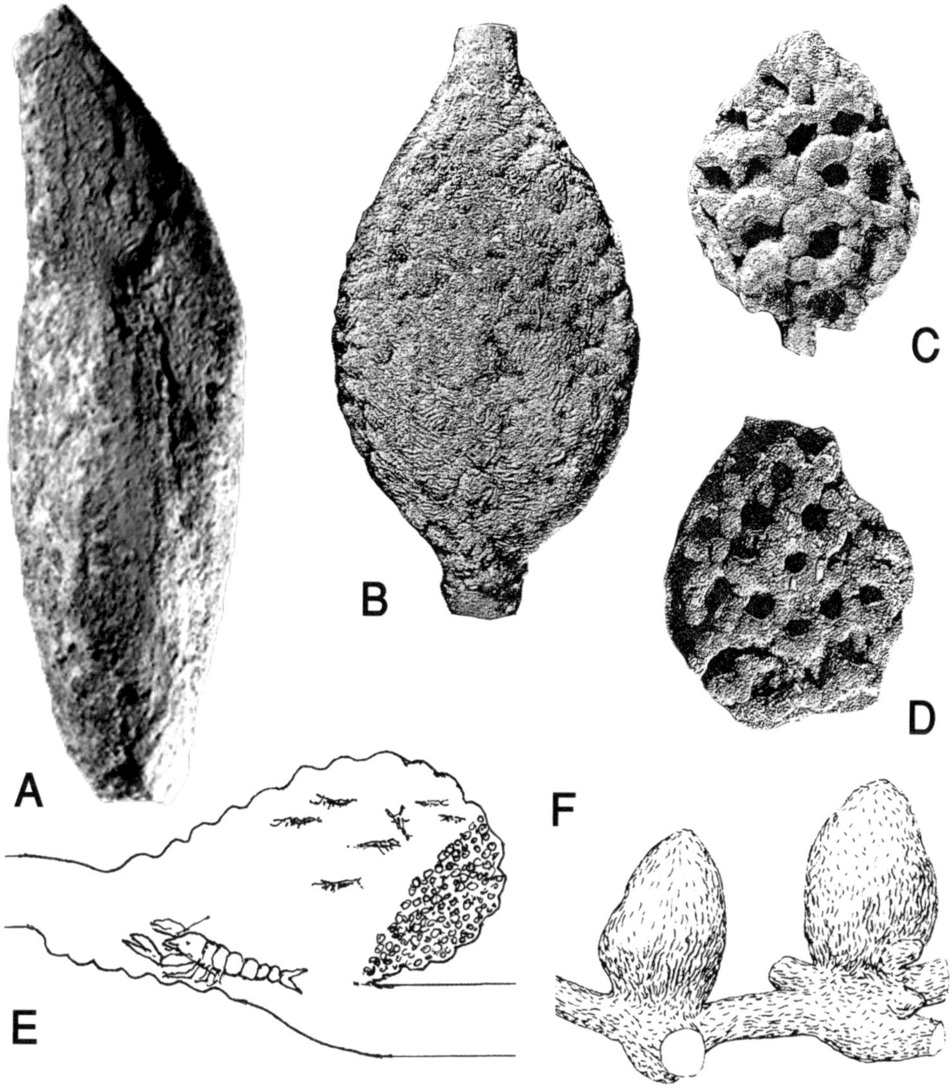


Fig. 6. A-D: Phosphatic molds of the two types of chambers in Campanian crustacean burrow system (samples GSI 5597). A & B: Nursery and storage chambers. A: Side view of vertically enlarged chamber; B: The pattern of the chamber wall consisting of low, circular to ovoid, blister-like elevations in irregular orientation of their length as indicated by fine transversal scratches in different directions. C & D: Casts of chamber floor comprising rings of pits arranged in a honeycomb pattern, leaving alternating non-pitted parts of the floor, which enables crustacean to cross the chamber without stepping on the pit's content. The pits were interpreted as protected sites for individual large eggs in a brood chamber. E: Sketch of the single type of chamber of Lower Eocene burrowing crustaceans suggesting the position of the egg-mass and the hatched young in relation to adult crustaceans crossing the chamber within the burrow network. H: Sketch of Pleistocene *Spongeliomorpha sicula* (Sicily, Italy; D'Alessandro & Bromley, 1995). All figures in 80% natural size.

the room, shifted in the Lower Eocene single chamber type to a new position. One gallery tube enters into the middle of the floor and the other tube exits from the upper narrow end of the chamber (Figs. 3, 4C). In that way, a large part of the space is not disturbed while crustaceans cross the chamber. In most chamber casts, the horizontally narrow part is also vertically smaller than the opposite side, which forms a greater space undisturbed by crustaceans crossing the chamber mainly over the floor. This constant chamber configuration attests to its special functions which required digging all chambers in the same design. When compared to the functions of the two Campanian chambers, the single Lower Eocene type of chamber probably fulfilled the necessary functions of brood chamber and nursery room. The lack of special sites protecting the eggs does not mean that the brood was released into the water and the carefully constructed chambers served for storing food and gardening. It is more reasonable to assume that in this case the whole brood mass laid by several females was concentrated in one place in the chamber, protected underground from potential predators and undergoing minimal brood-care. Accordingly total recovery was not expected and the brood consisted of numerous, hence small eggs which in previous Late Cretaceous times would not have been preserved. This breeding strategy reflects the early stages of r-type mode of breeding which in its fully developed stage the brood of numerous small eggs laid by each female was shed into open water. Thus, a small number of hatchling survived to maturity enough to maintain the species' population size relative to the associated organisms. The Lower Eocene hatched young probably developed in a corner where the eggs were concentrated. These free-swimming young swam within the chamber above the egg mass on the bottom of the broader part of the chamber (Fig. 6E).

The communal organization concluded for the Campanian burrowing crustaceans (Lewy & Goldring, 2006) can be extended to the Lower Eocene ones, which were constructed of one type of chamber where the brood was of several females was assembled and the hatching young were cared for in another corner.

## 2.4 Possible cause for the change in breeding strategy

The Campanian brood chambers were constructed to host each of the selected eggs in individual pits, where they were protected from incidental damage by the crustacean crossing the chamber while swimming through the burrow network. The size of the pits (D=4 mm) suggests that the largest (yolk-rich) eggs were selected from the brood of the females whereas small eggs were not included in the processed brood of every breeding season. Each brood chamber comprised a limited number of eggs, yielding 60-70 hatchlings under full recovery. This limited number of young probably maintained the community, reflecting optimal living conditions under 'luxurious' stable ecological settings. The disappearance of the brood chamber indicates that their function lost its previous significance, though underground protection of the brood continued in the Lower Eocene single type of chamber. What probably rendered these brood chambers useless seems to have been a change toward more eggs in every breeding season, which would require less attention and will supply offspring in the quantity required to maintain the size of a living community. It is suggested that the small eggs which were previously discarded were now gathered into an egg-mass placed at a corner of the underground chambers. These continued to provide a connecting passage within the burrow system and probably served as nursery and storage rooms as well (Fig. 6E). This multifunction may have involved some damage to the free-lying eggs, but their large quantity assured the maintenance of the population size and even an increase in the number of hatchlings. The unlimited number of eggs that could be obtained in the Eocene brood seasons is in sharp contrast to the Campanian brood in which the number of eggs was limited by the number of pits in the brood chambers, and hence the number of hatchlings even if all the eggs were fertile. The interpreted reduction in egg size did not affect the size of the mature crustaceans as attested to by the similar range of diameters of the Campanian (D=14-17 mm) and Lower Eocene (D=10-15 mm) galleries. This reflects a change in breeding strategy that is neither the result of changes in body or population size, nor of food shortage. It reflects a transition to a more economic mode of life to which the Lower Eocene (or earlier) crustacean population had to adapt under ecological pressure within the same pelagic habitat of their Late Cretaceous ancestors. The reduction in egg size increased the number of eggs that each female laid and hence the overall quantity of eggs in each breeding period. It probably increased the number of hatchlings despite some egg loss due to reduced brood-care. This interpreted need for more hatchlings must have compensated for the loss of individuals being killed while swimming outside the underground tunnel network to look for food. This fatal threat of predation has profoundly increased during the Late Cretaceous as evidenced by other faunal groups (discussed herein).

The evolutionary trend expressed by changes in crustacean burrowing systems can be extended to Early Pleistocene times as exemplified by a burrow system of *Spongeliomorpha sicula* D'Alessansro & Bromley (1995) from Sicily, Italy. Cylindrical vertical shafts and horizontal galleries about 10 mm in diameter bear longitudinal fine ridges replicating scratches. Plum-shaped chamber casts 30 mm high and 25 mm in diameter with similar longitudinal striations occur at gallery junctions every few centimeters (Fig. 6F). They are associated with much shorter cylindrical inflations. The plum-shaped chambers were interpreted as microbial gardening sites (D'Alessandro & Bromley, 1995). Following the study of the Campanian and Lower Eocene burrow systems, we are inclined to refer these chambers to brood and nursery chambers as in the Lower Eocene example. However, in the Pleistocene example each chamber has its own entrance from the lower side, whereby the chamber content is not jeopardized by chamber-crossing crustaceans as in the Eocene example. The associated sediments indicate shallow marine environments which were rich in food sources and should not require production of special nourishment in specially constructed gardening chambers. On the other hand, eggs and larvae shed into the shallow marine water were subjected to rapid consumption by many predators. Thus keeping the hatchlings until they were capable to defend and feed themselves would have been needed to protect the species. This interpretation coincides with the evolutionary trend in this group of burrowing crustaceans from the Late Cretaceous to almost present times. The interpreted care for the brood and the young attests to communal organization at least until the Early Pleistocene and probably might be detected in extant *Spongeliomorpha* species. It seems that these burrowing Crustaceans have maintained communal organizations from times when they inhabited deep water bottoms exposed to predators, and continued to experience the benefits of this co-operation in shallower marine environments.

## 3. The ecological affinities of the upper part of the Cretaceous Period

MacLeod (2005) summarized the characteristic affinities of the Cretaceous Period which ranged between 145.5-65.5 Myr and is generally divided at the Albian-Cenomanian boundary (99.6 Myr) into the Lower and Upper Cretaceous. The warm equable climate (warmer than today) extended into the high latitudes, and the poles were probably without ice cover most of the year (Hay, 2008). High sea-levels with their peak during the Turonian characterized the general transgressive trend of the oceans throughout the Upper Cretaceous and the Lower Tertiary. The continental plates continued to move toward their present-day position whereby the Atlantic Ocean further opened perpendicular to the Tethys Ocean improving the north-south water circulation. This global plate movement triggered local tectonic movements which differentiated plate margins and intensified the reduction of the previous Albian-Cenomanian-Turonian broad (50-300 km wide) shelves, where rudistid bivalves thrived together with sessile ostreids and chondrodentids. These could survive temporary exposure or cover by sediment, in contrast to hermatypic corals which occupied protected regions in this shallow marine ecosystem. A broad intertidal flat extended landward of this rudistid reefal belt, where calcareous detritus (bioclasts) accumulated and was partly dolomitized under the high salinity of the lagoon and sabkha settings. The abundance of calcitic mollusk conchs highly increased during the middle part of the Cretaceous Period, represented by the shallow marine rudists, oysters and chondrodonts, with Inoceramidae inhabiting the deeper water. The expansion of pelagic environments during the Upper Cretaceous was associated with an increase in the diversity and abundance of planktonic foraminifera and calcareous nannoplankton (Coccolithophorida) all having calcitic endoskeletons. This abundance of biologically precipitated calcite suggested a very low Mg/Ca ratio in Cretaceous seawater (MacLeod, 2005). Keeled planktonic foraminifera diversified from the uppermost part of the Lower Cretaceous (Albian) onward throughout the Late Cretaceous, and were associated with globular forms. Their calcitic tests accumulated as foraminiferal ooze in the outer shelf and deeper marine bottoms forming chalk characterizing the Upper Cretaceous, such as the Lower and Upper Chalk in northwest Europe. The gradual increase in the plankton bloom in the broad oceans triggered a gradual rise in marine productivity evidenced in the later Upper Cretaceous (Campanian-Maastrichtian) by extensive accumulations of organic-rich ('bituminous') chalk with chert (from dissolved diatoms and radiolarians) and phosphorite beds mainly in the Tethys ocean (Lucas & Prévot-Lucas, 1996). These optimal living conditions are corroborated by the increasing diversity of the marine fauna and the development of gigantic organisms. The largest ammonite *Parapuzosia seppenradensis* (Landois) with a diameter of about 2.50 m (Summesberger, 1979) and *P. bradyi* Miller & Yongquist (D=1.37 m) from Wyoming, USA (Larson et al., 1997) are from the Lower Campanian. Large inoceramid bivalves with an axial length of 1 m, and occasionally over 2-3 m in size of the genus *Platyceramus,* occur in the Santonian-Lower Campanian of Colorado (USA) (Kauffman et al., 2007). Marine reptiles related to plesiosaurs and mosasaurs grew to a length of 9-15 m (MacLeod et al., 1997). Their flying relatives (Pterosauria) reached in latest Cretaceous time wide wing spans up to 11-12 m in *Quetzalcoatlus* (Langston, 1981). The marine high productivity extended into early Cenozoic times (Lower Eocene) despite the profound change in zoo- and phytoplankton composition, suggesting that the Late Cretaceous marine physical and chemical properties were neither affected by the Deccan volcanism (India) nor by the asteroid impact. These marine conditions were controlled by the continuing movement of the plates and the general transgressive trend of the widening oceans during the Late Cretaceous and Early Tertiary. Thereby the size of the shelves and the neritic habitats were considerably reduced. Organisms living within the 'reefal' habitat of the rudistid and ostreid buildups decreased in abundance and disappeared from many shrinking shallow marine environments, surviving only in restricted regions. The expansion of the pelagic habitats increased the abundance of the nektonic organisms which comprised the top predators and thus changed prey-predator relationships. Micropaleontological analyses of Late Maastrichtian pelagic sediments detect short-term paleotemperature fluctuations (Li & Keller, 1998). A progressive cooling trend between ~66.8-65.45 Myr was followed by rapid extreme warming 400-200 Kyr before the end of the Maastrichtian, which was succeeded by a cooler climate during the last 100 Kyr of this stage (Abramovich & Keller, 2002; Abramovich et al., 2010).

## 4. Victims of disturbed prey-predator relationships under climatic and ecological instability

Despite the overall flourishing of the Upper Cretaceous marine fauna and flora, the local balance of prey-predator relationship was very fragile, threatening this global paradise. Optimal living conditions on land increased animal diversity, populations size and the dimensions of individuals (gigantism), as in the marine environments. However, any temporary ecological disturbance might have reduced reproduction, resulting in much less young, which played a significant role in the food-chain. The further collapse of the food chain, as the result of increased predatory stress, is examined herein, by comparing the affinities of the organisms which became extinct at the end of the Cretaceous Period to those which survived the biological crisis. The present study builds on the detailed analysis of the geological record of most faunal and floral groups around the K-T boundary, carried out by a large team of experts (MacLeod et al., 1997). They pointed out the terrestrial and extraterrestrial factors which affected life on Earth during a long period as well as short-catastrophic processes close to the K-T boundary. However, the control of the extinction-survivorship pattern was not defined. Re-evaluation of the characteristics of representative marine and terrestrial faunal groups will demonstrate that all those which became extinct at the end of the Cretaceous, were at some stage of their life unable to avoid their predation, thus being victims of temporary extreme predatory stress. This selective over-predation of the vulnerable organisms was caused by the collapse of the food chain as the result of climatic and ecological instability. These are partly reflected by paleotemperature fluctuations of the oceans surface water and deeper levels (Li & Keller, 1998; Abramovich & Keller, 2002), as well as reduction in oxygen content in seawater and dwarfing in marine calcareous planktonic microorganisms. All these phenomena can be related to fluctuations in the intensity of the volcaniclastic dust screening the sunlight, hence affecting photosynthetic activity of the flora and disturbing the biological clock of animals in the sea and on land.

### 4.1 Vertebrates: Reptiles and fishes

The disappearance of the dinosaurs close to the end of the Cretaceous Period is presented in scientific and popular publications and films as evidence to the most impressive catastrophic event in Earth history. These reptiles ruled over the land while their flying relatives (pterosaurs) and the marine ones (e.g., plesiosaurs and mosasaurs) were the top predators in the sky and the sea. They diversified during the Late Cretaceous and many of their species grew to giant dimensions (9-15 m long mosasaurs; 11-12 m wing-span of the pterosaur *Quetzalcoatlus*). Their apparent simultaneous disappearance from over the whole world was puzzling. Whatever caused it did not kill the related crocodiles and did not harm the sensitive frogs and salamanders. This selective elimination of the most skillful predators resulted from their early ontogenetic stage. Dinosaurs and pterosaurs were oviparous, laying 1-7 eggs (or more) in nests on a rather flat land, such as sea-shores (Sanz et al., 1995), tidal flats (López-Martínez et al., 2000) and beside estuaries, lagoons, marshes and fluvial plains (Vianey-Liaud & Lopez-Martinez, 1997). Despite brood-care by the adults, the eggs and the hatchlings were frequently exposed to potential predators and could be snatched from the nest or consumed on site after the distraction of the parent. Dinosaurs and pterosaurs living in this region were probably involved in the killing of their kind. This common process among birds was suspected to have occurred in dinosaurs as indicated by the name *Oviraptor*='egg stealer', given to a dinosaur situated in an egg nest. Other oviparous reptiles such as crocodiles, land lizards and sea-turtles hid the brood and only experienced predators knew were to search for them, whereby most of the eggs hatched and the young quickly looked for shelter. The marine reptiles gave birth to young which were likewise vulnerable to predation at this stage as hinted by bones of small, probably young mosasaurs in Upper Campanian sediments in southern Israel. The dinosaurian branch of birds survived this predatory threat thanks to their rather small body, and hence their egg size. These could be laid in nests high above the ground hidden in trees, in bushes or at inaccessible sites such as on cliffs. The threat to the brood was thus minimized and restricted to the few predators which could discover and reach these breeding sites. The extant large ostrich exemplifies the mode of breeding of the dinosaurs on open ground whereby some of the 10-12 eggs might be consumed by predators.

MacLeod et al. (1997) summarize the record of the Upper Cretaceous cartilaginous **fishes** many of which survived into the Tertiary like many of the bony fishes. Among those which became extinct are *Enchodus* and *Stratodus*. *Enchodus* species reached 1 m of length whereas *Stratodus* species were over 3 m long (Lewy et al., 1992). These rather large bony fishes, as well as most of the cartilaginous ones seem to have swum as individuals in contrast to present-day small fish, forming vortex-like swarms, hence confusing predators. Thus the surviving potential of these small fish is higher than of individual large ones, despite their skills as vicious predators which could be overcome by larger sharks and by big marine reptiles.

### 4.2 Cephalopods

**Ammonites** are another example of a large group seemingly to suddenly disappear at the end of the Cretaceous Period like the dinosaurs. These conch bearing cephalopods diversified during the Upper Cretaceous, providing excellent biostratigraphic markers. Ammonite species gradually disappeared during the Upper Maastrictian, with 12 reaching close to the K-T boundary in the section exposed in northern Spain, in which other species disappear in groups or individual species during the Upper Maastrichtian (Marshall & ward, 1996). This is probably the most complete latest Cretaceous sedimentary sequence with ammonites. The simultaneous disappearance of twelve species close to the K-T boundary gives the impression of a catastrophic event that killed all ammonites in this area and seems to corroborate the total elimination of the order Ammonoidea throughout the world.

Lewy (1996, 2002a, b) analyzed the functional morphology of ammonites (ammonoid conchs), especially of heteromorphs, but also of the planispirally coiled ones. Most of the heteromorph ammonites (except for the Baculitidae) developed a U-shaped terminal whorl with an upward facing aperture, which in some species was partly occluded by the previous whorls. Some planispiral ammonites changed the shape of the last whorl, inflating it and constricting the terminal aperture. Others added apertural appendages. All of these modifications in the last growth stage limited the mobility of the ammonoid by the constricted or upward oriented aperture, complicated nourishment, and in some cases must have resulted in death of starvation. These rather fatal modifications could not have been intended to live further in a different way (Westermann, 1990) and were interpreted to have served the last and most important biological duty of breeding by providing protected brood chambers (Lewy, 1996). The female was situated beside numerous tiny eggs (detected in fossil ammonites) being drifted by currents across the ocean while the eggs developed. This drifting process is corroborated by the wide distribution of many of these non-streamlined heteromorphy ammonite species along the Tethys Sea, which could not have been explained by swimming. Such mode of breeding in cephalopods is known in extant octopods which carry out two breeding strategies. The common one is laying large, yolk-rich eggs in bundles attached to submarine substrates ('stationary' mode of breeding), in which both parents care for the brood during several months without eating. They die of starvation close to when the young hatch, each with some yolk for their nourishment during the first hours of free swimming. In the single group of argonautids the female *Argonauta* (larger than the male) secrets from the expanded edge of two of its tentacles a thin calcitic, widely-coiled shell. It situates itself in this boat-shaped shell and lays numerous tiny spherical eggs about 1 mm in diameter, similar in shape and size to spheres detected in fossil ammonites (Lewy, 1996). Together they drift over the sea while the eggs develop ('pelagic' mode of breeding) and the young are shed into the water to cope with life. The shapes of all of the argonautid brood chambers (several fossil and extant species) are identical to latest Cretaceous ammonites. This fragile conch cannot protect its content and merely carries the female and the brood. For this purpose, a smooth boat-shaped shell would be sufficient and the ammonite-like complex sculpture cannot be explained by convergent evolution. Argonautid egg-cases occur since the Late Oligocene (Saul & Stadum, 2005). It is unlikely that only then the 'pelagic' mode of breeding was introduced into octopod breeding strategy. It is more reasonable to explain the first occurrence in Late Oligocene times of fossils of ammonite-like argonautids by the preservation of their calcitic shell. This brood-case might have been made in earlier times of an organic substance (conchiolin) which disintegrated close to after burial in the marine sediment. Octopods have neither an external conch nor an internal hard feature, which renders their geological record sparse. The earliest fossil octopod is an imprint from the Middle Jurassic, but the development of these cephalopods must have been earlier. The few octopod fossils preserved in restricted environments of unusual burial conditions resulted in the dispute over their systematic position among cephalopods. The comparison between the anatomy and physiology of extant octopods and the functional morphology of ammonites suggests close genetic relationships between these two cephalopod groups, one of which survived the end-Cretaceous biological crisis (Lewy, 1996, 2002).

The straight (orthocone) baculitids are heteromorph ammonites which did not modify the last growth stage and probably had another breeding strategy. All baculitid species seem to have formed local evolutionary lineages and were hence indigenous species like several planispiral ammonites, which characterize biogeographic provinces in contrast to the 'cosmopolitan' distribution of most heteromorphs. The 'pelagic' mode of breeding resulted in the wide distribution of these species, whereas in the 'stationary' mode of breeding the hatchlings remained in the area where they hatched and formed indigenous species. This means that the two modes of octopod breeding occurred in ammonoids and controlled their distribution. The similarity of the argonautid brood cases to Upper Cretaceous ammonites strengthens these ammonoid-octopodid relationships, suggesting phylogenetic connections. Lewy (1996) suggested that octopods descended from ammonoids in which the conch degenerated until total loss like in opistobranch gastropods. This evolutionary trend can be explained by the diversification of fish, sharks, marine reptiles and belemnites (Lewy, 2009) predating, among others on slow swimming ammonites which their conch did not protect anymore from these skillful hunters. On the other hand, the conch limited the expansion of the mantle cavity and hence the expelled water jet which controlled swimming speed. The lack of an external conch overcame these restrictions and improved maneuverability, while additional strategies improved octopods' escape from predators. The earliest octopods descended from several ammonoid groups, whereby some carried out the 'stationary' mode of breeding and the others- the 'pelagic' mode. These conchless creatures were physiologically required to carry out the two modes of breeding in which the 'pelagic' one a floating egg-case was needed. Empty ammonites floated for some time over the Jurassic and Cretaceous seas before sinking into the depths. These were occupied by the relevant octopods and amended into suitable egg-cases. In Late Cretaceous times, the common ammonites *Hoplitoplacenticeras*, *Jeletkytes* and *Phylloceras* were amended into floating egg-cases by the breaking off of the terminal part of the conch and its extention in an uncoiling shape, enabling the octopod female to enter and care for the brood. This added part was probably made of conchiolin secreted from glands developed at the end of two tentacles as reflected by extant argonautid octopods. The disappearance of floating conchs in earliest Cenozoic times forced the surviving octopods to produce the whole brood-case, which they did in the shape of the Late Cretaceous ammonites, which their ancestors had learned to construct. The short longevity of extant octopods (1-3 years), when applied to ammonoids provided an explanation to the function of the fluted margins of ammonite septa (Lewy, 2002a), ammonoid high evolution rate and other phenomena in ammonoids (Lewy, 2002b), corroborating the deduced ammonoid-octopodid genetic relationships as hinted by other common characteristics (Lewy, 1996). Accordingly, the order Ammonoidea did not completely disappear at the K-T boundary, but only the conch-bearing ones. In this respect the Ammonoidea are comparable to those dinosaurs from which the birds descended.

The endoskeleton of the cephalopod order of **Belemnitida** has been suggested to balance the horizontal orientation in the water while the belemnoid preys and swallows skeletal fragments (Lewy, 2009). These were mainly made of calcareous composition of a specific gravity twice that of flesh, being temporarily stored in the frontal crop. The change in weight in the anterior side through the accumulation and regurgitation of these fragments was balanced by water-gas exchange in the phragmocone. The rapid evolution of the belemnites since the Early Jurassic was associated with the appearance of calcitic opercula (aptychi) in ammonites. This calcification of the pair of 'wings' of the lower jaw in the shape of the aperture was intended to protect the ammonoid by preventing crustacean claws, belemnite tentacles and other means of predation from penetrating into the conch. Most aptychi are found associated with belemnites suggesting prey-predator relationships. The same ammonite genera (e.g., *Baculites*) in regions without belemnites, lack any associated aptychi plates (Lewy, 2009). Some Late Cretaceous belemnites reduced the size of the guard (rostrum) up to complete disappearance (e.g., *Naefia*, *Groenlandibelus*) suggesting a change in their diet comprising less skeletal parts- a fact which was attributed to the profound reduction in the abundance of ammonites as prey. These latest Cretaceous belemnites share common morphological affinities with Early Tertiary sepiids (Coleoidea) suggesting an evolutionary transition across the K-T boundary rather than belemnoid extinction (Lewy, 2009).

Five Late Cretaceous **nautiloid** genera crossed the K-T boundary (Kummel, 1964). Many had a spherical shape which was not easy to catch and crush, despite the fact that the ammonoids and the nautiloids were predated by mosasaurs (Kauffman, 2004). All cephalopod groups descended from ancestral nautiloids and the extant ones have similar anatomical features, except for nautilids. The latter have a primitive eye structure, two pairs of gills and numerous small tentacles in contrast to vertebrates-like eyes, a single pair of gills and ten or eight tentacles as seen, for example in cuttlefish, squids and octopods. These anatomical and physiological differences suggest that the extant nautilid adapted to darkness and oxygen deficiency, such as exists in the deep ocean where they are found today, restricted to the southwestern Pacific Ocean (Kummel, 1964). Mesozoic fossil nautiloids occur in shallow and deep marine sediments. It is reasonable to assume that nautiloids swimming in open marine waters were attacked by sharks, large fish, large octopods and squids. The slow swimming nautiloids escaped into deeper marine environments already millions of years ago during which their anatomy and physiology considerably changed and therefore cannot be applied to Mesozoic and older nautiloids (Lewy, 2000). This trend explains how nautiloid genera survived the Late Cretaceous biological crisis which affected their associated cochleate ammonites. This crisis was not caused by acid rain (Prinn & Fegley, 1987) which would have killed most nektonic organisms, but reflects an increase in predation pressure (as reflected by other faunal groups) from which nautiloids escaped into deeper water and less menacing habitats.

### 4.3 Bivalvia

Most bivalves are burrowers into the sediment and are thus hidden from predators, in contrast to epifaunal species. Among the few groups which did not survive into the Cenozoic are the sessile, epifaunal, gregarious **incoceramid** bivalves, which thrived in large communities on rather deep marine bottoms of calcareous shale and chalk. These sediments preserved organic matter in some places, suggesting temporary reduced oxygen content and hence living conditions unfavorable to other organisms (Kauffman et al., 2007). The decimeter to over a meter long bivalves are found up to the base of the Upper Maastrichtian, with questionable relics at higher levels. Their small relative *Tenuipteria* survived to the end of the Maastrichtian (Dhondt, 1983; Marshall & Ward, 1996). This selective extinction can be explained by increased predation by sharks and mosasaurian reptiles (Kauffman, 1972), which dived into the deep bottom for the easy prey due to the fragile nature of the prismatic shell structure of these sessile, rather large bivalves, which until the beginning of the Upper Maastrichtian coped with the usual predation rate.

**Rudists** were individual marine bivalves attached to substrates or reclining on the soft sediment. Whether or not they hosted photosynthesizing zooxanthellae, they concentrated in shallow water where food supply and aeration were optimal. They probably had a short larval stage and could not have drifted far from their ancestral rudists before settling down and undergoing metamorphism. Therefore the young rudists are found attached beside, or on top of the previous generation accumulating into wide thickets forming the carbonate-platform framework, or building elongated or lenticular biogenic buildups (bioherms) with or without hermatypic corals, stromatoporoids, calcareous algae and other attached faunal groups. Some rudists reclined on the bottom in the low-energy neritic zone. The general narrowing of these neritic habitats during the Late Cretaceous reduced rudistid abundance (e.g., central-eastern Mediterranean and Middle East region; Steuber and Löser, 2000) and the associated 'reefal' communities. Late Cretaceous rudist buildups prevailed in restricted regions of the Tethys Sea, such as in the Caribbean province (e.g., Jamaica; Mitchell et al., 2004) where a few genera reached the K-T boundary and became suddenly extinct probably as the result of the asteroid impact at the Yukatan Peninsula in the same region (Steuber et al., 2002). The following initiated tsunami waves might have broken and killed the rudists or covered them by sediment (e.g., Scasso et al., 2005; Bralower et al., 2010; with references). However, the dating of these turbulence-induced deposits relative to the age of the asteroid impact and the K-T boundary are still controversial (Keller et al., 2007).

Most **oysters** are attached to substrates in shallow marine environments tending to concentrate and form oyster banks. A few genera recline on soft bottoms in low-energy environments (e.g., *Gryphaea, Pycnodonte*). The 'tribe' Exogyrini (Stenzel, 1971) was highly abundant in the Cretaceous neritic zone and their calcitic shells are well preserved in carbonate platform sediments beside the long-ranging *Ostrea*. Their attached (left) valve first grew in a spiral pattern which opened and straightened into an elongated or rounded cup-shape in which the posterior margin stretched over the substrate and the anterior margin was raised, whereby the flattened upper valve was inclined to the substrate. This mode of growth subjected these oysters to penetration of sedimentary particles in between the valves as well as total cover by sediment. The disadvantageous growth orientation in shallow marine environments added to possible exposure at low tide or predation, all of which resulted in the extinction of this group at the end of the Cretaceous. Thereby they differ from the subfamilies Gryphaeinae and Pycnodonteinae in which the lower valve grew in a nearly planispiral curvature into a cup-shape, whereby the valve commissure (margins) was elevated above the substrate and the flat upper valve was in horizontal orientation (Stenzel, 1971). The larvae of these oysters had to attach before undergoing metamorphism. The shallow marine habitats of the Upper Triassic-Jurassic Gryphaeinae consisted mainly of friable sediment such as sand and marl. Because these sediments lacked large firm substrates, any grain or small fragment served as attachment site as evidenced by the small attachment scar at the oyster beak. With further growth the small substrate lost its anchorage function and the oyster was tilted, raising the substrate above the ground whereby the ventral margins of the oyster nearly sunk into the friable sediment. To avoid the penetration of sedimentary particles, the oyster increased the upward growth of the lower valve whereby the oyster balance changed and required further tilt and upward growth. The resulting planispiral curvature increased the living space in between the two valves beyond the size of the mollusk which was compensated by secondary deposition of shell material on the inner surface of the lower shell (Lewy, 1976). However, the precipitated calcitic foliated shell structure increased the oyster total weight and enhanced its sinking into the sediment. Thereby the curvature and thickness of the Gryphaeinae lower valve reflect the plasticity of the sediment on which it reclined. The crucial effect of this secondary deposit in the lower valve was partly solved in Cretaceous times by changing the compact structure into a vesicular one which characterized the similarly looking Pycnodonteinae. These oysters thickened their valves by layers of light vesicular structures in between layers of foliated structures minimizing the weight of the secondary fill. Thereby the Pycnodonteinae could inhabit very soft bottoms and thrive on marl and planktonic foraminiferal ooze forming chalk in the Upper Cretaceous. Thanks to their adaptation to rather deep marine environments, the Pycnodonteinae survived the end-Cretaceous biological crisis. This is in contrast to the Exogyrini tribe which lived attached to firm substrates and therefore inhabited mainly shallow marine environments such as gregarious oysters which were subjected to local exposure by long low-tides, being covered by sediment from terrestrial runoff, as well as predation. There they were associated with the oyster-like *Chondrodonta* which could withstand wave impact and thus thrived in high-energy environments and disappeared before the Exogyrini during the Upper Campanian (Stenzel, 1971).Perhaps *Chondrodonta* attained larger dimension and had a thin fragile shell in contrast to *Exogyra* species. The pectinid **Neithea** Group thrived in Upper Cretaceous neritic and continental-slope sediments and probably disappeared at the end of the Mesozoic (MacLeod et al., 1997). After an early byssate stage they reclined on the sea bottom and occasionally leaped for a short distance, thus being exposed most of the time to diving predators, subject to over predation and extinction at the end of the Cretaceous Period.

### 4.4 Gastropoda

Most gastropod families crossed the Cretaceous-Tertiary boundary nearly unaffected except for the **Nerineidae** and **Actaeonellidae**. These two families comprised rather large gastropods which thrived on the Tethyan warm-water carbonate platforms and their marginal mainly low-salinity zones (Sohl & Kollmann, 1985). The elongated nerineids had a thick external shell reinforced by folds of the inner shell layer. Actaeonellids likewise had a thick shell and most of them formed ovate conchs which probably slipped through the teeth of predators, increasing the resistance of both gastropod groups to predation. Though these gastropods were mobile, they formed layered concentrations and in places lenticular structures. These accumulations suggest that these gastropods lived close below (shallow burrowers) or on the bottom and were thus subjected to exhumation and concentration by turbulent water in the neritic zone. Generally their representative species survived up to the end of the Cretaceous although some disappeared earlier from many provinces (Sohl & Kollmann, 1985, fig. 14) probably as the result of the reduction of the neritic zones and predation of previously untouched organisms.

## 5. Marine microorganisms with symbiotic zooxanthellae

Marine floral and faunal microorganisms flourished and diversified throughout the Late Cretaceous. Therefore the disappearance of most of them at the end of the period seemed catastrophic (MacLeod et al., 1997). Close to the K-T boundary the calcareous tests of **nannoplankton** and the **planktonic foraminifera** reduced their size (dwarfing) and the assemblage became dominated by low-oxygen-tolerant small heterohelicid foraminifera and the disaster opportunist nannofossil *Micula decussata* (Abramovich & Keller, 2002; Keller & Abramovich, 2009). These latest Maastrichtian affected microfossils occur in Indian Ocean drilling samples with volcanic sediments attributed to the Deccan volcanism, hinting to a connection between the intensive volcanism and the deterioration of the marine ecological systems (Tantawy et al., 2009).

## 6. The end-Cretaceous biological crisis caused by the Deccan volcanism

The main volcanic phase, comprising ~80% of the total Deccan Trap volume, occurred around the K-T boundary and is interpreted as being active during a short time interval in the middle of the paleomagnetic chron 29r (Keller et al., 2009, fig. 5). Oxygen isotope analyses (Li & Keller, 1998) and the response of microorganisms to water temperature (e.g., Abramovich & Keller, 2002; Abramovich et al., 2010) reflect fluctuations in surface and intermediate depth ocean water temperature during the Late Maastrichtian, especially in the latest 0.5 Myr. Fluctuating cool temperatures (average degrees of $9.9^0$ C intermediate and $15.4^0$ C surface water) during 66.85 and 65.52 Myr were followed by a short-term warming between 65.45 and 65.11 Myr which increased intermediate water temperatures by $2\text{-}3^0$ C, and decreased the vertical thermal gradient to an average of $2.7^0$ C (Li & Keller, 1998). A previous study by Stüben et al. (2003) on hemipelagic sediments of Tunisia differentiated three cool periods (65.50-65.55, 65.26-65.33, 65.04-65.12 Myr) and three warm periods (65.33-65.38, 65.12-65.26, 65.00-65.04 Myr). Tantawy et al. (2009, p. 85) point out that "**the biotic effects of volcanism have long been the unknown factors in creating biotic stress. The contribution of the Deccan volcanism to the K-T mass extinction remained largely unknown, although recent investigations revealed that the main phase of Deccan volcanism coincided with the K-T mass extinction**". Keller et al. (2009, p. 723-4) refer to "**the dust clouds obscuring sunlight and causing short-term global cooling**" as the result of the volcanic eruptions, but **"how Deccan volcanism affected the environment and how it may have led to the mass extinction of dinosaurs and other organisms in India and globally is still speculative**". The direct cause for seawater temperature fluctuations during the Maastrichtian last half million years and the dwarfed microfossils in this time interval (Keller, 2008) are herein related to sunlight screening by volcaniclastic dust from the Deccan volcanism, suggesting that its main activity extended over the same period.

Most Late Maastrichtian planktonic microorganisms reduced their size (dwarfing) at about 65.4 Myr (Keller, 2008) reaching sexual maturity at smaller dimensions and probably more rapid than their normally-sized ancestors. This assemblage became dominated by low-oxygen-tolerant small heterohelicids (Keller & Abramovich, 2009). All these globally detected abnormal morphological and ecological aspects of the latest Cretaceous marine microfossils attest to the deterioration of the ecological conditions, as the result of sunlight screening and darkening of the Earth to various extents and periods. Global darkening of the atmosphere by fine volcaniclasts decreased photosynthetic activity of the symbiotic zooxanthellae in extreme cases these useless symbionts were digested by their host. The dwarfing of the latest Maastrichtian microfossils was artificially demonstrated by the elimination of these symbiotic dinoflagellates from within the planktonic foraminifer *Globigerinoides sacculifer* (Bé et al., 1982). The loss of symbionts resulted in early gametogenesis (at small size), short life span of the foraminifer and its smaller shell size at sexual maturity (dwarfing), exactly as described from the latest Maastrichtian planktonic foraminifera and calcareous nannoplankton. When the tested live foraminifers were reinfected by zooxanthellae they resumed normal shell growth and size as before the removal of the symbiotic zooxanthellae (Bé et al., 1982). The lack of planktonic microfossils of normal size in the latest Maastrichtian 0.5 Myr indicates that solar radiation was, during this period, too low to resume symbiotic relationships between these photosynthesizing dinoflagellates and the microorganisms. The drastically reduced photosynthetic activity lowered the oxygen content in the upper water column as attested to by the increased abundance of low-oxygen-tolerant small heterohelicids and the blooming of the disaster opportunist *Guembelitria* (Keller & Abramovich, 2009). The shading of sunlight reduced the depth of the marine euphotic zone and affected water temperature. A thick cover by ash dust probably lowered sea-water and Earth surface temperature. A less dense screen may have resulted in a greenhouse effect, keeping the warmth of partially penetrating sun-light from escaping into the atmosphere. Paleotemperature fluctuations in the latest Maastrictian marine environments (Li & Keller, 1998) are accordingly related to fluctuations in the intensity of the Deccan volcanic eruptions and world-wide dispersal of volcaniclasts.

The polyps of reef-building hermatypic **corals** house symbiotic zooxanthellae which are involved in the precipitation of the calcareous skeleton and in other physiological processes, but they can also become part of the coelenterate diet. Unlike in planktonic foraminifera these symbiotic relationships observed on extant corals in the Great Barrier Reef of Australia can be stopped for a while (coral bleaching) during which coral growth slows down while the polyps feed on other algae and microorganisms, organic debris and bacteria (Vernon, 1993). This may explain the survival of some hermatypic coral groups, though a great deal disappeared during the latest Cretaceous (MacLeod et al., 1997).

The fatal influence of the Deccan volcanism on the latest Cretaceous marine planktonic microorganisms applies to marine and large terrestrial creatures as well. The darkening of the atmosphere by dispersed volcaniclasts blurred the distinction between the annual seasons controlling plant growth and blooming, as well as the biological clock of animal reproduction, which provides a significant food source to carnivores after months of near starvation. Seasonality controls the timing of sperm and egg spawning into the water, most of which is consumed by predators awaiting this process. Mating and reproduction among larger animals is coordinated with availability of food supply (plant and meat) which will assure the survival and development of the young. Long-ranging darkness confused the instincts and physiology of animals, reducing birth rate and hence food supply crucial for predating mammals to feed the young as well as for reptiles and birds. The reduction in birth rate (including the laying of eggs) immediately reduced the food supply. The aggressive and large predators (mainly among the reptiles) were forced to consume part of the prey that smaller predators used to eat, whereby the 'normal' food chain collapsed, resulting in the intensive predation of the temporarily unprotected ones. These were dinosaurs and pterosaurs eggs in nests on flat-land and their hatched young, as well as mature ones sitting on the eggs, and other creatures which for a moment were careless. The over-predation of this easy prey reduced the size of the victim's population which gradually diminished until the remaining ones could not preserve the species, leading to extinction. The organisms which were not affected by the collapse of the food chain were small creatures which could escape and hide themselves or their brood such as crocodiles and turtles which covered their brood, birds which laid the small eggs in between plants, small mammals which could hide underground or among bushes, and amphibians and fishes capable of hiding in aquatic environments. The darkening effect of the Deccan volcaniclasts must have slowed down the metabolism of cold-blooded reptiles, among which were probably some large dinosaurs. During the severe darkening they were completely unable to defend themselves even from small predators. The selective elimination of the temporary vulnerable ones is an extreme example of **natural selection** as the result of catastrophic changes in the regular pattern of the long-operating ecological system in which organisms and plants lived in harmony. The additional destructive effect of a single or multiple asteroid impacts (Keller, 2008) would have had little contribution to the gradual collapse of Earth's biological systems.

## 7. Early Tertiary biological recovery

The Deccan volcanic activity extended into the lowermost Tertiary (~64.8 Myr; Keller et al., 2009) and thereafter the climatic and ecologic systems began their recovery. Small globigerinid planktonic foraminifera survived the end-Cretaceous biological crisis and appear in Early Paleocene sediments (e.g., Orue-Etxebarria & Apellaniz, 2000). Their trochospiral coiling with 4-7 nearly globular chambers in the last whorl resembles other associated species such as *Parvularugoglobigerina eugobina* (Olsson et al., 1999). This general shape and size has neen observed in one of the earliest planktonic foraminifer from the Lower Jurassic (probably Hettangian; ca 190 Myr) of Hungary (Görög, 1994). Keeled planktonic foraminifera appeared in the Upper Albian (*Rotalipora*) about 90 Myr later (Leckie, 1987). The earliest Tertiary keeled (pseudo-keel) planktonic foraminifer [e.g., *Morozovella angulata* (White)] appeared at the base of the Upper Paleocene (Thanatian, 58.7 Myr) about 6 million years after the recovery of the ocean ecological setting. This rapid introduction of the keel structure among pelagic foraminifera suggests that the survivors in a 'primitive' appearance preserved in their genome the ability to secrete keels and other morphologies under suitable conditions. The relative quick recovery of planktonic microorganisms after their near elimination at the end of the Cretaceous Period explains the similar recovery of life in the marine and terrestrial bioprovinces. Many Late Cretaceous species retreated to small-restricted niches protected from the side effects of the Deccan volcanism. They continued living in these numerous small habitats, adapting to the restricted ecological settings and thereby gradually changing their physiology, anatomy and the skeleton. With the recovery and stabilization of the ecological systems all of these 'hidden' communities tried to enter and adapt to the physical and chemical conditions of the open-large habitats and share the environment with other communities. Only those which succeeded to accommodate themselves in these extensive bioprovinces in large populations were discovered. The fossil record of all earlier small communities which lived in restricted areas is still missing, giving a misleading impression of a big hiatus in taxa ranges and sudden first appearance of new ones. These are actually members of evolutionary lineages, of which the earliest Paleocene ancestors have not been yet discovered. This all took part in an evolutionary biological continuum from the latest Mesozoic into the Cenozoic. The technical comparison of taxa names between these eras intensified the apparent catastrophic aspect of the end-Cretaceous biological crisis, being erroneously referred to a **mass extinction**.

## 8. Conclusion

Sunlight screening by volcaniclast dust from the Deccan volcanic eruptions blurred the distinction between annual seasons and disordered the biological clock of organisms and plants on land and in the sea. Flowering plants produced less fruits for vegetarians, reducing their birth-rate. The disturbed sexual cycle of carnivores likewise lowered their birth rate and drastically reduced the amount of food (eggs and young born), on which adults depended for feeding themselves and their young ones after a long period of near starvation. Food shortage resulted in intensive predation of those which could not escape or hide. The comparison between Campanian and Lower Eocene crustacean burrow systems into pelagic chalk suggest a change in the mode of breeding from a few large eggs specially treated (K-type) to numerous tiny eggs partly cared for (transition to r-type) to compensate their over-predation. Dinosaur and pterosaur eggs were laid in nests on open-flat land such as estuaries, tidal-flats and shores. They and the few successfully hatched young provided easy prey, most probably to carnivorous reptiles living in the same region. The hatched young of marine reptiles were vulnerable to predation by other reptiles, sharks and large fishes. Reptiles which hid their brood (e.g., crocodiles, sea-turtles), birds laying their small eggs among plants or at sites inaccessible to non-flying organisms were not much affected by the predatory stress. The small mammals of that time could hide in the underground and in hidden places where they survived the predatory threat. The detected fluctuations in seawater temperature during the Cretaceous last half million years (Li & Keller, 1998) resulted from variations in the amount of volcaniclastic dust released into the atmosphere by the Deccan volcanic eruptions of different intensities and duration. A thick, long-lasting dust screen blocked the solar radiation resulting in the cooling of Earth's surface land and ocean-water. A thin volcaniclastic screen created a 'greenhouse' effect raising the temperature on the Earth. The associated darkening reduced the metabolism and the activity of cold-blooded reptiles, whereby the large ones living on land could not withstand even small predators. Darkening reduced and stopped the photosynthetic activity of the symbiotic zooxanthellae in planktonic foraminifera and calcareous nannoplankton lowering the oxygen content in the reduced euphotic zone as reflected by an increase in abundance of microorganisms tolerating low-oxygen conditions (e.g., Tantawy et al., 2009). The lack of these symbionts lowered the rate of calcium-carbonate precipitation, as attested to by smaller test sizes (dwarfing) of the planktonic microfossils during the last 0.5 Myr before the K-T boundary (Keller, 2008), being demonstrated in laboratory experiments on extant planktonic foraminifera (Bé et al., 1982). The recovery of most of the Late Mesozoic life forms during the Early Cenozoic suggests that all those organisms and plants survived predation thanks to their capabilities as well as by retreating to restricted and protected habitats. There they adapted to the local and changing ecological settings during a few million years until most of them succeeded in returning to the open-large marine and terrestrial habitats. Thereby they re-appeared in the fossil record, some in a new shape as the result of adaptation to changing settings during a few million years of the recovery of Earth's ecological systems. All of these seemingly new taxa took part in continuous evolutionary lineages ranging across the K-T boundary and during the aftermath of the biological crisis. They passed most of this period in hitherto undiscovered sites and therefore these intermediate evolutionary stages do not appear in the fossil record. The resulting different nomenclature of taxa between the Late Cretaceous and the Early Tertiary was erroneously referred to the mass extinction of the Cretaceous species. The end-Cretaceous biological crisis was actually an extreme example of natural selection caused by the Deccan volcanic activity.

## 9. Acknowledgment

We thank Michail Kitin (GSI) for the technical assistance in the field and in the laboratory; to Chana Netzer-Cohen and Nili Almog (GSI) for their graphic work.

## 10. References

Abramovich, S. & Keller, G. (2002). High stress late Maastrichtian paeoenvironment: inference from planktonic foraminifera in Tunisia. *Palaeogeogr., Palaeoclim., Palaeoecol.*, Vol. 178, pp. 145-164, ISSN 0031-0182

Abramovich, S.; Yoval-Corem, S.; Almogi-Labin, A. & Bejamini, C. (2010). Global climate changes and planktic foraminiferal response in the Maastrichtian. *Paleoceanography,* Vol. 25, PA2201, pp. 1-15, ISSN 0883-8305

Bé, A.W.H.; Spero, H.J. & Anderson, R. (1982). Effect of symbiont elimination and reinfection on the life process of the planktonic foraminifer *Globigerinoides sacculifer. Marine Bioloby,* Vol. 70, pp. 73-86, ISSN 0025-3162

Bralower, T.; Eccles, L.; Kutz, J.; Yancey, T.; Schueth, J.; Arthur, M. & Bice, D. (2010). Grain size of Cretaceous-Paleogene boundary sediments from Chicxulub to open ocean: Implications for interpretation of the mass extinction event. *Geology,* Vol. 38, No. 3, pp. 199-202, ISSN 0091-7613

D'Alessandro, A. & Bromley, R.G. (1995). A new ichnospecies of *Spongeliomorpha* from the Pleistocene of Sicily. *Journal of Paleontology,* Vol. 69, pp. 393-398, ISSN 0022-3360

Dhondt, A.V. (1983). Campanian and Maastrichtian inoceramids: a review. *Zitteliana,* Vol. 10, pp. 689-701, ISBN 0373-9627

Görög, A. (1994). Early Jurassic planktonic foraminifera from Hungary. *Micropaleont.,* Vol. 40, pp. 255-260, ISSN 0026-2803

Hay, W.W. (2008). Evolving ideas about the Cretaceous climate and ocean circulation. *Cretaceous Research,* Vol. 29, pp. 725-753, ISSN 0195-6671

Kauffman, E.G. (2004). Mosasaur predation on Upper Cretaceous nautiloids and ammonites from the United States Pacific Coast. *Palaios,* Vol. 19, pp. 96-100, ISSN 0883-1351

Kauffman, E.G. (1972). *Ptychodus* predation upon Cretaceous *Inoceramus. Palaeontology,* Vol. 15, No. 3, pp. 439-444, ISSN 0081-0239

Kauffman, E.G.; Harries, P.J.; Meyer, C; Villamil, T.; Arango, C. & Jaecks, G. (2007). Paleoecology of Giant Inoceramidae (*Platyceramus*) on a Santonian (Cretaceous) seafloor in Colorado. *J. Paleont.,* Vol. 81, No. 1, pp. 64-81, ISSN 0022-3360

Keller, G. (2008). Cretaceous climate, volcanism, impacts, and biotic effects. *Cret. Res.,* Vol. 29, pp. 754-771, ISSN 0195-6671

Keller, G. & Abramovich, S. (2009). Lilliput effect in late Maastrichtian planktic foraminifera: response to environmental Stress. *Palaeogeogr., Palaeoclim., Palaeoecol.,* Vol. 284, pp. 47-62, ISSN 0031-0182

Keller, G.; Sahni, A. & Bajpai, S. (2009). Deccan volcasnism, the KT mass extinction and dinosaurs. *Journal of Biosciences,* Vol. 34, pp. 709-728, ISSN 0250-5991

Krenkel, E. (1924). Der Syrische Bogen. *Centralb. Mineral. Geol. Palaeontol.,* Abh. B 9, pp. 274-281, 301-313, ISSN 0372-9338

Kummel, B. (1964). Nautiloidea-Nautilida, In: *Treatise on Invertebrate Paleontology,* Part K, Mollusca 3, R.C. Moore (Ed.), K383-K457, Geol. Soc. Amer. & Univ. Kansas Press, ISBN 08137-30112

Langston, W. Jr. (1981). Pterosaurs. *Scientific American,* Vol. 244, pp.122-136, ISSN 0036-8733

Larson, N.L.; Jorgensen, S.D.; Farrar, R.A. & Larson, P.L. (1997). *Ammonites and the other Cephalopods of the Pierre Seaway.* Black Hills Institute of Geological Research, 148 pp., ISBN 0-945005-25-3

Leckie, R.M. (1987). Paleoecology of mid-Cretaceous planktonic foraminifera: a comparison of open ocean and epicontinental sea assemblages. *Micropaleontology,* Voll. 33, pp. 164-176, ISSN 0026-2803

Lewy, Z. (1976). Morphology of the shell in Gryphaeidae. *Israel Journal of Earth Sciences*, Vol. 25, pp. 45-50, ISSN 0021-2164

Lewy, Z. (1996). Octopods: nude ammonoids that survived the Cretaceous-Tertiary boundary mass extinction. *Geology*, Vol. 24, No. 7, pp. 627-630, ISSN 0091-7613

Lewy, Z. (2000). The erroneous distinction between tetrabranchiate and dibranchiate cephalopods. *Acta Geologica Polonica*, Vol. 50, No. 1, pp. 169-174, ISSN 0001-5709

Lewy, Z. (2002a). The function of the ammonite fluted septal margins. *J. Paleontol.*, Vol. 76, No. 1, pp. 63-69, ISSN 0022-3360

Lewy, Z. (2002b). New aspects in ammonoid mode of life and their distribution. *Geobios*, Vol. 35 (M.S. No. 24), pp. 130-139, ISSN 0016-6995

Lewy, Z. (2009). The possible trophic control on the construction and function of the aulacocerid and belemnoid guard and phragmecon. *Revue de Paléobiologie*, Vol. 28, No. 1, pp. 131-137, ISSN 0253-6730

Lewy, Z. & Goldring, R. (2006). Campanian crustacean burrow system from Israel with brood and nursery chambers representing communal organization. *Palaeontology*, Vol. 49, Part 1, pp. 133-140, ISSN 0081-0239

Lewy, Z.; Milner, A.C. & Patterson, C. (1992). Remarkably preserved natural endocranial casts of pterosaur and fish from the

Late Cretaceous of Israel. *Geological Survey of Israel, Current Research*, Vol. 7, pp. 31-35, ISSN 0195-6671

Li, L. & Keller, G. 1998. Abrupt deep-sea warming at the end of the Cretaceous. *Geology*, Vol. 26, pp. 995-998, ISSN 0091-7613

López-Martínez, N.; Moratalla, J.J. & Sanz, J.L. (2000). Dinosaur nesting on tidal flats. *Palaeogeog., Palaeoclim., Palaeoecol.*, Vol. 160, pp. 153-163, ISSN 0031-0182

Lucas, J. & Prévôt-Lucas, L. (1996). Tethyan phosphates and bioproduction. In: *The Ocean Basin and Margins. The Tethys Ocean*, Vol. 8, A.E.M. Nairn, L.-E. Ricou, B. Varielynck & J. Decourt (Eds.), 367-391, ISBN 0306451565, Plenum Press, New York

MacLeod, N.; Rawson, P.F.; Forey, P.L.; Banner, F.T.; Boudagher-Fadel, M.K.; Bown, P.R.; Burnett, J.A.; Chambers, P.; Culver, S.; Evans, S.E.; Jeffery, C.; Kamiski, M.A.; Lord, A.R.; Milner, A.C.; Milner, A.R.; Morris, N.; Owen, E.; Rosen, B.R; Smith, A.B.; Taylor, P.D.; Urquhart, E. & Young, J.R. (1997). The Cretaceous-Tertiary biotic transition. *Journal of the Geological Society, London*, Vol. 154, pp. 265-292, ISSN 0016-7649

MacLeod, N. (2005). Cretaceous, In: *Encyclopedia of Geology*, R.C. Selley, L.R.M. cocks & I.R. Plimer, (Eds.), 360-372, ISBN 012- 636-3803, Academic Press, London.

Marshall, C.R. & Ward, P.D. (1996). Sudden and gradual molluscan extinctions in the Late Cretaceous of Western European Tethys. *Science*, Vol. 274, (22 November 1996), pp. 1360-1363, ISSN 0036-8075

Mitchell, S.F.; Stemann, T.; Blissett, D.; Brown, I.; Ebank, W.O.; Gunter, G.; Miller, D.J.; Pearson, A.G.M.; Wilson, B. & Young, W.A. (2004). Late Maastrichtian rudist and coral assemblages from the Central Inlier, Jamaica: towards an event stratigraphy for shallow-water Carribean limestones. *Cretaceous Research*, Vol. 25, pp. 499-507, ISSN 0195-6671

Olsson, R.K.; Hemleben, C.; Berggren, W.A. & Huber, B.T. (1999). *Atlas of Paleocene Planktonic Foraminifera*, Smithsonian Contribution to Paleobiology, No. 85, pp. 1-252, ISSN 0081-0266

Orue-Etxebarria, X. & Apellaniz, E. (2000). Modification of the original stratigraphic distribution of *Globigerina hillebrandi*

Orue-Etxebarria, 1985 and its inclusion in another genus: one more planktic foraminifer species surviving the mass extinction of the K/T boundary. *Bulletin de la Société géologique de France*, Vol. 171, pp. 291-293, ISSN 0037-9409

Prinn, R.G. & Fegley, B. Jr. (1987). Bolide impacts, acid rain, and biospheric traumas at the Cretaceous-Tertiary boundary. *Earth and Planetary Science Letters*, Vol. 83, pp. 1-15, ISSN 0012-821X

Sanz, J.L.; Mortalla, J.J.; Diaz-Molina, M.; López-Martinez, N.; Kälin, O. & Vianey-Liaud, M. (1995). Dinosaur nests at the sea shore. *Nature*, Vol. 376, (31 August), pp. 731-732, ISSN 0028-0836

Saul, L.R. & Stadum, C.J. (2005). Fossil Argonauts (Mollusca: Cephalopoda: Octopodida) from Late Miocene siltstone of the Los Angeles Basin, California. *Journal of Paleontology*, Vol. 79, No. 3, pp. 520-531, ISSN 0022-3360

Scasso, R.A.; Concheyro, A.; Kiessling, W.; Aberhan, M.; Hecht, L.; Mrdina, F.A. & Talge, R. (2005). A tsunami deposit at the Cretaceous/Paleogene boundary in the Neuquén Basin of Argentina. *Cret. Res.*, Vol. 26, pp. 233-297, ISSN 0195-6671

Sohl, N. & Kollmann, H.A. (1985). Cretaceous Actaeonellid Gastropods from the Western Hemisphere. *Geological Survey Professional Paper* 1304, pp. 1-104, U.S. Government Printing Office, Washington, ISSN 1044-9612

Stenzel, H.B. (1971). Oysters, In: *Treatise on Invertebrate Paleontology*, Part N, Mollusca 6, Bivalvia N953-N1224, Geol. Soc. Amer. & Univ. Kansas, ISBN 0-8137-3026-6, Univ. Kansas Press

Steuber, T. & Löser, H. (2000). Species richness and abundance patterns of Tethyan Cretaceous rudist bivalves (Mollusca: Huppuritacea) in the central-eastern Mediterranean and Middle East, analysed from a palaeontological database. *Palaeogeog., Palaeoclim., Palaeoecol.*, Vol. 162, pp. 75-104, ISSN 0031-0182

Steuber, T.; Mitchell, S.F.; Buhl, D.; Gunter, G. & Kasper, H.U. (2002). Catastrophic extinction of Caribbean rudist bivalves at the Cretaceous-Tertiary boundary. *Geology*, Vol. 30, No. 11, pp. 999-1002, ISSN 0091-7613

Stüben, D.; Kramer, U.; Berner, Z.A.; Meudt, M.; Keller, G.; Abramovich, S.; Adatte, T.; Hambach, U. & Stinnesbeck, W. (2003). Late Maastrichtian paleoclimatic and paleoceanographic changes inferred from Sr/Ca ratio and stable isotopes. *Palaeogeog., Palaeoclim., Palaeoecol.*, Vol. 199, pp. 107-127, ISSN 0031-0182

Summesberger, H. (1979). Eine obersantone Ammonitefauna aus dem Becken von Gosau (Oberösterreich), *Annalen des Naturhistorischen Museum in Wien*, Vol. 82, pp. 109-176, ISSN 0255-0091

Tantawy, A.A.A.; Keller, G. & Pardo, A. (2008). Late Maastrichtian volcanism in the Indian Ocean: effects on calcareous nannofossils and planktic foraminifera. *Palaeogeogr., Palaeoclim., Palaeoecol.*, Vol. 284, pp. 63-87, ISSN 0031-0182

Vernon, J.E.N. (1993). *Corals of Australia and the Indo-Pacific*. University of Hawaii Press, 644 pp., ISBN 08248-1504-1

Vianey-Liaud, M. & Lopez-Martínez, N. (1997). Late Cretaceous dinosaur eggshells from the Tremp Basin, southern Pyrenees, Lleida, Spain. *Journal of Paleontology*, Vol. 71, No. 6, pp. 1157-1171, ISSN 0022-3360

Westermann, G.E.G. (1990). New developments in ecology of Jurassic-Cretaceous ammonoids, In: *Fossili, Evoluzione, Ambiente*, G. Pallini, F. Cecca, S. Cresta & M. Santantonio (Eds.), 459-478, Atti del Secondo Convegno Internazionale, Pergola 1987, Tecostampa, Otra Vetere, Italy